CREATIVE HOMEOWNER®

ULTIMATE GUIDE

HOME REPAIR
AND IMPROVEMENT

**Technical Editor for 3rd Updated Edition:
Charles T. Byers**
Assistant Professor, Thaddeus Stevens College of Technology, Lancaster, PA

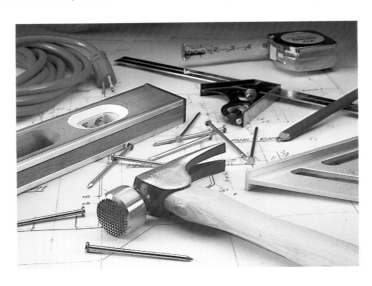

Copyright © 2000, 2006, 2011, 2017, 2022 Creative Homeowner

CREATIVE HOMEOWNER®

This book may not be reproduced, either in part or in its entirety, in any form, by any means, without written permission from the publisher, with the exception of brief excerpts for purposes of radio, television, or published review. All rights, including the right of translation, are reserved. Note: Be sure to familiarize yourself with manufacturer's instructions for tools, equipment, and materials before beginning a project. Although all possible measures have been taken to ensure the accuracy of the material presented, neither the author nor the publisher is liable in case of misinterpretation of directions, misapplication, or typographical error.

Creative Homeowner is a registered trademark of New Design Originals Corporation.

ULTIMATE GUIDE TO HOME REPAIR AND IMPROVEMENT
Managing Editor: Fran J. Donegan
Senior Editor: Mike McClintock
Associate Editor: Paul Rieder
Assistant Editor: Craig Clark
Proofreader: Sara M. Markowitz
Contributing Editors: John Wagner, Roy Barnhardt, Joseph Gonzalez, Michael Morris, Ken Textor, Laura Tringali, Bruce Wetterau
Photo Coordinator and Digital Imaging Specialist: Mary Dolan
Photo Researchers: Robyn Poplasky, Craig Clark, Dan Lane, Amla Sanghvi
Editorial Assistants: Laura DeFerrari, Dan Lane, Stanley Sudol
Thanks to: Mark Arduino, Juan Calle, Richard L. DeJean, Craig Fahan, Joseph L. Fucci, David Geer, Rafael Lian, Felix Nieves, Andrew Parsekian, Ann Parsekian, James Parsekian, Neil Soderstrom
Indexer: Sandi Schroeder/Schroeder Indexing Services
Senior Designers: Glee Barre, Kathryn Wityk, David Geer, Diane P. Smith-Gale
Designers: Maureen L. Mulligan
Design Assistants: Virginia Wells Blaker, Susan Hallinan
Senior Photographer: John Parsekian
Staff Photographer: Brian C. Nieves
Staff Illustrators: Vincent Alessi, Clarke Barre
Contributing Illustrators: Tony Davis, Ron Hildebrand, Greg Maxson, Thomas Moore, Ian Worpole

3RD UPDATED EDITION
Technical Editor: Charles T. Byers, Assistant Professor, Residential Remodeling Technology, AAS, AST, Thaddeus Stevens College of Technology
Cover Design: David Fisk
Managing Editor: Colleen Dorsey

ISBN 978-1-58011-868-2

Library of Congress Control Number: 2021940915

We are always looking for talented authors. To submit an idea, please send a brief inquiry to acquisitions@foxchapelpublishing.com.

Printed in China

Third Printing

Creative Homeowner®, www.creativehomeowner.com, is an imprint of New Design Originals Corporation and distributed exclusively in North America by Fox Chapel Publishing Company, Inc., 800-457-9112, 903 Square Street, Mount Joy, PA 17552.

Technical Consultants
Association of the Wall and Ceiling Industries, Intl., Lee Jones
Brick Industry Association, Brian E. Trimble, CDT, Director, Technical Services Engineering and Research
California Redwood Association, Charles Jourdain
Carpet and Rug Institute, R. Carroll Turner
Carrier Corp.
Concrete Foundation Association, J. Edward Sauter, Executive Director
Merillat Industries
National Association of Homebuilders, David DeLorenzo
National Association of the Remodeling Industry, Brett S. Martin
National Concrete Masonry Association, Dennis W. Graber, P.E., Director of Technical Publications
National Oak Flooring Manufacturers Association, Mickey Moore
North American Insulation Manufacturers Association, Charles Cottrell
Plumbing Manufactures Institute, David W. Viola, Technical Director
Plumbing, Heating, Cooling Contractors National Association, Robert Shepherd
Roofing Industry Education Institute, Richard L. Fricklas, Founder
Ross Electrical Assessments, Joseph A. Ross, former Chief Editor of the NFPA-NEC Handbook
Scotts Training Institute, John Marshall, Instructor
Southern Forests Products Association, Richard Wallace
Tile Council of America, Duncan English
Western Wood Products Association, Frank Stewart, Director, Technical & Product Support
Window and Door Manufacturers Association, Alan J. Campbell, President

Safety

Though all the designs and methods in this book have been reviewed for safety, it is not possible to overstate the importance of using the safest construction methods you can. What follows are reminders—some do's and don'ts of work safety. They are not substitutes for your own common sense.

- Always use caution, care, and good judgment when following the procedures described in this book.

- Always be sure that the electrical setup is safe, that no circuit is overloaded, and that all power tools and outlets are properly grounded. Do not use power tools in wet locations.

- Always read container labels on paints, solvents, and other products; provide ventilation; and observe all other warnings.

- Always read the manufacturer's instructions for using a tool, especially the warnings.

- Use hold-downs and push sticks whenever possible when working on a table saw. Avoid working short pieces if you can.

- Always remove the key from any drill chuck (portable or press) before starting the drill.

- Always pay deliberate attention to how a tool works so that you can avoid being injured.

- Always know the limitations of your tools. Do not try to force them to do what they were not designed to do.

- Always make sure that any adjustment is locked before proceeding. For example, always check the rip fence on a table saw or the bevel adjustment on a portable saw before starting to work.

- Always clamp small pieces to a bench or other work surface when using a power tool on them.

- Always wear the appropriate rubber or work gloves when handling chemicals, moving or stacking lumber, or doing heavy construction.

- Always wear a disposable face mask when you create dust by sawing or sanding. Use a special filtering respirator when working with toxic substances and solvents.

- Always wear eye protection, especially when using power tools or striking metal on metal or concrete; a chip can fly off, for example, when chiseling concrete.

- Never work while wearing loose clothing, hanging hair, open cuffs, or jewelry.

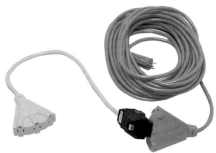

- Always be aware that there is seldom enough time for your body's reflexes to save you from injury from a power tool in a dangerous situation; everything happens too fast. Be alert!

- Always keep your hands away from the business ends of blades, cutters, and bits.

- Always hold a circular saw firmly, usually with both hands so that you know where they are.

- Always use a drill with an auxiliary handle to control the torque when large-size bits are used.

- Always check your local building codes when planning new construction. The codes are intended to protect public safety and should be observed to the letter.

- Never work with power tools when you are tired or under the influence of alcohol or drugs.

- Never cut tiny pieces of wood or pipe using a power saw. Always cut small pieces off larger pieces.

- Never change a saw blade or a drill or router bit unless the power cord is unplugged. Do not depend on the switch being off; you might accidentally hit it.

- Never work in insufficient lighting.

- Never work with dull tools. Have them sharpened, or learn how to sharpen them yourself.

- Never use a power tool on a workpiece—large or small—that is not firmly supported.

- Never saw a workpiece that spans a large distance between sawhorses without close support on each side of the cut; the piece can bend, closing on and jamming the blade, causing saw kickback.

- Never support a workpiece from underneath with your leg or other part of your body when sawing.

- Never carry sharp or pointed tools, such as utility knives, awls, or chisels, in your pocket. If you want to carry such tools, use a special-purpose tool belt with leather pockets and holders.

Table of Contents

About this Book — 8

1 HOME EMERGENCIES — 10
- Weather-Related Emergencies — 12
- Emergencies Around the House — 14

2 SAFETY & SECURITY — 16
- Security Basics — 18
- Door & Window Locks — 20
- Strengthening Doors — 22
- Security Systems — 24
- Fire Safety — 26
- Childproofing — 28
- Environmental Hazards — 30
- Universal Design — 32

3 REMODELING GUIDE — 34
- Planning — 36
- Hiring — 38
- Making Contracts — 40
- Supervising — 42
- Dealing with Disputes — 44

4 TOOLS — 46
- Basic Power Tools — 48
- Basic Hand Tools — 50

5 FASTENERS & ADHESIVES — 54
- Nails — 56
- Screws — 58
- Bolts — 60
- Fasteners — 62
- Framing Hardware — 64
- Adhesives & Caulk — 68

6 MASONRY — 70
- Tools & Materials — 72
- Concrete — 74
- Formwork — 78
- Block — 82
- Brick — 86
- Stone — 90
- Masonry Repairs — 92

7 ELECTRICAL — 94
- Power Distribution — 96
- Wiring Systems — 98
- Safety Measures — 100
- Circuit Breakers — 102
- Boxes & Connectors — 104
- Conduit — 106
- Wiring Basics — 108
- Wiring Paths — 110
- Adding a New Circuit — 114
- Wiring Receptacles — 118
- Ground Fault Circuit Interrupters — 126
- Wiring Switches — 128
- Lighting Fixtures — 134
- Specialty Wiring — 142
- Phone Wiring — 144
- Wiring Appliances — 146
- Safety Devices — 154
- Outdoor Lighting — 156

8 PLUMBING — 164
- Basics — 166
- Tools — 170
- Working with Pipe — 172
- Sinks & Faucets — 180
- Faucet Repairs — 182
- Installing Sinks & Faucets — 194
- Toilets — 206
- Tubs & Showers — 228
- Water Heaters — 238
- Pumps — 250
- Appliances — 252
- Water & Waste Treatment — 256
- Sump Pumps & Wells — 262
- Fixture Improvements — 266
- Solving Problems — 268

9 INSULATION — 272
- Thermal Protection — 274
- Types & Applications — 276
- Installation — 278
- Blown-in — 282
- Cold Spots — 284

10 HEATING — 286
- Basic Systems — 288
- Basic Installation — 290
- Basic Maintenance — 292
- Space Heaters — 294
- Stoves & Fireplaces — 296
- Humidifying — 298

11 COOLING — 300
- Systems — 302
- Window Units — 304
- In-Wall Units — 306
- Maintenance — 308
- Indoor Air Quality — 310
- Dehumidifiers & Thermostats — 312

12 VENTILATION — 314
- Foundations — 316
- Baths — 318
- Kitchen & Laundry — 320
- Whole-House Ventilation — 322
- Roofing — 324

13 FLOORS & STAIRS — 328
- Preparing Old Floors — 330
- Solid Wood — 332
- Engineered-Wood Flooring — 334
- Wood-Floor Repairs — 336
- Refinishing — 338
- Resilient Flooring — 340
- Tile Floors — 342
- Finishing & Trim — 346
- Carpet — 348
- Stairs — 352

14 WALLS & CEILINGS		**356**
Materials		358
Paneling		360
Drywall		364
Painting		370
Wallpaper		376
Wall Tile		378
Glass & Mirrors		382
Soundproofing		384
Ceilings		386

15 TRIMWORK		**388**
Basics		390
Crown Molding		394
Base Trim		396
Door & Window Casings		398
Wall Frames		402
Wainscoting		404
Chair Rails & Plate Rails		406

16 CABINETS & COUNTERS		**408**
Tools & Materials		410
Basic Techniques		412
Cabinet Construction		414
Installing Cabinets		416
Doors		418
Drawers		420
Laminate Counters		422
Solid Counters		424
Finishing		426
Resurfacing Cabinets		428
Cabinet Repairs		430
Cabinet Specialties		432

17 SHELVING & STORAGE		**434**
Basic Assembly		436
Shelf Supports		438
Basic Built-Ins		440
Closets		442
Utility Storage		444
Storage Options		446

18 ROOFING		**448**
Tools & Equipment		450
Design Options		452
Flat Roofs		454
Asphalt Shingles		456
Slate		460
Masonry		462
Wood Shingles & Shakes		464
Metal		466
Skylights & Roof Windows		468
Flashing		470
Climate Control		472
Gutters & Leaders		474

19 SIDING		**476**
Tools & Materials		478
Panel Siding		480
Clapboard Siding		482
Shingles & Shakes		484
Wood-Siding Repairs		486
Brick & Stone		488
Stucco		490
Masonry-Siding Repairs		492
Vinyl & Metal		494
Vinyl & Metal Repairs		496
Re-siding		498
Finishing New Siding		500
Fixing Paint Problems		502
Repainting		504

20 WINDOWS & DOORS		**506**
Window Basics		508
Energy Efficiency		510
Installing New Windows		512
Window Trim		514
Replacement Windows		516
Basic Repairs & Improvement		518
Storms & Screens		522
Door Basics		524
Installing Doors		526
Weatherproofing		528
Interior Doors		530
Door Repairs		532
Door Hardware		534
Garage Doors		536

21 DECKS, PATIOS & WALKS		**538**
Decks		540
Walk Design & Layout		552
Concrete Walks		554
Bricks & Pavers		558
Stone & Interlocking Systems		560
Gravel & Surface Systems		562
Drives		564

22 UNFINISHED SPACE		**568**
Attics		570
Basements		574
Garages		578
Small Spaces		580

23 CANADIAN CODE		**584**
Plumbing & Electrical		586
General Construction		588

Glossary	590
Index	593
Photo Credits	607
Metric Conversion	608

To find over 300 how-to projects, turn the page.

Table of Projects

2 SAFETY & SECURITY 16
   Fixing Common Lock Problems 18
   Installing a Peephole 20
   Strengthening Frames 22
   Installing a Dead Bolt 23
   Hard-Wiring a Detector 26

5 FASTENERS & ADHESIVES 54
   Toenailing 57
   One-Hand Nailing 57
   Cat's Paw Pulling 57
   Adding Wood Bracing 66
   Adding Metal Bracing 66

6 MASONRY 70
   Pouring a Patio 74
   Testing Concrete 75
   Forming a Curved Corner 76
   Repairing Cracks 77
   Building Formless Piers 78
   Building Formed Footings 80
   Cutting Block 82
   Laying Block 82
   Patching Block 84
   Buttering Blocks 84
   Replacing Block 84
   Cutting Brick 86
   Mortaring Bricks 86
   Laying Brick 88
   Laying Face Stone 91
   Shaping Stone 91
   Laying Full Stone 91
   Patching Steps 92
   Patching Stucco 92

7 ELECTRICAL 94
   Cutting off Power 100
   Fixing a Cord 101
   Testing Fuses 102
   Stripping Cable Sheathing 108
   Fastening Cable 108
   Attaching Wires 109
   Capping Wires 109
   Fishing Connections 110
   Installing Surface Wiring 111
   Fishing Cable from Below the Floor 112
   Fishing Cable Across a Ceiling 113
   Installing an Outlet 120
   Wiring a GFCI Outlet 126
   Wiring a Three-Way Switch 128
   Single-Pole Switch 128
   Wiring a Single-Pole Dimmer Switch 131
   Installing a Ceiling Box 135
   Installing a Fluorescent Fixture 135
   Installing a Ceiling-Mounted Fixture 136
   Installing Track-Mounted Lighting 137
   Installing a Chandelier 138
   Installing Vanity Lighting 139
   Installing a Recessed Light Fixture 141
   Installing a Transformer 142
   Fixing a Doorbell 142
   Replacing a Thermostat 143
   Wiring a Telephone Jack 145
   Direct-Wiring a Dishwasher 146
   Wiring a Waste-Disposal Unit 147
   Ceiling Fan or Light 148
   Ducted Range Hood 149

   Radiant Floor Heating 150
   Installing a Schluter Underlayment System 151
   Installing a Dryer Receptacle 153
   Wiring Smoke Detectors 154
   Hardwiring Carbon Monoxide (CO) Detectors 156
   Installing a Floodlight 156
   Extending Power Outdoors 160
   Installing UF Cable 160
   Installing an Outdoor Receptacle 160
   Installing Low-Voltage Wiring 163

8 PLUMBING 164
   Installing Pipe through Joists 172
   Running Pipe 172
   Connecting Plastic 174
   Connecting Copper 174
   Making Flared Fittings 175
   Tying Into Old Cast Iron 178
   Unblocking a Cleanout 179
   Installing a Sink 180
   Fixing a Seat Washer 182
   Replacing an O-Ring 183
   Repairing a Packing Washer 183
   Repairing Ball-Type Faucets 184
   Repairing a Kitchen Cartridge Faucet 185
   Repairing a Two-Handle Cartridge Faucet 186
   Repairing a Single-Handle Cartridge Faucet 187
   Fixing a Leak in a Ceramic-Disk Faucet 188
   Repairing a Two-Handle Ceramic-Disk Faucet 189
   Replacing a Spray Attachment 189
   Repairing a Two-Handle Faucet Spout 190
   Repairing a Single-Handle Faucet Spout 191
   Reaching Recessed Faucets 191
   Fixing Single-Handle Tub & Shower Faucets 192
   Removing the Drain 194
   Installing a New Bathroom Faucet & Drain 196
   Removing the Wall-Hung Bathroom Sink 198
   Installing the Cabinet & Top 199
   Installing a Sink in a Plywood Top 200
   Attaching Drain Fittings to the New Sink 201
   Installing a Metal-Rim Sink 202
   Attaching the Faucet 203
   Installing a Freestanding Laundry Sink 204
   Connecting the Water 205
   Upgrading a Water Closet 206
   Fixing a Running Pressure-Assisted Toilet 208
   Cleaning a Bacteria-Clogged Toilet 211
   Removing Grit from the Diaphragm 212
   Replacing a Tank Ball 213
   Replacing a Flapper 214
   Replacing a Flush Valve 216
   Replacing a Fill Valve 218
   Flush-Valve Cartridge 220

   Taking Up and Resetting a Toilet 225
   Installing a New Toilet 226
   Fixing a Faucet 228
   Replacing a Tub Drain Assembly 230
   Replacing a Tub-Shower Faucet 232
   Installing a Cast-Iron Tub 235
   Installing a Shower Stall 236
   Installing an Instant Hot-Water Dispenser 242
   Changing an Anode Rod 243
   Maintaining the Burner & Thermocouple 245
   Troubleshooting the Wiring & Thermostat 246
   Replacing a Gas-Fired Water Heater 248
   Installing an Electric Water Heater 249
   Installing a Recirculating System 250
   Removing an Old Dishwasher 252
   Installing a New Dishwasher 254
   Installing a Sediment Filter 256
   Removing a Waste-Disposal Unit 258
   Installing a Waste-Disposal Unit 260
   Restarting a Jammed Waste-Disposal Unit 261
   Installing a Sump Pump 262
   Installing a Water Softener 263
   Replacing a Leaking Pressure Tank 265
   Replacing a Showerhead 266
   Reglazing a Tub 266
   Tub Surrounds 266
   Relining a Tub 267
   Anti-Scald Faucets 267
   Installing an Anti-Freeze Faucet 268
   Temporary Repairs 268
   Quieting Noisy Pipes 269
   Clearing a Waste-Line Clog 270
   Clearing a Tub Clog 271

9 INSULATION 272
   Insulating Foundation Exteriors 278
   Insulating Foundation Interiors 278
   Insulating Crawl Spaces 278
   Insulating Walls 280
   Insulating Ceilings 280
   Insulating Roofs 280
   Insulating Existing Walls 282
   Insulating Attics 283

10 HEATING 286
   Baseboard Convectors 290
   Furnace Maintenance 292
   Installing Toe-Space Heaters 294
   Installing Wall Heaters 294
   Installing a Masonry Fireplace 297
   Installing a Humidifier 298

11 COOLING 300
   Installing a Window Unit 304
   Installing an In-Wall Unit 306
   Basic Cleaning 308
   Cleaning Ducts 310
   Servicing a Dehumidifier 312

12 VENTILATION 314
   Installing Foundation Vents 317
   Installing Bath Vents 318

MONEY-SAVING PROJECTS ARE IN BLUE.
ENVIRONMENTALLY FRIENDLY PROJECTS ARE LISTED IN GREEN.

Installing Timer Switches	319
Installing Dryer Vents	321
Installing Whole-House Fans	322
Installing Ceiling Fans	322
Installing Strip-Grille Vents	325
Installing Roof Vents	326
Installing Gable Vents	326
Installing Ridge Vents	326

13 FLOORS & STAIRS 328

Removing Old Trim	330
Installing Solid-Wood Flooring	333
Preparing Floors	334
Installing Floating Floors	334
Laying Parquet Floors	335
Removing Stains	336
Replacing Boards	336
Six Ways to Stop Squeaks	336
Plugging	337
Refinishing Wood Floors	339
Installing Sheet Flooring	340
Repairing Tile Floors	341
Installing Tile	342
Replacing Tiles	344
Installing Baseboards	346
Installing Carpet	348
Patching Carpet	350
Spot Patching	351
Tuft Patching	351
Adding Pull-Down Stairs	352
Carpeting Stairs	353
Tightening Balusters	355
Replacing Balusters	355

14 WALLS & CEILINGS 356

Installing Panels	360
Replacing an Interlocked Plank	362
Patching Paneling	362
Scribing Joints	363
Installing Drywall	364
Cutting for Outlets	365
Finishing Panel Seams	366
Finishing Corners	366
Fixing Small Holes	368
Fixing Large Holes	368
Fixing Corners	369
Preparing Walls	372
Painting Walls	372
Repainting Trim	373
Spot-Painting Patches	375
Making Spot Repairs	376
Stripping Old Wallpaper	376
Hanging Wallpaper	377
Installing Tile	379
Replacing Tile	380
Restoring Bath Tile	381
Installing Glass Blocks	382
Mounting Mirrors	383
Making Sound-Absorbing Walls	384
Installing a False Beam	386
Installing a Suspended Ceiling	387

15 TRIMWORK 388

Biscuit Joinery	390
Back-Cutting Joints	390
Coping Joints	391
Installing Molded Trim	392
Installing Crown Molding	394
Installing Three-Piece Base Trim	396
Making a Scarf Joint	397
Assembling a Jamb	398
Installing Simple Colonial Casing	399
Installing Wall Frames	402
Installing Sheet Paneling	404
Installing Wainscoting	404
Milling and Installing a Plate Rail	406
Installing a Chair Rail	406

16 CABINETS & COUNTERS 408

Cutting & Edging Plywood	412
Doweling	412
Cutting Dadoes	412
Making Biscuit Joints	413
Wall Preparation	416
Installing Wall Cabinets	416
Installing Base Cabinets	416
Installing Metal Guides	420
Cutting Dovetails	420
Installing Wood Guides	420
Installing Post-Form Counters	422
Installing Sinks	422
Installing Laminate	423
Installing Tile	424
Retrimming Bases	428
Refacing Cabinets	429
Installing a Pullout Platform	432

17 SHELVING & STORAGE 434

Making Wide Plywood Shelves	436
Reinforcing Shelves	436
Hanging Wood-Framed Standards	438
Plugging	440
Filling Holes	440
Constructing a Built-In Shelving Unit	441
Framing a Closet	442
Attaching Casters	444
Assembling a Wall System	445

18 ROOFING 448

Repairing a Flat Roof	455
Installing Asphalt Shingles	456
Shingle Repair	457
Preparing for Reroofing	458
Installing Double-Layer Roll Roofing	459
Working with Slate	460
Repairing Slate	461
Repairing Clay Tile	462
Repairing Concrete Tile	463
Installing Wood Roofing	464
Repairing Shingles & Shakes	465
Two Ways to Repair Metal Roofs	467
Installing a Skylight	468
Installing Step Flashing	470
Flashing Repairs	471
Repairing Gutters	475

19 SIDING 476

Installing Panels	480
Installing Clapboard	483
Installing Shingles	484
Repairing Bows	486
Replacing Shingles	486
Repairing Panels	487
Repairing Clapboards	487
Setting Veneer Stone	488
Setting Face Brick	488
Installing Stucco	491
Replacing Brick	492
Repairing Stucco	492
Installing Vinyl	494
Repairing Vinyl	496
Repairing Aluminum	497
Painting Aluminum	497
Re-siding over Existing Siding	498
Flashing Windows	499
Finishing New Siding	500
Repainting	505

20 WINDOWS & DOORS 506

Installing a Window in New Construction	513
Installing Window Trim	514
Installing an Upgraded Replacement Window	517
Cutting Glass & Plexiglas	518
Replacing Glass	518
Removing an Old Sash	520
Improving Windows by Using Friction Channels	520
Installing Heat-Shrink Plastic	522
Replacing Screens	522
Screening Frames	523
Installing a Prehung Door	527
Installing a Door Shoe	528
Replacing a Threshold	528
Installing an Interior Door	531
Repairing a Slider	532
Adjusting a Closet Door	532
Refinishing Metal	533
Installing a Privacy Lock	534
Installing a Garage-Door Opener	537

21 DECKS, PATIOS & WALKS 538

Installing Girders	544
Setting Joist Hangers	547
Installing Stairs	548
Installing Railings	551
Forming Curves	554
Forming Walks	554
Finishing Concrete	555
Edges & Joints	556
Cracks & Breaks	556
Installing a Paver Patio	559
Laying Stone Pavers	560
Forming a Gravel Walk	562
Building a Boardwalk	562
Installing In-Ground Steps	563
Replacing Expansion Joints	566
Cleaning Concrete	567
Repairing Asphalt	567
Sealing Asphalt	567

22 UNFINISHED SPACE 568

Strengthening Attic Floor Joists	573
Sealing a Floor	574
Covering Basement Walls	575
Fixing Water Leaks	577
Installing a Garage Entry Door	578
Patching & Leveling Floors	579

About this Book

Think of the *Ultimate Guide: Home Repair and Improvement* as an owner's manual for your house—a comprehensive guide that you can turn to any time to help you with day-to-day home repair and maintenance. It is also a valuable source for information on larger, more involved home improvements, such as kitchen and bathroom upgrades.

Inside, you'll find 600 pages of helpful information that covers your house from top to bottom, inside and out. We've laid out the full spectrum of home repair and improvement in 23 chapters that cover tools and materials, home remodeling, outdoor living areas, and every part of the building, including the foundation, the framing, the mechanical systems—the works.

Home improvement is a big subject, and experienced do-it-yourselfers know it can sometimes be complicated. *Ultimate Guide: Home Repair and Improvement* helps seasoned homeowners make sense of the subject and further their skills, but you don't need to be an accomplished do-it-yourselfer to use it. You'll find the information you need in sensible text that takes the time to cover the basics. It also offers money- and time-saving tips, explains your options, and shows you what to do using more than 3,300 photographs and illustrations.

There are 325 step-by-step projects that run the gamut from simple repairs, such as fixing a stuck window, to major improvements, such as installing replacement windows. In addition, you'll find help with plumbing, heating, cooling, and electrical systems with how-to photo sequences that focus in on the information you need. This latest edition includes updated information on energy-efficient products and building techniques as well as updates on plumbing and electrical work.

Ultimate Guide: Home Repair and Improvement conveys all of this information in a number of informative and enter-

Main Features and Other Elements

Remodeling Guide

Sequence Photos

It looks like the drawings—sort of.

At last: some big windows and a view.

There used to be a really nice lawn here.

Deck-Post Anchors

J-bolts and post anchors are attached to the concrete footing, and the anchor bracing is nailed or screwed to the post. Soak post ends in preservative before attaching.

Exploded Illustrations

Material Options

Tip Boxes

◆ **CONCRETE SLABS** make strong, long-lasting driveways as long as the ground underneath is relatively stable and has been prepared thoroughly. The material can be slightly commercial-looking and without much visual interest—unless it is stamped with a pattern when poured. Light-colored concrete will show oil stains. Concrete generally costs more to install than asphalt.

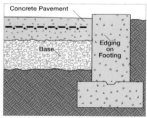

House-Mounted Ledgers

Cutaway & Schematic Illustrations

When attaching ledger boards to masonry, predrill the ledger, and use holes as guides for masonry holes. Place board onto the masonry wall, level it, and use a marker to indicate drill points. Then drill and insert shields or anchors. Don't confuse masonry with stucco; ledgers on stucco must be bolted into the house's frame. On other exteriors, strip siding to expose the sheathing; bolt the ledger into the frame. To preserve this crucial timber, use pressure-treated wood and cover the top edge with flashing.

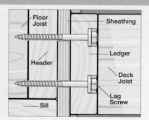

MONEY SAVER

GROUT REPAIRS
Water can puddle on tile all day without leaking, but even a hairline crack in grouted seams lets water through. If you catch cracks early, you can scrape out existing joints (a can opener works well) and regrout, which will save you the cost of retiling and prevent damage to the tile substrate. You may have to experiment with grout samples to get a good color match—aging affects the color—even if you use a leftover supply of the original material.

If adjacent tiles are loose, new grout will not hold them in place.

Money Saver Tips & Projects

taining ways—many of which are shown below—including step-by-step projects, detailed drawings, and informative photographs.

Throughout the book you will find "Green Solutions," which are tips and projects to help you make earth-friendly choices when selecting products and using techniques to improve your home. "Money Savers" alert you to the opportunity to make a repair or improvement that will save you money in the long run—even if a pro does the work for you. For example, repairing a few damaged roofing shingles to stop a leak is a lot cheaper, and quicker, than a total reroofing project. You will find both highlighted in either green type for "Green Solutions" or blue type for "Money Savers" in the "Table of Projects," pages 6–7 and at the beginning of each chapter.

You will also find a few special sections that you won't find in most home how-to books, but which cover important aspects of owning a home. One is the "Remodeling Guide," beginning on page 34, a special photo-illustrated section that covers the ins and outs of contracts, contractors, building codes, payment schedules—everything you need to know to successfully manage your own remodeling project.

Throughout the book you'll also find reminders on working safely—a key part of any project. A glossary on page 590 helps explain terms used in repair and remodeling.

All in all, the *Ultimate Guide: Home Repair and Improvement* is about how a house works, what you can do to improve it, what can go wrong, and what you can do about fixing it. You're sure to find yourself turning to it again and again as you make your house a home. Best of luck with your projects!

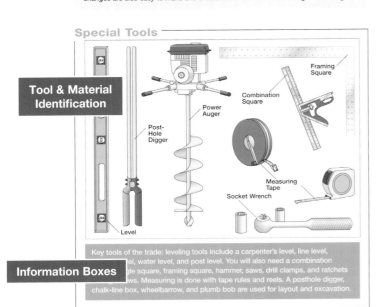

1 Home Emergencies

12 WEATHER-RELATED EMERGENCIES
- Preparedness
- Severe Weather
- No Power, No Problem?
- Water, Water Everywhere

14 EMERGENCIES AROUND THE HOUSE
- Fire! Are You Safe?
- Making Accidents Preventable
- Making Wiring Safer
- Dealing with the Aftermath

1 Home Emergencies

Preparedness

Not every emergency can be prepared for, but if you live in an area prone to hurricanes, floods, earthquakes, or tornadoes, you should have basic emergency supplies on hand, and your family should be aware of what steps to take when disaster strikes.

Hurricanes. The National Hurricane Center recommends that those living in low-lying areas have an evacuation plan. Find out about the best routes from your local police or Red Cross chapter. Also plan for emergency communication, such as contacting a friend out of the storm area, in case family members are separated. Listen to the radio or TV for warnings, check your emergency supplies, and fuel the car. Bring in outdoor objects such as lawn furniture, and close shutters or install plywood before the storm arrives. Unplug appliances, cut off the main circuit breaker, and turn off the main water-supply valve. If time permits, elevate furniture to protect it against flooding.

Tornadoes. Have a place ready where you can take shelter—if you don't have a basement, find a windowless spot on the ground floor, such as a bathroom or a closet under stairs. As tornadoes usually happen with little warning, each family member should know the danger signs, where your emergency supplies are, and what to do in case of a power outage or gas leak.

Earthquakes. If you live in an earthquake zone, have all shelves fastened securely to your walls, and store heavy or breakable items close to the floor. During an earth-quake, the safest place in your home (according to FEMA) is under a piece of heavy furniture or against an inside wall, away from windows or furniture that may topple.

Survival Tips. If you plan to ride out a storm, have basic emergency supplies on hand, including flashlights and extra batteries, a battery-operated radio, first-aid kit, extra non-perishable food and water, essential medicines, and a cell phone. Turn the refrigerator to its coldest setting, and open and close it only when necessary. Store drinking water in jugs and bottles—and in clean bathtubs.

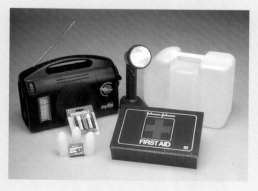

Your basic emergency kit should include a first-aid kit, flashlights and extra batteries, bottled water, and a portable radio.

Severe Weather

Severe storms are quantified by their power and the damage they do—like the Richter scale for the force of an earthquake. Hurricanes, for example, are measured according to the Saffir-Simpson scale, which predicts five levels of damage you can expect as storm winds rise from 74 miles per hour (when hurricane warnings are issued) to over 155 mph. Often, it's a combination of wind and rain that causes damage, particularly to roofs.

Temporary Roof Repairs. It's natural to try to patch an active leak but unwise to work on a wet roof in bad weather. There are exceptions: mainly, if the house has a low-sloped or flat roof that you wouldn't roll off even if you slipped. When you can work safely, temporarily stem roof leaks with roof cement (not roof coating). On standard shingles, flashing, roll roofing, and even built-up flat roofs, pry apart the leaking seam, and fill the opening with the thick tar. Then, push the shingle seam or flashing edge back in place, and add another thick layer of tar on top. If a shingle tab (the exposed section) has blown off, cover the area with tar, particularly exposed nailheads on the shingle layers below, and weave in a cover layer—if you don't have spare shingles, a piece of tarpaper or even a plastic bag will work—to maintain the system of overlapping edges that shed water.

Clearing Bottlenecks. To help prevent damage, it pays to regularly check and clear gutters and downspouts, particularly the S-shaped offset fitting that directs water from roof overhangs back toward the building leader board. These fittings typically are held in place with sheet-metal screws, which you need to remove to gain access for cleaning.

De-icing. To prevent gutters and drains from becoming laden with ice during a winter storm, you can install UL-approved electric heat cables equipped with built-in thermostats that trigger a power flow when temperatures drop to the freezing point. Once the drainage system freezes, ice dams can form on the roof edge.

An early storm warning can allow time to pack up essentials and batten down the house with plywood or boards.

WEATHER-RELATED EMERGENCIES

No Power, No Problem?

The best way to know what you'll need is to remember what you most missed last time there was an outage. For example, in a house with a well and only a small holding tank, you might miss water more than lights, which can be replaced temporarily by candles.

Conserving Heat. In winter, conserve heat during an outage by making only the quickest entries and exits through exterior doors and opening drapes and blinds to winter sun for solar heat gain during the day. If power goes out at night, drape blankets over windows to provide more insulation. Stay in the warmest room, normally on the south side of the house, and insulate the space from colder areas by hanging blankets over hallways.

Automatic Lighting. To avoid a maddening search for a flashlight with good batteries, use a recharging flashlight. Leave it plugged into an outlet, and when the power goes off, the light will turn on automatically so that you can see where it is. Remove the unit, and use it as a portable flashlight.

Portable Generators. A home generator is handy for an area with frequent power outages or for anyone who relies on electricity for their water or heat. To prepare a generator to run your lights and outlets, have an electrician install a transfer panel at your main service panel (the main circuit-breaker box where electricity enters the house). With this kind of hookup, you can supply limited power through your normal house wiring. Without a transfer panel, you need to string extension cords from the generator and plug the refrigerator, well pump, and a few lights directly into the unit. Remember, never run a gas-powered generator in the house or garage. Its exhaust fumes can be lethal.

Restarts. Before you resume the normal operation of appliances after a power outage has ended, check the manufacturer's restart instructions. Some, particularly older furnaces and water heaters, may require a specific sequence in order to restart safely.

Install a battery-powered flashlight at stairwells. It charges when power is on and lights automatically when power fails.

Water, Water Everywhere

The natural impulse after your house is flooded is to remove as much water as quickly as you can. But after a major flood, you should resist the impulse, and drain the water slowly.

Pumping Out Water. The hidden danger is that the ground outside the foundation wall is saturated and pressing against the masonry with the potential force of a mudslide. In extreme cases, several feet of water inside the wall pressing in the opposite direction may be the only thing preventing a collapse. According to the Federal Emergency Management Agency (FEMA), you should wait until water on the ground outside begins to drain away before pumping out the basement. Even then, you should reduce the level only 2 or 3 feet the first day. Remember, don't use a gasoline-powered generator or pump inside the house because it releases deadly carbon monoxide fumes.

Sump Pumps. Check your sump pump; it can prevent major damage from flooding. Many models turn on when a float rises along a wire as water rises in the sump hole. If the sump hasn't kicked in recently, the float can seize in place. Run it up and down a few times to make sure that the sump, and everything else in the basement, won't wind up submerged.

Foundation Repairs. Interior surface patches won't work on foundations because leaks have a wall of water behind them—sometimes massive hydrostatic pressure from a yard of compacted dirt that has turned to mud. But there is one material, hydraulic cement, which has the potential to stem an active leak through masonry. The dense cement mix should be forced into wet cracks, packed in layer after layer, and held in place with a cover board. Even if water continues to flow, the mix will harden and swell as it sets up. If you pack the crack tightly, the swelling mix fills nooks and crannies and can stop the leak.

On many building sites, flood waters from heavy rains can fill basements and rise close to window height.

PREPAREDNESS 13

1 Home Emergencies

Fire! Are You Safe?

The most important fire protection is a working smoke detector. Next is a fully charged ABC-rated extinguisher you can use against any type of home fire. For fireplaces and stoves, use a special chimney extinguisher. Most look like a road flare. You remove a striking cap, ignite the stick, and toss it into the fireplace or wood stove. It can suppress a fire in the chimney by displacing oxygen needed for combustion with a large volume of noncombustible gas.

Smoke Detectors. If your smoke detectors are battery-powered, change the batteries on a set schedule. There are also hard-wired smoke detectors that run off house current (with battery back-ups). Many building codes require hard-wired detectors. Install at least one smoke detector on every level of the home, and one in an open area near bedrooms.

Chimneys. Have a chimney sweep inspect chimneys, even if you use a fireplace only occasionally. Sweeps have the tools to dislodge hardened creosote, a by-product of incomplete combustion that can reignite and start fires. You can make an unlined flue safer with one of the proprietary masonry mix systems that forms a fire-safe shell inside the flue or by running code-approved stainless steel exhaust duct through the chimney.

Extinguishers. Mount extinguishers near points where fires may start—say, one in the kitchen and one at the entrance to the utility room that houses a gas-fired furnace, water heater, and clothes dryer. Check the pressure dials to make sure extinguishers are fully charged.

Escape Routes. For maximum safety, particularly with children in the house, make sure you establish an evacuation plan with two ways out of every room, and walk children through the routes so they know what to do in an emergency.

Emergency Numbers. Post telephone numbers of local fire, police, and emergency services. Use an extinguisher against small, spot fires, but don't try to fight large, developing fires; leave the house, and call the fire department.

Your best protection against property loss and injury from fire is a smoke detector. Push the test button to check it.

Making Accidents Preventable

Every year, 18,000 people are killed by poisonings, falls, and other common household accidents. More than 13 million others are seriously injured at home in preventable accidents. "Preventable" means that you can correct the conditions that lead to accidents—for instance, by storing chemicals, medicines, and other potentially hazardous products in locked cabinets where children can't reach them. One million children under five years old are injured by unintentional poisoning every year. You can also prevent the most dire consequences of threats that you can't eliminate—for example, reducing the possibility of being injured in one of the 800,500 reportable home fires every year by installing smoke alarms.

Falling Hazards. Reduce the chance of falling by improving visibility at night with low-wattage night lights near bedrooms, baths, and stairs. Eliminate dark paths to exterior lights with fixtures triggered by timers or motion sensors. Install nonslip mats or tack strips in bathtubs to provide better footing; also install grab bars and handrails. Sand-finish polyurethanes are available for traction on wooden stairs. Brush-finishing concrete improves traction on exterior walks.

Safety Glazing. Be sure that shower doors and all large glass panels in the home are made of safety glass, which pebbles when broken. A safety-glass mark is permanently etched into the lower corner of every panel.

Cutoff Valves. Locate and check the operation of the cutoff valves that control the flow of natural gas or propane to appliances such as furnaces and stoves.

Gas and CO Detectors. As a backup to regular maintenance, install both natural gas and carbon monoxide detectors that can detect leaks of methane and propane. If you smell gas, the safest course is to leave the house immediately and report the leak. All gas utilities provide a 24-hour emergency number.

One easy way to ensure a supply of clean indoor air is to periodically clean heating and cooling equipment.

EMERGENCIES AROUND THE HOUSE

Making Wiring Safer

There are two basic ways to reduce risks when you work on wiring. First, make a circuit map and post it at the main service panel so that you know which breaker to trip. Second, double-check wires and outlets with a neon tester. If the power is off, the tester bulb won't light.

Built-in Shock Protection. Safety is provided from the point where electricity enters the house and through the network of wiring to appliances it powers. At the service panel, there is a main cutoff, usually a double toggle at the top of the box, that shuts off all power. Next in line are individual circuit breakers in rows beneath the main cut-off. Each controls a loop of wiring that services a specific part of the house. Some circuits feed several lights and outlets, while others only a single appliance that uses a lot of electricity like a kitchen range.

GFCIs. More protection against electrical shock is provided by ground-fault circuit interrupters (GFCIs) at electrical outlets that are close to sources of water because they are most likely to produce a shock. GFCI outlets are more sensitive than standard circuit breakers and trip instantaneously. GFCIs are required by the National Electrical Code in all new baths, kitchens, laundries, and exterior outlets. Some electrical appliances, including hair dryers manufactured since 1991, are equipped with appliance leakage current interrupters (ALCIs) or immersion detection circuit interrupters (IDCIs), which give extra protection against shock when an appliance is accidentally dropped in water.

Extension Cords. Permanent wiring systems have many built-in safety features, but extension cords, which are widely used, do not. Check the UL label, and you'll find that there are different types (for inside versus outside use) and different wattage ratings. A standard cord is fine for a lamp with a 100-watt bulb. But plug in a room heater, and the cord can heat up and start a fire. To be safe, the extension cord wattage rating should be 1.25 times the rating of the appliance.

Dealing with the Aftermath

When a severe storm causes damage to your house, you may have to make many temporary fixes, such as covering a leaking roof with a tarp, before the weather improves and you or your contractor can work safely to make permanent repairs.

Temporary Roof Repairs. When you can work safely, use roof cement to stem leaks. It has a thick consistency and won't run on sloped roofs. Use it to fill punctures from tree limbs and to cover nailheads and exposed courses where shingle tabs have been damaged or blown away.

Releasing Leak Reservoirs. To safely release leak reservoirs in a ceiling, put a big basin under the area, and pop the bulge with one small hole away from the center of the bulge. As that hole drains, make another hole closer to the center to release the water gradually.

Clearing Iced Gutters. A propane torch or a heat gun will melt small blockages in frozen gutters. The most drastic solution is to pull frozen downspouts off the wall. Remember, if gutters and leaders are filled with ice, they will be extremely heavy. Once you pry the downspout loose, stand clear, and let it fall to the ground like a tree. That way, water trapped on the roof and in the gutter can begin to drain away without waiting for the giant icicle in the downspout to melt.

After a Storm. If you return to a damaged home after a storm, enter with caution; beware of animals that were driven to higher ground by flooding. If you smell gas, open a window, leave immediately, and report the leak. If you see sparks or broken electrical wires in a flooded house, do not enter; have the problems fixed by an electrician. Be wary picking through a structurally damaged building. FEMA reports that after one of the most devastating hurricanes of the 1990s (Hurricane Andrew), 18 of the 54 deaths attributed to the storm occurred after the weather cleared, when some people fell in unsafe buildings or were struck by falling debris.

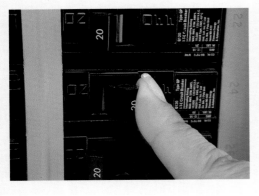

Circuit breakers trip automatically when there is a problem. Reset them once, but if they trip again, call in an electrician.

Sometimes a simple job such as heating a frozen water pipe is all you need to do to restore essential services.

FIRE! ARE YOU SAFE? 15

2 Safety & Security

18 SECURITY BASICS
- A Sense of Security ◆ Keys & Combinations
- Key Locks ◆ Fixing Lock Problems
- Security Programs ◆ Vacation Security

20 DOOR & WINDOW LOCKS
- Door Locks ◆ Installing a Peephole
- Window Locks ◆ Sliding-Door Locks

22 STRENGTHENING DOORS
- Doors & Jambs ◆ Strengthening Frames
- Security Hardware ◆ Installing a Dead Bolt

24 SECURITY SYSTEMS
- System Layout ◆ Common Components
- Auto-Dialers ◆ Remote Sensors

26 FIRE SAFETY
- Preventing & Detecting
- Battery-Powered Detectors
- Typical Detector Locations
- Hard-Wiring a Detector
- Home Fire Extinguishers
- Safety Checklist

28 CHILDPROOFING
- Dangers at Knee Level ◆ Built-In Features
- Openings ◆ Soft Surfaces
- Hazardous Materials ◆ Electricity

30 ENVIRONMENTAL HAZARDS
- Bad Air & More ■ LEAD PAINT ■ LEAD IN WATER
- ■ RADON ■ ASBESTOS ■ GAS

32 UNIVERSAL DESIGN
- Access ◆ Kitchens ◆ Baths
- Special Fixtures

2 Safety & Security

A SENSE OF SECURITY

No home can ever be made absolutely burglar-proof: if a burglar really wants to get in, even elaborate electronic security systems probably won't stop him. It's a question of degree: not enough protection can be foolhardy, while too much can be overbearing. What makes one person feel safe at home may leave another fearful.

You can achieve a comfortable level of security in many ways, ranging from taking such commonsense precautions as not advertising your absence, to installing expensive electronic alarms linked to a security station that monitors them. But some of the most beneficial security measures are relatively simple. Begin by strengthening the most basic defenses you already have, including door locks, window latches, and lights, before adding to your home security arsenal. That way you can make your home more secure without disrupting your day-to-day life—or denting your checkbook.

Locking Up

There's a reason castles were built with moats around them: limiting intruders to one point of entry makes a building easier to defend. The average home, however, provides a dozen or so points of entry: all your doors and windows.

You need a good lock on the front door, of course, but no burglar will waste time on double dead bolts if you've left a first-floor or basement window open a crack for ventilation. It makes more sense to build in a reasonable amount of protection at every point of entry. Windows need locks, and you have to remember to use them when you go out for the evening. If you want to leave a window open a crack for ventilation in hot weather—just install a second lock in the cracked-open position as detailed under "Locking an Open Window," p. 20.

Basement windows are especially easy targets for burglars; they are so low to the ground it's hard to see anyone breaking in. To make them more secure, you can install scissor-type gates and a lock on the inside of the window or custom-made iron grilles on the outside. A less prison-like solution would be to use glass blocks to replace the windows—you'd lose the ventilation, but not the light. (See "Glass & Mirrors," pp. 374–375, for installation information.)

Keys & Combinations

While a standard lockset is relatively easy to break through, the dead bolt above is not because it locks the door to the frame.

This type of keyless lock has a combination cylinder on the dead bolt, and a regular key-lock in case you forget the combination.

Key-Lock Cutaway

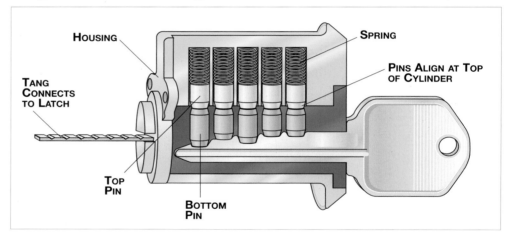

Fixing Common Lock Problems

When a lock sticks or is slow in responding, it may be clogged with dirt. Lubricate the cylinder with penetrating oil.

If you can't push in a key because of ice in the lock, thaw it out with a hair dryer, or heat the key with a match and work it in.

SECURITY BASICS

Many people enter through the garage, where computer chips in modern remotes can set new code combinations every day.

The cylinders of most locks have the same basic design. Two opposing rows of spring-loaded pins are cut at different lengths so that they align (and you can open the door) only when the pins are shifted into position by a particular key. The system provides reasonable security and convenience. Unfortunately, many burglaries today are kick-ins where the door and jamb are smashed. Long screws that join jamb and house frame will help; so will a dead bolt. Some codes may not allow dead bolts with an inside key (instead of a thumb latch) because you may have to search for the key in a fire emergency.

When a key breaks inside a lock, lift up the broken end with a narrow piece of metal, and remove the stub with pliers.

Security Programs

In some areas, there are so many false alarms from security systems that police reduce their level of response—or stop responding altogether. Electronic systems are a deterrent, but there is no substitute for high-quality locks on windows and doors, good exterior lighting so that burglars can't conceal themselves, and good neighbors who watch out for each other.

A crime watch program, like a security company sticker, may not stop a determined thief but may deter a casual one.

Vacation Security

If you have piles of cash in the house, burglars may fight through a pack of rottweilers to get it. But if your cache is like most people's—TVs, stereos, and such—most burglars are likely to break in only if they're sure you're not there. Below are several ways to make your home look like a bad risk by simulating normal activity with light timers and other devices—even if you're away on vacation.

A stuffed mailbox and a pile of news-papers is a clear signal no one's home—have deliveries held while you're away.

VACATION CHECKLIST

- **Don't close up** — Leave signs of normal activity, like a rake on the front lawn
- **Stop deliveries** — Don't let mail or newspapers accumulate while you're gone
- **Phone calls** — Leave your answering machine on (and clear the messages from your vacation spot) or have your calls forwarded
- **Outside lights** — Put outdoor lights on a timer, photoelectric switch, or motion-sensor switch
- **Inside lights** — Mimic your normal schedule by putting upstairs and downstairs lights on automatic timers
- **Trigger activity** — Put some indoor lights on special switches that turn on when they detect noise or motion—a good idea for lamps located near your front porch

A SENSE OF SECURITY

2 Safety & Security

WHAT LOCKS DO

Good locks help keep the honest people honest and provide you with a sense of security on a dark and rainy night. Of course, burglars know plenty of ways to defeat even the best locks. But jimmying a locked door or smashing a window can be noisy or take too long to do—either of which may convince an intruder to look elsewhere for a less risky target.

Door Locks

The easiest type of standard door lock for a burglar to open is a key-in-knob lock. These are the locks that can be opened with a credit card: just slip the card between the strikeplate and the spring latch to pop it open. Some of these locks have a separate tongue on the latch that makes this more difficult. But the lock can still be easily stripped out with a screwdriver, or simply removed from the door.

To give would-be intruders more trouble, attach a separate dead-bolt latch above the existing lockset, or remove the old knob and reinstall a stronger lockset, one with a full mortise dead bolt or spring-latch rim lock. The rim lock is easier to install than a dead bolt and somewhat stronger but has a clunky appearance. The dead bolt is hidden from view.

Locking an Open Window

On windows, permanent locking clips and screws are no help if you want an occasional breath of fresh air; window locks need to be secure yet allow at least some ventilation. You can solve this problem with proprietary hardware or some do-it-yourself installations.

On a double-hung window, the plan is to lock the two sashes together in a partly opened position—say, with the upper sash cracked 2 inches. To make your own lock, set the sash in a vented position, then drill a hole through the frame of the inner sash and three-fourths of the way into the outer sash where they overlap. The two frames then can be joined solidly but temporarily by a dowel or a common nail.

Several manufacturers also offer more attractive alternatives, such as elegant brass-capped sash bolts. A small, threaded receptacle fits into, but not through, the outer sash. It accepts a 2¼-inch-long threaded bolt that slips through a corresponding hole in the inner sash.

Basic Door Locks

Passage locksets that have no key are commonly used on interior doors where there is no need for security.

Standard keyed locksets used on exterior doors have a key cylinder on the outside and a thumb latch inside.

Installing a Peephole

USE: ▶ power drill/driver ▶ peephole

1 Use a sharp bit to drill a centered, eye-level hole the same diameter as the cylinder of the peephole.

2 A typical peephole has an eyepiece attached to a cylinder that goes through the door and screws to a cover plate.

Window Locks

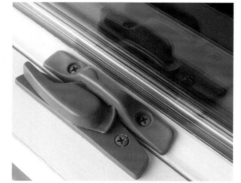

This rotating lever lock slides under the adjacent housing on the outer sash to join the two sections together.

This rotating cam lock is harder to pick from the outside because it clips around the housing on the outer sash.

DOOR & WINDOW LOCKS

Combination locksets may have a separate latchset and dead bolt, or they can be connected on the same face hardware.

Dead-bolt locks make any key-lock door more secure by connecting the door to the frame with a long-throw bolt.

Exterior keyed locks come with many types of handles, including lever types that are easier to use for people with disabilities.

Sliding-Door Locks

The typical in-line hook lock on sliding doors can't offer the degree of security that a lockset provides on a swinging door. The best bet is a bar lock. One end is hinged to the far jamb and can fold down out of the way when you're using the door. The other side folds down into a U-shaped bracket on the sliding panel and is pinned there with a small key. You can also install a sliding bolt lock that ties the movable panel to the fixed panel. Fixed panels should be permanently clipped to the door frame.

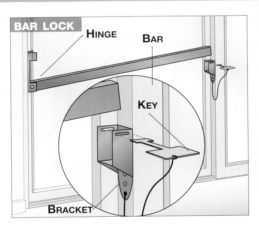

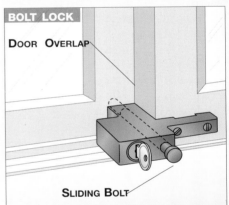

Keyed window locks are secure, but can be inconvenient if you need to find and use a key every time you want some ventilation.

This window lock increases security with a small stop that prevents the lock from turning unless you squeeze the handle.

This sash lock has a stub that travels in a slotted bracket. In this position, the stub is out of the way so the sash opens.

WHAT LOCKS DO

2 Safety & Security

DOORS & JAMBS

A good place to start making your home more secure is the point that a burglar is most likely to attack: the door. Invest in strong locks, but remember that locks only make a connection between exterior doors, which are pretty solid, and doorjambs, which aren't. An intruder may not bother to pick or drill through an expensive dead-bolt lock when one swift kick can break loose the doorjamb that holds the dead bolt keeper. The entire assembly may stay securely locked while swinging into the room with the jamb. You can fix this weak link in your household security by making the door frame part of the building frame, as shown at right in "Strengthening Frames."

Sliding Glass Doors

This method of strengthening the jamb won't work on a sliding glass door, and their locks tend to be very small. Here's what you can do to beef up sliding-door security. First, replace the screws that came with the door with ones that reach several inches into the structural house framing. Then, install a special dead-bolt lock or commercial security bar (as described in "Sliding-Door Locks," on the previous page). You could also cut a piece of broomstick or a 2x4 to place in the track between the door and the side jamb instead. Then your door is secure, short of someone smashing the glass—a step many intruders won't take because it makes such a racket.

Strengthening Frames

USE: ▶ pry bar • power drill/driver ▶ shims • wood screws

1 Prevent kick-in entries where burglars crash in the door and jamb with the lock intact. First, remove the trim.

2 Insert blocks of wood at several points into the gaps between the door casing and the house wall framing.

3 Remove the stop or weatherstripping so you can drive screws through the door casing and blocks into the house frame.

4 Use screws long enough to reach at least 1 in. into the nearest wall stud. Replace the stop to conceal screwheads.

Security Hardware

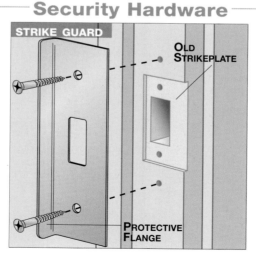

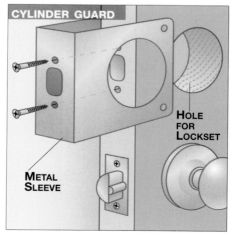

STRENGTHENING DOORS

Installing a Dead Bolt

USE: ▶ power drill/driver • hole saw • spade bit • wood chisel • utility knife • screwdriver • pencil or scratch awl ▶ dead-bolt kit

1 Use the paper template provided with the dead bolt to mark the center point of the holes you will need to drill.

2 Bore a hole in the door face using a hole saw. When the tip breaks through, drill from the other side to prevent tearout.

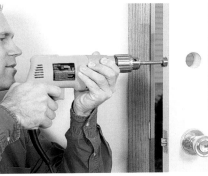

3 Bore a hole in the door edge for the latch bolt. To keep the drill (and the hole) level, it helps to sight along the drill.

4 Set the latch bolt in its hole, and make a tracing of the latch plate on the door edge with a utility knife.

5 Chisel out a mortise to match the size and depth of the latch plate. Turn the chisel over to clean the bottom of the mortise.

6 Insert the latch bolt (this one being for a new dead bolt) on the door edge, and fasten it in place with the screws.

7 Install the lock cylinder by sliding its metal extension bar, called a tang, through the latch mechanism.

8 To line up the bolt keeper on the jamb, color or chalk the end of the bolt, close the door, and turn the lock to make an imprint.

9 For maximum security, install a heavy-gauge keeper using screws long enough to reach into the house framing.

DOORS & JAMBS

2 Safety & Security

WIRED FOR SECURITY

For most people, alarm systems should be considered only a last resort. A good system is very expensive, often requiring monthly monitoring fees, and you will still need other security measures, such as window locks.

Too often, elaborate alarms have also one of two undesirable by-products: they either produce a false sense of security because no single system can keep out a determined burglar, or they become a nuisance—because of all the arming, disarming, and false alarms. That only makes you overly security-conscious and more fearful than you reasonably need to be.

Types of Alarm Systems

Alarm systems are either wired directly into your house's electrical system or are radio-controlled. A radio system uses battery-powered transmitters to send alarm signals to the master control unit. For a wired system, you need to loop wiring to and from each component of the system. A radio-controlled system is much easier to install but more expensive, and the batteries must be checked periodically.

Sensors installed at entry points in your home feed signals to the master control panel. Typical sensors are magnetic switches set on door-jambs or first-floor window sashes; trap switches that string across an air-conditioner or casement window; metal-foil alarm tape that detects movement in a window, motion sensors, and sensors that can detect the sound of breaking glass.

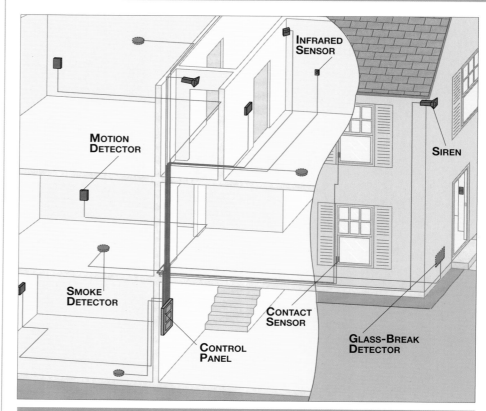

Basic System Layout

A complete home alarm system can combine many different security functions: monitoring entry at windows and doors, sensing motion inside rooms, and reacting to signals from a variety of sensors such as smoke alarms.

Common Components

A whole-house security system may include dozens of components, including inside and outside sensors and alarms.

Sensor packages at a window can be wired or remote. When a connection between units is broken, the alarm sounds.

Remote sensors also work on doors. A wired alternative, a plunger switch in the jamb, releases when the door is opened.

SECURITY SYSTEMS

Auto-Dialers

Some security systems offer off-site monitoring or some other way to respond to an emergency, even if you're not home. When an alarm is triggered at your house, it shows up at the security company's monitoring station. One alternative to this is an auto-dialer. When a security system sensor is triggered, the auto-dialer automatically calls the telephone numbers that you programmed it to call. Any auto-dialer that uses regular phone lines (that is, nonwireless) won't work when lines are down.

Some security systems tie the entry sensors and other components in your house to a central monitoring location.

CAUTION

▶ Before investing in an expensive security system that automatically reports trouble to local authorities, check with local police about their policy on false alarms. In some areas, the police may not have the manpower or the budget to cover every alarm. There may be a penalty for repeated violations. In some cases, police may not respond at all after there have been a certain number of false alarms from the same address.

Remote Sensors

Remote sensors are like wireless phones. They do the same job as standard security-system components, but they broadcast trouble to the central control panel instead of relaying it by wires strung through your house. Remote sensors make installation easy in an existing home where it may be difficult to conceal wiring. Most types of sensors are available as remote units, including motion detectors and sensors that monitor glass breakage and basement flooding.

A remote sensor has circuitry to monitor entry at windows and doors, for example, and battery power to signal the entry.

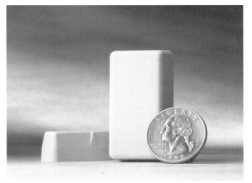

Many remote sensors are barely noticeable (only twice the size of a quarter) and do not require hard-wire connections.

Many whole-house systems include a remote trouble switch—a panic button—that can trigger an alarm from any room.

One economical alternative to a detector at every window is a centrally located audio unit that detects glass breaks.

Motion detectors are installed inside to signal movement in rooms and outside to trigger lights. Their range is adjustable.

WIRED FOR SECURITY 25

2 Safety & Security

PREVENTING & DETECTING

Many house-fire tragedies that make the news could have been prevented. But because the possibility of one's own house burning down seems so remote, many homeowners don't take even the most basic preventive measures. Many are surprisingly simple and inexpensive, yet very effective at saving both lives and property.

Smoke Detectors

Smoke detectors may be the most cost-effective consumer product on the market. Just consider what it costs (about $25 for a battery model) and what it can do (provide a warning in enough time to save lives). Detectors should be installed on every level of a home, high on the walls or on ceilings in open areas like hallways. Because deaths are most likely from fires that start at night when everyone is asleep, it's important to install detectors in halls just outside bedrooms. Many building codes now require hardwired units (with battery backup) for new construction and major remodeling.

Heating Equipment

Regular checkups are the best preventive measure for your heating system. Annual tune-ups are recommended for oil-fired furnaces; once every 3 years for gas-fired units. Electric units, which do not produce any combustion by-products, normally do not need regular tune-ups.

If you burn wood or coal regularly, have the flue cleaned annually by a chimney sweep. Wood and coal combustion in a stove is dirtier than other types of heating—you need the sweep because the worst hazards are out of sight: creosote, a gummy and flammable product of wood combustion that collects inside the chimney, and cracks in the chimney liner or bricks, which could let smoke and fire escape.

Escape Routes

Fire departments call it an alternative means of egress—a second way out of a room. On the first floor you could climb out a window. On second stories, you may need a portable safety ladder with metal arms that hang on the window sill and steps that unroll to the ground below. It's important to go over escape routes with children and practice using the escape routes.

Battery-Powered Detectors

Each year, more than 3,700 people die in over 400,000 residential fires. The best way to prevent property damage and injury is to install smoke detectors. About 90 percent of U.S. households have at least one, but up to 16 million detectors don't work, due mainly to dead or missing batteries. You should test battery-powered units monthly, and replace batteries that are low on power. Many building codes require hardwired units that are equipped with battery backup.

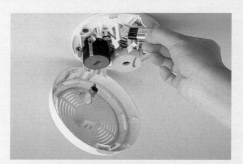

Most safety organizations recommend that you change the batteries in your smoke detector at least once a year.

Typical Detector Locations

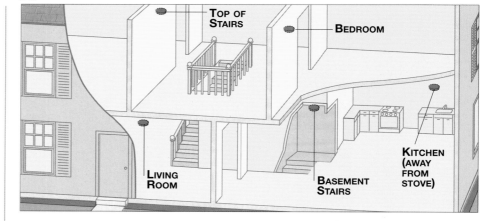

Hard-Wiring a Detector

USE: ▶ circuit tester • combination tool • screwdriver • drywall saw • pliers ▶ hard-wired smoke detector •

1 The most convenient power source is a junction box mounted to a ceiling joist. Cut power to the box before opening it.

2 Check your local codes before running a new supply line from connectors in the junction box to the detector mounting box.

FIRE SAFETY

Home Fire Extinguishers

Most homes need at least two extinguishers: a small unit in the kitchen and a larger, wall-mounted unit (generally installed in a closet) to use elsewhere. To avoid confusion in an emergency, choose A-B-C-rated units, which work on all types of fires. To use an extinguisher effectively, remember the acronym P A S S—Pull (the pin), Aim, Squeeze, and Sweep.

Use an all-purpose A-B-C extinguisher against paper, grease, and electrical fires. Aim at the base of the fire.

Typical Clearance Codes

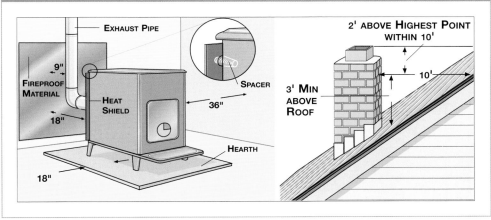

SAFETY CHECKLIST

▶ **Don't overload:** Do not plug more than one heat-producing device into an outlet.

▶ **Maintain smoke detector:** Replace battery and vacuum the unit annually; test a detector monthly, and replace it every 10 years.

▶ **Provide safe egress:** Have two ways out of every room—a door and a code-compliant egress window—including rooms in finished basements.

▶ **Fire-safe security:** Don't use security locks, bars, or devices that make it difficult to escape a fire.

▶ **Clean your chimney:** Have wood-burning chimneys inspected annually and cleaned as needed.

▶ **Store inflammables safely:** Store inflammable liquids in original containers with tight-fitting lids. Keep them away from heat sources or flames, preferably in a shed.

▶ **Be prepared:** Keep an extinguisher handy to stop a small fire from spreading. In other cases, call 911.

cable • wire connectors

3 Most hard-wired detectors have a surface mounting plate that attaches to the electrical box above the drywall.

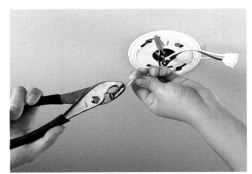

4 Following the instructions supplied by the manufacturer, join the detector leads to the power-supply cable.

5 The leads are attached to a harness that plugs into a receptacle on the detector. Twist the detector onto the mounting plate.

PREVENTING & DETECTING

2 Safety & Security

DANGERS AT KNEE LEVEL

Don't put off babyproofing your house until the little one can walk—a fast-crawling baby can find plenty of trouble. Crawl around your house yourself, and you'll discover many dangerous things at infant eye level: outlets, rickety TV stands, appetizing potted plants. And when a toddler learns to stand, he or she will hold onto anything to get upright, including the tablecloth dangling from a table full of bone china.

Storing Hazardous Items Safely

Any potentially harmful item should be either locked away or kept in cabinets and drawers with childproof safety latches. This includes knives and the obvious poisons such as medicine and cleaning products, but also mouthwash, shaving cream, perfume, and deodorant, which can be harmful to curious toddlers who like to put things in their mouths. Even high counters are not necessarily safe places; you need to put away that decorative knife rack because toddlers will figure a way to get up on a counter well before you think they can. It's a good idea to have one cupboard or drawer full of safe distractions, such as plastic containers or wooden bowls, that a toddler can get to.

Leave small appliances unplugged, and store them as far out of reach as possible. Plastic bags and plastic wrap also need to be kept in a high place. When discarding plastic bags from dry cleaning, recycle them at the dry cleaner's or tie them into knots before tossing. Buy garbage cans with secure lids that kids can't open.

Keeping Rooms Safe

Get safety gates to close off stairs or any room where you don't want the baby to go and you don't have a door to lock. Guards or gates are also needed to keep them away from fireplaces, wood-burning stoves, or space heaters.

One door lock that kids figure out quickly is the push button on bedroom and bathroom doors. For bathrooms, an adult-height bolt would be a better option for a house with young children. You won't have time to look for a bobby pin to unlock the door if your child slips in the bathtub.

Built-in Features

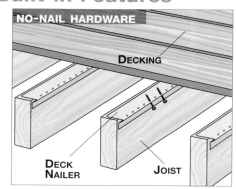

Remove one source of accidents on decks (raised nails) with hardware that allows you to fasten boards from below.

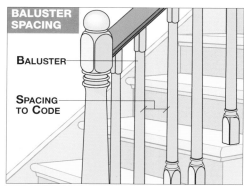

Codes control spacing (often only 4 in.) between parts of stairs and railings so that children can't get caught between them.

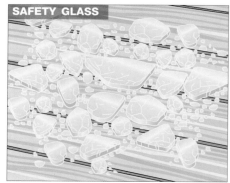

Unlike standard sheet glass that breaks into razor-sharp shards, tempered safety glass breaks into pebble-like pieces.

Openings

Safety grates can prevent falls. Building codes will not allow locks if the window is a potential fire-escape route.

Safety gates can prevent accidents on stairs and wall off rooms. This model has mesh panels that won't trap children.

Reduce the risk of accidents with landscaping tools and materials by walling them off with a hinged lattice gate.

CHILDPROOFING

Soft Surfaces

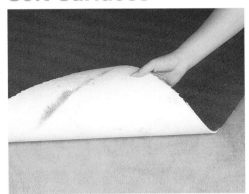

To reduce noise transmission through the floor and take the edge out of falls, install wall-to-wall carpet over a thick pad.

This cushioned chair rail for a child's room has thick foam stapled around a strip of plywood and covered with fabric.

On furniture where you can't create cushioned surfaces, you can at least reduce hard edges using a roundover bit and router.

Hazardous Materials

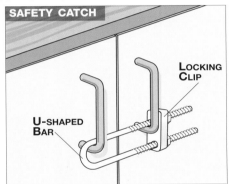

When you can't remove all hazardous materials from children's reach, lock up the cabinets that contain them.

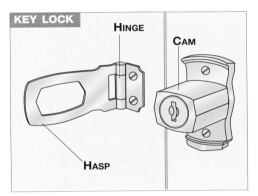

There are locks to fit every type of door and cabinet combination, including hasp locks that can't be opened without a key.

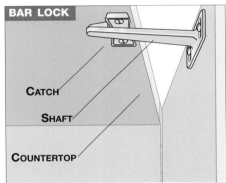

Where only minimal security is needed, this under-counter spring lock will keep a door from opening fully.

Electricity

Short cords are inconvenient on counter-top appliances, but they keep the wires from hanging within the reach of children.

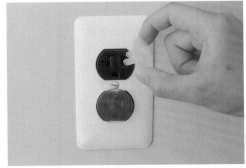

To eliminate shock hazards when an electrical outlet is not in use, plug a plastic insulator cap into the receptacle.

Required by code in many locations, circuit-breaker outlets (ground-fault circuit interrupters) reduce shock hazards.

DANGERS AT KNEE LEVEL

2 Safety & Security

BAD AIR & MORE

If you are a nonsmoker or allergic to dust, you can tell as soon as you walk into a room that these common pollutants are present. But some air pollutants can be harder to identify; fumes from volatile organic compounds (VOCs) and high levels of carbon monoxide are prime examples.

VOCs

VOCs are released as gases from many ordinary products, including wood finishes, paints, adhesives, rug and oven cleaners, dry-cleaning fluids, furnishings, and office equipment. Some (not all) VOCs have a distinct odor, and some (not all) products containing VOCs come with caution labels.

Health problems resulting from exposure to VOCs include skin rash, upper respiratory irritation, nose bleed (from formaldehyde glues), headache, nausea, vomiting, fatigue, and dizziness. Because VOC emissions are greatest in new materials and gradually dissipate, symptoms are likely to be triggered during or shortly after remodeling and cleaning work.

At home, there are several things you can do to reduce your exposure. Meet or exceed label cautions for ventilation when using products that emit VOCs. Don't store opened containers of paint and other materials containing VOCs in the house. To reduce emissions from composite boards used in some cabinets, seal the interior surfaces with two coats of polyurethane.

Carbon Monoxide

Carbon monoxide (CO) gas is made whenever fuels such as gas, oil, kerosene, wood, or charcoal are burned. A properly used (and maintained) stove or heating appliance won't leak a significant amount of the gas. But CO poisoning is tricky to detect, and the initial symptoms (dizziness, headache, nausea, shortness of breath) are easily mistaken for other illnesses. The best preventative measure is to have heating appliances inspected annually by a professional. CO detectors are widely available but should only be used as a backup for yearly maintenance. Performance of these detectors varies widely, and because CO is colorless and odorless, it's easy to think the real thing may be a false alarm.

Lead Paint

Many house paints made before 1978 contain lead, which is a threat to children and can cause permanent brain damage, behavioral problems, and other serious health problems. If you live in a pre-1978 home, you should contact the EPA National Lead Information Center (800-424-LEAD or epa.gov/lead) for free information on testing and safety precautions and for guidelines on whether the paint should be left alone, covered, or removed.

You can test existing paint for lead with a simple kit. Following instructions, scrape the surface, apply the activator, and wipe.

GREEN SOLUTION

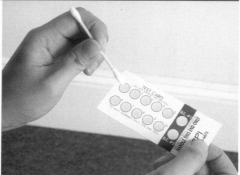

The activator makes a liquid sample on the swab that you then apply to the test card, to find out the lead-content reading.

Lead in Water

Sources of lead in drinking water include lead pipes (common until around 1930), brass faucets or fittings that contain some lead, or copper pipes soldered with material containing lead. If you suspect lead in your water, have the water tested. To reduce the lead you may be consuming, use cold water for consumption (because it doesn't sit in pipes for long), and run the tap 1 or 2 minutes before you drink. You may have to replace old pipes.

To take a sample of water for testing, first use the flame from a match to burn off impurities on the faucet head.

GREEN SOLUTION

Fill a small, clean container with a sample, which can be tested by some town health departments or a private lab.

ENVIRONMENTAL HAZARDS

Radon

Radon (a colorless, odorless gas) is the second-leading cause of lung cancer. This naturally occurring gas comes from the ground, well water, and some building materials. Nearly one out of every 15 homes contains high levels. It's easy and inexpensive to test your home or well for radon. Indoor levels of 4 picocuries per liter or more need to be fixed. Contractors can install an air-pumping system that vents radon from the ground under your house to the outside. (For more information, contact epa.gov/radon.)

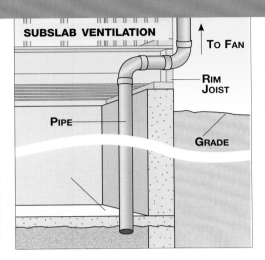

GREEN SOLUTION

A radon test kit consists of a small canister that you leave in your living areas. You mail it to a lab for results.

Asbestos

Asbestos is often found as insulation and fire protection on pipes. You also may find asbestos in old cement roofing and siding shingles, insulation (in houses built between 1930 and 1950), walls and floors around wood-burning stoves, and hot-water or steam pipes in older houses. The safest solution is usually to leave it undisturbed—asbestos material in good shape won't release fibers. If you need to remove it, hire a state-licensed abatement contractor.

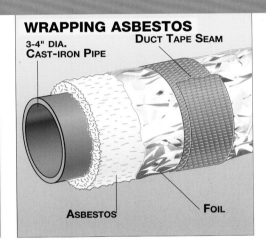

GREEN SOLUTION

Asbestos was often used to insulate heating pipes in older homes. You should test a sample before deciding on removal.

Gas

A natural- or propane-gas leak is detectable due to mercaptan, an additive in gas that has a rotten-egg smell. If you suspect a gas leak, the safest course is to leave the house immediately and report it to 911 and the utility's emergency number. If you suspect problems in a gas-fired appliance but don't smell the aroma of a major leak, you can turn off the gas valve near the appliance or the main valve at the gas meter.

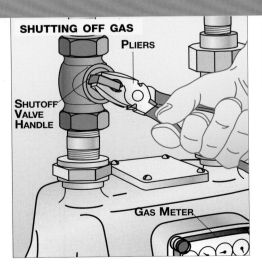

GREEN SOLUTION

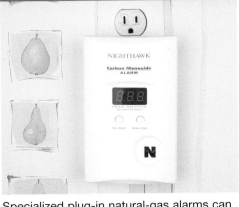

Specialized plug-in natural-gas alarms can detect small amounts of leaking methane and propane in your house.

2 Safety & Security

UNIVERSAL DESIGN

Making homes accessible for disabled people was once considered an extra that added to the cost of building it. But the trend in residential design today is to include features that make a building accessible for a disabled person and easier and more convenient for anyone else, too. It's called accessible or universal design. Here are some of the basic principles that can be incorporated into new construction and remodeling projects.

Entrances & Floor Level

Even in single-story homes, level changes are common at the entrance because the floor level is generally higher than ground level. There are a few good ways to eliminate this barrier without sticking a wood ramp on the front of the house. One is to create a gradually bermed, or earth-ramped, entrance with landscaping timbers.

In new construction, you can lower the foundation or floor level. Typically, the foundation is several inches aboveground; the first floor is a foot or so higher than that. But when building a home, you stop the foundation at grade and use a combination of pressure-treated plywood sheathing and a waterproofing barrier to prevent leaks at the critical transition between foundation and frame. Another way to reduce floor height is by creating a ledge in the top inside of the foundation equal to the depth of the floor framing. To reduce the chance of damage from wood rot if water seeps in, the joists can be set onto metal hangers attached to a pressure-treated ledger board bolted to the foundation.

Other Features

There are other universal-design alternatives to standard architectural details. For example, a round, easily grasped stairway banister makes a safer, more convenient handrail than the typical 2x6 wood cap piece that is too wide to hold onto firmly. Another design defect is common to by-passing closet doors and cabinets: inset pulls. These pulls put even more stress on finger joints than round knobs. Instead of applying the strength of your whole arm, these pulls focus all the stress on the ends of your fingers. D-shaped pulls with at least 1½ inches of clearance to the door surface make the job easier.

Access

HALLWAYS

RAMPS

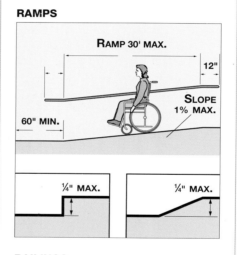

RAILINGS

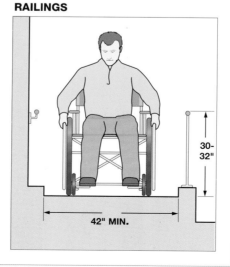

Kitchens

STOVE CLEARANCES

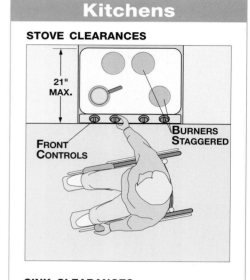

SINK CLEARANCES

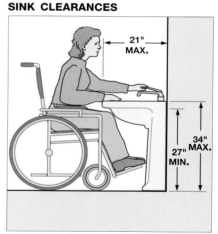

TABLE CLEARANCES

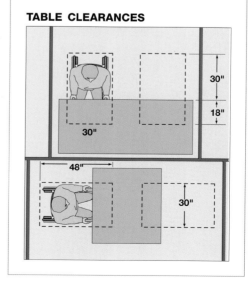

UNIVERSAL DESIGN

Baths

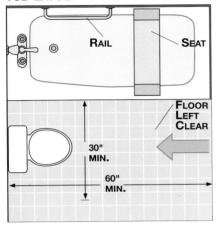

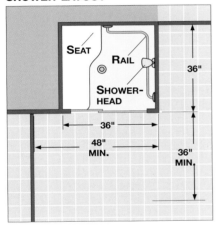

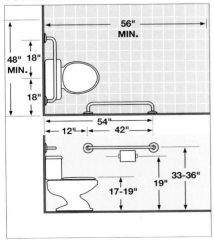

Universal design seeks to make houses easy to use and accessible for almost everyone. Overall, rooms don't look much different, although this space has a raised dishwasher, lowered dish cabinets, open-plan sink, pull-out steps and stools, and many similar features.

Special Fixtures

This sink has flexible plumbing connections and an electric motor to raise and lower the sink platform.

Special motorized cabinets that raise and lower also are available to make storage more accessible in kitchens.

Counters are more versatile when they have several levels and pull-out extensions near appliances.

To reach higher shelves, the door on this bin-storage cabinet is equipped with a sturdy (four-legged) fold-down stool.

3 Remodeling Guide

36 PLANNING
- Planning a Remodeling Job
- Design Professionals
- Do-It-Yourself Plans
- When to Call for Help

38 HIRING
- Management Options
- Hiring a Contractor
- Watching the Budget

40 MAKING CONTRACTS
- Checking Estimates
- Contracts
- Starting & Completion Dates
- Payments

42 SUPERVISING
- Following Codes
- Supervising a Project
- Changes & Extras

44 DEALING WITH DISPUTES
- Settling Disputes
- Material Substitutions
- Time Problems
- Resolving Disputes

3 Remodeling Guide

BEFORE

AFTER

Every home is different, but most successful remodeling jobs follow a similar course. Here is the process from a winter start to a fall finish, with pictures from a major overhaul.

Planning a Remodeling Job

When you're lounging in the new family room, finally free of whining saws and paint fumes, it may seem incredible that such a complicated remodeling project began on the back of a napkin. But that's the way many jobs start—with a simple sketch. The question is how to develop the idea into a plan that balances what you want, what you need, and what you can afford.

Not every homeowner is up to the task of developing home-improvement dreams into working blueprints. But the more information you gather and the more details you give a remodeling contractor, the more likely you are to get a reasonable version of what you want at a reasonable price. And that's the final measure of a successful project. The work may cost more and take longer than you imagined. But if you can transform the heart of your first sketch into long-awaited extra living space, you'll find that the hassles tend to fade.

There are some nightmare jobs. But talk to friends and neighbors who have lived through a major project, and you'll find that most report their own take of the same basic experience. They may grumble about the delays and disruptions—and the money, of course—but then they'll want to take you on a guided tour through the new addition. To reach that position, your ideas must be on paper in the form of measured drawings. You need them to get realistic estimates, to get a building permit, and to pin down construction details for your contractor. Although there are many variations, you have three basic options: hire a design professional, work with a design-build con-struction firm, or draw your own.

Design Professionals

Both interior designers and architects must meet standards and follow guidelines set by professional societies. They work for you and can represent your interests with workers and suppliers. They can come up with a plan, help you analyze estimates, select a contractor, and follow through with regular checkups on materials, schedules, payments, and job quality.

Although there are exceptions, interior designers, true to their title, are likely to give more attention to the surface than the structure, while architects take it in reverse. This built-in bias is reflected in their drawings: full-color perspectives from interior designers and measured blueprints from architects. That means an interior designer may be the best choice if you need a lot of ideas about remodeling an existing space, and an architect may be a better choice if you need to reshape the building.

Despite some controversy about who can legally use the word "design" to describe their services, design-build firms are an attractive option because they offer a complete package. On a kitchen remodeling job, you may be able to view dozens of designs on a computer screen and get detailed plans, plus cabinets and installation. Some firms have an architect or engineer on staff for more comprehensive jobs, while others work with code-approved plans that can be modified to suit your needs. Whatever the circumstance, you deal with one company instead of an array of design professionals and independent contractors. That's good, because one party is responsible for every phase of the job. But if problems develop, there is no intermediary to sort things out.

Do-It-Yourself Plans

It's a good idea to develop your ideas in detail, even if you turn to an architect or design-build

PLANNING

firm to complete the final version. This preliminary planning will refine your best ideas and weed out the worst variations. You can work from stock designs in plan books and, in some areas, use preapproved plans supplied by the building department for basic projects such as decks. Steal ideas from the pages of shelter magazines that show materials and designs you like, and survey displays in local home centers. It helps to see materials firsthand instead of in a catalog where the scale can be deceiving.

Sketch out a basic floor to scale on graph paper, and use models to approximate furniture—or use home-design computer software to work up your plans. Even basic programs allow you to try different furniture arrangements and change materials with the click of a mouse. But it pays to test your plan at full scale—for example, by laying out the lines of a new deck with stakes and strings. Seeing even the barest outline helps to guard against a chronic planning problem that can occur no matter who draws the plans—the underwhelming moment when the new area fills with people and furniture and seems smaller than the ample space you drew months ago on the back of a napkin.

When to Call for Help

Once you have plans for a remodeling project, you need to find contractors to carry them out—or maybe not. You can save money by handling part of the work yourself—maybe 30 percent of the job price if you act as the general contractor of the project. But you can run into pitfalls, too, mainly when you get in over your head on jobs that require skills you don't

Your contractor doesn't want to hear...

> "*But you're going to the landfill anyway.*"
>
> This is one of the many twists to the as-long-as-you're-there idea in which a client stretches the boundaries of the project to include all kinds of little favors and accommodations that amount to freebies, and acts wounded if the contractor doesn't oblige.

have. You may start with the best intentions but wind up getting in the way of pros who are accustomed to the daily regimen—and work weekdays when you're at that other job.

There are no set rules about which jobs to tackle and which to leave to the pros. Do-it-yourself skills, interests, and budgets vary so widely that you might do an excellent job on part of the project your neighbor wouldn't touch. But there are sensible guidelines you can apply. Number one is "if in doubt, don't." Don't plunge into a project unless you have a realistic idea of the tools, skills, time, and money involved. Then you can make common-sense decisions, including the most important one: whether or not you can do the work safely.

So when you consider the array of jobs on a major remodeling project—from foundation work to roofing—the best approach is to avoid work that is inherently dangerous. Naturally, that depends to some extent on your idea of danger. For one do-it-yourselfer, laying shingles on a low-slope roof may be a snap. But if you start sweating halfway up a ladder, even the simplest roof improvement can be hazardous. You don't need construction expertise to sort this out. There is a good reason painting is at the top of the do-it-yourself list and wiring is at the bottom: paint can't kill you.

You can't rule out everyday accidents. Some scrapes and bruises are inevitable when you handle a lot of tools and building materials. But it's wise to steer clear of jobs where problems due to lack of experience could result in structural defects or serious injury.

As to structure, that means in most cases you should leave foundations, framing, and similar work to the pros, and concentrate on finishing trades such as drywalling, trimming, and painting that use the structural systems for support. Foul up on the surface and you may make a mess, but nothing is likely to collapse.

As to injury, that means in most cases you should let pros handle mechanical systems and equipment—heating and cooling, electrical, and plumbing. Don't take a shot at installing a new furnace that could burn down the house if it's installed incorrectly. Plumbing is a bit of a gray area because on some pipes, the worst you'll get from a mistake is a leak, and you can fix leaks—or call a plumber. But making the wrong kind of connection on other piping, such as gas lines and sewer lines, can have serious consequences.

A surprise extra: removing an old oil tank.

Blocks delivered for the foundation.

The hole in the ground takes shape.

3 Remodeling Guide

Blocks rise, and the floor is ready to pour.

The truck only crushed a few bushes.

First-floor framing—and the new porch.

If you don't know enough about a project to have a reasonable idea about its potential dangers and the crucial installation steps that prevent them, leave it alone. Also rule out installation work that could void a warranty and jobs that require special licenses or inspections. Removing asbestos is one good example; wiring a new addition is another. With the proper permits, you can pour a bump-out foundation and nail up framing on your own house, although a building inspector may pick your work apart. But in many areas there are jobs, such as installing a septic system and wiring, that won't be approved unless you are licensed to do the work.

Management Options

If you decide to have others do most of the work, you may want to consider hiring a remodeling manager to monitor the job. They have no standardized qualifications and no professional society, so finding one is often a matter of word of mouth. Some provide supervision on the job site; others help you select materials and deal on your behalf with architects, designers, contractors, and even mortgage bankers. You may pay a flat fee or a percentage of the job price for what amounts to a hand-holding service provided by a manager who may be very helpful but doesn't design or build anything. But some homeowners need a friendly voice on the other end of the phone who will do some of their negotiating—someone who doesn't have a stake in the design or a hand in the construction.

> Your contractor doesn't want to hear...
>
> **"*You're not eating dinner, are you?*"**
>
> It's aggravating to be constantly on call, and hand-holding a client day and night, Saturdays and Sundays, even though the questions can wait a few hours and both of you are done with work for the day and at home with other things to do.

The next rung up from hiring a remodeling manager is managing—acting as your own general contractor. On the plus side, you can save at least some of the contractor's markup buying materials in bulk. However, you probably won't get the best price that suppliers reserve for repeat customers (year-round contractors), and you may not save anything on a small, one-time order.

The most taxing part of being a general contractor is hiring and managing several subcontractors. To start with, you probably won't get their best price, either. They might trim down a bid for the general contractor who keeps them working job to job—and who knows where the excess is—but not for you. You will save the general contractor's salary, of course, but only if you have the time to spend on site and figure your time is free. Don't try to run a remodeling job from your other job over the phone unless you're clairvoyant. You need to be there to see what's happening and have enough general construction knowledge to understand what you're seeing. You don't have to know as much about electricity as the electrician. But you will need to be familiar with the basics of every trade so you aren't taken advantage of, or simply stumped by conversations about the project.

The best jobs to manage yourself are short in duration, relatively inexpensive, straightforward, and self-contained—like adding new siding. Whole-house remodeling that involves plumbing, electrical, heating, cooling, and structural systems throughout the building are more difficult, particularly kitchen and bathroom jobs that cram a lot of mechanical and finishing work into a small space. As you hire more subs with different expertise, interests, and schedules, the job becomes exponentially more difficult to control.

You may save money when you're the boss, but you'll be shouldering a lot of responsibility. It can be a difficult spot to be in—when the buck stops with you, and it's your own buck.

Hiring a Contractor

If every homeowner with a remodeling job gets a top contractor who comes highly recommended, puts in 10-hour days, and always finishes on time, then who hires all the other guys? And when the economy is strong, loan rates are low, and a lot of homeowners are remodeling, how can you snag these top contractors? The search can be time-consuming and frustrating, and many contractors you call may be unavailable.

Locating the right contractor takes legwork: developing leads, tracking down referrals, verifying recommendations, conducting face-to-face interviews, and keeping track of the

HIRING

one who seemed so forthcoming and the one who rubbed you the wrong way.

The ideal shortcut is to hire a contractor recommended by a trusted friend or neighbor who has just finished a project like the one you have in mind—a situation where you'll be able to find out about costs, schedules, attitudes, and see the results firsthand. Short of that ideal, the best bet is to leave a lot of time for the search and to start with the biggest possible pool of qualified contractors. You can get names from many sources that generally fall into two categories: referrals and recommendations. Referrals come from advertisements—for instance, the side of a truck or a card posted at the lumberyard. There is no screening involved; they are just names.

Recommendations carry more weight than referrals because they include an evaluation—say, from a friend or neighbor, your home-owners' insurance agent, mortgage banker, real estate agent, or someone else in the housing business who knows what you want.

Before you start weeding out candidates, also consider the Internet. In a typical Web search, you'll get lost in lists of contracting firms halfway across the country. But keep looking and eventually you will hit trade associations, such as the National Association of the Remodeling Industry (NARI, at www.nari.org), and a growing number of Internet-based contractor-referral services. Several on-line referral services are dead ends; either too new or too disorganized to use. You scroll through snappy-looking option screens (and ads to click on, of course), but when you finally start to zero in, it turns out that only a few states and counties can be searched—and yours is not among them.

But fully operational sites, such as www.improvenet.com, one of the largest referral services with a national database of 600,000 contractors, help to make a match. You submit the particulars of your job and receive a list of screened contractors that omits anyone who has been in business less than three years and does not have a clean legal and credit record.

The service is free to consumers; contractors pay for the referrals. It's a good way for them to get leads and screen out homeowners who aren't really serious about starting a project in the near future and are just shopping for ideas. A typical response, via e-mail, may give you four local names within 48 hours. But there is no guarantee that these contractors will be agreeable, work for the price you have in mind, or produce high-quality results. All that is for you to discover.

By whatever method, gather as many names as you can, and make the basic checks with local consumer protection agencies and the Better Business Bureau to find out if the contractor has a license, plus a record that is free of consumer complaints.

Finally, the contractor must be available to do the job—but not immediately without a good explanation. You can't be the only one to find the guy. If he's good, a lot of people want him, and you may have to get in line.

Watching the Budget

Soliciting bids from remodeling contractors is like shopping for any consumer service—just more drawn out and confusing. The main mystery is that even though the product—your project—is painstakingly defined with plans and specifications, everyone who looks at it sees a different price tag. Few will be as low as the one you envision, and some will be so far from others that it will be hard to imagine the contractors are estimating the same job.

Variation can stem from basics like the law of supply and demand, of course. When loan rates are low, the weather is warm, and a lot of homeowners are looking for estimates, availability falls and prices rise. If you solicit a highly sought-after contractor who doesn't really need the work, the estimate may soar.

> Your contractor doesn't want to hear...
>
> **"So what do you think about these nine kinds of tile?"**
>
> This syndrome occurs when clients in the thrall of a major remodeling project come to believe that their house is endlessly fascinating and fail to realize that rehashing every possible feature is like showing their most recent 200 baby pictures to a pediatrician.

At last: some big windows and a view.

There used to be a really nice lawn here.

It looks like the drawings—sort of.

3 Remodeling Guide

It looks much too big from the outside.

From the inside the rooms seem small.

Have to pick paint colors by next week.

If your job involves a lot of demolition work, and no one can know ahead of time how much rebuilding will be required, the bid may rise to cover the worst scenario. Bids also can increase if you impose extra conditions on the work such as a tight schedule, if your house is full of custom details and odd angles—and if you come off as difficult as your house.

After a few meetings to check the site and talk about the plans, an experienced contractor can tell the accommodating client from the one who will check for paint blemishes with a magnifying glass—and adjust the bid accordingly. But don't dismiss a bid just because it's high. Find out why it's higher and if the increased labor and material estimate is for features that offer durability or convenience others don't provide.

Checking Estimates

This part of the project involves a continuation of the sorting and sifting, first through references and recommendations of many contractors to find the bidders, and then through the sometimes complex estimates of your short list of candidates. But if you take the time to solicit and analyze bids thoroughly, you can discover a lot more than price. A detailed breakdown can reflect the contractor's professionalism, attention to detail and overall attitude, and help you make the final selection.

The first key is to be sure that you are getting real prices, which means that you should be prepared to hire any of the bidders. You can try to get a bid or two purely for comparison. But when a contractor gets the idea that you want an estimate only for negotiations with another contractor who basically has the job, the bottom line may not be realistic.

The second key is to make sure that every contractor bids on the same job, based on a complete description of the project. If three contractors have different ideas about your deck, ask them to price the same basic platform, and list their plans for floating stairs or built-in benches separately. If you're not sure yet about some of the materials in your project—say, exactly what type and brand of window—pick one for the bids so you won't be stuck comparing apples and oranges. Whenever possible, and particularly on big projects, submit plans that show the job pictorially, as well as specifications, which list every door and floor tile and appliance by type, size, and model.

Because labor often costs more than materials and has the greatest effect on overall job quality, it's also crucial to know whose labor you'll be getting for the money: who will be the main job supervisor on-site every day. Occasionally, firms send an estimator to lock up the bid—someone you never see again who may forget to tell the contracting crew all the things you were promised.

Some firms start off with a general contractor who seems to be the supervisor but only visits the site from time to time and leaves most of the work to employees with less experience and subcontractors who only know about their part of the project. You ask why the door will swing one way instead of the other, and no one knows for sure except the general contractor, who may respond to his pager in the next few minutes—or not until the end of the day when the door is already installed. The worst scenario is paying for the reputation of a particular contractor or crew and getting the second team stocked with apprentices.

To help you analyze and compare bids, ask for an itemized estimate. Bottom lines and square-foot prices will give you an overall view. But itemized bills allow you to decide between differently priced materials, isolate parts of the job you might do yourself, and find key steps one bid includes and another leaves out or glosses over. Even on modest remodeling jobs, it's helpful to have bids broken down by labor and materials for four major operations: foundation and framing, closing in (installing windows, doors, siding, and roofing), mechanicals (heating, cooling, plumbing, and electrical), and interior finishing.

To be practical, every bid should predict true and complete costs, which means the best bids may contain some bad news—mainly, charges for remodeling related jobs, including fees; permits; inspections; preliminary demolition;

> Your contractor doesn't want to hear...
>
> **"How could it possibly cost so much?"**
> No one likes a whiner, and this is whining. Every remodeling product and service has a cost, and before a client orders a top-of-the-line commercial stove or triple-glazed windows, it's easy enough to find out what the bill would be.

MAKING CONTRACTS

carting, which can loom large if several rooms are gutted; relandscaping to cover the tracks of heavy equipment or a new sewer line; and upgrading piping and wiring in the old part of the house to meet the demands of modern systems in remodeled rooms. Leaving them out can make a bid look enticingly low—until you start paying for all the extras.

Contracts

No legal language can make dishonest people honest or change slow, sloppy carpenters into expert woodworkers. But on the up-side, contracts put the job into words that describe its components in enough detail that you don't have to rely on memory or good will.

A good contract also allows for misunderstandings with a provision for third-party arbitration—a last-chance to maintain job progress on-site where work is done instead of in court. Nobody installs roof flashing in court. On commercial projects with millions of dollars at stake, it may pay to pursue a settlement. On many residential remodeling jobs, even when you're sure that you're right and the contractor is wrong, it rarely pays to sue. It takes too much time, too much money, and prolongs the life-disrupting overload of aggravation.

To guard against problems, use a contract that begins with basic provisions, the who's-who and what's-what: details of names, addresses, the contractor's license and insurance, starting and completion dates, and a detailed description of what the job entails. To simplify the contract, it can refer to blueprints instead of listing every design feature, and to a separate document called the specifications, which lists wood types, appliances, paint colors, and everything else in enough detail to avoid substitutions and misunderstandings. Here is a look at some key provisions.

Starting & Completion Dates

To have any teeth, dates should be described as "of the essence of the contract." That phrase will give more weight to your case in court. But on the job what helps most is a detailed schedule you and your contractor agree to before work begins. It should include target dates for acquiring permits, ordering a dumpster for debris, initial demolition, and the first material deliveries.

The more detailed your schedule, the better your chance of avoiding serious trouble—of almost any kind and at any stage of the project. Track progress in small, detailed steps that can be corrected before they mushroom into complex problems that can seem insurmountable. Delays don't count if they are generally beyond the contractor's control—including weather and your decision to change to windows that don't arrive for two extra weeks.

To coerce a projected completion date into reality, some contracts contain a bonus clause for finishing early and a penalty clause for finishing late—bad ideas for three reasons. First, many older homes ripe for remodeling contain too many unknowns to pick a reliable date months in advance. Second, you don't want anyone to cut corners in search of a bonus. Third, end-of-job penalties intended to encourage completion can have the opposite effect as the final payment erodes to the point where the contractor will be financially better off starting another project than finishing yours.

Payments

The best course with remodeling work is to pay for services as you receive them, typically in stages tied to job progress—with two exceptions: a down payment that gives the contractor a financial head start, and a final payment that gives you some leverage with the checklist. Percentages vary (and they're negotiable), but should be large enough to count—say, 15–20 percent of the total at each end of the job.

The rest can be split into four equal payments: for the structure, for closing in, for mechanicals (plumbing, heating, cooling, and electrical work), and for finishing work. No reputable contractor needs 40–50 percent up front

> Your contractor doesn't want to hear...
>
> **"Lend me a saw; I'll help."**
> It sounds so cooperative—like we'll all pitch in and raise the barn frame together—but it can strain the relationship, slow up the job, and raise hackles when a client doesn't have enough sense to know that you shouldn't inflict a part-time hobby on a full-time professional.

White trim but natural wood railings.

Saws run all day, inside and outside.

In place of the wobbly old steps.

3 Remodeling Guide

Inside now, it sounds like a train wreck.

The antique appliances leave for good.

The kitchen: like camping out inside.

to cover costs. They have credit at lumberyards, and don't pay their subcontractors in advance. A contract also should direct suppliers and subcontractors on the project to provide a release called a Waiver of Mechanic's Lien Rights. Once signed, the plumber can't sue you for payments you provided to the general contractor that were not passed on.

Following Codes

With a plan in hand and a contractor in the wings, there is a moment of calm before the remodeling storm when you can clear out rooms, prepare for material deliveries, allocate parking space, and arrange the details of a subject that's rarely mentioned in contracts: bathroom facilities for a crew of contractors. Most of the pre-job jobs are organizational chores that contractors will eventually tend to if you don't. But there are two subjects that won't tolerate a haphazard or last-minute approach: building codes and zoning ordinances.

Building codes are unlike everything else about remodeling because you can't change them. There is no fussing or fudging the way there is with paint colors. The building inspector says your foundation trench is 2 inches too shallow; you start digging. Inspectors have the final say on the size of girders, the locations of electrical outlets, the thickness of drywall, and just about everything else. And they can show up unannounced to tour the job and set you straight. It's not that you want to be bowing and scraping when the inspector pulls up, but it pays to be accommodating.

To start the relationship on a positive note, it's a good idea to visit the building department with your preliminary plans, particularly on small jobs where you sketch the drawings yourself. Ask questions, including the most obvious one: do I need a permit? Generally, you do if the project involves structural work such as extending a foundation or breaking through bearing walls. But you also need a permit to change the use of a structure—say, to convert an attached garage or storage attic into living space.

The most common problems stem from incomplete plans—failures of omission such as showing joists in the right places but failing to specify their sizes, and showing a new second story without accounting for the extra load on the old foundation. But details that aren't caught in planning probably will be on site, where you can expect at least four inspections: one to verify the depth of footings, one to examine framing, one to check mechanical systems, and one to make a final check before granting a Certificate of Occupancy.

Some homeowners try to avoid the supervision and do the job without a permit, generally to keep the value of the job from increasing their real estate taxes. But this omission leaves you vulnerable to substandard work, unsound materials—and a disgruntled neighbor who doesn't like the construction noise. One call to the building department and you could be liable for fines on top of the costs and hassles of rebuilding to code.

You have to observe building codes that govern the details of construction and zoning regulations that control what you can do with your site. The main zoning categories are commercial and residential, but there are many subcategories. For instance, in one area you can remodel a large house into a two-family, but a few blocks away in a different zone you can't. Single-family houses on one street may need 2 acres of land but only ¼ acre on another. In historic districts, your plans may have to be altered to include certain styles of windows, siding, and even paint colors.

But the main stumbling blocks for remodelers expanding the building footprint are two restrictions on property: the percentage of land that can be covered by buildings and how the uncovered property can be distributed among front, side, and backyards. For example, the code may specify at least 100 feet of side yards, with a distribution percentage no greater than 80-20. That means the narrowest yard must be at least 20 percent of 100 feet—in theory, to prevent next-door neighbors from

> Your contractor doesn't want to hear...
>
> "*I changed my mind.*"
> This is fine in the planning stage. It would be okay at any stage if the client were willing to pay the full tab of extra time and materials for switching in midstream from, for example, casements to double-hungs—and didn't grumble later about delays caused by the indecision.

SUPERVISING

building nearly on top of each other. The rules can box you in, but there may be a way out. It's called getting a variance, and each jurisdiction has its own, generally time-consuming, procedure for granting variances.

Typically, you submit copies of your blueprints (including a detailed plot plan) to an appeals board, and present your case at a hearing after posting notices of the hearing up and down your street so your neighbors have a chance to object. You can present the case yourself—relying as much as possible on facts and figures instead of what you want—or hire a real estate attorney to present a complex case where you anticipate objections from adjacent property owners.

Supervising a Project

How do you keep track of a complicated process undertaken by people you don't know who use techniques that may be unfamiliar and report their progress in terms you may not understand? You could turn over every detail of job management to the general contractor. But that's too much like falling asleep in the barber's chair. If there aren't any problems, you can wake up to a happy ending. But if there are—and you can count on more than a few glitches on most projects—the problems may be time-consuming, expensive, or even impossible to correct.

The best way to safeguard your investment is to establish a schedule for meetings on job progress where you can divide the project into manageable pieces, watch for signs of trouble and plot the progress of work done and money paid. End-of-week meetings are best, even if everyone stops work a half-hour early to attend. A slight dip in productivity is a good trade for up-to-date information. Include the general contractor and any major subcontractors, such as a kitchen firm hired by the G.C., as well as the architect, designer, or decorator. Review the previous week's work (mainly things on the schedule that didn't get done); the upcoming week, including materials in hand or on order; and an update of payments made and due.

The idea is to uncover potential problems and small delays, although you may have to probe for the bad news because many contractors tend to minimize trouble. They know you don't want to hear it and may honestly think that they can fix the wound before it begins to fester.

To minimize the disruption that a major remodeling job can inflict on a household, also solicit advance warnings of any significant demolition or cutoffs—for instance, periods when the power or water supply will be shut down and inevitable disruptions, such as when you'll have to vacate rooms while two coats of polyurethane take two days to dry. The meetings are likely to expose a few surprises. Three of the most troublesome are subcontractor scheduling, material and design changes, and debates about work that you think is part of the contract and the contractor thinks is an extra.

On many projects, even with a supervising general contractor who seems to take part in every operation, most of the work is handled by subcontractors such as roofers and plumbers.

The main stumbling block with them is coordination—a logical sequence that prepares foundations for framers and framing for drywallers. If the electrician disappears before the rough wiring is complete, the general contractor may not be able to call for an inspection, which means a delay for drywallers, finish carpenters, painters, and other trades.

To make sure that there aren't any gaps—and to avoid a few days of Keystone Kops where everyone shows up at the same time—you should know whom to expect next on the job, roughly what they are supposed to do, and how long it will take them to do it.

Changes & Extras

Another potential monkey wrench, changes in the plan, can occur because materials weren't available or because someone forget to order the windows; it happens. And sometimes the

> **Your contractor doesn't want to hear...** "*What do you mean it's not in the contract?*"
> It may seem reasonable for a contractor who hangs drywall in the new garage to tape and finish the seams—but not for free. Experienced contractors specify in the contract for all to see exactly what work will be done and tag any steps or finishing touches that won't be with the initials N.I.C (not in contract).

Didn't even scratch the paint; amazing.

It's almost there; ready for finishing.

Another three days with no furniture.

3 Remodeling Guide

The new family room.

The new kitchen.

One of the new bedrooms.

change is an improvement everyone wants and no one foresaw. Whatever the reason, any significant deviation from the blueprints needs a little last-minute planning because it wasn't scheduled initially.

On big jobs, these alterations are committed to paper with a change order. You can use an official form, or at least note what's being changed and whether it effects the job price. That's raises one of the touchiest subjects:

Before: the old house in the snow; and after, finally, the new addition.

extras. It's often contentious because contractors tend to see most changes as extra work, particularly toward the end of a job when payments for materials and subs leave a smaller than expected share of the job price as profit.

Across the table, homeowners tend to see most changes as part of the contract—simple replacements. There are obvious points of agreement at both ends of the spectrum. If the basic cooktop you specified is out of stock, and you want to substitute a six-burner restaurant stove, that's an extra. If the contractor can't locate the off-white paint you wanted, substituting a slightly whiter white is a non-issue. But there is so much gray in between, and there's not much you can do about it except talk out your differences. Be firm on any material or work that is either written into the job specs or drawn in the blueprints. That includes things that are inescapably part of an explicitly detailed job, such as hardware for doors—unless they are marked N.I.C. (not in contract) in the contract. Be prepared to negotiate the cost of unexpected work, such as installing a beam to beef up joists that turned out to be undersized only when the job was underway and the fault was exposed.

You could argue, and some homeowners do, that a contractor should be liable for any unexpected conditions. But no G.C. or architect has X-ray vision. The only other option is to draw a contract that includes charges for every foreseeable extra, which will prove to be a bad deal when most don't materialize. The goal is to keep the job moving, and that won't happen if you try to take advantage of a miscalculation. You need to find a middle ground on extras—and try to balance them with a give-back on jobs or materials you can live without.

DEALING WITH DISPUTES

Settling Disputes

A few parts of your remodeling job will be different than the picture you had in mind. Some problems can't be helped—for instance, a week of rain that bogs down the excavating work. Some problems can be settled by compromises and negotiations. But when it comes to major differences about materials, time, and money, discussions too often deteriorate into ugly disputes that can bring the job to a halt.

Material Substitutions

Effective contracts include an "or equal" clause to deal with situations where you can't get what you planned. It means that a substitute window must be similar to the one already planned—not much more expensive, which wouldn't be fair to the contractor, or a bargain model, which wouldn't be fair to you.

If you disagree about what's equal, get together with the contractor and a copy of the material specifications (or blueprints or contract) that list the windows by manufacturer and model number. Price one, and only agree to a replacement that matches the cost and basic characteristics of the original.

This process does not require specialized construction knowledge. If the plan calls for a window with low-E glazing—the central feature of the window—don't accept a less efficient glazing system in a substitute. If you upgrade, expect an extra charge. If you downgrade, insist on a refund of the reduced cost.

Time Problems

The number-one consumer complaint about building and remodeling projects—that they don't finish on time—is unlike most other job problems, such as unpainted walls or leaking roofs, which are obviously right or wrong. Time problems are rarely that clear-cut.

Bathroom windows won't arrive for a few days, so the tile contractor is going to another job for a few days. Add on others, such as special order trim that hasn't arrived, a plumber with the flu (the possibilities are almost endless), and you have a full-blown delay.

The job schedule is so disrupted that other contractors can't be on site when you need them. Materials delivered in a timely fashion pile up and have to be moved again and again because they're in the way. Everyone spends more and more time doing things to work around the delay, which keeps the job moving but makes it take even longer.

To catch time problems at an early stage, start by asking the contractor who delivers the bad news about a hard date for returning to schedule. Even if the delay is an unexpected deluge that floods the new foundation, it's up to the contractor to make an adjustment to fix the problem. Don't leave a delay up in the air; deal with it immediately.

Resolving Disputes

Standard practice is to draw up a checklist (or punch list) at the end of the job to cover work not done or done incorrectly. But this collection of molehills can seem like a mountain with only a week to go. It's more sensible to check details in stages—for framing, mechanical systems, windows, and doors—and leave the final list for finishing touches. This also helps to keep one problem from creating a negative chain reaction and exposes problems that require extra money to fix. If a problem develops that can't be sorted out in discussions on site, document the difficulty in a registered letter to the contractor. Explain why you are entitled to a solution (listing a provision of the contract, for example), and offer a reasonable time frame for a response to the letter and the problem. Also include your ideas about a solution. A dispassionate letter may bring a positive response by offering a way to make right what is wrong without financial penalty. If it doesn't, an arbitration service or attorney will probably want to see some record of your attempt to solve the dispute before they get involved.

If you do wind up in front of a third party, the best policy is to present facts, not a hand-wringing case of what you wanted and how terrible it is that you didn't get it. Use records of estimates, payments, meetings, and material orders to corroborate your complaint. The best bet is to include in your contract a provision for binding arbitration that can prevent serious job problems from becoming disastrous.

GREEN SOLUTION

WHAT IS GREEN REMODELING?
The goals of green remodeling include providing a home that uses less energy and water than the original house, that makes use of products and building materials originating from sustainable sources, that does not result in waste that must be placed in a landfill, and that provides a healthy environment for the people living the house. Not every remodeling project will address all of these goals, but you will be surprised at how close you come by simply making the right decisions, especially during the planning stages of a project. Some decisions include:

♦ Exceeding the insulation levels of the local energy code
♦ Replacing old appliances with Energy Star-rated models
♦ Using recycled building materials and products where possible
♦ Choosing low VOC paints, caulks, and sealants

You'll find more Green Remodeling Tips throughout the book, but here are some additional sources:

♦ **ENERGY STAR:** (www.energystar.gov) For listings of energy-efficient appliances, including heating and cooling systems, kitchen and laundry appliances
♦ **NATIONAL ASSOCIATION OF THE REMODELING INDUSTRY:** (www.greenremodeling.org) For general information and listings of certified remodeling contractors
♦ **NATIONAL ASSOCIATION OF HOME BUILDERS:** (www.nahb.org) For general information on green remodeling

4 Tools

48 BASIC POWER TOOLS
- Tool & Shop Safety
- Saws
- Drills
- Cutters
- Sanders
- Pneumatic Tools

50 BASIC HAND TOOLS
- Practical Tools
- Wrenches
- Pliers
- Scrapers
- Prying Tools
- Hammers
- Saws
- Cutters
- Drivers
- Extras

4 Tools

TOOL & SHOP SAFETY

The how-to tools of a first-time homeowner may fit in a kitchen drawer, but the collection of hammer, screwdrivers, and duct tape is bound to grow. Wherever you keep them and use them, consider these measures that can help make your workplace safe.

Hazardous materials. You can read label cautions and safely use hazardous materials, but unless your shop is a separate, locked room, don't assume that everyone else will, particularly children. Protect them by designating one cabinet for dangerous materials. Lock it, and hang the key out of sight high up where only an adult can reach it.

Secured power tools. Most manufacturers build in safeguards, including key-lock start-stop switches. But the best bet is to unplug power tools when you're not using them. Also be sure that your shop wiring is up to code.

Operating power tools. Owner's manuals contain such a long, legal-sounding list, including a few ridiculous warnings, that people don't take the cautions seriously. But there are a few guidelines worth noting. Don't remove built-in and sometimes cumbersome safety features such as a blade guard on a saw. You could wind up hurting yourself, damaging the tool, and voiding the warranty. Make a quick check of high-speed power tools before you turn them on to be sure there is nothing in the cutter path and that adjustment wrenches or chuck keys have been removed. Don't use excessively dull blades, bits, and cutters. They can make wood and other materials chatter and jump unexpectedly—and won't produce clean cuts.

Personal protection. Use appropriate personal protection—safety glasses when cutting hard or splintering materials and a respirator mask when you're doing a lot of power sanding, for example. Don't wear anything loose like a tie (or loose sleeves or jewelry) that can become tangled and pull you into machinery. Include a well-stocked first aid kit as standard shop equipment. And if your shop has stationary power tools and a stock of wood, install a smoke alarm and an ABC-rated extinguisher that can be used against any type of fire.

Abundant lighting. Install lighting that is more than adequate. It will make your work safer and help you produce better results.

Saws

A 7 ½-in. circular saw with a combination blade is the most practical model for most DIY building projects.

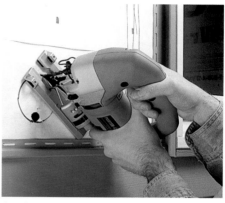

A saber saw is the right tool for many furniture and hobby projects where you need to cut curved shapes in wood.

A reciprocating saw is valuable mainly for demolition work such as removing an old stud wall. It makes rough cuts.

A power miter saw can be the right choice to increase production if you don't have space for a table or radial-arm saw.

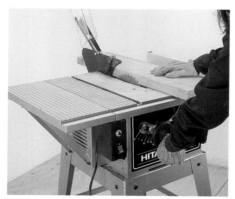

A table saw adds tremendous capability to a home shop because you can rip sheets of plywood and boards to suit.

A radial-arm saw can crosscut and, with the cutter rotated, can work as a rip saw to size lumber.

BASIC POWER TOOLS

Drills

A ¼- or ⅜-in. drill will handle DIY projects, and a cord will save you the trouble of recharging batteries.

Cordless drills are convenient in tight spots. Also consider a keyless chuck that makes bit changing easier.

Cutters

A router can cut dadoes and create decorative edges on your woodworking and furniture projects.

A cutout tool can save a lot of time making holes in drywall or paneling, such as for outlets and lighting fixtures.

Sanders

Depending on the grit of the sanding belt, you can use the sander to remove wood or finish a surface.

A random-orbit pod sander is probably the most practical tool overall for finishing wood surfaces.

Pneumatic Tools

Power nailers speed the work and eliminate hammer marks. Shown is a brad nailer (left), a finishing nailer (right).

Pneumatic impact wrenches make short work of tightening and loosening stubborn nuts and bolts.

Air compressors provide the power for pneumatic tools. For convenience, make sure the hose is at least 25 ft. long.

TOOL & SHOP SAFETY

4 Tools

PRACTICAL TOOLS

Some DIYers need a basement full of tools, while others get by with the basics. It depends not only on the work you want to do but also on how often you use tools, how expert you are at handling them, and how much you want to spend. Here are some of the key points to consider if you're starting or adding to a collection.

Durability. It's nice to buy the best, but the high-priced, long-lasting version often isn't necessary. One example: the throw-away brush, which is fine for slapping some stain on a fence post. Generally, it's wise to buy better quality in tools you'll use often—basics like a hammer and saw, a set of screwdrivers with comfortable handles, or chisels with steel-capped instead of plastic heads. Don't pay top dollar for a tool you rarely use or a very specialized tool that you can rent.

Precision. The truth is that inexperienced do-it-yourselfers don't get professional results by using professional tools. Skill comes from the hand that holds them, not from the tools themselves. Most DIYers should stick with DIY tools instead of the super-heavy-duty professional model designed to run for hours every day.

Strength. Look for hammers, wrenches, and pry bars that are drop-forged instead of cast metal. Casting traps air bubbles in molten metal, creating weak spots that can break under stress. (Cast tools often are painted.) Drop-forging removes more bubbles and makes the metal stronger and safer. When manufacturers take the time and money to drop-forge a tool and machine-grind its surface, they generally leave the metal unpainted.

Feel. Try the tool in the store to see if it feels controllable, too heavy, or too light. It can be difficult to compare tools you can't normally test on the spot, such as power saws—but some tools you can test, such as levels. Check three or four on the store floor or counter, and stack them on top of each other to catch the one whose bubble is out of line with the others. If a tool feels bulky or clumsy in the store, it's likely to feel even worse plugged in and running.

Price. The best bet is to avoid the most and least expensive models. The top end often has more capacity than you need, and the bottom end often has fundamental flaws that make work difficult even for an expert.

Wrenches

Use ¼- or ⅜-in.-drive ratchets for household and automotive jobs. Larger, ½-in. drives are truck tools.

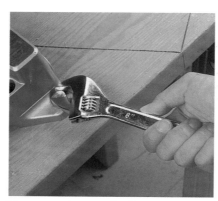

There are dozens of adjustable wrench designs, but the classic cres-cent will handle small and large jobs.

For heavy-duty plumbing work, a Stilson wrench with serrated jaws provides the most turning power.

Pliers

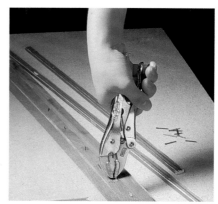

Use locking pliers (such as Vise-Grip or Robo-Grip brands) to securely grab what you're turning.

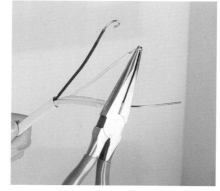

The needlenose pliers is mainly an electrician's tool but is handy for detail work in tight places.

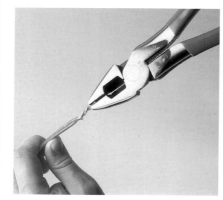

An electrician's or lineman's pliers is the most useful for wiring work. It can twist wires and cut through cable.

BASIC HAND TOOLS

Scrapers

Use a scraper to remove paint and old caulk from siding and to prep old surfaces for refinishing.

A sharp block plane is the right tool when you need to trim wood just a little but not enough to use a saw.

For shaping and trimming, use a wood rasp and a metal file. Use a mill bastard file to touch up saw blades.

Prying Tools

Use a pry bar to pry off old moldings and move materials more efficiently than you can with a hammer claw.

One step up from a pry bar is a crowbar, with more leverage for heavy-duty moving and demolition work.

Need to pull out a nail? Drive a cat's paw under the head, and pry it up with minimal damage to the wood.

Hammers

The basic claw hammer is the most versatile model for DIY repairs and home improvement projects.

Use a drop-forged 2-lb. hammer for heavy-duty jobs such as splitting brick pavers and concrete blocks.

A sledgehammer, generally 10 lbs. or heavier for demolition work, is known in the trade as a persuader.

PRACTICAL TOOLS

4 Tools

THE BASICS

Hammers. The most versatile is a curved-claw nail hammer. Many stores offer at least three sizes, determined by weight: 13, 16, and 20 ounces. The governing principle is to use the heaviest hammer you can while maintaining control. More weight drives nails with fewer blows; too much weight causes muscle fatigue and makes it harder to hit nails on the head.

As for handles, ten carpenters will give you ten different reasons why one type is better than another. Wood is durable, absorbs shock, and transmits a good feel for the work. Fiberglass is supposed to be stronger than wood but may provide less feedback. Rubber-covered steel tends to have more rebound than other types—and sometimes transmits the zing of metal on metal. They also can become slippery. If a wood handle gets slippery, try the carpenter's trick of scuffing up the sides of the handle with rough sandpaper or a finishing saw.

Saws. The most practical handsaw is a crosscut model with a tapered blade about 24 inches long that ends in a big wooden handle. Crosscut saws have 7 or 8 teeth per inch. More teeth make a finer cut but work more slowly; fewer teeth move faster but leave rougher edges. You may also want a backsaw with 12 or 13 teeth per inch to make smooth-edged cuts on trim.

If you prefer power, try a 7½-inch model circular saw. (Size goes by blade diameter.) Look for one with good balance that is easy to adjust for angle and depth, and make sure it has a comfortable handle position that lets you push instead of drag through the work.

Drills. A portable power drill has long been a DIY staple. Now, most are cordless, and some pack 18 volts of power. Among a slew of features and a wide range of power capacities, you will not be disappointed with a ⅜-inch model that reverses, has variable speed (which makes the drill more versatile), and has a keyless chuck. A torque limiter is handy but not essential; it keeps you from ruining screwheads and surrounding wood. It's nice to have one charger and two battery packs so that one is always ready to go, but it's not necessary if you remember to store the pack in the charger instead of the drill. Most DIYers can handle projects and repairs with a 12-volt model, while 14–18 and higher voltages are good for big projects.

Saws

There are specialized trim saws and rip saws, but crosscut saws with 7 or 8 teeth per in. are the most versatile.

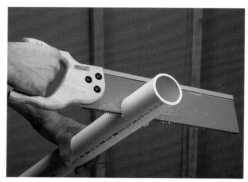

Short, stiff backsaws have fine teeth for detail and trim work. The teeth are not splayed, so the saw kerf is very narrow.

Cutters

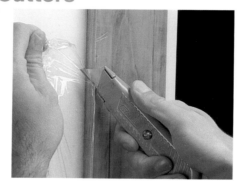

A utility knife handles everything from cutting drywall and batts of insulation to trimming the edges of a door mortise.

You'll need metal-cutting shears to cut the materials a utility knife can't handle, including flashing, downspouts, and more.

Drivers

In the basic screwdriver collection, you need a flat tip and a Phillips tip. Two sizes of each will handle most screws.

The nut driver, a pint-sized alternative to a full set of socket wrenches, will handle nuts and boltheads for most appliances.

BASIC HAND TOOLS

The keyhole saw's narrow, pointed blade is handy for making small cutouts. Drywallers use a stubby version to cut openings.

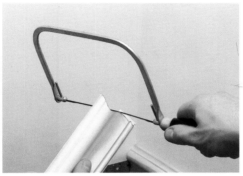
Once a home shop staple, the coping saw now is used mainly to join curved-profile moldings. Modern saber saws do the rest.

The fine teeth of the hacksaw will handle bricks, nails, old pipes, and other materials too tough for wood saws.

Handle most woodworking jobs with a ¼-in. chisel and a larger ¾- or 1-in. model. Steel caps stand up under pounding.

A hardened cold chisel can score and cut concrete blocks and bricks. Always wear eye protection when you use one.

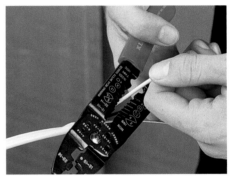
Use this specialized cutter-stripper instead of a utility knife to cut wire and safely strip its insulating sheath.

Extras

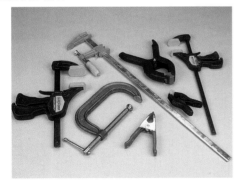
You can't have too many clamps. Among the most versatile are quick-clamps; you just squeeze the handle.

For paint and touch-up work, keep a brush, a roller and pan, a pair of drywall knives, and a caulking gun.

Basic safety equipment is a must, including work gloves, safety glasses, and a respirator mask.

THE BASICS

5 Fasteners & Adhesives

56 NAILS
- Applications
- Nail Sizes
- Nail Types
- Weights
- Nail Tips

58 SCREWS
- Heads & Washers
- Screw Sizes
- Screw Types
- Screw Tips

60 BOLTS
- Applications
- Bolts & Nuts
- Threaded Fasteners
- Locking
- Loosening
- Removing
- Splitting

62 FASTENERS
- Frame-Wall Fasteners
- Fastening
- Masonry-Wall Fasteners

64 FRAMING HARDWARE
- Hardware Types
- Sills & Girders
- Rafters
- Posts & Caps
- Code Compliance
- Special-Use Hardware
- Bridging
- Adding Wood Bracing
- Adding Metal Bracing

68 ADHESIVES & CAULK
- Common Adhesives
- Hot Glue
- Contact Cement
- Common Caulks

5 Fasteners & Adhesives

APPLICATIONS

Pound for pound, nails are probably the least expensive, most available, and easiest fasteners to use for all manner of construction. There are many different shapes and sizes, and more than one way to use most of them. Spend some time on a few different construction sites, and you'll find that one carpenter will toenail a stud with an 8d (eight-penny) nail, while another will start a little higher on the stud and use a 10d nail. In general, of course, there are finishing nails for trim, common nails with large heads and increased holding power for framing, and several other types for specialized purposes, such as concrete nails for when you need to connect wood to masonry.

For DIYers working on home repairs and improvements, the best bet is to keep a small supply of several types on hand. You can buy a handful of the most common types or choose a prepackaged kit. If you need to buy nails in quantity for a project, be prepared for the slightly confusing practice of ordering nails by pennyweight (notated by the small letter d) instead of by length. For example, if you want 3-inch common nails for framing, standard procedure is to order 10d nails by weight. The chart on p. 57 gives a general guide as to how many nails you get per pound.

Nailing Specs

As a rule, it's wise to use the largest possible nail. The larger the nail, the more holding power you get. But using too large a nail can cause two problems. First is splitting, particularly where the thick shank of a nail is driven close to the end of a board. The other problem is that sharp points may protrude through boards. That won't decrease the strength of the connection but could cause an accident during construction. For example, if you are spiking together two 2x10s to make a header, the sharp points of 10d (3-inch) nails could poke through because the header is just 3 inches thick. You could angle the nails slightly, or use 8d (2½-inch) nails instead. But if you are building the header with a ½-inch plywood spacer between boards, 10d nails will suit. Nailing specs are rarely listed on plans or blueprints, but a building inspector is likely to spot inadequate nailing on-site.

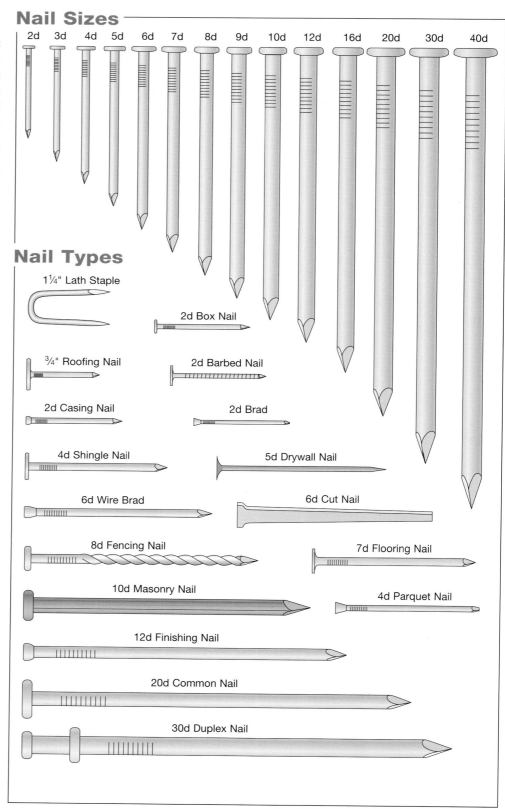

NAILS

Weights

Type of nail	Nails/lb.
3d box (1¼")	635
6d box (2")	236
10d box (3")	94
4d casing (1½")	473
8d casing (2½")	145
2d common (1")	876
4d common (1½")	316
6d common (2")	181
8d common (2½")	106
10d common (3")	69
12d common (3¼")	63
16d common (3½")	49
2d roofing (1")	255
6d roofing (2")	138

SAFETY

Pneumatic tools used by DIYers and professionals have built-in safety features. The most important is a lock-out device that won't let you fire the tool unless the head is firmly against a board. Still, these tools require maximum caution.

Air-powered nailers have a clip of banded nails that feed into the gun and fire with a trigger squeeze.

Nail Tips

Toenailing

Start your nail at an angle, up an inch or so from the joint. As the tip goes in, steepen the angle slightly.

In some cases, driving a toenail may be the only fastening option—it's also a good way to move a nailed board over slightly.

One-Hand Nailing

When you don't have the helper or the extra clamps you need, try this carpenter's trick of one-hand nailing.

Hold the head squarely against the side of your hammer to set the point; then grab the handle and drive the nail home.

Cat's Paw Pulling

Remove nails with minimal damage to surrounding wood by driving the forked end of a cat's paw under the nailhead.

Rotate the cat's paw handle to raise the nailhead, and finish pulling it out with the claw of your hammer.

APPLICATIONS 57

5 Fasteners & Adhesives

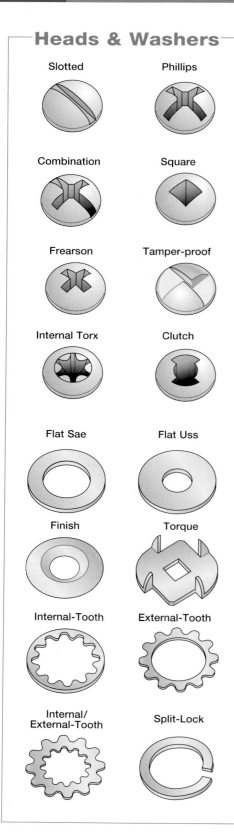

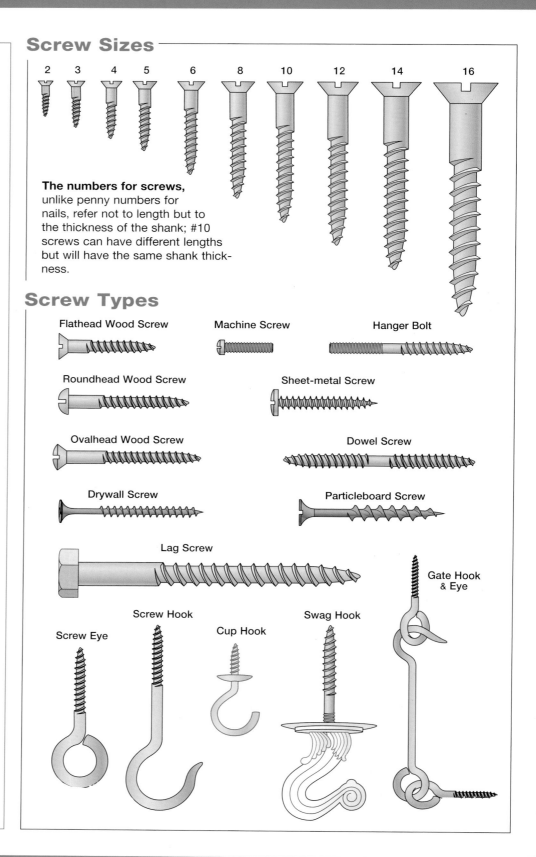

The numbers for screws, unlike penny numbers for nails, refer not to length but to the thickness of the shank; #10 screws can have different lengths but will have the same shank thickness.

SCREWS

Screw Tips

Holders

Cordless power drill/drivers have become a standard DIY tool. They're good for drilling but sometimes hard to control driving screws. That's where an extension holder will help. These tube-shaped attachments fit into the drill chuck and can be fitted with different driving tips. To keep the tip engaged and drive screws in a straight line, the tube extends over the screw head and along its shank. The tube withdraws as you seat the screws.

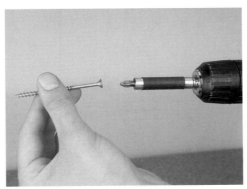

Tighten the driver attachment in the drill chuck, and fit one of the interchangeable driving tips onto the screwhead.

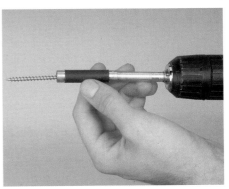

You can let go of the screw because the driver head is magnetized; then slide the guide tube over the shank.

Countersink

A countersink is a shallow, conical hole that allows a screwhead to sit flush in the wood surface. This is a standard feature of screwed connections, and typically requires two bits—and a time-consuming bit change every time you drive a screw. First, you need a pilot hole for the screw shank. Second, you need a countersink for the screw-head. Countersink bits do both jobs in one pass. You can adjust drilling depth with a collar on the bit.

A combination bit has a blade tip that carves a hole for the screw and a secondary cutter to make a countersink.

Most countersink bits have an adjustable collar so that you can control the depth and size of the recess.

Extract

When you need to remove a screw and the turning slot is damaged or stripped bare, most DIYers resort to pliers. Sometimes you can grab just enough metal to start backing out a screw. When you can't, use a screw extractor bit. The idea is to bore a small hole in the screwhead, and turn in a specially spiraled extractor bit that tightens when you turn it to the left instead of to the right. Once it seats, continue turning to back out a stuck screw.

To extract a screw with a damaged head, drill a pilot hole in the head. Its diameter should match the extractor bit's diameter.

You can turn the extractor bit with a drill or by hand. When its reverse threads take hold, the damaged screw backs out.

5 Fasteners & Adhesives

APPLICATIONS

On most DIY repairs and improvements around the house, you'll use nails and screws instead of bolts and nuts. But there are a few places where the extra strength of bolts is valuable. One is along the foundation where there is a critical change in materials from masonry to wood. The great weight of the building should be enough to keep the sill, typically a 2x6, in position. But to prevent any shifting on top of the concrete or blocks, it is standard practice to attach the sill with long, J-shaped anchor bolts instead of nails, screws, or other kinds of fasteners. The bottom of each bolt is embedded in solid concrete, and the threaded end is attached to the sill with a washer and nut.

Also consider bolted connections on decks to secure posts and railings—another location where you can't afford to have loose screws or nails that might give way. You can use roundheaded carriage bolts with large washers to permanently pin 2x4 or 4x4 posts to the solid framing of the deck platform. Recess the nuts and washers for improved appearance.

Fine Threads

The difference between standard threads and fine threads is significant in some applications—where you need maximum holding power between metal components in cars, for example. If you want the highest possible strength on critical projects, fine threads provide it by creating more interlocked surfaces between male and female threads.

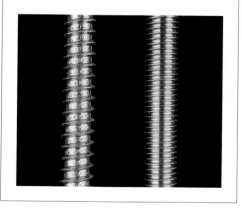

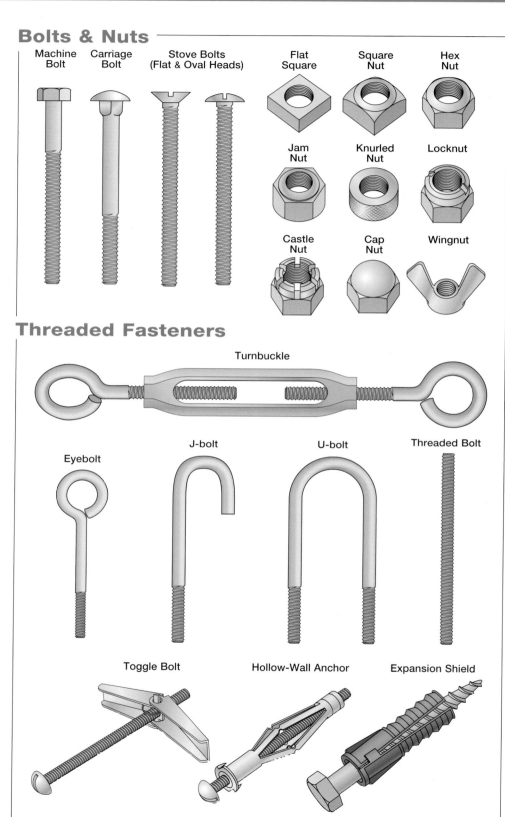

BOLTS

Locking

In most cases you can tighten nuts enough with a wrench or ratchet. But on installations subject to regular use and vibration, such as garage door tracks, there are two ways to add security. Either coat the bolt threads with an adhesive agent such as Locktite untightening the nut later if need be, or coat the nut with silicone caulk, which helps prevent corrosion.

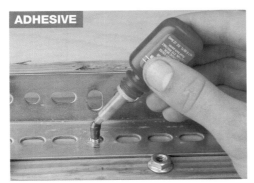
ADHESIVE

SILICONE

Loosening

Some nuts and bolts won't come apart no matter how much leverage you apply, particularly if the connection is rusted. Try breaking the corrosion with a penetrating lubricant. Another option is to break the rust bond by impact, with one hammer below the nut so the bolt won't bend, and another striking from above.

OIL

TWO HAMMERS

Removing

When lubricants and hammering won't budge the nut off a bolt, remove it with a hacksaw. Instead of trying to saw through the bolt shank (and possibly damage the surface underneath), cut through one of the facets on the nut. This weakens the nut enough so that you can twist off the remaining section with a wrench.

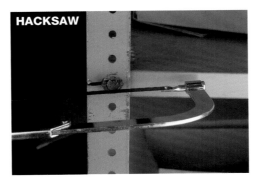
HACKSAW

WRENCH

Splitting

As a last resort, you can free a frozen nut—the kind you might find on a rusted bracket or license plate—with a nut splitter. Fit the head of this hardened tool over the nut, and tighten down its splitting wedge. Then, use a wrench or ratchet to drive the wedge into the side of the nut. This pressure will crack the nut.

SPLITTER

RATCHET

APPLICATIONS

5 Fasteners & Adhesives

Frame-Wall Fasteners

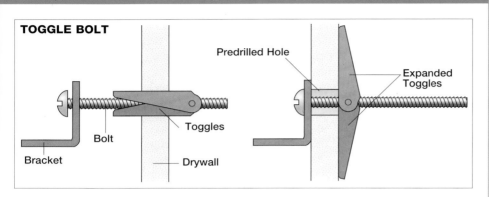

TOGGLE BOLT

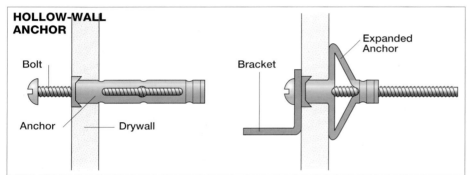

HOLLOW-WALL ANCHOR

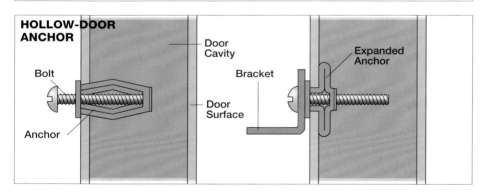

HOLLOW-DOOR ANCHOR

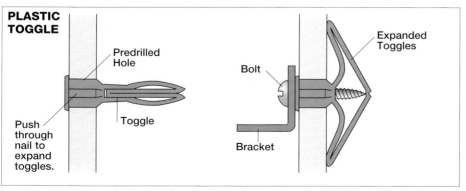

PLASTIC TOGGLE

Fastening

Your standard ¼-in. drill will handle most drilling and fastening jobs. But for wood, an old fashioned brace and bit provides good control on slow-speed drilling, and a hand-cranked rotary drill does the same on smaller holes. A compact, ratchet-action pin drill is handy for drilling pilot holes in moldings. There are many options when it comes to drilling holes for fasteners in masonry. For drilling by

AUGER

ROTARY DRILL

PIN DRILL

FASTENERS

hand, use a star drill and hammer, rotating the tip of the tool with each blow. With a drill, use masonry bits that have a wide carbide tip to do most of the cutting. For maximum production, use a hammer drill. It has a cam that drives the bit back and forth into the masonry with a hammering action as it rotates. It's important to wear eye protection, particularly when drilling in masonry.

STAR DRILL

MASONRY BIT

HAMMER DRILL

Masonry-Wall Fasteners

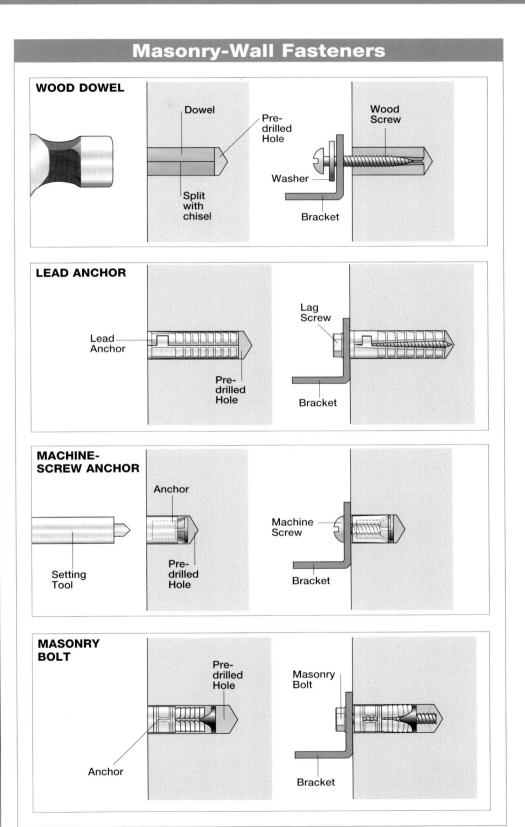

FRAME-WALL FASTENERS 63

5 Fasteners & Adhesives

HARDWARE TYPES

How much strength does your house really need, and how much protection does it pay to build in against storms? In a theoretical design for a nuclear power plant off the coast of New Jersey, the U.S. Army Corps of Engineers built in protection against the worst storm that could be anticipated in 1,000 years—a double hurricane producing 50-foot waves. It's not economically practical for homeowners to anticipate such freak events. But there are ways to make a building more storm-resistant, particularly in regions where hurricanes and tornadoes are common.

Corner Bracing

After corner posts are plumbed, add let-in bracing. A 1x4 runs from the top of the corner down at approximately a 45-degree angle across several studs to the bottom of the wall frame. Cut a series of pockets (¾ inch deep) so the brace will sit flush with the outside of the wall. Installing two let-in braces angled away from each corner will tie walls together and strengthen the corner.

Rafter Bracing

Use a similar system to strengthen the outermost rafters in a roof. Diagonal bracing, or 2x4 blocking (wide side down), can be set at approximately 45 degrees between the outside rafter near the top and bottom of the rafter run. These braces provide more resistance to movement and may keep high winds from raising the roof edge and getting into the attic where it can push up at the entire frame.

Tie-Down Strapping

Most buildings are not designed to deal with an uplift force. They rely on gravity and the weight of construction materials, pinned together with nails, to hold them down, which is more than enough in most areas most of the time. But you can easily build in extra protection by adding tie-down straps at critical connections: where rafters join the walls and where walls join the foundation.

Several manufacturers offer metal hardware preformed to fit around different framing combinations—a wing-shaped piece called a ridge strap, which fits across the top of a ridgeboard with two extensions that fasten along the rafters on either side, for example. The hardware is prepunched. You just set it in place, and then nail through the holes.

Stirrups

The most common construction hardware is a U-shaped bracket, often called a joist hanger, or stirrup. It makes framing connections stronger, for instance, where deck framing joins a ledger along the house wall. Hardware also speeds construction. You can nail up all the hangers, drop joists into the U-shaped pockets, and nail through the prepunched holes to secure them.

Rafters

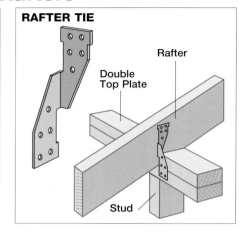

Posts & Caps

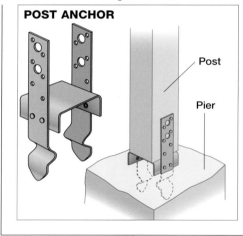

Sills & Girders

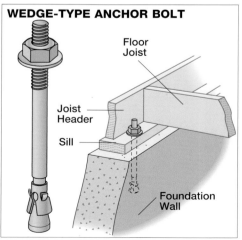

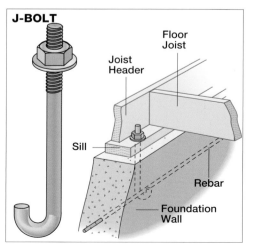

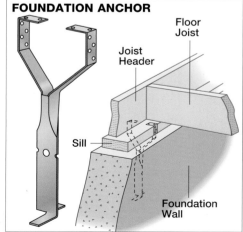

FRAMING HARDWARE

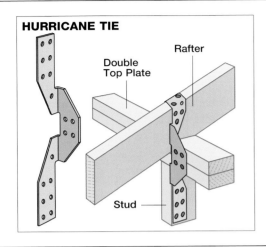

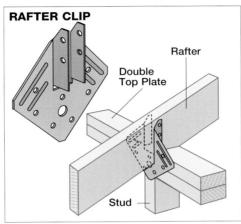

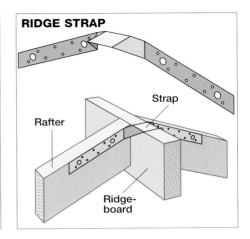

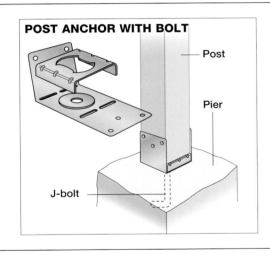

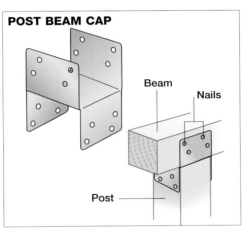

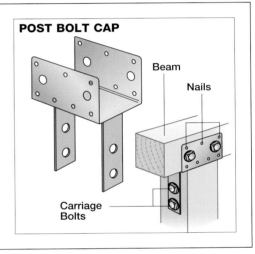

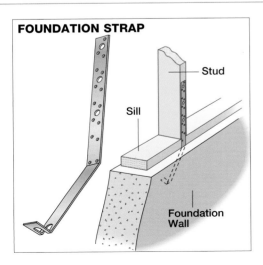

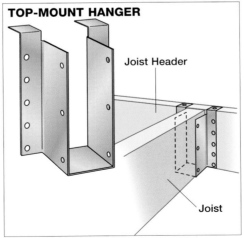

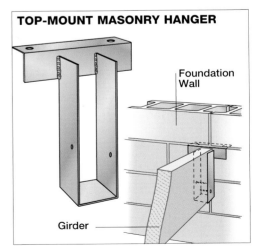

HARDWARE TYPES 65

5 Fasteners & Adhesives

CODE COMPLIANCE

The building inspector is unlike other housing professionals on a major remodeling or construction project. He can come and go at will, call a halt to work if he sees something wrong, and make you or your contractor peel off siding or drywall that may have been installed prematurely without his okay.

There are cases where inspectors seem to go overboard and insist on details, such as the number of nails in a sheet of drywall, that do not impact on safety. But overall, the inspection process provides a valuable double-check of your plans and oversight of your contractor.

You might not be on site when the walls are clad with drywall and the details of framing and insulation vanish from sight. But at key stages of the project, an inspector will get a look before important components such as the foundation, framing, and insulation are hidden. It's important to remember that you have to schedule these inspections and give the inspector some advance warning. In the middle of a busy building season, it may take a few days to get the inspector on site. In some regions, you may also deal with more than one inspector, even though one person handles most of the oversight work. For example, you may deal with one inspector who specializes in septic systems and another who handles wiring.

If you build without a permit or manage to conceal features such as an extra lavatory during construction, you could save on permit fees and reduce increases in your real estate assessment. But you could be in for trouble later on. A future inspection could uncover the non-permit work and make you liable to fines. Your home insurance might not cover damages to work found to be done illegally. And you could have problems when it comes time to sell your house and the current footprint doesn't match the one on record at the town tax office.

Special-Use Hardware

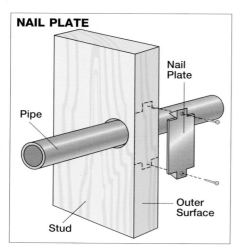

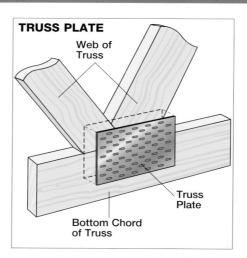

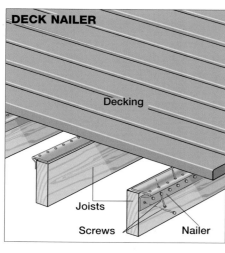

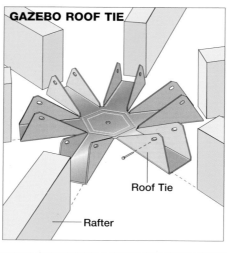

Adding Wood Bracing

USE: ▶ circular saw • wood chisel • hammer

Tack a 1x4 at an angle across the studs near the corner. Mark along the edges of the brace where it crosses the studs.

Adding Metal Bracing

USE: ▶ chalk-line box • circular saw • hammer

Many DIYers will find metal bracing easier to install than wood bracing. Start by snapping a chalk-line guide at the corner.

FRAMING HARDWARE

Bridging

Bridging is a standard feature of most floor construction. There is some controversy about how much it actually helps to stiffen the structure. But even a floor with large joists and plywood decking seems a little more secure when the joists are secured to each other with bridging. The traditional braces are short pieces of the joist material set at midspan. But you can use other systems instead. The most common is an X-pattern of wooden 1x3 or metal strapping nailed to the tops and bottoms of each joist. Connectors called Z-clips are used to anchor braces between I-joists.

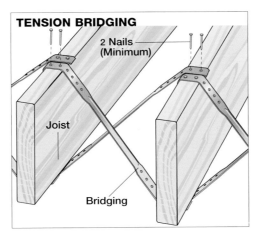

TENSION BRIDGING

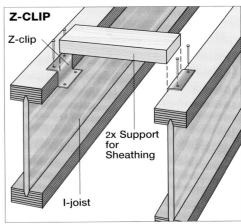

Z-CLIP

- pencil • eye protection ▶ 1x4 bracing • nails

Set your saw to cut ¾ in. deep (the thickness of the brace), and cut along the two sets of pencil marks.

Use a hammer and chisel to remove the wood between cuts and form a recess for the brace. Extra sawcuts make the job easier.

Set the 1x4 brace back in position, this time nailed into the recesses of each stud, to stiffen and strengthen corners.

- eye protection ▶ metal bracing • nails

As you do with wood bracing, set your saw to cut only as deep as the brace, and make one cut along the chalk-line guide.

Metal let-in bracing is L-shaped. You simply set one edge into the straight line of cuts angled from the corner to the sill.

The metal bracing is perforated, so it's easy to nail it in place at each stud. Installing two braces strengthens corners.

CODE COMPLIANCE

5 Fasteners & Adhesives

Common Adhesives

White Glue Also known as PVA or polyvinyl acetate, and perhaps best known by the brand name Elmer's Glue-All, white glue is the one kind of adhesive most people keep around the house. It's useful for making quick repairs to furniture, woodwork, ceramics, and paper. Because it's water-soluble, never use it in a place that might get wet.

- **Solvent:** soap and warm water
- **Curing:** sets in 1 hour, cures in 3–8 hours

Yellow Glue Often called carpenter's glue or aliphatic resin, yellow glue is a good general-purpose woodworking glue. Like white glue, it dries clear and is often used to repair furniture and indoor woodwork. It sets more quickly than white glue, usually within an hour. Also like white glue, it isn't waterproof and shouldn't be used outside.

- **Solvent:** warm water
- **Curing:** sets in 1 hour, cures in 3-8 hours

Acrylic Acrylic adhesives come in two parts—either powder and liquid, which must be mixed before use, or liquid and paste, which must be separately applied to the opposing surfaces being joined. Acrylic is used for quick-bonding adhesion of metal, glass, and wood. It is waterproof and dries to a light brown color.

- **Solvent:** acetone
- **Curing:** sets in 5 minutes, cures overnight

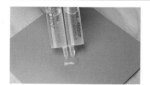

Epoxy Epoxy is particularly good for joining together dissimilar materials, such as glass to metal. It's also useful for bonding ceramics, wood, and many other materials. Epoxies come in tubes, in syringes, or as a mixable putty. It provides a very strong bond, is waterproof, and dries clear to brownish.

- **Solvent:** acetone
- **Curing:** set times vary widely, cures 3–72 hours

Hide Glue Hot hide glue is prepared from granules (made from animal hides) and water. You can adjust its curing time, making it useful for complex projects such as kit furniture. It has superior strength, a powerful grip when setting (handy for gluing veneer). Joints won't "creep" once they dry, and dry glue can also be "reactivated" with steam. Premixed hide glues are also available.

- **Solvent:** warm water
- **Curing:** set time varies with formula, cures in 24 hours

Resorcinol Resorcinol is an extra-strong adhesive particularly adaptable to laminating wood, making outdoor furniture, and boatbuilding. It will cure at 70°F and up, but no lower. It is available in a liquid with a separate (usually powdered) catalyst—just mix up the amount that you need. It is waterproof, very durable, and dries to a dark red.

- **Solvent:** cool water
- **Curing:** sets and cures in 10 hours at 70°F, 6 hours at 80°F

Cyanoacrylate Most commonly known as super glue, this is a powerful, fast-curing adhesive that can be used to bond most metals, plastics, ceramics, vinyl, and rubber; the gel form of the glue can be used on wood. It is water-resistant and dries clear. Always use extra caution not to get any on your skin.

- **Solvent:** acetone
- **Curing:** sets in 10–30 seconds, cures ½–12 hours

Construction Adhesive Sometimes called mastic adhesive, this is used to bond wood and concrete. Many brands also can be used on acoustic tile and other materials. While not a substitute for thorough nailing, it can improve the overall strength and rigidity of face-to-face connections—where you sister one joist onto an existing joist, for example. It can be spread with a notched trowel but is commonly applied with a caulking gun from a tube.

- **Solvent:** usually mineral spirits
- **Curing:** sets in 15 minutes to 1 hour, cures in 4–24 hours

ADHESIVES & CAULK

Hot Glue

Glue guns are excellent for home-improvement projects because they dispense a versatile range of adhesives that bond quickly. The solid glue or caulk cartridges are heated and melted in the gun. Once dispensed, the glue usually sets within one minute.

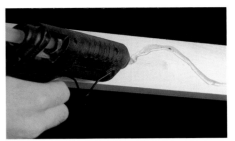

Contact Cement

To bond plastic laminate and wood—on a countertop, for example—use contact cement. A full coat is spread on each piece and allowed to dry until it feels dry but tacky. Use wood strips to keep the components from bonding until they are in position.

Common Caulks

Acrylic Latex Acrylic latex caulk (slightly longer-lasting than similar, cheaper vinyl latex) is inexpensive and easy to apply but degrades in direct sunlight and adheres poorly to porous surfaces.
- **Drying skin:** ½ hour ◆ **Curing:** 1 week
- **Life:** 5–10 years

Butyl Also called butyl rubber, this caulk has better adhesion and stretching ability than acrylic but costs more and takes longer to cure. It also degrades in sunlight.
- **Drying skin:** 24 hours ◆ **Curing:** 6 months
- **Life:** 5–10 years

Polyurethane Polyurethane caulks are expensive and more difficult to apply than latex and butyl, but they last longer, can cover a wider gap (up to ¾ inch), and will stretch further.
- **Drying skin:** 24 hours ◆ **Curing:** 1 month
- **Life:** 20+ years

Silicone Silicone—not to be confused with paintable siliconized acrylic—has good stretching ability and can cover a 1-inch gap, but it can't be painted and adheres poorly to plastic and wood.
- **Drying skin:** 1 hour ◆ **Curing:** 1 week
- **Life:** 20+ years

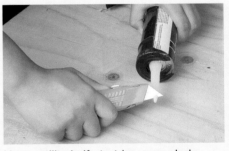

Use a utility knife to trim an angled opening in the cartridge tip.

To keep adhesive from hardening in the tip between jobs, insert a common nail.

6 Masonry

72 TOOLS & MATERIALS
- Handling Materials ◆ Handling Heavy Loads
- Weights ◆ Safety Gear ◆ Special Tools

74 CONCRETE
- Concrete Basics ◆ Pouring a Patio ◆ Testing Concrete ◆ Concrete Finishes ◆ Ready-Mix
- Estimating ◆ Placing Concrete ◆ Edging & Jointing ◆ Forming a Curved Corner
- Reinforcing ◆ Diagnosing Problems
- ■ REPAIRING CRACKS

78 FORMWORK
- Concrete Forms ◆ Form Types ◆ Building Formless Piers ◆ Building Wall Forms
- Building Formed Footings
- Pipe-Sleeve Blockouts ◆ Window & Door Blockouts

82 BLOCK
- Building with Block ◆ Cutting Block
- Laying Block ◆ Estimating
- Mortar Types ◆ Concrete Block
- ■ REPAIRING BLOCK WALLS ■ PATCHING BLOCK ◆ Control Joints ◆ Garden Block
- ■ REPLACING BLOCK

86 BRICK
- Brick Basics ◆ Cutting Brick ◆ Mortaring Bricks
- Mortars ◆ Materials ◆ Brick Maintenance
- Cleaning Brick ◆ Laying Brick ◆ Tooled Joints
- Diagnosing Brick Problems

90 STONE
- Working with Stone ◆ Stone-Wall Patterns
- Laying Face Stone ◆ Shaping Stone
- Laying Full Stone

92 MASONRY REPAIRS
- Cleaning Masonry ■ PATCHING STEPS
- ■ PATCHING STUCCO ■ REMOVING STAINS
- ■ EFFLORESCENCE

6 Masonry

HANDLING MATERIALS

A rustic-looking rock wall for the garden, a flagstone patio, curving brick walkway, or much-needed retaining wall can do wonders for the appearance of your yard. When used as part of an overall landscape design, masonry adds a touch of permanence and solidity. Ditto for more down-to-earth uses, such as foundations and floors, except here it's truly essential. Laid properly, concrete and block foundations will last for a lifetime—and more—with little or no maintenance.

The reason masonry looks so solid is that it really is. If you've never picked up a concrete block or mixed an 80-pound bag of cement before, you're in for a surprise. The stuff is heavy—and the larger the project, the more weight you have to deal with. The bending and lifting that always come with masonry projects can easily get to your back and shoulders if you are not careful, so break the job up into manageable chunks. Wear a lower-back support if you need to, and always lift with your legs, not your back. Remember that cement is caustic and will burn your skin after prolonged contact. Wear sturdy work gloves, a long-sleeved shirt, and rubber boots (because you may have to stand in the wet mix to spread it).

Excavation

Moving dirt is the one part of improvement projects that just about everyone underestimates. The work is messier, more expensive, and more time-consuming than it seems beforehand. On big jobs, such as excavating a foundation for a patio, you will probably need to hire a backhoe.

Save your energy on small jobs by loosening dirt with a pick before removing it with a shovel. If you're digging a hole for concrete footings, try to leave the bottom undisturbed. If the soil seems soft, compact it with a metal tamper or by pounding it with the end of a 2x4. A concrete footing poured over soft, uncompacted fill will settle, and so will anything resting on it, such as a foundation wall.

Moving Dirt

Unless you're looking for exercise, it's wise to move the dirt as little as possible. Before you dig, decide where the leftover dirt will go in the end—into a flower bed or a low spot that puddles in heavy rain. You could load dirt straight from the hole into a wheelbarrow, and take breaks from digging with the comparatively easy work of carting each load away.

After the concrete is poured and it's time to backfill your excavation, add dirt in stages, tamping to compress the loose fill as you go. Though it's always tempting to bury construction debris in the backfill, have it carted away instead. It's okay to fold in rocks or chunks of concrete, but not wood, paper, and other biodegradable job scraps, because wood and paper make excellent termite food, and you don't want to be spreading appetizers around the house. In addition, as the material biodegrades, the dirt above gradually compresses and fills in the space, creating a water-catching trench around the foundation or pier.

Handling Heavy Loads

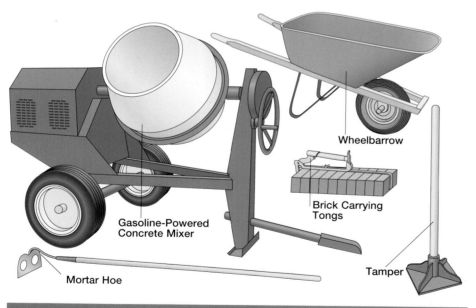

Many of the same tools are needed whether you are working with concrete, brick, or stone. Often you can rent tools such as these, so check before buying a tool you may not use that often. Some tools—such as wheelbarrows, shovels, and hoes—you probably already own.

Weights

◆ **CONCRETE**
 Plain: 90-144 lbs./cu. ft.
 Lightweight: 35-105 lbs./cu. ft.
 Reinforced: 111-150 lbs./cu. ft.
 1-inch slab: 6-12 lbs./sq. ft.
 Ready-mix: 50-96 lbs./bag. An 80-lb. bag + water weighs near 100 lbs.

◆ **BRICKWORK**
 4-inch-thick wall: 40 lbs./sq. ft.
 Mortar: 116 lbs./cu. ft.
 Bricks: about 3–5 lbs. each

◆ **CONCRETE BLOCK**
 8-inch-thick wall: 55 lbs./sq. ft.
 8-inch-thick wall (lightweight): 35 lbs./sq. ft.

◆ **STONE (ASHLAR)**
 Granite: 165 lbs./cu. ft.
 Limestone: 135 lbs./cu. ft.
 Marble: 173 lbs./cu. ft.
 Sandstone/bluestone: 144 lbs./cu. ft.
 Slate: 172 lbs./cu. ft.

TOOLS & MATERIALS

Safety Gear

Always wear a dust mask, gloves, and protective eyewear when mixing masonry products or cutting bricks, blocks, or concrete. Prolonged contact with caustic cementitious materials can also cause irritation, so protect skin with gloves and rubber boots.

Special Tools

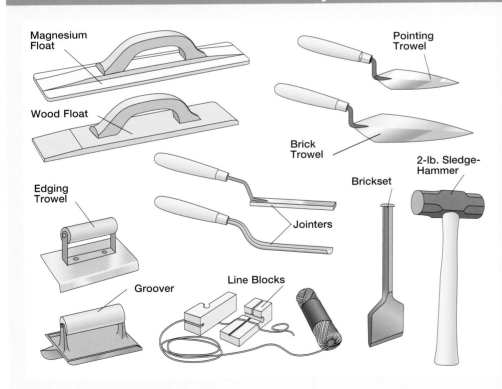

Important tools for concrete-block construction include a line level and line blocks. Mason's twine is strung between line blocks and checked with the line level to help you keep a structure straight, level, and plumb.

Concrete finishing tools include a bull float to level and smooth the surface of wet concrete, an edging trowel to separate concrete from its forms, a wood or magnesium float to apply a smooth finish, and a steel finishing trowel for a final pass. An edging trowel also creates a rounded edge that is safe and durable. A jointing or grooving trowel is used to form control joints.

In brickwork, a brick hammer and brick set are used to split bricks, while various other tools are used to make different profiles in brick mortar joints. (See page 89.)

HANDLING MATERIALS 73

6 Masonry

CONCRETE BASICS

Fresh concrete is a semifluid mixture of portland cement, sand (fine aggregate), gravel or crushed stone (coarse aggregate), and water. As the cement chemically reacts with water (a process called hydration), the mix hardens to a stonelike consistency. Properly mixed and cured concrete creates strong structures that will weather the extremes of summer heat and winter cold with little maintenance.

Formwork and steel reinforcement are needed to build concrete structures. Formwork is generally made from wood, and may be a simple square or complex shape. Reinforcing steel (most often reinforcing bar, or rebar) can be light or heavy, depending on size and strength requirements. Welded or woven wire mesh is also used to reinforce concrete slabs on grade.

Mixing Concrete

For large-scale projects such as a patio, concrete is sold by the cubic yard and delivered in a ready-mix truck ready to pour. For smaller jobs—say, steps for the patio—you can mix your own by purchasing dry ingredients in bags and adding water. But don't get too ambitious: one wheelbarrow-sized batch is less than 3 cubic feet. You would need about nine batches to make just one cubic yard. For mid-sized jobs, it makes sense to rent a portable power mixer.

You may be tempted to adjust the mix proportions—say, by adding more water to make the concrete easier to mix and pour. However, as water content can drastically affect strength, the best policy is to order concrete ready-mixed or follow directions on the dry ingredient bags.

The standard proportion of water to cement produces concrete with a compressive strength of about 3,000–4,000 pounds per square inch (psi). Adding less water makes mixing more difficult but increases concrete strength. Concrete that dries too quickly can be a problem, however.

Even the best mix may dehydrate in hot weather, robbing the concrete of the water it needs to harden. In extreme cases, a steady hot, dry breeze can even accelerate evaporation so that the masonry surface begins to set before it can be smoothed. Concrete begins to harden as soon as it is mixed, and can support your weight within a few hours. Most of the curing takes place in the first two weeks, but it takes a month to reach maximum hardness.

There are some solutions for extreme cases, such as adding flaked ice or cooling down the aggregates with a sprinkler before adding them to the mix. To eliminate the risk of wasting your efforts on a job that doesn't last, don't pour in temperatures over 90°F; if you must, start very early in the morning to beat the heat.

It's also important to remember that hot surfaces contacting the mix can burn off moisture. For instance, it's wise to spray some cool water on forms that are sitting in the sun, as well as on the reinforcing bar, which can get quite hot to the touch.

Concrete Finishes

- A wooden float finish is slightly rough, just enough to be slip-resistant and glare-free.

- A steel trowel finish should be perfectly smooth, and is best suited for interior applications.

- A broomed surface is rough and slip-resistant, ideal for outdoor steps and patios.

- Exposed aggregate is smoothed into the concrete while drying for an attractive, rough surface.

Pouring a Patio

USE: ▶ compactor • darby • line level • mason's hoe • mason's string • measuring tape • rebar chairs • screed • shovel • sledgehammer • wheelbarrow

1 Your crew may not bring one this big, but most jobs begin with a bulldozer that cuts away the sod and levels the ground.

2 Once the perimeter is established and the forms are in place, the ground should be compacted.

3 To strengthen the concrete, welded wire is laid near the bottom of the slab, generally on short supports called chairs.

CONCRETE

Testing Concrete

- A typical mix is 11% portland cement, 26% sand, and 41% crushed stone, plus 16% water and 6% air.

- If concrete is too wet, ridges made in the mix with a trowel won't hold their shape.

- If concrete is too dry, you won't be able to make ridges, and it will be difficult to work.

- When mixed correctly, the ridges will hold most of their shape; only a little water will be visible.

Ready-Mix

There are several advantages to ordering ready-mixed cement, aside from the fact that you don't have to mix yards of the stuff by hand. The chute can extend and swivel to pour concrete where it's needed. Ready-mix trucks can deliver concrete at temperatures that make it possible to pour during a heat wave. Also, ready-mix concrete is available with an additive that produces microscopic air bubbles in the mix—air-entrained concrete that is more resistant to cracking than the concrete you can mix on-site.

Estimating

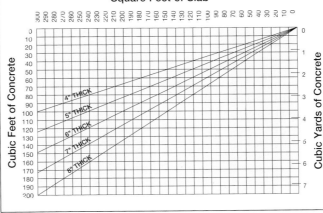

To figure out how much concrete to mix or order, use the chart at left; or total up the volume inside the forms in cubic feet (length x height x width); then, divide this figure by 27 to convert into the ordering standard of cubic yards. Some contractors build in a reasonable excess factor of about 8 percent by changing the connversion factor to 25.

• edging trowel • float trowel • jointing trowel • broom ▶ concrete mix • double-headed nails • formboards • gravel • stakes • welded or woven-wire mesh

4 As concrete pours from the ready-mix truck, a straightedge guided by a screed board on each side levels the mix.

5 The rough surface left by screeding can be smoothed with a float or textured with a broom finish for better traction.

6 Control joints that prevent cracking can be formed into the pour, although some crews cut them after the mix hardens.

CONCRETE BASICS

6 Masonry

PLACING CONCRETE

Steel reinforcement helps control the cracking associated with the natural shrinkage of concrete as it dries. The two basic types are rebar (reinforcing bar) and welded or woven wire mesh. Rebar ranges in size from ¼ to 1 inch in diameter, and it is ridged for a better bond with the concrete or smooth for nonbonding control joints. Rebar is stronger than wire mesh; use it for concrete that will carry a heavy load, such as footings and piers. Wire mesh is made from steel wire in a grid of squares and is sold in rolls and mats. Use it in flat slabs on grade, such as patios and walks. Cut wire mesh with fencing pliers, and flatten it out before use. Fill large areas by overlapping sections of mesh by at least 3 inches and tying them with wire.

Avoiding Problems

Pouring the concrete is a simple matter; the crucial stage begins as the mix sets and begins to cure. Curing is a long-term process during which concrete continues to gain strength. If raw concrete is left exposed to the wind and sun, it may dry too quickly and may not attain half its potential strength. You could still walk on it, but the slab would be likely to crack.

Tooling control joints into the surface of concrete slabs helps make settling cracks break at planned locations. Control joints weaken the concrete surface, causing cracks to occur at the bottom of the joints, where they are inconspicuous and will not spread.

Edging & Jointing

Edging is tooled into fresh concrete as soon as the water sheen disappears after the first floating. Run an edging trowel along the entire perimeter of a slab. Control joints can be hand-tooled into fresh concrete with a jointing trowel, cut into cured concrete with a circular saw fitted with a masonry blade, or (usually with isolation joints) preformed with fixed divider strips of hardboard, cork, rubber, plastic, or felt paper.

EDGING

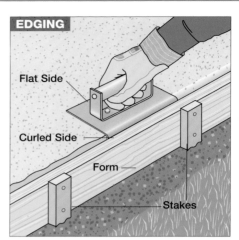

CONTROL JOINTS

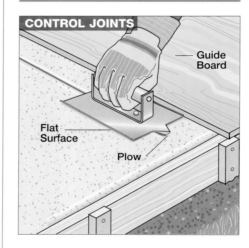

ISOLATION JOINTS

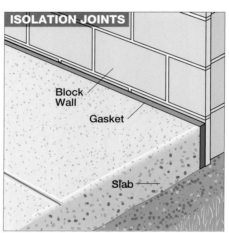

Forming a Curved Corner

USE: ▶ clamps • hammer • saw • screw gun or drill driver • small sledgehammer ▶ common nails • 1x4 or 1x6 boards • ¼-inch plywood • screws • 2x4 stakes

1 To make a curved form, use a flexible material, such as thin hardboard, that you can bend in a large radius.

2 Center the hardboard in the corner; clamp the ends in position; and secure them to the forms with screws.

3 A flexible form needs at least one supporting stake to prevent the radius from distorting when concrete is poured.

CONCRETE

Reinforcing

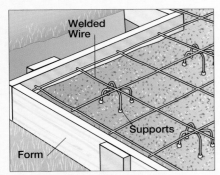

Welded wire mesh must be fully embedded in concrete for maximum strength. Supports called chairs hold it off the ground.

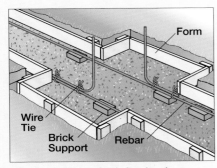

Short sections of reinforcing bar turned up into the foundation at each corner will add support to pilasters on a brick or masonry wall.

Diagnosing Problems

Crazing—many minute, shallow cracks—is caused by over-floating. The surface should be cleaned and sealed to stop further crazing, or entirely resurfaced for a permanent fix. Spalling is also the result of overfloating, and indicates faults in the concrete—requiring resurfacing. Oxidizing iron causes rust-colored stains; these can be cleaned with a solution of oxalic acid crystals, water, and ammonium bifluoride. Always wear a mask when mixing an acid solution.

CRAZING

OXIDATION

SPALLING

Repairing Cracks

USE: ▶ cold chisel • small sledgehammer • putty knife • masonry trowels • wire brush ▶ concrete patch

MONEY SAVER

1 To make the most secure patch in concrete, use a cold chisel to undercut the surface on each side of the crack.

2 To ensure that your patch material bonds with the old concrete, wire-brush and dust out the crack.

3 Use a trowel to force cement into the undercut areas; then, fill the remaining gaps, trying to match the surrounding surface.

PLACING CONCRETE

6 Masonry

CONCRETE FORMS

Wet concrete is poured into molds called forms, which hold and shape the concrete until it hardens. Usually made of lumber and (in the case of wall forms) plywood sheathing, forms must be level, plumb, and strong enough to withstand the weight of the concrete pushing against them. Forms for edges of slabs and continuous wall footings are typically made of 2x4, 2x6, or 2x8 lumber, depending on the thickness of the slab or footing.

Form boards should be free of holes, cracks, loose knots, and other flaws that might weaken them or mar the concrete's surface. The boards are set on edge, perpendicular to the subgrade, and are braced every 3 to 4 feet with wood stakes driven firmly into the ground. To make the forms easier to remove, use double-headed nails to fasten the form boards to the stakes. Wherever two boards butt together, screw a plywood gusset across the outside of each joint.

Forms for footings and slabs are usually built on site. The more complicated forms for perimeter wall foundations are either site-built or constructed using prefabricated panels rented from concrete form suppliers.

Continuous Footing Forms

Foundation walls rest on concrete footings set below the frost line to avoid frost-heave damage. The footing height should be the same as the thickness of the wall, or a minimum of 8 inches. The footing width should be twice that of the wall, or a minimum of 16 inches. The batter boards set up to establish the building's outline are also used to establish and mark the width of the footings.

If the soil you're building on can hold an edge without crumbling, an earth trench can serve as the form. However, concrete for wall footings is usually poured into wooden forms anchored by 1x4 stakes driven into the ground and braced with 1x4 spreaders nailed across board tops every 4 to 6 feet. A chamfered 2x4 suspended from the spreaders down the center of the form creates a depression, or keyway, in the footing that will help secure a poured concrete wall.

Wall Foundation Forms

Install basement wall formwork after the footings have cured. The forms are usually made of smooth, knot-free plywood sheathing supported by 2x4 studs, horizontal members, called rangers or wales, and braces. Wire ties hold the plywood walls together; wood spreaders keep them a fixed distance apart.

Pier Forms

Square or cylindrical concrete piers support structures ranging from decks to houses. To make forms for piers, use either a simple wooden box constructed on-site or a prefabricated fiberboard hollow tube. The simplest type of pier form is a hole in the ground with or without an above-grade box form.

Form Types

Formwork for a wall can be a simple stud form, consisting of stud wall frames sheathed with plywood on their inside faces. Wire ties hold facing walls together, while 2x4 spreaders keep them a fixed distance apart. A stud-and-ranger form has the added support of horizontal members, called rangers, and braces. You can also rent pre-fabricated, reusable box formwork from concrete form suppliers. In this type of formwork, both the inside and outside faces of the stud walls are sheathed with plywood.

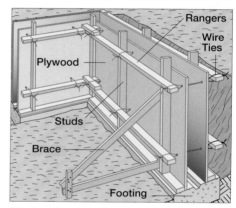

Stud-and-ranger forms are reinforced with horizontal rangers, or wales.

Building Formless Piers

USE: ▶ posthole digger • measuring tape • shovel • wheelbarrow • spirit level • hammer ▶ concrete • steel reinforcement • post • post base or steel dowel

1 Use a posthole digger to excavate a hole for a formless pier. The earthen walls of the hole serve as the formwork.

2 Measure the hole's depth carefully to make sure that the base of the pier will rest on soil below the frost line for your region.

3 Fill the hole with the required concrete mix and steel reinforcing. The bottom of the hole should be undisturbed soil.

FORMWORK

Box forms for rectangular piers are easy to construct on-site. Cut and assemble four boards to make a box with inside dimensions equal to those of the pier. To keep the box form from sticking and make it easier to remove after the concrete sets, coat its interior with motor oil or form oil, a release agent. To make removal even easier, nail a pair of 2x4 handles to the form above grade level. Lower the box into the excavated hole. Use wire ties to suspend an anchor bolt in the center of the pier, and then pour the concrete.

Stay-in-place tube forms are usually made of high-quality, moisture-resistant fiber that is spiral-wound and laminated with heavy-duty adhesives. Other types of permanent tube forms are made of molded fiberglass, clay tile, or concrete pipe. Also available are fiberboard tube forms that can be peeled off after the concrete has cured. To form a concrete pier, simply lower the tube form into the hole; suspend an anchor bolt in the center of the form with wire ties; and then pour the concrete into the form.

A surface form is a short box form placed over a hole; it serves as the main formwork for a concrete pier. The surface form simply gives a clean-edged finish to the above-ground portion of the pier. After the hole is dug, the box form is installed over it. Temporary stakes driven into the ground and nailed to the form hold it in place. Use wire ties to suspend a steel pin or anchor bolt in the center of the form. Pour the concrete mix into the hole up to the top of the surface form.

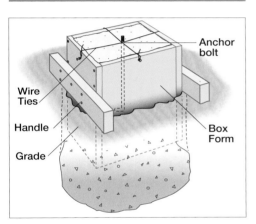

A box form is easily built on-site. Above-grade handles make it easier to remove.

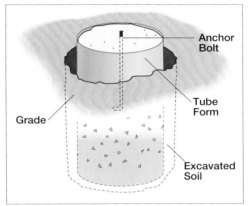

Tube forms, designed to stay in place, need only be positioned and filled with concrete.

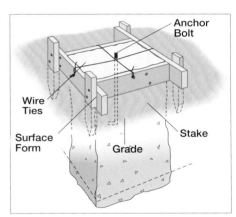

A surface form gives a smooth finish to the exposed part of a formless pier.

4 Insert a metal post base or steel anchor into the concrete when the mix is firm enough to hold it but not yet hardened.

5 Adjust the post base so that it is level, plumb, and properly oriented to support the structural post to be installed later.

6 Install the structural post by nailing it into the metal post base after the concrete pier has hardened.

6 Masonry

BUILDING WALL FORMS

Wet concrete exerts a great deal of pressure on any form, but especially on tall wall forms. A simple stud form consisting of plywood sheathing over stud walls held together by wire ties might be sufficient for a small garden wall. But a crawl space or basement wall requires a more securely braced stud-and-ranger form.

Assembling the Forms

After the footings have set and their forms have been removed, build 2x4 stud wall frames using 16d nails and spacing the studs 16 inches on center. Drill holes in the plywood panels for wire or snap ties. The spacing of studs and ties depends on the size of the wall and the pour rate, the speed at which concrete is poured. In a typical scenario, ties are spaced 16 inches on center between studs and in horizontal rows 12 inches from the top and bottom of the form and every 2 feet in between.

Nail the plywood to the studs with 8d nails. Snap chalk lines on the footings to mark the edges of the finished wall. Raise the outer walls first, nailing adjacent panels to butting studs. Shore up panels with braces running from panel studs to stakes driven in the ground. Insert snap ties into the predrilled holes. The outer stems of the ties are sandwiched between two horizontal rangers held together by U-shaped brackets that attach to the ties. Overlapping rangers nailed together at outside corners provide added support. (See page 78.)

When the outer walls are up, make sure they are level and square, and fasten the bottom plates to the footings with masonry nails. Spray the inside plywood faces of both inner and outer walls with form oil to prevent sticking.

Raise the inner walls, threading the free ends of the snap ties through the predrilled holes in the plywood panels. Install and secure walls and braces. Fasten bottom plates to footings.

Wall Openings

Once the forms are built, you'll need to install blockouts to accommodate pipes, electrical and cable lines, windows, and doors after the concrete is poured. A blockout can be a short length of plastic pipe or a plywood box. You may also have to excavate and build formwork for a well around a basement window.

Rebar & Anchors

To keep cracks from spreading in the finished wall, place steel reinforcing bars (#4 rebar) horizontally 1 foot from the top and bottom of the wall and at intervals in between (depending on the wall's height). Rebar is fastened with wire ties to the snap ties holding the form walls together and to vertical rebar set in the footing.

After pouring the concrete into the forms and smoothing and leveling it, set anchor bolts into the concrete 1 foot from the ends of each wall and 6 feet on center. The bolts will be used to secure the sill plate that anchors the wood house frame to the foundation.

MATERIALS

Formwork for concrete foundation walls is generally constructed with the following materials:

- **LUMBER**
  Formwork studs, rangers (horizontal supports), diagonal braces, and stakes are made of relatively lightweight 2x4s that resist splitting when nailed.

- **PLYWOOD**
  For formwork sheathing, use smooth, water-resistant ¾-inch exterior plywood. To keep the plywood from sticking to the hardened concrete, spray it with motor oil or form oil.

- **FORM TIES**
  Wire ties hold form walls together while wood spreaders keep them a fixed distance apart. Snap ties perform both functions. A snap tie is a steel wire with a bolt head and bracket on each end to grab the outsides of facing form walls and hold them together. Two plastic cones on the wire press against the inside of form walls, keeping them a fixed distance apart.

Building Formed Footings

USE: ▶ power drill/driver • mason's twine • shovel • 4-ft. level • tape measure • hammer • short sledgehammer ▶ concrete • gravel • 2x formwork boards

1 Set up batter boards and string lines to locate the footing excavation. On this wall project, the footing is near grade.

2 Begin excavating the soil within your string lines. Remember, your footings should rest below the frost line.

3 Drive in the stakes that will support the outside form boards. You can build them from 2x4s with an angled tip.

FORMWORK

Pipe-Sleeve Blockouts

Before pouring concrete into foundation wall forms, you need to install blockouts, or barriers, to accommodate electrical conduits and gas and water pipes. The simplest way to create this type of blockout is to drill holes into the formwork and slide a piece of PVC pipe through the openings. You can leave the pipe sleeves long, and trim them flush with the concrete once the mix is set and the forms are stripped. Pipe diameters should be slightly greater than whatever has to pass through them.

Use a drill (or a drill and a saber saw) to cut holes matching the pipe diameter through the faces of the formwork.

Insert a sleeve of plastic pipe through the holes in the formwork to provide access for utility lines.

Window & Door Blockouts

Openings for doors, windows, and crawl-space vents must also be blocked out inside formwork before any concrete is poured. These rough-opening forms are plywood boxes nailed to the inside of the wall forms. If you are placing a window in a basement wall, you'll also need to excavate a well outside the window and build forms for the walls of the well (as shown at far right).

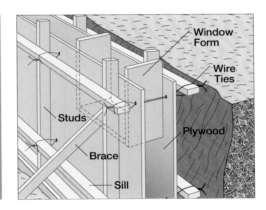
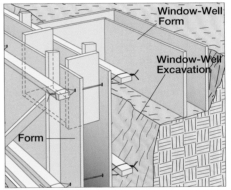

• 2x4 batter boards • 1x4 stakes • double-headed nails • galvanized screws • plywood gussets

4 Use a clamp to fasten boards to the stakes temporarily. Check for level; adjust the boards; and then nail or screw them in place.

5 Secure butt joints in the formwork with plywood gussets screwed in place across the exterior face of each joint.

6 Use 1x2 spreaders to bridge the forms so that the weight of the concrete pour will not cause the forms to bulge.

BUILDING WALL FORMS

6 Masonry

BUILDING WITH BLOCK

Concrete block, like lumber, is not really the size it's labeled. A standard 8x8x16-inch block is actually 7⅝ x 7⅝ x 15⅝ inches, allowing for the ⅜-inch-thick mortar joints. Blocks consist of an outside shell with a hollow center that is divided by two or three vertical webs. The ends of a unit have flanges that accept mortar and join with the adjacent block (except blocks intended for corners and the ends of walls). There are many types of blocks, such as solid, load-bearing, and nonload-bearing. Heavier blocks are made with sand, gravel, and crushed stone mixed with cement; lighter blocks have lighter aggregates, such as coal cinders.

Building with masonry block can be taxing. Standard concrete blocks weigh more than 40 pounds each. It's difficult enough to sling one of the rough-surfaced blocks up into position, but it's even tougher when you have to lower it gingerly onto a bed of mortar.

Ties, Flashing & Reinforcement

Most structures are single-wythe—the thickness of a single row of blocks. Multiple wythes of masonry must be tied together with header bricks, cap blocks, and metal ties. Masonry flashing (made of metal, rubberized asphalt, or other material) is used to control moisture by keeping the top of the wall dry. To laterally reinforce joints between wythes, place wire wall ties in the mortar beds at 16 inches on center vertically; foundations and pilasters require rebar to strengthen them.

Cutting Block

Usually you can get by with store-bought, preformed half-sized blocks, but there may be times when you need to cut a slightly smaller or larger partial block. Concrete block, like all other types of masonry, often does not break smoothly. Scoring the block, as shown here, helps improve the odds of getting an even break when using a hammer and brickset. But if you plan to cut even a few, order extra.

You can cut through concrete block with a circular saw, but only if it has a special masonry blade.

To cut a block by hand, score each face with a brickset, making several passes with light pressure along each cut line.

You can keep scoring until a piece breaks free, or turn the block on edge and apply more force to the brickset.

Laying Block

USE: ▶ bricklayer's hammer • brick trowel • brickset • line level • mason's twine • pointing trowel • spirit level • work gloves • jointing trowel ▶ concrete block

1 To embed the first course of concrete blocks, trowel on a liberal amount of mortar in two rows under the block edges.

2 Carefully level the block, particularly on the first few courses. Use the end of your trowel to make small adjustments.

3 You may want to build from one end of your wall to the other; masons often build up a few courses at the corners first.

BLOCK

Estimating

For standard 8x8x16-in. concrete block, calculate 113 blocks per 100 sq. ft. of wall area and add 10 percent for breakage. For mortar, figure 8.5 cu. ft. per 100 sq. ft. of wall area. A contractor would order the components separately and make the mix on-site, but to keep your proportions accurate, buying mortar premixed makes more sense.

Mortar Types

TYPE M— A high-strength mortar for masonry walls below grade and walls subject to high lateral or compressive loads or severe frost heaving.

TYPE S— A medium-high-strength mortar for walls requiring strength to resist high lateral loads.

TYPE N— A medium-strength mortar for most masonry work above grade.

TYPE O— A low-strength mortar for interior nonbearing partitions.

TYPE K— Low-strength lime-sand mortar used for tuck pointing.

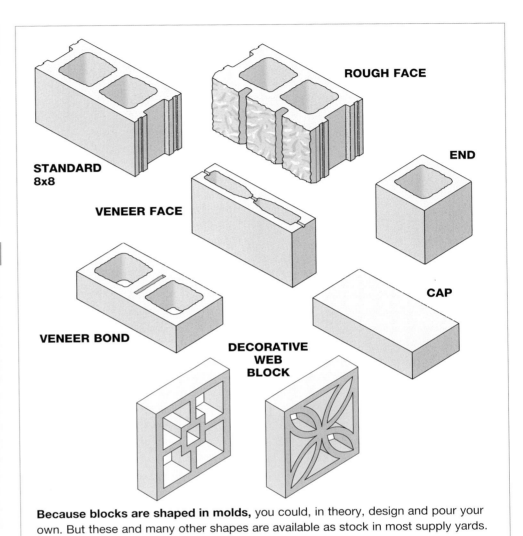

Because blocks are shaped in molds, you could, in theory, design and pour your own. But these and many other shapes are available as stock in most supply yards.

- mortar mix • steel reinforcing bars • truss-type horizontal joint reinforcement

4 Masons can toss the right amount of mortar in the right place; you may want to form the mortar bed one small bit at a time.

5 After spreading mortar on the final course of full-size blocks, finish the wall with cap blocks.

6 Use a jointing tool to smooth the mortar into a water-shedding, slightly concave shape.

6 Masonry

REPAIRING BLOCK WALLS

Usually, the first repair you'll need to make to a block structure will be repointing—using fresh mortar to replace mortar joints that have failed. (This process is the same for brick; see "Maintaining Mortar," on page 88.) Other simple cosmetic repairs to a block structure involve reinforcing a damaged area with a patch of repair compound, or coating the block with paint, stucco, or mortar.

Foundation Wall Repairs

If a block or brick foundation has cracked and chipped on the surface but is still structurally sound, there are several repair options, including repointing the mortar joints, replacing badly chipped units, and covering with stucco. (See "Patching Stucco," on pages 92–93.) If the bricks or blocks don't wobble in and out of line too much, you can use ½-inch pressure-treated plywood or foamboard to bridge the nooks and crannies that let in water and moisture. Even small openings can eventually ruin the foundation's appearance and sap its strength.

To make a complete seal and create a neat line near the ground, clear away enough dirt so that the bottom few inches of the new covering can be buried, and tuck the upper edge into a liberal bed of caulking below the house siding. Although pressure-treated wood will resist rot, you might refill the small trench with gravel to encourage drainage before replacing the top layer of sod. The only tricky part of this simple operation is nailing the covering panels into the brick. Get an assist by lacing the backs of the covering panels with construction adhesive; then use several hardened masonry or cut nails to lock each sheet in place. Nailing into the mortar between bricks is easiest if you can gauge the courses of brick.

For a neutral, masonry-like finish, coat the covering material with heavy-bodied gray exterior stain. From a few feet away, it will look like stucco. While you're at it, consider adding a layer of rigid foam insulating board behind the plywood to insulate the exposed portion of the foundation.

Painting Block

If you plan to finish masonry basement walls, there are special waterproofing paints that roll or brush on thickly, like wet plaster. First, you should patch any open or leaking cracks with a cement-based patching compound or hydraulic cement. Even on rough walls, surface applications look better and last longer over sound, uncracked surfaces.

For painting, opt for a heavy-napped roller, called a "bulldozer" by professional painters because it can be loaded with thick paint and it will push the excess along the wall just as a bulldozer pushes dirt. The thick, ragged nap is essential when working on rough masonry. It helps to work the viscous paint into surface crevices and the joints between courses of concrete foundation block.

Patching Block

USE: ▶ cold chisel • hawk • pointing trowel • small

1 To make your patch material bond securely to the block, chip away cracked edges with a cold chisel.

Control Joints

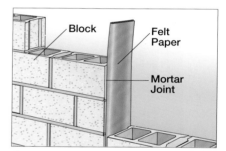

Control joints, made by setting felt paper in continuous seams, isolate wall sections from each other.

Buttering Blocks

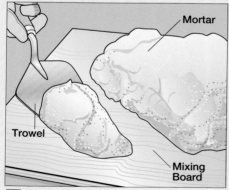

1 Using a trowel, slice a wedge of mortar from your mixing board or hawk.

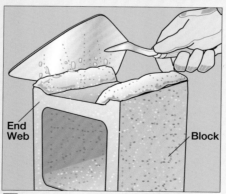

2 Tap the trowel against the raised edge of the block to lay the mortar.

Replacing Block

USE: ▶ power drill with masonry bit • safety

1 For many repairs, you don't have to remove the entire block, only the face. To start, drill a series of holes.

BLOCK

MONEY SAVER

sledgehammer • wire brush • work gloves ▶ cement patch • water

2 Because dust and loose particles prevent the patch from adhering, sweep the damaged area using a wire brush.

3 Hold a board or trowel loaded with cement at the base of patch area, and force the material into the hole.

4 To smooth out small patches, use the side of a mason's trowel with each end riding on the undamaged surface.

Garden Block

Garden block, or screen block, is used to create walls with open patterns. Walls can be made entirely of garden block, or a few courses can top a solid wall or be worked into the middle. Build them as you would a regular concrete block wall. The joints are smaller, so you will have to be a little neater, but you won't need to mix as much mortar. The work goes faster because the units are spaced far apart.

MONEY SAVER

goggles • brickset • jointing trowel • mason's hoe • mortar box • pointing trowel • small sledgehammer • work gloves ▶ mortar mix • replacement block • water

2 Use a brickset to chip away the interior webs of the block. Always wear gloves and eye protection when chiseling block.

3 Split a replacement block; butter the edges with mortar; and then set the new block face in position.

4 Because there isn't room to place mortar above and below the block, force mortar deeply into joints before tooling.

6 Masonry

BRICK BASICS

Brick is made from clay fired at high temperatures. Brick textures vary depending on the molding process. Most bricks made today are dense, hard, and durable. But it is important to choose the right brick for the project.

Types of Brick

Face brick is used where a consistent appearance is required. Face brick (FB) is produced in three varieties. FBX (select) brick have the tightest limits on size variation and flaws. Sharp edges and crisp outlines give them a contemporary look. FBS (standard) has wide color ranges but only slight variations in size. FBA (architectural) has no limits on either size variations or the cracks or chips that are permitted because it is intended to look like historic brick. **Building bricks** (or common bricks) are rough in appearance but structurally sound. The chips, cracks, color variations, and slight deformations in building brick create a rustic look. **Paving brick** (always solid, unlike other types, because the widest faces will be visible) are pressed into molds, and a longer baking time reduces how much water they absorb—critical for bricks placed on the ground, which will have to withstand freeze-thaw cycles and heavy traffic. Paving brick classes are similar to face brick: PS (standard), PA (architectural), and PX (select). **Firebrick** are a dull yellow, highly heat-resistant type used for fireplaces and ovens.

Don't feel too bad if your concrete patio is sinking. The S.S. Atlantus is an entire concrete ship that's sinking, off the coast of Cape May Point, NJ. Built as an experiment during World War I, it was, as the onshore marker reads, "proven impractical because of weight."

Brick Grading

There are three grades each for face brick and building brick; the grades are based on how well a brick resists damage from freezing and thawing. Grade MW (moderate weathering) can be used when bricks will be exposed to moisture but not saturated. Grade SW (severe weathering) should be used when bricks will be frozen when saturated. Grade NW (no weathering), available only in building brick, is for indoors only. Paving brick is divided into similar grades: SX (for highest freeze-thaw resistance), MX, and NX.

Cutting Brick

To cut bricks, score all around the brick with a brickset and a brick-layer's hammer. A firm blow on the score line will split the brick. To cut pavers, use a circular saw equipped with a masonry blade. Cut several half-bricks before starting a project.

To cut bricks, score all four sides with a brickset and hammer.

Deepen the score line with repeated blows until the brick splits.

Mortaring Bricks

Mortar is the glue that holds all brick walls together. There are different mortar mixtures to suit different conditions, but every mortar mix should be prepared keeping the same three pointers in mind. First, don't add excess water to make mixing easier; it will weaken the mortar. Second, follow proportions in every small batch to maintain a uniform mix throughout the wall. Third, keep the mix clean, which means periodically cleaning your mixing board of hardened particles from past batches.

USE: ▶ brick trowel • brickset • hammer • hawk • jointing trowel • mason's hoe • mortar box • pointing

1 Mix just enough mortar to use before the material sets up, cutting off one trowel's worth at a time.

2 Carry the mortar on one side of the trowel, and set it in place by sharply striking the clean side against the masonry.

BRICK

Mortars

Much like block mortar, brick mortars are mixes of portland cement, lime, sand, and water. You can mix your own or purchase factory-blended dry mortars, which you then combine with sand and water.

Mortar Proportions by Volume

Type*	Portland Cement	Hydrated Lime	Mason's Sand
Portland cement & lime mortar:			
N	1	1	6
S	1	½	4½
Factory-blended masonry cement mortar:			
N	1	n/a	3
S	1	n/a	3

*See page 83 for Mortar Types.

Materials

Modular brick styles are simpler to work with than non-modular ones because they make for easier planning and estimates. For example, with mortar joints, one modular brick's length is the same as two bricks laid widthwise or three bricks stacked. This makes it easier to turn corners or vary brick patterns in a double-wythe wall.

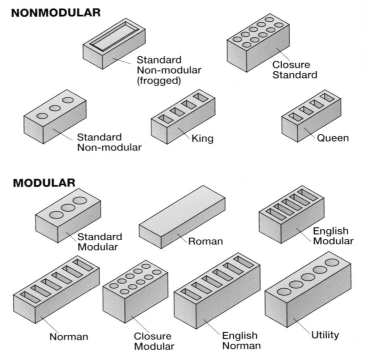

NONMODULAR: Standard Non-modular (frogged), Closure Standard, Standard Non-modular, King, Queen

MODULAR: Standard Modular, Roman, English Modular, Norman, Closure Modular, English Norman, Utility

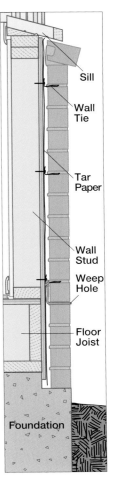

Sill, Wall Tie, Tar Paper, Wall Stud, Weep Hole, Floor Joist, Foundation

---

trowel ▶ metal ties • mortar mix • truss-type horizontal joint reinforcement • water

3 Turn the trowel over, face side in, to form the outer edge of the mortar bed. You may need several passes to make it even.

4 To make sure the bricks are fully supported along their edges, make a shallow furrow along the center of the mortar bed.

5 If furrowing pushes mortar off the bricks, slice away excess with a trowel. Don't reuse mortar from the ground.

BRICK BASICS

6 Masonry

BRICK MAINTENANCE

To evaluate a brick wall, each component must be considered. Large cracks, called faults, in the overall structure are usually the most obvious problems and the most costly to repair. Most faults can be traced to uneven settling: when soil under one section compacts more than at another, the foundation and the wall above it can crack under the strain.

Cracking & Curving Walls

The signs of fundamental structural problems are staircase-pattern cracking along many courses of brick, and either large-scale convex or concave cupping of the walls. You can check for this curving, which can be difficult to detect over a large surface, by using line blocks and string.

Don't write off a wall's soundness just because it has some cracks. They may be only cosmetic or from settling that occurred long ago. Old, stable cracks (which usually are somewhat weathered, dirty, and may contain bits of leaves, dirt and debris, or spider webs) can be patched and sealed against the weather. New, unstable cracks (which are usually clean, with the masonry a lighter color than that on the surrounding wall) indicate that the building is still in motion. Such a wall will have to be watched carefully because if it is moving quickly, the foundation is unstable, and it may be in danger of collapse. The only fix for a wall with a poor foundation is to either prop it up with braces (an unsightly, temporary, and potentially dangerous solution) or demolish it and begin again from the foundation.

Maintaining Mortar

The maintenance program for your brick wall should also include a careful check of the mortar joints, the source of the most potential danger next to the foundation. Any mortar that is loose, spongy, and easily scraped away needs to be repointed. This process includes excavating the mortar to a depth of approximately two times the width of the joint, cleaning away all loose dust and debris, and then refilling the joints with mortar. Without repointing, the process of deterioration accelerates, particularly in winter (due to freezing and thawing). You can expect to repoint brick homes built before 1900, but not because of age alone. The mortar commonly used at that time was lime-based, softer, and very porous.

Bricks are generally more durable than the surrounding mortar, although leaks and condensation can sometimes harm them unex-pectedly. Water absorption through the brick can lead to fragmentation and flaking of the exterior brick face. This splitting off of the exterior face of bricks—a condition called spalling—can result in the need for expensive repairwork. To avoid the potential problems associated with spalling, keep water out of your masonry walls, and repair leak-prone brick and mortar joints as soon as possible.

Cleaning Brick

Cleaning old brick walls by sand- or water-blasting might do more harm than good. Stresses from high-pressure cleaning could seriously erode if not shatter nineteenth-century brick, which may not tolerate more than 100 psi. While sand hits the brick with uniform pressure, old brick isn't uniform; leading to pitting and channeling. Chemical cleaning may produce good results with less damage.

Laying Brick

USE: ▶ chalk-line box • brick trowel • bricklayer's hammer • brickset • hawk • jointing trowel • line level • mason's hoe • mason's line • measuring tape

1 Snap a chalk line to establish the outer boundary of the wall; then trowel on the embedding layer of mortar.

2 Professionals can tap the brick into proper position by eye; you may want to check the first few courses with a level.

3 String a line from corner to corner, and use it as guide to keep the wall flat. Make fine adjustments by tapping with the trowel.

BRICK

Tooled Joints

Proper tooling seals masonry joints, but only some types are weather-resistant. Horizontal joints should be tooled first, then move on to the vertical.

WEATHER RESISTANT
- Concave
- V-Shaped
- Weathered

NONWEATHER RESISTANT
- Flush
- Raked
- Struck

Diagnosing Brick Problems

Eroded mortar joints, caused by old or poorly made mortar, are fixed by repointing. Staircase-pattern cracking is a sign of major structural problems, resolved only by removing the structure and improving its subbase. Spalling occurs when freezing water has caused a brick to fracture—usually the face of the brick cracks off. The entire brick should be replaced, and if possible, water should be redirected away from this area.

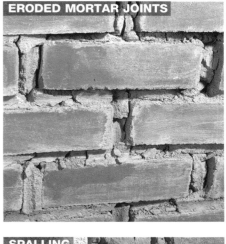
ERODED MORTAR JOINTS

STAIRCASE CRACKING

SPALLING

• mortar box • pointing trowel • spirit level ▶ mortar mix • water

4 When you've set the brick in proper position, use your trowel to scoop away excess mortar that oozes from the joint.

5 Regularly check your wall for level. Every few courses, take a measurement from the footing at several points on the wall.

6 Unlike rough block joints, which often are concealed by stucco, brick mortar requires more uniform tooling.

BRICK MAINTENANCE

6 Masonry

WORKING WITH STONE

Although many types of stone are available throughout the country, only a few can be used for building. Suitable stones must satisfy the requirements of strength, hardness, workability, durability, and density.

Stone can be described by its shape or the form in which it is used (such as rubble, ashlar, or flagstone), by its class or mineral composition (such as granite, limestone, sandstone, or slate), or by the way in which it is obtained (fieldstone or quarry stone). Rubble stone is irregular in size and shape; fieldstone is a type of rubble, naturally rough and angular. Ashlar is stone that has been cut at the quarry to produce relatively smooth, flat bedding surfaces that stack easily. Flagstone is designed to be used for paving and has been cut into flat slabs. Flagstone ranges from ½ to 2 inches thick and may be irregularly shaped or cut into geometric patterns.

Buying Stone

Stone is sold by the cubic yard at quarries and stone suppliers. Cut stone is more expensive than fieldstone or rubble because of the labor involved. To estimate how many cubic yards you will need, multiply the length times the height times the width of your wall in feet, and then divide by 27. For ashlar stone, add about 10 percent to your order for breakage and waste; if using rubble, add at least 25 percent.

Dry-Stacking a Stone Wall

Dry-stacked stone walls are built without mortar: friction, gravity, and the interlocking stones hold the wall together. The wall is flexible enough to absorb some frost heave and is usually built without footings when less than 3 feet tall. Taller dry-stacked walls are not recommended because of the lifting involved and code restrictions. Stone may require cutting and shaping for a good fit. If you are gathering fieldstone, look for angular shapes rather than round stones; larger stones should be kept for the base course. A dry-stacked wall consists of two wythes, with a space between filled with small rubble. The wythes are tied together with bond stones that span the width of the wall. A 3-foot-high wall should have a base about 2 feet wide that tapers at the top.

Stone-Wall Patterns

RANDOM RUBBLE

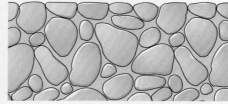

Rubble wall construction uses stones that are irregular in size and shape. Fieldstone is a type of rubble; quarried rubble comes from fragments left over after stonecutting. Random rubble walls are usually dry-laid but can also be mortared. Use this pattern for an informal garden wall or short dry-laid boundary walls.

COURSED RUBBLE

Coursed rubble walls have a neater appearance than random rubble walls but are more difficult to construct, and require a large selection of stone. Rubble stones can also be roughly squared with a brick hammer to fit into place more easily. Coursed rubble walls can be used for foundations and structural walls as well as garden and retaining walls.

MOSAIC

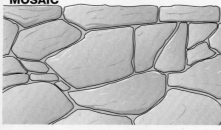

Mosaic, also called web, is a tighter version of a random rubble wall. Large and smaller stones are fit together much more tightly than with a random rubble wall. To ensure that all of the pieces fit together without large gaps, the stones are first laid out on the ground, face down, and test-fitted in the order in which they will be installed.

RANDOM ASHLAR

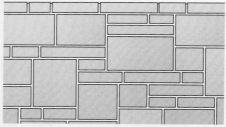

Ashlar has been cut at the quarry to produce smooth, flat bedding surfaces that stack easily. It is generally cut into small rectangles and has sawed or dressed faces that can be either smooth or slightly rough. Ashlar patterns are not really random; as with brick, a variety of bond patterns are used alone or combined for different effects.

COURSED ASHLAR

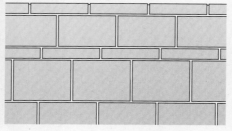

Coursed ashlar has a more formal appearance than random ashlar, and it requires more precisely cut stone. Ashlar mortar joints are sometimes used as a decorative element. They may be a color that complements the stone, raked concave like block joints, or filled and dressed to have an extruded appearance that stands out from the stone.

STONE

Laying Face Stone

USE: ▶ brick trowel • bricklayer's hammer • brickset • jointing trowel • line level • mason's line • pointing trowel • spirit level • work gloves ▶ mortar mix • water

1 Manufactured face stone (thinner and lighter than full stone) is applied like brick, using a string line to keep the wall flat.

2 Because the stones are slightly irregular to create a less manufactured look, periodically check the tops for level.

3 After the stone is embedded on the wall, fill the joints between courses, and tool the joints to suit.

Shaping Stone

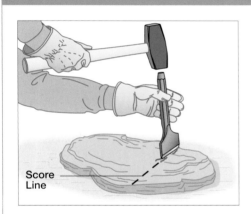

You will often have to cut and shape stones to make them fit—especially when dry-laying a wall. Place the stone on flat, solid ground rather than a hard surface such as concrete. Use a heavy hammer and cold chisel to score the cut line. The stone may break at the line before you finish scoring. If not, strike one sharp blow with the hammer. You can also split flagstone after scoring by placing it over a section of pipe and striking it. Always wear heavy gloves and goggles when cutting stone.

Laying Full Stone

USE: ▶ brickset • hawk • hoe • line level • mason's line • mortar box • small sledgehammer • trowels • work gloves ▶ mortar mix • rubble • stone • water

1 The rocks you use help design the wall you build. But try to use square-edge rocks at corners and flat-faced rocks on the sides.

2 To build a rock wall in manageable pieces, use modest-size stones with flat faces, and fill the hollow core with rubble.

3 Bind the rocks together in stages, first with an embedding coat on the foundation, then by filling the rubble core.

6 Masonry

CLEANING MASONRY

There are four ways to clean concrete and stone: with chemicals, water, steam, or by sandblasting. Sandblasting takes away surface and embedded dirt—and often some of the masonry, too. Chemical- and steam-cleaning contractors can tailor their mix of chemicals to the job at hand—removing algae, for example. Always be careful when working with acid cleaners. Water cleaning is a job you can do yourself, either with bucket and brush or with a pressurized sprayer (power washing).

Plant life can be destructive to stone walls. When ivy roots start growing into cracks in mortar joints, you should cut the roots as close to the wall as possible and treat the ends with ammonium phosphate paste to kill the plant. Mold and mildew also may take hold on stone not exposed to enough sunlight. To test discoloration, drop a small amount of bleach on the area. It will whiten mildew and have no effect on dirt. To clear the mildew, scrub the area with a solution of one part bleach to one part warm water, and then rinse.

Stains from iron can be removed with a solution of oxalic acid. Mix about 1 pound of the crystals in a gallon of water with ½ pound ammonium bifluoride; brush the mix over the stained area; then rinse. Stone may also become stained with asphalt and tar from a roof. Remove as much tar as possible. The remaining stain can be cleaned with a solvent such as benzene.

Patching Steps

USE: ▶ chisel • formboards • goggles • sledgehammer • trowel • work gloves ▶ mortar mix

1 Wire brush the damaged area to remove all loose debris, and then build a plywood form to square the step repair.

3 Nails driven halfway into the old step are covered by fresh concrete, smoothed out to be level with the form.

MONEY SAVER

2 To help a patch adhere, coat the area with a bonding agent, or set a few concrete nails to reinforce the repair.

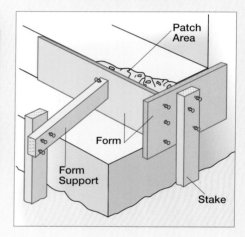

Patching Stucco

USE: ▶ metal snips • brick trowel • hawk • pointing trowel • sponge trowel • spray mister • wire brush • work gloves ▶ 15-lb. felt paper • galvanized-wire lath

1 To check for damage beneath cracked and broken stucco, cut away bent reinforcing mesh and tar paper.

2 Install new lath over the damaged area, and trowel on a base coat of patch material that nearly fills the hole.

3 Rake the surface of the base coat before it hardens. Even a homemade tool (nails through a board) will do the job.

MASONRY REPAIRS

MONEY SAVER
Removing Stains

For stubborn spot stains, mix cleaners into a paste with flour or talc, which keeps the cleaner on the wall.

Stain	Solution
Oil	For brick, emulsifying agent; for concrete/block, automotive degreaser
Iron	1 lb. oxalic acid crystals, 1 gal. water, ½ lb. ammonium bifluoride
Paint	2 lbs. TSP to 1 gal. water
Smoke	Scouring powder with bleach; or poultice of trichlorethylene and talc; or alkali detergents and emulsifying agents

Efflorescence

Efflorescence is a white powdery deposit of soluble salts leached to the surface of masonry or mortar by moisture from within. Residue is left when the moisture evaporates. For a quick fix, dry brush the deposit using a stiff fiber brush; then saturate the masonry with water. For stubborn deposits, use a 1-to-10 solution of muriatic acid and water, following label cautions, and then rinse thoroughly. Of course, the best solution is to find the source of the leak and to stop it.

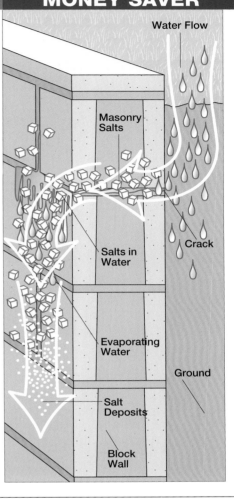

MONEY SAVER

MONEY SAVER

• masking tape • 1¼-inch roofing nails

4 Keep the finish coat flush with the surrounding wall by using the stable edges of the damaged area as a guide.

5 Texture the thin top coat (about ⅛ in. thick) with a trowel, sponge, broom, or anything that helps the patch blend in.

6 To harden and cure the stucco, moisten the patch at least twice a day for the first two days.

7 Electrical

96 POWER DISTRIBUTION
 ◆ How Electricity is Provided ◆ Transmission
 ◆ Power Generation and Delivery

98 WIRING SYSTEMS
 ◆ Service-Entrance Panel ◆ Circuit Voltage
 ◆ Wire Types ◆ Wire Sizes ◆ Wire Colors
 ◆ Ground Faults on Appliances

100 SAFETY MEASURES
 ◆ Electrical Cautions ◆ Testing ◆ Testers
 ■ FIXING A CORD ◆ Cutting off Power
 ◆ Socket Safety

102 CIRCUIT BREAKERS
 ◆ Testing Fuses ◆ Breaker Types

104 BOXES & CONNECTORS
 ◆ Electrical Boxes ◆ Electrical Box Types
 ◆ Wire Connectors ◆ Wire Connector Ratings
 ◆ Metal-Box Markings ◆ Using Crimping Ferrules

106 CONDUIT
 ◆ Conduit and Accessories ◆ Raceway
 Components ◆ Wiring Capacities for Metallic
 Tubing ◆ Bending and Cutting Conduit

108 WIRING BASICS
 ◆ Basic Materials ◆ Wire Terminals
 ◆ Fastening Cable ◆ Stripping Cable

110 WIRING PATHS
 ◆ New Wiring Options ◆ Fishing Connections
 ■ INSTALLING SURFACE WIRING ◆ Fishing
 Cable: Walls and Ceilings

116 ADDING A NEW CIRCUIT
 ◆ Creating New Circuits ◆ Installing a Circuit
 Breaker

118 WIRING RECEPTACLES
 ◆ Typical Outlet Wiring ◆ Grounding Boxes
 ◆ Receptacle Types ◆ Installing an Outlet
 ◆ Typical Outlet and Wiring Layouts ◆ High-
 Voltage Receptacles ◆ Wiring High-Voltage
 Receptacles ◆ Appliance Receptacles

126 GROUND-FAULT CIRCUIT INTERRUPTERS
 ◆ GFCI Receptacles ◆ Wiring a GFCI Outlet

128 WIRING SWITCHES
 ◆ Switch Types ◆ Single-Pole Switch ◆ Wiring a
 Three-Way Switch ◆ Wiring Dimmers ◆
 Dimmer Types ◆ Switch Types ◆ Testing a Switch
 ◆ Splicing Grounding Wires ◆ Typical Switch
 and Wiring Layouts

134 LIGHTING FIXTURES
 ◆ Bulb Types ■ COMPACT FLUORESCENTS
 ◆ Lamp Repair ◆ Boxes ◆ Installing a Ceiling
 Box ◆ Installing a Fluorescent Fixture
 ◆ Installing a Ceiling-Mounted Fixture
 ◆ Track-Mounted Lighting ◆ Chandeliers
 ◆ Vanity Lighting ◆ Quartz Halogen
 ◆ Recessed Lighting

142 SPECIALTY WIRING
 ◆ Low-Volt Wiring ◆ Installing a Transformer
 ◆ Fixing a Doorbell ◆ Replacing a Thermostat

144 PHONE WIRING
 ◆ Telecommunications ◆ Wiring a Phone Jack

146 WIRING APPLIANCES
 ◆ Range Receptacles ◆ Direct-Wiring a
 Dishwasher ◆ Waste-Disposal Units
 ■ CEILING FANS ■ RANGE HOODS
 ■ ELECTRIC RADIANT FLOOR HEATING
 ◆ Dryer Receptacles

154 SAFETY DEVICES
 ◆ Smoke Detectors ◆ Hard-Wiring Smoke
 Detectors ◆ Carbon Monoxide Detectors

156 OUTDOOR LIGHTING
 ◆ Installing a Floodlight ◆ Outside Power
 ◆ Weatherproofing ◆ Conduit and Boxes
 ◆ Receptacles and Switches
 ◆ GFCI Outdoor Power
 ◆ Extending Power Outdoors
 ◆ Installing UF Cable
 ◆ Installing an Outdoor Receptacle
 ◆ Installing Low-Voltage Lighting

7 Electrical

HOW ELECTRICITY IS PROVIDED

Utility companies generate electricity in a variety of ways. One of the most common methods uses the energy of running water to power a generator. Electrical power created in this way is termed hydroelectricity. To harness the energy of flowing water on a scale this enormous, a dam may be built across a narrow gorge in a river or at the head of a man-made lake. Water backed up behind this dam, in what is called the forebay, is then allowed to flow through a submerged passage, or penstock, in a controlled release. The massive force of this elevated water spins the generator's giant turbines as it falls, producing electricity. Electrical power produced in this way is called AC power, or alternating current.

Transmission

Once a utility company produces electricity, it must then transmit it through a distribution system for use by its customers. For ease of transmission, the electrical power is raised to many thousands of volts and conducted over high-voltage transmission lines to the utility company's regional switching stations, where it is then stepped down to a lower voltage for transmission to local substations. A typical transmission starts at 230,000 volts, is stepped down to 69,000 volts at a switching station, then is stepped down further at a substation to 13,800 volts for direct distribution to a local area. Once at your home, this is again reduced, to 240 volts. All homes are wired for 240 volts, which can be stepped down to 120 volts.

Point of Use

To be stepped down, the electricity that arrives at your home must first pass through a utility transformer. It then leaves this transformer via three terminals, mounted on its side, which are connected to three wires. These wires constitute the service drop that leads to your house service entrance. They include two insulated hot wires, or legs, and a grounded neutral. The two hot wires can each provide 120 volts or supply 240 volts of power between the two hot legs. The neutral conductor is usually bare on overhead and insulated in underground service laterals.

A glass-domed meter is connected to the two hot wires leading from the utility transformer. This meter, generally mounted on the outside of your house, is provided by your utility com-

Power Generation and Delivery

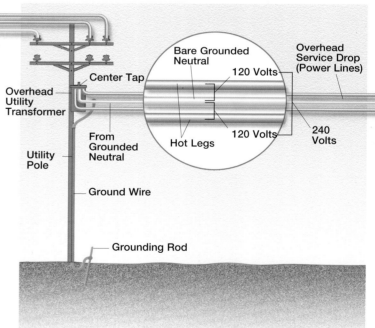

At a hydroelectric plant, the massive kinetic force of elevated water that drops down through a dam penstock in a controlled release turns giant turbines that generate electricity. Electrical current travels over high-voltage power lines to a step-down utility transformer near your house. The current then leaves the transformer, enters your home as available voltage for use (120-volt and 240-volt), and then returns to the transformer. It is transmitted to your main

POWER DISTRIBUTION

pany to measure the amount of electrical energy in kilowatt hours consumed by your household. This is the rate of energy consumption in kilowatts multiplied by usage in hours. Directly from the utility meter, the two hot wires and the grounded wire continue on to a service-entrance panel (SEP), which distributes power throughout your house. The service panel also contains circuit breakers or fuses that will open if a short circuit or overload occurs in the system.

Service-Entrance Panel

It is the service-entrance, or main, panel that controls the flow of power to individual circuits within your home. These circuits may be 120-volt, 240-volt, or both (120/240-volt). All 240-volt devices pull current from both of the hot insulated legs. At any given moment, electricity is exiting from one terminal on the utility transformer and returning by the other. Current flows from one terminal, travels through the service drop to the house, and then down the service-entrance conduit or cable into the meter base. From here it flows through the meter into the main panel and is then distributed to each of the circuits within your home, flowing through the main panel via one (or two) insulated hot leg(s), or wire(s), and returning to the panel via another insulated wire—directly through the utility meter and back to the transformer. The final result is that you never actually "consume" electricity—you just borrow it (although you transform much of its energy, which is what you pay for).

All 120-volt devices draw from one of the two hot insulated wires going to the device and use the grounded white wire as the return. The grounded conductor is connected to ground (via a ground electrode) at both the transformer and at the main service-entrance panel. All 120/240-volt appliances draw from both of the hot insulated wires as well as using the grounded conductor as the return. An electric clothes dryer, for example, uses 240 volts to heat the element but also uses 120 volts for the timer, motor, and alarm circuits. Such circuits carry current on all three wires at the same time. A 120/240-volt appliance, like a clothes dryer, needs two insulated hot wires, one insulated grounded wire, and one grounding conductor. A 120-volt duplex outlet needs an insulated hot wire, an insulated grounded wire, and a bare or green grounding wire. A 240-volt-only appliance needs just two insulated hot wires and a grounding wire.

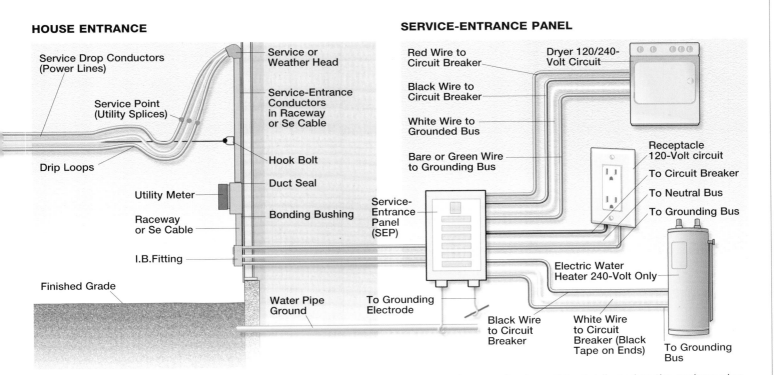

service-entrance panel (SEP) by means of two insulated hot conductors and a bare grounded neutral wire. Power that enters your home must first flow through the utility company's electric meter to be measured. The electricity then goes to your service panel, where it is distributed to the various electrical circuits in your home. Though power may be generated by other means, it is delivered through this same system.

HOW ELECTRICITY IS PROVIDED 97

7 Electrical

Service-Entrance Panel

Also called the circuit-breaker panel, the main service-entrance panel (SEP) is the distribution center for the electricity you use in your home. Incoming red and black hot wires connect to the main breaker and energize the other circuit breakers that are snapped into place. Hot (black or red) wires connected to the various circuit breakers carry electricity to appliances, fixtures, and receptacles throughout the house. White and bare-copper wires connect to the neutral and grounding bus bars, respectively. (Representative 120-volt and 120/240-volt circuits are shown.)

Aluminum Wire. Some older homes have aluminum wiring, identified by the silvery color and an AL stamp. Aluminum wire expands and contracts at a different rate than copper, and this can loosen the copper-to-aluminum connection. So, use aluminum-compatible connectors, marked CO/ALR or CU/AL.

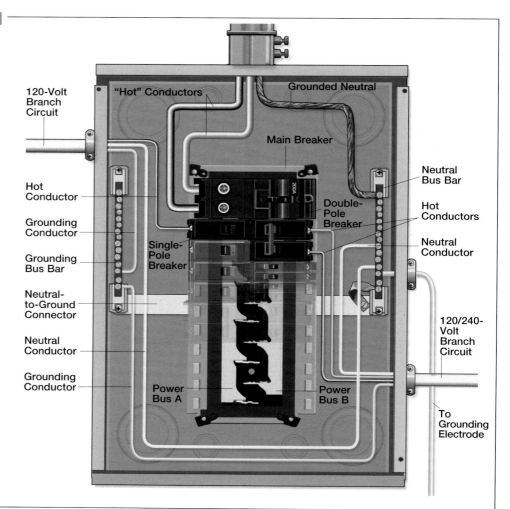

Circuit Voltage

Power rated at 120 volts travels from a black wire on a single-pole circuit breaker to the device. It returns through a white wire from the device to the neutral bus. In a 120/240 volt circuit, 240-volt power flows from one pole of a double-pole circuit breaker to the appliance and back to the second pole on the breaker. Additionally, 120-volt power that runs the lights, clock, and timer travels through a hot wire and back on a neutral white wire.

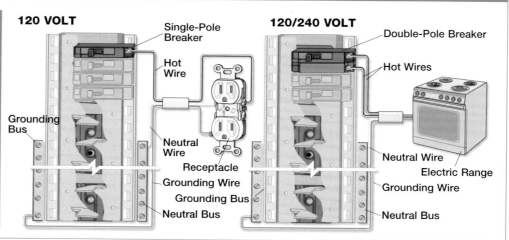

98 ELECTRICAL / WIRING SYSTEMS

WIRING SYSTEMS

Wire Types

Single wires can be insulated to carry electricity or bare for grounding. Most household wiring is contained in cable, inside flexible metal (such as BX, pictured at top), or plastic insulation (such as NM, 2nd from top). Cords (such as lamp cords, 3rd from top) are stranded wires in plastic insulation, not to be used as fixed wiring; low-voltage wire (bottom) is used to wire doorbells and thermostats.

Wire Sizes

Wires have size numbers based on the American wire gauge (AWG) system, which expresses wire diameter as a whole number. For example, No. 14 wire is 0.064 inches in diameter; No. 12 is 0.081 inches. Smaller numbers indicate larger diameters that can carry more power. The National Electrical Code requires a minimum of No. 14 wire for most house wiring.

Wire Colors

Wires have color-coded plastic insulation to indicate their function in your house's wiring system. Hot wires carrying current at full voltage are usually black, red, or white with black marks (marker or bands of electrical tape), but can be other colors. Neutral wires carrying zero voltage are white or gray. Ground wires can be bare copper or copper clad in green plastic insulation.

Ground Faults on Appliances

A ground fault can occur in an appliance any time that excess or misdirected current causes the appliance to become energized. A grounding system, including an equipment grounding conductor, is intended to provide a low-resistance path for current back to its source to prevent an electric shock or possible electrocution. In a properly grounded system, the current remains within the wiring system and trips the affected circuit breaker at the panel, rather than being directed to a grounding rod. In an ungrounded or improperly grounded circuit, the appliance itself may become energized, which means the ground-fault current will not be sufficient to trip the affected circuit breaker at the panel.

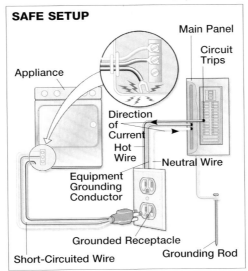

Grounded Appliance. The ground-fault current will return to the service-entrance panel through the equipment grounding conductor.

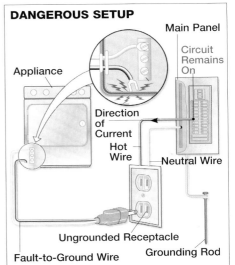

Ungrounded Appliance. The appliance frame can become electrically powered, and anyone touching it and a grounded surface may be electrocuted.

7 Electrical

ELECTRICAL CAUTIONS

Electricity can be dangerous, but if you use common sense, you can work with it quite safely. The most important thing to remember is to always, without fail, turn off the power at the main service panel before working on a circuit. Only use one hand to disconnect or reactivate a fuse or circuit breaker, and keep the other hand in your pocket or behind your back. Before starting work, check the circuit with a voltage tester to make sure that it is powerless. If you follow this rule, you will never suffer an electrical shock.

Confine your projects to outside the main service entrance. Do not go in the fuse/breaker box to add new circuits or into a transfer panel for a backup generator unless you have the professional know-how. You can wire in new circuits, repair old ones, and make countless other improvements, but call a licensed electrician when it is time to hook up the project to the entrance panel. The cost is not prohibitive, and the professional will check your work.

Codes

All electrical procedures and materials are governed by local building or electrical codes. They may prohibit the use of a certain type of cable or require a particular size wiring or minimum number of circuits, for example. The codes are for your protection. You may need a permit before beginning some projects; always consult with a municipal building inspector.

Testing

Testing tools let you know that wiring is safe. Use a neon circuit tester to see whether power is present, an important safety step even when you have tripped a breaker or removed a fuse. Touching the probes to a hot circuit causes an indicator to light. Use a continuity tester when a circuit is turned off to check whether an electrical path is uninterrupted. A multitester, which has a voltmeter on its face, performs both functions and is essential for measuring low voltages.

Cut off power at the service panel, and insert the metal probes of the tester into each slot of the receptacle.

Testers

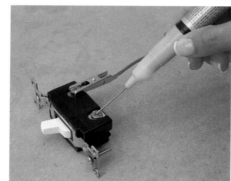

Use a continuity tester on wiring and appliances to pinpoint trouble by determining if a complete circuit exists.

Some analyzers can test for power, reversed wire connections, and other conditions of your electrical system.

Cutting Off Power

Touch only the insulating rim when removing a fuse. An overload melts the fuse's ribbon; a short discolors the window.

A tripped circuit breaker can be in the off position or between on and off. Reset it by switching it to off, then pressing it back on.

Ground-fault circuit-interrupter receptacles cut power automatically. They are required in many locations.

SAFETY MEASURES

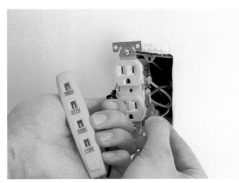

Touch brass and silver terminals with the probes; if the bulb glows, you have to shut off the circuit at the service panel.

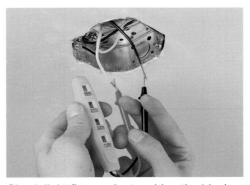

Check light fixtures by touching the black wire and the box or grounding wire, or by touching the black wire and white wire.

Touch the grounding wire of a switch with one probe; then touch each terminal with the other. No glow, no power.

Fixing a Cord

USE: ▶ cutting pliers • screwdriver ▶ new plug

MONEY SAVER

1 Cut a frayed cord below the damage; expose the individual wires inside; and strip ¾ in. of insulation from each one.

2 Choose a new plug rated for your wire. Some need an Underwriter's knot to prevent wire stress; some have a built-in clamp.

3 Hook exposed leads clockwise on the terminal screws (the way they are tightened), and clamp the cord.

Socket Safety

Three-slot outlets are built to accept a standard two-wire plug with a third grounding leg to reduce shock hazards.

Place protective plastic caps onto receptacles near the floor to keep young children safe from electrical shock.

A receptacle cover closes over plug and outlet to shield connections and prevent accidental pull-outs.

7 Electrical

CIRCUIT BREAKERS

Circuit breakers have replaced fuses as the preferred type of circuit protection. Technically, they are called molded-case circuit breakers, or MCCBs. Circuit breakers use a two-part system for protecting circuit wiring. When a small overload is on the circuit, a thermal strip will heat up and open, or trip, the circuit. When a massive amount of current comes through very quickly, as in a ground fault or short circuit, an electromagnet gives the thermal strip a boost. The greater the amount of trip current, the faster the breaker will trip.

The most important advantage circuit breakers have over fuses is that they can be easily reset; you don't have to buy a new one every time an appliance draws excessive current. When a breaker is tripped, it won't work unless you throw it all the way to the off position before you turn it back on again. Another characteristic of circuit breakers is that they are air-ambient-compensated—the hotter the air around them gets, the sooner they will trip. For example, if all the circuit breakers around a specific 20-amp breaker are running hot because of an excessive flow of current, the 20-amp breaker may trip at only 18 amps.

Residential circuit breakers typically range in size from 15 to 60 amps, increasing at intervals of 5 amps. Single-pole breakers rated for 15 to 20 amps control most 120-volt general-purpose circuits. Double-pole breakers rated for 20 to 60 amps control 240-volt circuits.

Standard circuit breakers are universal and have clips on the bottom that snap onto the hot-bus tabs in the panel box. Contact with the hot bus brings power into the breaker. Be aware, however, that some manufacturers make breakers with wire clips that mount on the side. These clips slide over the tab on the hot bus, requiring you to remove one or more of the other breakers to get at the one you want.

Common Breaker Types

In addition to single- and double-pole breakers, quad breakers, GFCI breakers, and surge-protection devices are also available. Single-pole breakers supply power to 120-volt loads such as receptacle and light circuits. A hot black or red wire is usually connected to the breaker. Single-pole breakers come full size or in a two-in-one configuration (twin). The latter type will only fit into a panel having a split-tab hot bus.

Double-pole breakers provide power to 240-volt appliances such as electric water heaters and dryers. If a standard NM cable is used as the conductors, both the black and the white wire are connected to the breaker. The white wire must be marked with black tape at both ends. Larger double-pole circuits have two black conductors in the circuit.

Specialty Breakers

A quad breaker falls within the half-size breaker

Breaker Contacts

Two contacts, or pressure clips, on the underside of a circuit breaker snap over a hot-bus tab in the main panel. These contacts bring power into the breaker.

family and can contain several configurations within one unit. It may, for example, contain two double-pole circuits, such as a double-pole 30-amp and a double-pole 20-amp circuit; it may have two single-pole circuits and one double pole; or it may provide power to some other combination of circuits. The advantage of a quad breaker is that it takes up half the space of a standard breaker. The panel, though, must be specially designed to accept quad

Testing Fuses

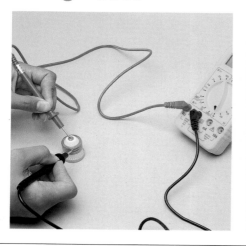

In homes with older wiring systems, you may need to test the conditions of fuses. To perform a continuity test on a glass fuse, left, touch one probe of a multitester to the center contact and the other to the screw shell. A zero reading means the fuse is working properly. Test a cartridge fuse, right, by touching one probe of a multitester to each end terminal on the fuse. The knob setting shown will give an audible signal as well as a meter reading.

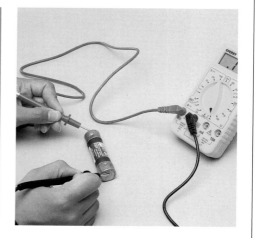

CIRCUIT BREAKERS

breakers. Furthermore, if the panel is too small, it may end up resembling a tightly interwoven nest of wires.

A GFCI circuit breaker fits into the main panel just like a standard circuit breaker. On its face is a test button but no reset. If properly installed, pressing the test button places a deliberate, preset current imbalance (6 milliamperes) on the line to verify that the breaker will trip when there is an unintended imbalance. When tripped, the breaker arm will go to a halfway off position, cutting power to the circuit. The circuit cannot be reset unless someone first turns the breaker arm completely off.

At first glance, a surge-protection device can be confusing. You will see a device that looks like a double-pole body. This type of device also has two lights that glow when power is applied to the panel. Nevertheless, a surge-protection device connects to the buses in the same way as any other circuit breaker.

Arc-Fault Circuit Interrupters

The National Electric Code requires an arc-fault circuit-interrupter (AFCI) breaker be installed to protect branch circuits to most living areas, with the exception of kitchens, bathrooms, and garages. The requirement applies to all new construction, but switching existing circuits will supply this protection to existing buildings as well.

Limits

Circuit breakers are limited in protecting wires, and therefore life and property. Breakers other than GFCI cannot prevent electric shock, for instance. Although breakers trip at 15 amps and above, it only takes about 0.06 amp to electrocute someone. Circuit breakers cannot prevent overheating of a fixture or appliance or other device, and they can't prevent low-level faults. For a breaker to trip, a fault must occur when enough current is being demanded to exceed the trip current of the breaker. Breakers cannot trip fast enough to completely block lightning surges from entering the house circuits. They cannot prevent fires within appliances. Circuit breakers are meant to save the wiring to the appliance—not the appliance itself.

Breaker Types

A single-pole breaker is the most prevalent type of circuit breaker found in residential use. It will power anything that requires 120-volt current.

A double-pole breaker is used with a 240-volt appliance, such as a 20-amp baseboard heater or a 30-amp clothes dryer.

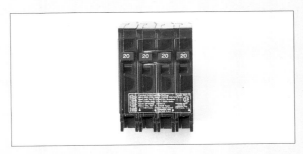

Use a quad breaker to serve two double-pole circuits in the same space as one standard double-pole breaker.

A GFCI circuit breaker will cut power to a circuit when it is tripped by an imbalance in current flow through the wires.

A surge-protection device provides protection for an entire service panel and simply installs in place of two single-pole breakers.

7 Electrical

ELECTRICAL BOXES

Electrical boxes are used for a variety of purposes, such as holding receptacles and switches, housing wire junctions, and supporting ceiling fans and lights. Many types of boxes are made for different purposes, and they can be metal or nonmetallic (plastic or fiberglass). Today, plastic boxes are the type most commonly used. However, metal boxes are still found in many homes. Check code requirements before installing electrical boxes. Plastic boxes have the advantages of low cost and simplicity of installation.

Electrical boxes come in standard shapes for each type of use. For example, they may be shallow for furred walls, wider for ganged arrangements, or waterproof for outdoor applications. Every box must be covered and accessible. It is also important not to use a box that is too small for the size and number of wires it will house. An electrical box can hold only a limited number of wires.

Although plastic boxes are labeled according to the number of wires they can house, and/or their size in cubic inches, metal boxes are not. Cubic-inch capacities for a variety of metal boxes are listed in the National Electrical Code (NEC) Table 314.16(A).

The safest way to overcome this problem is by purchasing the deepest box that will fit into your stud wall—one that is 3¼ to 3½ inches deep. For a single-gang box, this will provide 20.3 cubic inches of wiring space. Use common sense, and don't overcrowd any box.

Fuse Puller

A fuse puller is specially designed for the removal of cartridge type fuses that are found in older homes. The grips on one end of the puller enable you to remove cartridge fuses up to 60 amps in size, while those on the other end can pull fuses of greater capacity. The fuse puller is also used to insert the new or replacement fuse. Always be sure that the fuse box has been switched off before you pull a fuse.

To remove a cartridge fuse, grasp it firmly with a fuse puller, and pull it straight out.

Electrical Box Types

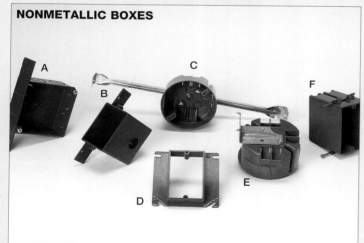

Nonmetallic Boxes:
(A) MP Bracket Square Box; (B) MP Bracket Switch Box;
(C) Adjustable Ceiling Box; (D) Raised Device Cover;
(E) JP Bracket Ceiling Box; (F) Receptacle and Switch Box

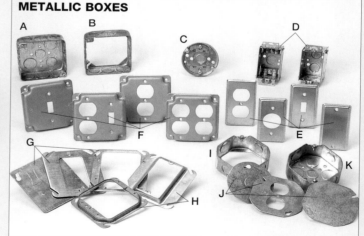

Metallic Boxes:
(A) Square Box; (B) Box Extension; (C) Ceiling Pan Box;
(D) Receptacle and Switch Boxes; (E) (F) Exposed Box Cover Plates; (G) Concealed Box Flat Plates; (H) Raised Device Covers; (I) Box Extension; (J) Concealed Box Flat Plates; (K) Octagonal Box

BOXES & CONNECTORS

Wire Connectors

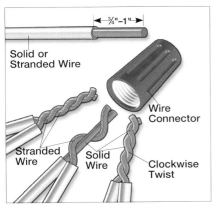

Use a wire connector to splice two or more wires securely together, adhering to the minimum and maximum wire capacity rating for the connector you are using. Hold the wires tightly; slip the connector over the stripped ends; and turn the connector clockwise to secure the wires together.

Wire Connector Ratings

Wire Connector	Color	Minimum		Maximum	
		Gauge	No. Wires	Gauge	No. Wires
	Orange	18	2	14	2
	Yellow	16	2	14	4
	Red	14	2	12	4
				10	3
	Green	Green wire connectors are used for grounding wires only.			

Metal-Box Markings

A depth gauge marked on an electrical box tells the installer where to align the box on a framing stud. The markings come in handy because they provide a quick way to make sure that the face of the box will be flush with the finished drywall.

Using Crimping Ferrules

Use bare ground crimps when a device box has multiple cables with multiple switches and/or receptacles. The circuit requires all grounds to be twisted together in a clockwise direction. Use lineman's pliers to accomplish this.

Each device requires a bare ground wire attached to each device's green ground terminal screw. Accomplish this by leaving individual bare copper conductors past the uninsulated ground crimp.

ELECTRICAL BOXES 105

7 Electrical

CONDUIT AND ACCESSORIES

Metal conduit, or tubing, is typically used to protect wires from damage and moisture in an exposed location, such as a basement or outdoors. If exposed to harsh atmospheric conditions, however, it must be corrosion-resistant. There are five basic types of metal conduit: EMT (electrical metallic tubing), the similar IMC (intermediate metallic conduit), rigid metal conduit, flexible metal conduit in a nonmetallic PVC cover (liquid-tight), and flexible metal conduit (helically wound). There are also two types of nonmetallic conduit generally used in residential work—electrical nonmetallic tubing (ENT) and rigid nonmetallic conduit (Schedule 40). These are made of polyvinyl chloride (PVC). Conduit sizes permitted by the NEC range from a minimum of ½ inch to a maximum of 6 inches in diameter, depending on the type and use of the conduit. Various accessories are used to connect conduit, just as with water pipes, including bends, couplings, compression and screw connectors, conduit bodies, and pipe supports. Check your local code and the NEC carefully before doing electrical work involving conduit. Note, for example, that no wire splices are permitted within conduit itself but only in electrical boxes or wherever wires remain accessible.

Wiring Conduit

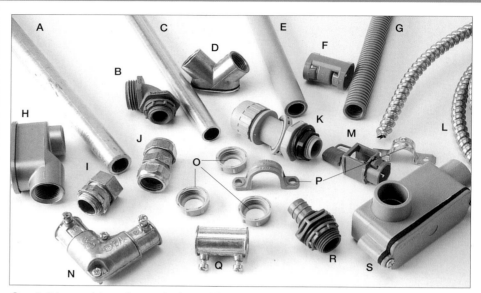

Conduit is used to safeguard wire cable where it is exposed to potential damage, such as in a basement workroom or outdoors. Shown here are some of the accessories used to extend and fasten conduit and protect cables wherever conduit changes direction. (A) Rigid Metal Conduit; (B) 90° Liquid-Tight Connector; (C) Electrical Metallic Tubing (EMT); (D) Metal Elbow Connector; (E) Rigid Nonmetallic Conduit; (F) Plastic ENT Coupling; (G) Electrical Nonmetallic Tubing (ENT); (H) SLB Fitting; (I) Raintight Compression Connector; (J) Compression Coupling; (K) Multiposition Liquid-Tight Connector; (L) Flexible Metal Conduit; (M) NM Cable & Cord Connection; (N) Metal Elbow Connector; (O) Plastic Bushings; (P) Pipe Straps; (Q) Setscrew Coupling; (R) Plastic Screw-in Connector; (S) Plastic LBL-Fitting

Raceway Components

Raceway components are available in metal or plastic and must be joined mechanically and electrically to protect wires. Raceway fasteners must be flush with the channel surface so that they don't cut the wires. Like metal, plastic raceways must be flame-retardant; resistant to moisture, impact, and crushing; and installed in a dry location. Components include channel sections, elbows, T-connectors, and electrical boxes.

Raceway wiring is connected in the same way as conventional wiring. A backing plate and extension fit over the electrical box, and the switch or receptacle mounts on the plate.

Typical raceway components include straight channel sections, elbows, T-connectors, extension boxes, plates, and covers.

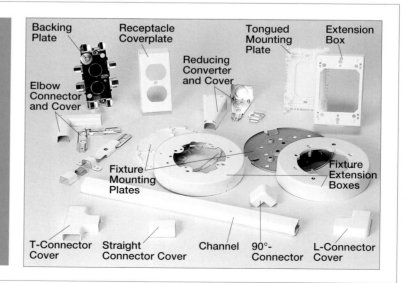

CONDUIT

Wiring Capacities — Electrical Metallic Tubing

The National Electrical Code limits the total number of individual wires of the same gauge that can simultaneously go in conduit, which can also be called a "raceway" or "electrical metallic tubing" (EMT). The code is based on common sense, as the number of wires in the conduit cannot be more than what can be installed or withdrawn without damaging the wires. Note that conduit can range in size from ½ inch to 1-½ inches. So, if you have more wires than code allows for your conduit, do not run fewer wires, simply run a larger conduit.

Wire Type	Gauge	Maximum No. of Wires Permitted in EMT				
		½ Inch	¾ Inch	1 Inch	1¼ Inch	1½ Inch
TW	14	8	15	25	43	58
	12	6	11	19	33	45
	10	5	8	14	24	33
	8	2	5	8	13	18
THW THHW THW-2	14	8	15	25	43	58
	12	6	11	19	33	45
	10	5	8	14	24	33
	8	2	5	8	13	18
	6	2	4	7	12	16
THHN THWN	4	1	2	4	7	10
	3	1	1	3	6	8
	2	1	1	3	5	7
	1	1	1	1	4	5

Bending and Cutting Conduit

Manual Conduit Bender. A manual pipe or conduit bender is used for bending metal conduit smoothly and efficiently. You can operate a pipe bender by hand or by using foot pressure to bend conduit to a 10-, 22½-, 30-, 45-, 60-, or 90-degree angle as marked on the conduit bender. This tool is essential for making accurate saddle bends, stub-ups, and back-to-back bends, as well as simple up and down bends, without crimping the pipe.

Hacksaw. You'll need a hacksaw to cut through metal pipe or conduit or metal-sheathed cable. The number of teeth in the blade determines the thickness of metal that can be cut. In general, thicker metals require coarser-toothed blades. A wing nut on the hacksaw handle allows you to remove the blade or adjust its cutting angle and tightness.

Conduit Connectors. Special types of connectors are needed to secure conduit at junctions and connection points.

Wrenches and Pliers. Many conduit connections, as with conventional plumbing pipes, consist of compression fittings. You will need pliers and sized or adjustable wrenches to properly secure these connections.

MANUAL CONDUIT BENDER

HACKSAW

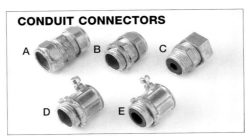

CONDUIT CONNECTORS

Conduit connectors including (A) straight compression coupling; (B) concrete-tight straight compression; (C) straight compression with insulated throat; (D) concrete-tight setscrew; and (E) setscrew with insulated throat.

WRENCHES AND PLIERS
- Combination Wrench
- Open-End Wrench
- Groove-Joint Pliers
- Adjustable Wrench
- Locking Pliers

7 Electrical

BASIC MATERIALS

House circuits are usually wired with nonmetallic sheathed cable, with metal-armored cable, or with insulated wires running through metal or plastic pipe called conduit. For most projects, you will be working with flexible nonmetallic sheathed cable known as NM (or by the trade name Romex). Armored cable is often called BX, also a trade name.

The markings on the cable tells you what type of wiring is inside. For example, the designation 14/3 WITH GROUND tells you that the cable contains three 14-gauge wires. "With Ground" means that a fourth bare copper or green insulated grounding wire is also contained in the cable.

Conduit, according to code, can be galvanized-steel pipe or plastic pipe. Metal conduit comes in an array of sizes and three types: rigid (often preferred for outdoor use), intermediate, and electrical metal tubing (EMT). Special tools are used to bend metal conduit, or shaped fittings are available to join sections.

Cutting Armored Cable & Conduit

To cut armor-clad cable, use a hacksaw to cut the flexible steel wrapper about 8 inches from the end. Make the cut diagonally across two metal ribs, and stop sawing as soon as the blade cuts through the metal or you will damage the wires. With your fingers, bend the cable back and forth until the metal snaps apart; then slide the armor off the cable.

Wire Terminals

Wires attach to screw terminals around threaded screws. Wire insulation should stop just short of the screw.

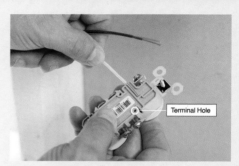

Push-in terminals feature molded-in gauges to ensure proper wire length. Strip the wires, and push them into the holes.

Fastening Cable

Fasten cable along framing with code-approved staples or plastic fasteners. Keep cable 1¼ in. back from the edge.

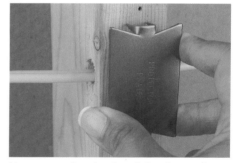

Hammer galvanized metal nail guards onto studs and joists to protect hole-threaded cable from nails or screws.

Stripping Cable Sheathing

USE: ▶ cable ripper • combination tool ▶ cable

1 Slide the cable ripper onto the cable, and squeeze it 8 to 10 in. from the end to force the point through the plastic sheath.

2 Grip the cable tool in one hand and the cable in the other, and pull the ripper toward the end of the cable.

3 Expose wires by peeling back the plastic sheathing and paper wrapping. A cable ripper won't damage individual wires.

WIRING BASICS

Attaching Wires

USE: ▶ combination tool • needlenose pliers • insulated screwdriver ▶ wire • switch or outlet

1 Clamp the wire in the proper gauge slot of a combination tool, and strip ¾ in. of insulation from each wire end.

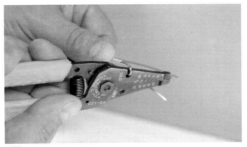

2 Use combination tool to form a clockwise half-loop at the end of each wire. Avoid making nicks that could weaken the wire.

3 Hook the looped wire end onto the screw terminal. The clockwise wire loop will close as you tighten the screw.

Capping Wires

USE: ▶ needlenose pliers • combination tool ▶ wire • wire connector

1 To join wires, strip ¾ in. of insulation; hold the wires parallel; and twist them together clockwise with pliers.

2 The twisted part should be long enough to engage the wire connector without exposing any bare wire when it is applied.

3 Screw a wire connector clockwise so that the exposed wires are covered. Use hand pressure only, not the force of pliers.

4 Use a combination tool's cutting jaws to trim away excess plastic sheathing and paper wrapping inside the sheath.

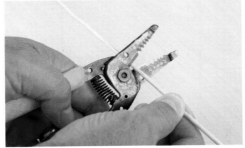

5 Use the cutting jaws of the combination tool to cut individual wires in the cable to length, if needed.

6 A combination tool has slots for different gauges of wire. The correct one will remove the sheath without crimping the wire.

BASIC MATERIALS

7 Electrical

NEW WIRING OPTIONS

If you are building new or remodeling old construction where the framing is exposed, wiring in new circuits and fixtures is easy—there's nothing in your way. Stringing wire or cable through existing walls and ceilings, however, is akin to fishing in a muddy creek: you can't see where the line is going.

Wiring in Open Walls

To add new wiring, you can drill holes in the studs for the cable using a ⅝- or ¾-inch bit. Holes should be at least 1¼ inches from the facing edge of the stud. If you can't leave that much space, attach a steel nailing plate on the outside of the stud to keep any drywall or paneling fasteners from being driven into the wires. You can also cut notches into the front of the studs, about ¾ inch deep and 1 inch apart. Once the cable is in position, cover the notch with a steel plate. If the ceiling framing is also exposed, cable can be run through the joists and down the wall, stapled to the studs.

Wiring in Closed Walls

If you have access to walls from an unfinished basement or attic ceiling, you can add new wiring by fishing the cable down or up (through a hole drilled in the top plate or soleplate) instead of across. Otherwise, you'll need to "fish" the cable through the closed walls by cutting additional openings in the drywall surface, such as behind the base molding, and then drilling holes in the studs. Pull the cable through using fish tape—wires with hooked ends that pull the cable between studs or joists.

Surface Wiring

Surface wiring—wiring installed in a raceway (a protective plastic or metal casing)—eliminates behind-the-wall cable fishing. The raceway is permanently attached to the wall, usually at the base molding or the ceiling. Raceway wiring can include outlets, switches, and ceiling fixtures. Special connectors turn corners and provide intersections for branches.

Edison produced the first practical lightbulb in 1879. He then had to build a power plant; by 1882, his New York City plant had 203 customers.

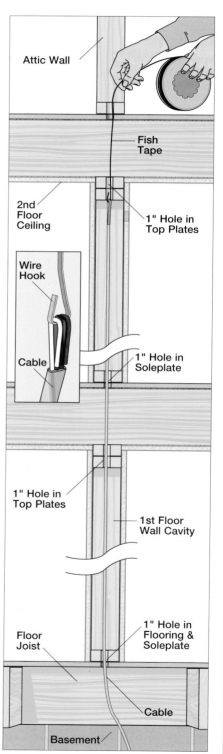

Fishing Routes

- Attic Wall
- Fish Tape
- 2nd Floor Ceiling
- 1" Hole in Top Plates
- Wire Hook
- Cable
- 1" Hole in Soleplate
- 1" Hole in Top Plates
- 1st Floor Wall Cavity
- Floor Joist
- 1" Hole in Flooring & Soleplate
- Cable
- Basement

Fishing Connections

1 Instead of trying to thread bendable cable through wall cavities, thread a more controllable steel fish tape.

2 Bend wire leads around the fish tape end hook; tape the leads down; and roll up the tape to pull the cable through.

WIRING PATHS

Installing Surface Wiring

USE: ▶ metal snips • hacksaw • screwdriver • combination tool • neon tester ▶ raceway channels, trim & clips • wiring & connectors • screws • boxes & fixtures

MONEY SAVER

1 Plastic surface wire channels, called raceways, are easy to cut—use metal snips on the base channel and a hacksaw on trim.

2 Screw the base channel to the wall with plastic anchors. At a tee, clip the edge to clear a path for wires.

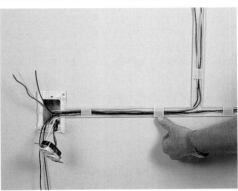

3 Install a box plate over an existing outlet, and extend wires from that circuit. Hold the wires with clips.

4 Once all wiring is in place, clip trim channel over the tracks. There are special connectors for T- and L-joints.

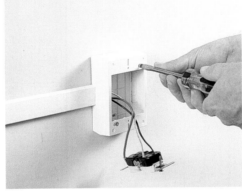

5 Where raceway wiring feeds power to a new outlet, you can mount a matching plastic outlet box on the wall.

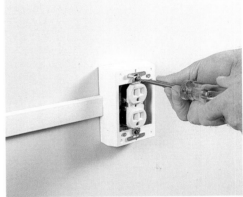

6 Strip the wire leads; connect them to the terminals on the outlet; and screw the outlet to the plastic wall box.

7 To feed power to a ceiling light, run a raceway up the wall and across to the fixture. Cover the corner with an L-clip.

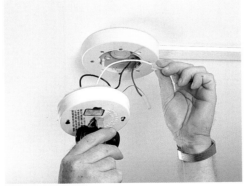

8 Raceway systems have surface mounts for outlets and switches—and ceiling fixtures. Connect the leads with wire connectors.

9 Once the wiring is complete, screw the fixture base to the raceway ceiling mount, and fasten the diffuser to the fixture.

NEW WIRING OPTIONS

7 Electrical

FISHING CABLE: WALLS

Retrofitting wire in an existing wall is always a challenge, because you do not have the advantage of open walls that grant access to stud bays. So, you have to identify the existing structure beneath the drywall wall or ceiling and fish your cable through special holes cut for that purpose.

Drill a Guide Hole

Using a keyhole or saber saw, cut an opening where the new receptacle or switch box will be installed in the wall. Directly in line with the opening for the new electrical box, drill a 1/16- to 1/8-inch hole at the base of the wall. If you have carpeting, cut a small X to prevent unraveling of the carpet fibers. If possible, remove the baseboard, and drill as close to the wall as you can; then insert a thin guide wire down the hole to the level below.

Locate the Guide Wire and Soleplate

Go down into your basement or crawl space; then locate the end of the guide wire you inserted through the floor. Use the guide wire to establish the location of the soleplate at the bottom of the stud wall above. The position of the plate should be conspicuous because of the protruding nails visible just a short distance away from the guide wire. Mark a spot for drilling along the centerline of the soleplate.

Using a 3/4-inch spade or wood bit, drill a hole up through the soleplate, about 2 inches behind the guide wire. The hole should extend into the wall cavity between framing studs.

Fish the Cable

Push the cable up into the wall. If you are installing a low wall receptacle, you should be able to reach the cable from above and grab it and pull through the hole. If you need to push the cable up high in the wall to reach a new switch location, use fish tape.

If there is no attic, basement, or crawl space, fishing wire may require that you cut access openings along the length of a wall. Although this method requires extensive repair work, you may have no alternative, unless you can run your wiring behind a baseboard.

Fishing Cable from Below the Floor

USE: ▶ keyhole or saber saw • 1/16 or 1/8-inch bit • fish tape (nonmetallic) • cordless drill • guide wire • 3/4-inch spade or wood bit

1 Outline the electrical box on the wall; then cut the drywall along the outline using a keyhole saw.

2 Drill a small hole in the floor below a new box cutout and drop a guide wire through the hole.

3 Find the guide wire in the ceiling below. This will help you locate the soleplate in the wall.

4 Mark the center of the soleplate—approximately 2 inches behind the guide wire.

5 Drill a hole up through the soleplate large enough to accommodate the passage of a cable connector.

6 Push the cable up into the wall, and fish it out through the box opening. Pull gently so the wire sheathing isn't scuffed.

WIRING PATHS

FISHING CABLE: CEILINGS

Using a keyhole or saber saw, cut a hole in the ceiling where you wish to locate the ceiling box. From the ceiling hole, make a visual reference line running perpendicular to the wall where the cable will turn downward; then mark a reference line on the wall at the point of intersection.

Cut Access Holes

At the point of intersection, use a utility knife to make adjoining 2x4-inch openings in the drywall, one in the ceiling and in the wall. From the ceiling cutout, bore a ¾-inch hole down through the double top plate, or use a chisel to notch the outer face of the top plates. Through the new opening that you cut for the ceiling box, feed a long fish tape through the joist space in the ceiling until it reaches across to the ceiling cutout made previously at the top of the adjoining wall. Tie a plumb bob or other weight to the end of the fish tape, and drop it down into the wall cavity.

Pull the Cable

Cut an access hole near the bottom of the wall; then grab and pull the fish tape through the hole. Fasten the cable to the fish tape, and pull the cable up through the wall, across the ceiling, and then out the hole that you cut previously to house the ceiling box. Cover the exposed cable with metal plates.

Fish Tape & Lubricant

Friction-reducing lubricant, usually in powder form, can make this job of manipulating the fish tape much easier.

Fishing Cable Across a Ceiling

USE: ▶ keyhole or saber saw • chisel • cordless drill • fish tape (nonmetallic) • wire shields • utility knife • pencil and straightedge • ¾-inch spade bit • electrical tape

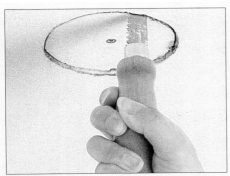

1 If your ceiling is not open from above, use the ceiling cutout for your new box as an access hole.

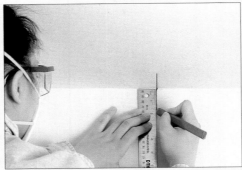

2 Site a line from the opening to the wall where the cable will turn down. Mark the intersection.

3 At the spot marked, cut adjoining access openings in both the ceiling and the wall.

4 Cut a ¾-in.-wide and 1-in.-deep notch through the top plate to serve as a raceway for the cable.

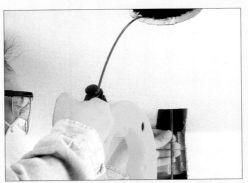

5 Feed the fish tape across the ceiling and down the wall to an access opening at the bottom of the wall.

6 Secure the cable to the fish tape at the bottom of the wall, and pull the tape back up to the ceiling hole.

FINISHING CABLE: WALLS

7 Electrical

CREATING NEW CIRCUITS

Remodeling plans often call for adding new electrical circuits to meet increased demands. In addition, appliances such as dishwashers and waste-disposal units often require a separate circuit to power the appliance. You will need to run wiring from the appliance to the main panel.

Working on a service panel is dangerous, so take all safety measures. If your installation is special in any way or if you are not confident of completing the hookup correctly, have a licensed electrician do the work at the panel after you have done the room wiring. Many local building codes require that a licensed electrician make the final hookups at the panel anyway.

Turn off the power to the house at the main breaker switch, which is usually mounted at the top of the service panel. Ideally, each breaker should be labeled. Remove the panel's cover plate, and note the breaker arrangement. Look for breakers that are not in use or spare slots for additional breakers. Use one of these for the new circuit.

Bringing the Cable to the Panel

Use a screwdriver to pry out a perforated knockout from the side or top of the panel box. Attach a cable clamp, and thread 12 or more inches of the cable through the connector, the hole in the box, and a locknut. Tighten the locknut with a screwdriver. Remove about 8 inches of the outer sleeve of the end of the cable, and strip the wire ends.

Insert the ends of the white (neutral) wire and the bare grounding wire into the holes along the bus bars at the side or bottom of the panel. Then tighten the setscrews, and note how the other circuits are connected. The neutral wire and the grounding wire should each have its own terminal.

Adding the Circuit Breaker

Place the breaker in the empty space, and engage the securing clip to the mounting bar of the panel board. Apply pressure to the end away from the terminals until the breaker seats flush with adjacent breakers. Loosen the setscrew on the breaker, and insert the black wire of the cable into the hole below. Then retighten the screw to secure the wire end. Screw the cover plate back onto the panel box, and record the new circuit on the panel door. To prevent a power surge, turn off the individual breakers; turn on the main breaker; and turn on the individual breakers one by one.

If you need to add a breaker for a 240-volt circuit, get a special breaker that occupies two slots in the panel box. The cable will have two hot wires—one black and one red. Insert one of the hot wires into each of two holes in the double breaker.

Maximum Wires in a Box

Electrical boxes that are overloaded with wires can be a hazard. Though a large amount of wires may cleanly fit through the knockouts (the holes that allow wires to enter a box), there may not be enough room within the box to allow the multiple physical connections and wire nuts required to make wire connections safe. Before tying into a junction box, be sure to trace the wires to check the voltage. Be sure you are not connecting a 120-volt outlet to a run of wire that is energized at 240 volts.

Whenever you work on a service panel, be sure the main breaker is turned off, and even then test the circuit you will be working on. In many houses, including older homes, multiple service panels can be providing power to the house. Never work alone when working with wires that could be live.

Box Type and Size	Maximum Number of Wires Permitted						
	18 GA	16 GA	14 GA	12 GA	10 GA	8 GA	6 GA
4"x1¼" Round or Octagonal	8	7	6	5	5	4	2
4"x1½" Round or Octagonal	10	8	7	6	6	5	3
4"x2⅛" Round or Octagonal	14	12	10	9	8	7	4
4"x1¼" Square	12	10	19	8	7	6	3
4"x1½" Square	14	12	10	9	8	7	4
4"x2⅛" Square	20	17	15	13	12	10	6
3"x2"x2" Device Box	6	5	5	4	4	3	2
3"x2"x2½" Device Box	8	7	6	5	5	4	2
3"x2"x2¾" Device Box	9	8	7	6	5	4	2
3"x2"x3½" Device Box	12	10	9	8	7	6	3
4"x2⅛"x1½" Device Box	6	5	5	4	4	3	2
4"x2⅛"x1⅞" Device Box	8	7	6	5	5	4	2
4"x2⅛"x2⅛" Device Box	9	8	7	6	5	4	2
3¾"x2"x2½" Device Box	9	8	7	6	5	4	2
3¾"x2"x3½" Device Box	14	12	10	9	8	7	4

ADDING A NEW CIRCUIT

Installing a Circuit Breaker

USE: ▶ insulated screwdrivers • cable stripper • multipurpose tool • flashlight • utility knife ▶ cable • cable clamps • circuit breaker

Before starting work, be sure to turn off the power to the house at the main breaker switch. The main breaker switch is usually mounted at the top of the service panel, but it may be contained in a smaller box nearby.

Because modern homes can often have two or more panels and subpanels, it is always worth checking to make sure a circuit is turned off before you begin work. Use a circuit tester to check whether the circuit you are about to work on is turned off.

1 Open the door to the panel box, and turn the main breaker to the off position. Unscrew the screws in the corners.

2 Each breaker in use is connected to a cable. A label on the door of the box should identify each circuit.

3 Check for any unused breakers. A breaker that is not in use will not have a cable attached to it.

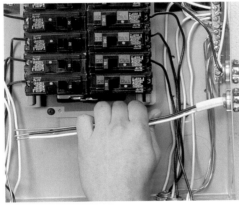

4 Pry out a knockout from the side or top of the box. Thread the cable through a cable clamp and then through the locknut.

5 When you have enough cable inside the panel, tighten the cable-clamp screws and the locknut with an insulated screwdriver.

6 Connect the white wire to the neutral bus bar, where other white wires connect. Connect grounding wire to the grounding bus.

7 A typical 120-volt circuit breaker has a clip that plugs into the hot bus and a hole in the side for inserting the black wire.

8 Screw the cover plate back onto the panel housing, and record the new circuit on the door. Turn the main breaker on.

CREATING NEW CIRCUITS 115

7 Electrical

GFCI PROTECTION IN ELECTRICAL CIRCUITS

Ground fault circuit interrupter protection was introduced and required by the NEC code starting in 1968 for underground swimming pool lighting fixtures. The mandate to protect homeowners in other wet areas around the home, such as kitchens and outdoor receptacles, began in 1971, and bathrooms added to the regulations in 1975. Updated location requirements in kitchens added all kitchen counter receptacles in 1996, then any receptacle within 6 feet of a kitchen sink, and then any sink, including laundry and utility sinks, in 2014. Unfinished basements were required to have a GFCI receptacle circuit to protect from electrical hazards with standing water on the floor in 1990, and now the same protection is required for finished basements as well.

The protection provided by the required GFCI receptacle and GFCI breakers use a complex circuitry that detects circuit imbalance between the black wire and the white (neutral) wire when current is flowing through the circuit. A properly balanced electrical circuit carries the same 120 volts on both the black and white wires of a circuit simultaneously. Any imbalance will trip the GFCI receptacle in the room or the GFCI breaker in the main service panel.

The ground (bare copper) wire also plays an important role in this circuit, as the white (neutral) wire is also grounded at the main service panel through the bus bar. This connection creates the direct path to ground for unwanted electrical currents, maintaining a safe electrical environment in the home.

GFCI receptacles and GFCI breakers are designed to work with specific amperages of 15 amps connected to 14-gauge wires and 20 amps connected to 12-gauge wires. Be sure the receptacle or breaker being replaced matches the correct installation.

Whether to install a GFCI breaker versus a GFCI receptacle is a decision made by the electrician and the homeowner when the electrical plan is being designed. Initially, it was beneficial to use a GFCI receptacle at the beginning of any required circuits and have the correct number of standard receptacles beyond the GFCI receptacle to meet the code requirements. The advantage of this circuit design is that when the GFCI receptacle trips and stops the flow of electrical current to all of the devices, resetting the circuit after the ground fault issue was corrected only requires pushing in the reset button on the GFCI receptacle.

With today's higher demand of electrical receptacles and required GFCI circuit locations, the use of the GFCI breaker is becoming more popular over the GFCI receptacle. In addition, the cost of GFCI receptacles continue to increase, especially with the self-test devices being introduced into the market. The disadvantage of the GFCI breaker is that it requires the homeowner to open the main service panel door, correctly identify the GFCI breaker that has tripped, and reset the breaker after the ground fault issue is corrected.

Both GFCI breakers and GFCI receptacles have a test button, and it is highly recommended that the test feature be performed each month to ensure the proper operation of the device. The newer self-test GFCI receptacles are capable of automatically performing this procedure every three seconds.

You must ensure that you do not connect a freezer or refrigerator/freezer appliance to a GFCI circuit, because when the GFCI circuit is tripped due to a ground fault, the current to these devices will also be interrupted; this may not be detected for a long period of time, which could allow the food to spoil or defrost.

Holiday outdoor lighting and extension cords plugged into outdoor receptacles can also create

AFCI/GFCI

This is an example of a dual-purpose AFCI/GFCI plug-on neutral breaker (side view and bottom view). This breaker eliminates the need for any GFCI receptacles.

ground fault issues that will cause the GFCI receptacle in the home to trip due to moisture issues from rain, snow, and ice. Take care to ensure that the electrical connections are not exposed to these elements. Many outdoor receptacles are connected to a first-floor bathroom or powder room GFCI electrical circuit.

Arc-Fault Circuit Interrupter Breakers (AFCI)

Arc-Fault circuit interrupter breakers, commonly labeled AFCI, have been in use since the 2002 National Electric Code (NEC) adopted their use in **dwelling bedrooms** requiring 120-volt 15 and 20 ampere outlets, including lighting fixtures, to provide protection for any and all new installations.

Additionally, the 2008 NEC increased the mandatory requirements to include, in addition to dwelling bedrooms, **dwelling family rooms, dens, recreational rooms, parlors, libraries, dining rooms, sunrooms, closets, hallways, and/or similar rooms** for any and all new installations. Virtually all single-pole breakers are now required to be AFCI type for any and all new installations.

AFCI breakers are designed to apply a specific intelligent algorithm to the flow of electricity in an attempt to differentiate normal acceptable arcing, such as from mechanical switches being turned on and off, from unwanted and unsafe arcing. Conventional circuit breakers cannot achieve this, as they are only designed to detect sustained current overloads and short circuit occurrences. Only an AFCI breaker can detect the presence of an arc, which can be attributed to erratic current flow; this will allow the control circuitry inside an AFCI breaker to trip. When this occurs, it will de-energize the circuit and eliminate further arcing conditions that could cause a fire in the dwelling.

In the earlier days, when extension cord manufacturers would produce a 6-foot-long, 18-gauge, two-prong extension cord with up to three or four female plug ends, it was common for residents to overload a circuit and repeatedly blow fuses or trip breakers. It was also very common in older dwellings to not have the necessary number of outlets in a room needed for an increase in electronic equipment usage,

ADDING A NEW CIRCUIT

Arc-Fault Circuit Interrupter Breakers (AFCI)

This is a typical wire-on neutral AFCI breaker. You will need to make sure that the neutral wire (white) that is provided with the breaker is attached to the main panel neutral buss bar (as shown).

Next, connect the branch circuit hot (black) wire and second neutral wire (white) to the ACFI breaker. The installer must ensure that the hot (black) wire is connected to the brass-colored screw and the branch neutral (white) wire is connected to the silver-colored screw for proper function.

This is a typical plug-on neutral breaker application for newer homes. This allows the installer to eliminate duplicate neutral wire connections.

Here you can see the plug-on neutral clip attached to the neutral buss bar of the main panel.

Dual-Function AFCI & GFCI Breakers

This is a dual-function plug-on neutral breaker for newer-style panels. Newer-style panels have specific neutral buss bar spacing that matches the manufacturer's breakers for each specific panel.

This is a dual-function wire-on neutral breaker for existing-style panels that provides easy adaptability to ensure proper neutral breaker connections. The attached neutral wire is long enough to meet existing neutral buss bar spacings.

so residents would use multi-socket power strips to plug in all of their devices.

The increase in electronic devices and needing devices to be charged has lead our society to have increased electricity needs. The branch circuits that these devices are connected to for operation and charging has led to increased safety and the use of the AFCI breaker that provides that protection.

House wiring, when being installed, can be damaged by many factors, and many of these will not be noticed until the circuit is energized during the final stages of construction. House wiring can be damaged by the installation of wire staples, sharp tips of framing nails, drywall screws, structural wood joint connectors, and even metal junction boxes. All of these can create an undesired arcing condition that can lead to unsafe electrical conditions—but an AFCI breaker will detect and shut down the circuit, letting the resident knows there is a potential hazard.

Some standard breaker panel boxes can be retrofitted to accept AFCI breakers, but you will need a professional electrician and inspector to verify the work, as it requires rewiring a portion of the neutral wire requirements to accept AFCI breaker. **Do not attempt this yourself unless you are qualified to do so.**

Dual-Function AFCI & GFCI Breakers

With new electrical wiring codes required by the NEC for laundry and kitchen locations, the use of a combination breaker, including a self-diagnostic breaker, is available as a dual-function breaker. Although the AFCI breaker has protection against unwanted arcing in the electrical circuit, it does not have the required sensitivity to interrupt the circuit that detects a ground fault condition.

The dual-function breaker now combines the class A 5mA GCFI sensitivity and the combination circuitry to detect the unwanted arc and can protect against AFCI and GFCI conditions independently in one breaker. The breaker will trip and shut off the flow of electricity to the entire circuit when protection has been compromised.

The dual-function breaker has many advantages, including a lower cost compared to installing separate GFCI and AFCI breakers to meet code requirements, a smaller device size, and the fact that both screw lugs are provided at the same angle to allow for easier wire installation. In addition, it provides both automatic and continuous self-testing that will ensure the breaker is performing properly. If the self-test determines the breaker has been compromised, it will automatically trip the breaker and stop the flow of electrical current.

7 Electrical

PLUGGED IN

Some outlets, or receptacles, are designed exclusively for use outside; some are made to handle heavy-duty appliances such as air conditioners, dryers, and ranges, and have distinctive "faces" that won't accept ordinary two- or three-prong plugs. The most common home receptacle is the duplex receptacle that is rated at 15 amps and 120–125 volts. A duplex receptacle has two outlets and accommodates two plugs. Ground fault circuit interrupter (GFCI) outlets may be code-required for kitchens, bathrooms, garages, crawl spaces, and other damp locations. GFCI outlets have a safety device that compares the amount of current flowing in the black and white wires of the circuit, and breaks the circuit if it detects a difference as little as 0.005 amp.

Note: According to the latest National Electrical Code, all 125-volt 15- and 20-amp receptacles used in new construction and renovation must be listed as tamper resistant. These receptacles prevent a child from inserting an object into one of the contact slots.

If you buy a replacement outlet, make sure it matches the circuit—the markings and ratings on the old and new outlets should be the same. Amperage and voltage ratings indicate the maximums that each outlet can handle. The type of current indicated (such as "AC ONLY") is the only one that outlet can use. The wire in your house must also match the outlet: either copper (CO or CU), copper-clad aluminum (CO/ALR), or solid aluminum (ALR).

WARNING

Older receptacles may be wired with the hot and neutral wires inserted in the push-in holes through the back of the receptacle. Replacement receptacles should have the wires connected to the corresponding screw terminals with a loop in a clockwise direction for proper connections when the screws are tightened.

Typical Outlet Wiring

Side-wired receptacles generally have two screws per side: one brass or black pair and one silver pair. The brass terminal connects to a hot (red or black) wire and the silver to a neutral (white) wire. (Some outlets have back-wire slots.) Newer outlets also have a green screw at the bottom for a grounding wire. Most outlets are always hot, but you can wire one or both receptacles to be controlled from a switch—say, to control a plugged-in lamp.

In the middle of a circuit, two cables with black, white, and bare copper wires enter a box. Power continues to other outlets in line.

At the end of a circuit run, power feeds the last outlet. In all cases, the grounding wires must be connected at the green terminal.

With a switch-controlled outlet, you can turn a lamp on and off from a switch, useful in a room without a ceiling fixture.

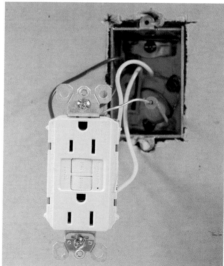

Ground fault circuit interrupters are wired uniquely but provide extra safety with a built-in circuit breaker.

WIRING RECEPTACLES

Grounding Boxes

All outlets must be properly grounded to prevent short circuits. Metal outlet boxes require that a grounding wire is pigtailed to the grounding screws of the outlet and the box. With plastic boxes, the cable grounding wire attaches directly to the outlet's grounding screw. Houses more than 50 years old may have metal-armored cable, called BX or Greenfield. In this case, the metal wrapping serves as the ground.

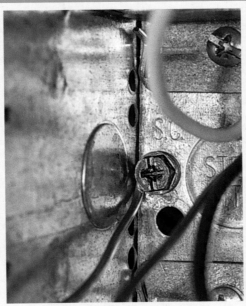

Most electricians now use plastic boxes. With older metal boxes (above), the box must be part of the ground system.

Grounds are not connected to plastic boxes, but rather are directly connected to each electrical device(s) in the box.

Receptacle Types

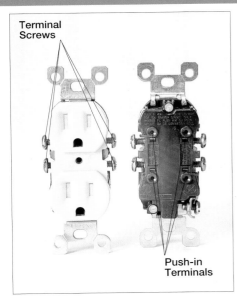

Tamper-proof receptacles have terminal screws for connecting wires. Avoid push-in terminals, as they are not always secure.

Older receptacles have only two slots. If both are the same size, the receptacle is neither grounded nor polarized.

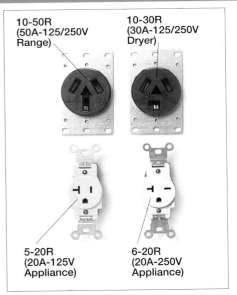

High-voltage appliance receptacles have specific slot configurations that accept only certain plugs.

PLUGGED IN 119

7 Electrical

Installing an Outlet

USE: ▶ measuring tape • pencil • hammer • cable stripper • combination tool • insulated screwdriver ▶ cable • outlet box with screws

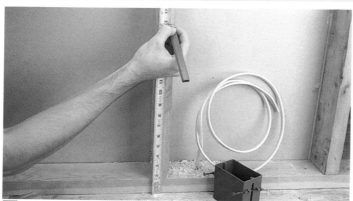

1 Bring your cable into the framing bay through the soleplate, and mark the box location 12–18 in. from the floor.

2 Most boxes have a slot through which you drive a mounting nail. Set the box ½–⅝ in. proud of the stud to be flush with drywall.

3 Pull the cable through the box; many have self-clamping inlets. Allow enough wire (at least 6 in.) to make connections.

4 Use a cable stripper to slit open the cable sheath and expose the individual leads inside without crimping the wires.

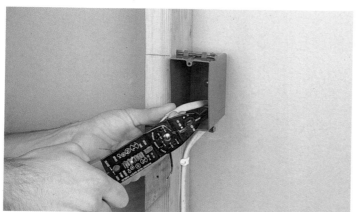

5 Use a combination tool to strip the insulation from individual wires. Match the tool stripping slot to the wire gauge.

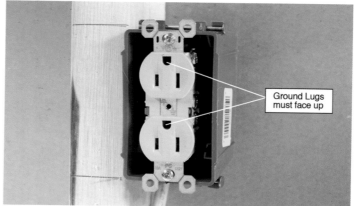

Ground Lugs must face up

6 Connect wire leads to the screw terminals; fold the wire bundle behind the outlet; and screw the outlet to the box.

WIRING RECEPTACLES

USB/120-Volt Dual-Purpose Residential Receptacles

These are available in both 20-amp and 15-amp 120-volt receptacles in the Decora-style (single rectangular opening) cover plate that will provide the required 3.6-amp USB charging current for all of you electronic devices. The advantage is the adapter plug to convert your USB cable to the three-prong outlet is no longer necessary in the event it gets lost or broken.

Be sure to match the amperage of the receptacle to the amperage of the branch-circuit breaker or fuse. To meet the latest National Electric Code, these dual-purpose receptacles are being produced in the tamper-proof style to prevent accidental electrical shock to children.

Note: This dual-purpose receptacle can be replaced where combination devices that are Decora-style are used in a junction box. Otherwise, it may be placed in a single-receptacle junction box with the new cover plate included.

USE (if adding a new receptacle to room): ▶ single gang old work box with mounting ears for drywall/plaster • correct length and wire gauge for extended outlet (14 AWG 15 amps, 12 AWG 20 amps) • wire nuts for pigtail wires in existing outlet to extend outlet circuit • USB/115-volt combination outlet with cover plate

1 To begin, disconnect the power to the receptacle by locating the fuse or breaker in the panel box. Test the receptacle by plugging in a device to be sure power is disconnected. With the grounding lugs facing the same way, up or down, detach one wire at a time and connect the wire to the matching screw terminal on the dual-purpose receptacle.

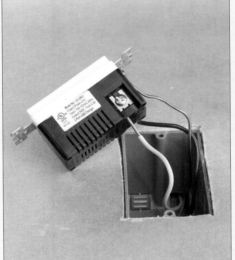

2 The new dual-use receptacle supports back wiring, as shown in the photos. Back wiring creates the fastest high-quality connection, but the device will also support side wiring under the connection screws (the silver screws on the gold plates visible on the sides in the photos). Follow the strip wire gauge on the back for proper length to ensure the best electrical connection.

3 Fasten the device screws to the junction box and install the Decora cover plate with the two supplied cover plate screws. Complete the installation by restoring power to the branch circuit, then test the receptacle with an electrical device.

INSTALLING AN OUTLET

7 Electrical

Typical Outlet and Wiring Layouts

Sequential 120-Volt Duplex Receptacles
Start- or middle-of-run receptacles are connected to all wires from both directions. End-of-run receptacles are the last on the circuit and have only two terminations and a ground connection.

Multiple 120-Volt Duplex Receptacle Circuit
On multiple 120-volt receptacle circuits, three-wire cable is used to connect all but the last receptacle. The white neutral wire is shared by both circuits.

Single-Pole Switch with Light Fixture and Duplex Receptacle
Wire a middle-of-run light fixture through a single-pole switch using two-wire cable to power the switch and three-wire cable from the switch to the fixture. Continue the circuit to an end-of-run receptacle using two-wire cable.

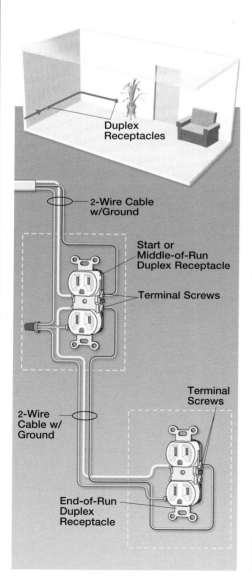
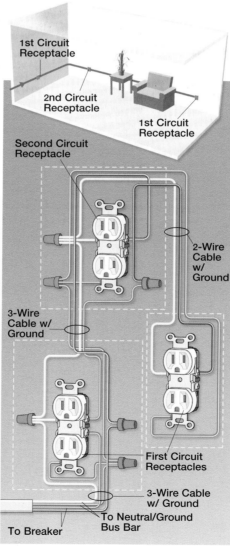
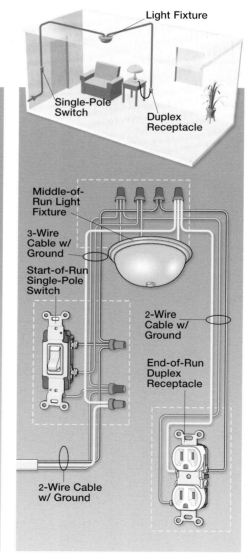

122 ELECTRICAL / WIRING RECEPTACLES

WIRING RECEPTACLES

Split Receptacle Controlled by End-of-Run Switch
In this configuration, one-half of a split receptacle (tab removed) is powered by a switch located at the end of the circuit run. The other half of the receptacle is constantly powered.

Duplex Receptacle with Split Receptacle Controlled by Start-of-Run Switch
To install a split receptacle controlled by a switch, remove the tab on the receptacle. Using two-wire cable, connect the switch to the power source. One hot wire from the switch 3-wire cable connects to each half of the receptacle.

Duplex Receptacle with Split Receptacle Controlled by End-of-Run Switch
In this combination, the split receptacle is located at the start of the cable run. One-half of the receptacle is controlled by the switch. The other half of the receptacle is always hot and feeds the remainder of the circuit.

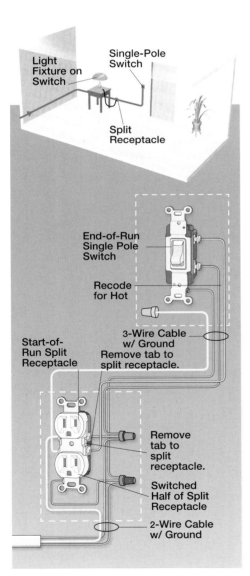
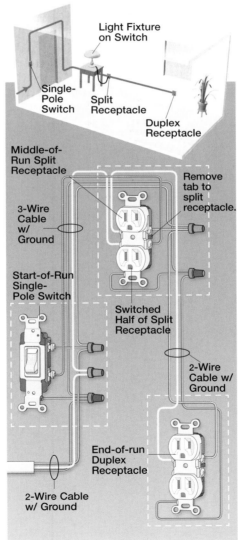
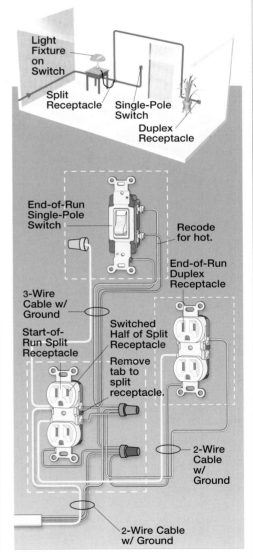

TYPICAL OUTLET AND WIRING LAYOUTS   123

7 Electrical

HIGH-VOLTAGE RECEPTACLES

Large appliances in a home often draw significantly more current than smaller appliances. For this reason, contemporary homes usually have two types of receptacles—one type provides low-voltage power (115 to 125 volts), and the other provides high-voltage power (220 to 250 volts). Appliances that are rated for 240 volts—such as cooking ranges, clothes dryers, and air conditioners—are required to be connected to a dedicated circuit. Though some high-voltage appliances incorporate their own electrical box and must be directly hard-wired, most are connected to either a flush- or surface-mounted receptacle box. A nonmetallic sheathed cable containing two hot wires, each carrying 120 volts, and a grounding wire typically form an end-of-run connection within the receptacle box, which must be located within the length of the appliance cord. Because no neutral wire is needed, the white wire is coded using black tape to indicate that it is hot. However, a high-voltage circuit that also requires 120-volt current to operate clocks, timers, and lights does need to have a white neutral wire connected to the receptacle—so that the appliance can split the entering current between 120 volts and 240 volts. These circuits use three-wire cable.

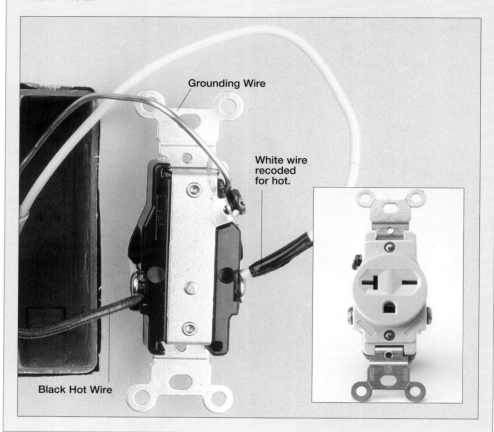

Wiring High Voltage Receptacles

A 240-(250) volt receptacle uses a two-wire cable with ground; the white wire is taped black. The grounding wire is pigtailed to the receptacle and the electrical box if metal.

Interpreting a Receptacle

The labels or markings that appear on a receptacle convey important information about safety and usage. A UL label, for instance, means that the device has been certified for safety by the American Underwriters Laboratories, while a CSA label indicates approval by the Canadian equivalent—the Canadian Standards Association. Also shown are amperage and voltage ratings, which state the maximum permitted for the device. You should be especially alert to the acceptable wire usage designation, which indicates what kind of wire is safe to connect to the receptacle. A CU label means that only copper wire can be used; CO/ALR indicates that aluminum wires are acceptable; and CU/AL specifies copper or copper-clad aluminum wires only.

The screw terminal colors on a receptacle also denote specific information. Use the brass screw terminals for black/red hot wires, the silver screw terminals for white neutral wires, and the green terminal screw for the grounding connection.

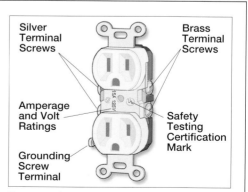

Receptacles must be clearly marked to indicate amperage and voltage, and the proper type of wire.

WIRING RECEPTACLES

Special Receptacles

RANGE RECEPTACLES

Older range receptacles had only three slots, combining the current-carrying neutral with the appliance's frame grounding.

Modern range receptacles have four slots, for three insulated wires and a grounding conductor or wire, either bare or insulated.

DRYER RECEPTACLES

An older-style dryer receptacle had two angled slots for hot wires and a third elbow-shaped slot for the grounding/neutral.

Newer dryer receptacles have four slots to provide independently for the two hot wires, the neutral, and the grounding wire.

High-Voltage Receptacle Layouts

240-Volt Appliance Receptacle
Wire a 240-volt appliance receptacle using a two-wire cable with a ground. Connect the white wire to one brass terminal and the black wire to the other. Tape the white wire black to indicate that it is hot. Pigtail the grounding wire to both the receptacle and box grounding screws.

120/240-Volt Appliance Receptacle
Red and black wires in a three-wire cable provide 240-volt power to this type of appliance receptacle. Either wire, combined with the white wire, provides 120-volt power. The receptacle has no ground terminal. Connect the grounding wire to the grounding screw in the receptacle box.

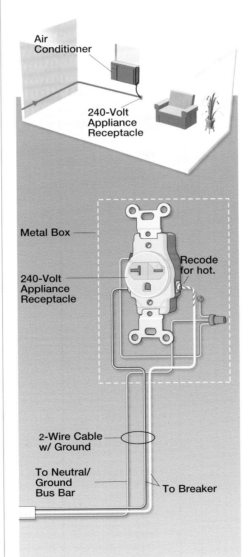
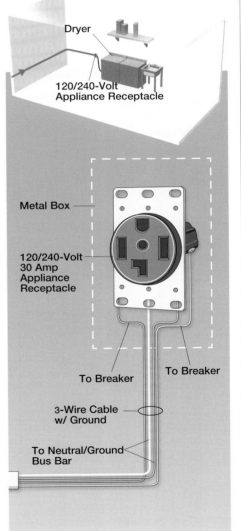

HIGH-VOLTAGE RECEPTACLES

7 Electrical

GFCI RECEPTACLES

A ground fault circuit interrupter (GFCI) is an electrical device that prevents electrocution caused by an accident or equipment malfunction. In a general-purpose, 120-volt household circuit, current moves along two insulated wires—one white and one black. Power is brought to the device or appliance by the black wire and returns from it by the white wire. As long as these two current flows remain equal, then the circuit operates normally and safely. However, if a portion of the return current is missing, or "faulted," a GFCI will immediately open the circuit in 1/25th to 1/30th of a second—25 to 30 times faster than a heartbeat. In this fraction of a second, you may receive a mild pinprick of a shock, rather than the dangerous or potentially lethal shock that would otherwise occur in a circuit without the protection of a ground fault circuit interrupter.

A GFCI receptacle, however, is not foolproof. For a ground fault circuit interrupter to succeed, a ground fault must first occur. This happens when current flows out of the normal circuit to a ground pathway, causing the imbalance between the black and white wires mentioned earlier. In this instance, if you place your body between the black and white wires, and you are not grounded, the GFCI will not function properly because it has no way of distinguishing your body from any other current-drawing device. The number of electrons entering the circuit is equal to the number of electrons returning from the circuit, except that they are passing first through the resistance within your body—causing your heart to go into fibrillation, beating erratically. If your heartbeat is not quickly restored to normal, then you will die. Even if the circuit is connected to a breaker panel, the breaker will not trip unless the internal current exceeds 15 or 20 amps—2,500 times more than is necessary to cause electrocution. A breaker or fuse is only designed to protect

GFCI Outlets and Breaker

Protecting multiple locations. If you want several receptacles farther down circuit, or downstream, from the GFCI receptacle to also have GFCI protection, then use the method of wiring for multiple locations. Wire the receptacle using a middle-of-run configuration, connecting the downstream hot and neutral wires to the screw terminals labeled LOAD. In this type of connection, any upstream receptacles will not be protected.

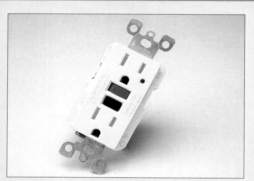

When a ground fault occurs or a test is made, the reset button will pop out. Press the button back in to reset the circuit.

An older GFCI circuit breaker (like this one) has a test button, but no reset button. To reset this old style breaker, turn it off and on again slowly. Refer to page 116 for more information about newer style dual-function AFCI/GFCI breakers for newer code applications.

Wiring a GFCI Outlet

USE: ▶ insulated screwdriver • cable ripper • wire connectors • neon circuit tester • long-nose pliers • copper grounding wire • 12/2g NM cable • diagonal-

Connect the GFCI Hot and Neutral Wires. After installing the receptacle box, pull and rip the cable, and strip the inside wires. Using a wire connector, splice the black hot wires; then pigtail them to the GFCI receptacle at the terminal screw labeled HOT LINE. Using another wire connector, splice the white neutral cable wires; then pigtail them to the GFCI receptacle at the terminal screw labeled WHITE LINE.

Connect the GFCI grounding wires, and install the receptacle. Splice the bare copper cable grounding wires together, and then pigtail them to the green GFCI receptacle grounding screw. Install the GFCI receptacle box and coverplate, and then turn on the power. Using a neon circuit tester, test the circuit for power. Press the test button to see whether or not the GFCI is operational; then reset it.

1 Label line and load wires. Line wires are connected to the breaker for correct wiring of a GFCI receptacle.

GROUND FAULT CIRCUIT INTERRUPTERS

your household wiring against excessive current—it is not designed to protect you.

Required GFCI Locations

Even though GFCI circuits are not foolproof, they are nevertheless required in certain locations within a dwelling unit, specified by the NEC (Section 210.8). These locations include, but are not strictly limited to, bathrooms, garages, outbuildings, outdoors, crawl spaces, unfinished basements, kitchens, and wet-bar sinks. A good general rule to follow is that if you are working in a damp or wet environment, then the receptacle you use should be GFCI-protected. If no GFCI receptacle is located nearby, then use an extension cord that has a built-in GFCI.

A GFCI receptacle resembles a conventional receptacle, except that it has built-in reset and test buttons. A GFCI can also be directly installed at the panel box as a circuit breaker. This type of ground fault circuit interrupter has a test button only; when tripped, the switch flips only halfway off to break the circuit. To reset the circuit, the breaker must be switched completely off and then flipped back on again. If it trips again immediately, there is still a ground fault in the circuit that needs repair. A GFCI receptacle is less expensive than a breaker-type GFCI and has the advantage of letting you reset a circuit at the point of use. Although ground fault circuit interrupters can be wired to protect multiple devices, they are most effective when limited to protecting a single location.

GFCI Breaker

To install the circuit breaker type of GFCI, simply insert the device into the panel box in the same way as a conventional circuit breaker; then connect the wires from the circuit you wish to protect. Connect the white corkscrew wire attached to the GFCI circuit breaker to the white neutral bus in the panel.

Ground Faults

If an electrical current flows through your body from a hot wire to a neutral wire, this completes an electrical circuit—just as though you were an appliance or fixture. In this case, a ground fault circuit interrupter cannot save you from being electrocuted because it cannot distinguish you from your microwave. If you hold only one wire, however, the resulting imbalance in current entering and leaving the circuit will trip the GFCI and protect you from serious shock or electrocution.

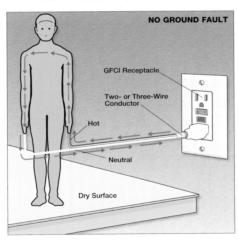

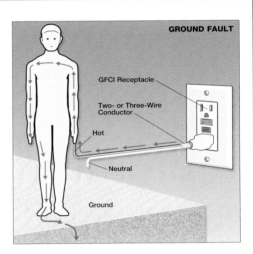

cutting pliers • wire stripper ▶ GFCI receptacle • receptacle box

2 Connect black wires to line and load brass terminals. Connect white wires to line and load silver terminals respectively.

3 Pigtail the grounding wires to the green GFCI receptacle grounding screw. This keeps connections neat and tight.

4 Check the reset button on the receptacle by pushing in the test button; the reset button should pop out when power is applied.

7 Electrical

SWITCH TYPES

Single-Pole Switches

An electrical switch controls the flow of power in an electrical circuit. It provides an open circuit in the off position and acts as a short, or closed, circuit in the on position. A switch having two screw terminals is known as a single-pole switch; it can control a circuit from one location only. Most residential switches are single-pole switches. Power is connected to one side of the switch at all times. When the switch is on, electricity flows from the wire attached to the powered screw terminal, through the switch, and into the fixture or appliance wiring connected to the other screw terminal. If the switch is at the end of a circuit, power will flow through the black hot wire and return through the white neutral wire, taped black to classify it as hot. (Neutral current in the white wire equals that in the black wire; in the on position, either wire can cause an electric shock.) If the switch is in the middle of a run, two black hot wires connect to the switch and the two white neutral wires are spliced together with a wire connector in the switch box. Splice together the bare copper grounding wires, and then pigtail them to the green grounding screw on the switch and in the box, if it is metal.

Three-Way Switches

Like a single-pole switch, a three-way switch controls the flow of power in an electrical circuit, but from two different locations instead of just one. Such switching requires special three-conductor or three-way switch cable with ground. This type of cable is usually round, rather than flat like conventional nonmetallic (NM) cable, and it contains an additional, insulated conductor—a red wire.

Three-way switches also differ from single-pole switches in that they have three screw terminals instead of two: a com terminal (dark screw), and two traveler screws to connect wires that run between switches. The switch does not have either an on or an off marked position because the com terminal alternates the connection between two different switch locations, allowing either position to potentially close the circuit.

You must consider three different cables when wiring a three-way switch: the feeder cable, the fixture cable, and the three-wire cable. The typical wiring method is to run the two-wire hot feeder cable into the first switch box, and then the three-way switch cable between the first and the second switch box. You can then run a second two-wire fixture cable between the second switch box and the fixture box. An alternative method is to run the hot feeder into one switch box; then run the three-way switch cable from the first switch box to the light fixture and then to the second switch box. Either method initially requires that you run the hot feeder to a switch box. It's also possible to run power first to the light fixture, but this method is not preferred because it's more difficult to troubleshoot if there's a problem in the circuit.

Single-Pole Switch

USE: ▶ insulated screwdriver • long-nose pliers

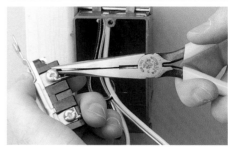

1 In a middle-of-run switch circuit, connect both of the black hot wires to the screw terminals on the switch.

Wiring a Three-Way Switch

USE: ▶ insulated screwdriver • cable clamps • wire stripper • diagonal-cutting pliers

A three-way switch installation requires two switches that control the same fixture. After you have installed the switch boxes, pull the cables into the boxes. Connect the black hot wire in the feeder cable—a two-wire cable with ground coming from the power source—to the "common" screw terminal on-the-first three-way switch. A three-wire cable with ground should connect the two switches. Splice the white neutral wire from the feeder cable to the white neutral wire in the three-way switch cable that goes to the second three-way switch. Connect the black and red traveler wires in the three-wire cable to the traveler terminals on the switches. Connect the black wire that goes to the fixture to the "common" terminal on the second switch. White neutral wires are not attached to the switches, simply use wire connectors to attach the segments in the boxes, but attach the white wire to the neutral wire in the fixture. Attach and pigtail all grounding wires as shown in the sequence.

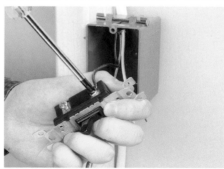

1 The feeder cable contains two wires and a ground; it supplies power to the first switch.

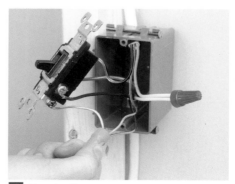

5 Feeder line and traveler switch grounding wires must be pigtailed to the grounding screw in the first switch.

WIRING SWITCHES

• multipurpose tool • cable ripper ▶ switch box and switch • 12/2g NM cable • grounding pigtail and screws • cable clamps • wire connectors

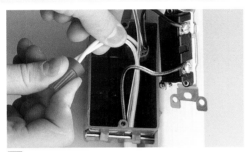

2 Splice together the white neutral wires in a middle-of-run switch circuit inside the switch box.

3 Braid the grounding wires together; then pigtail them to the switch and metal box grounding screw.

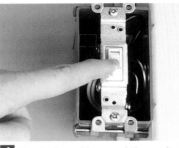

4 Push the wiring and the switch carefully into the switch box, screw the switch in place; then test the circuit.

• cable ripper • long-nose pliers (optional) ▶ three-way switches • switch box • two-wire cable • copper wire • wire connectors • three-way switch cable

2 The white neutral wires in a three-way switch circuit must run continuously through the entire circuit.

3 The black traveler wire maintains continuity of power between the first and second three-way switch.

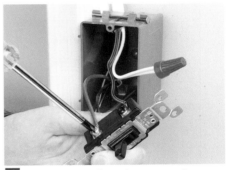

4 The red traveler wire connects a traveler terminal on the first switch to one on the second switch.

6 Fixture and traveler switch grounding wires must pigtail to the grounding screw on the second switch.

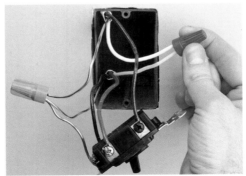

7 Continue neutral wiring through the second switch, connecting the traveler neutral to the fixture neutral.

8 Complete the circuit by splicing the hot and neutral wires from the second switch to those from the fixture.

SWITCH TYPES 129

7 Electrical

WIRING DIMMERS

A dimmer switch allows you to regulate luminosity, or brightness, of light emanating from a light fixture—either to set a mood or conserve energy. Dimmer switches can be single-pole or three-way switches. In a three-way configuration, only the dimmer switch regulates brightness, while the paired toggle switch merely turns the fixture on or off. Though not commonly used, dimmer switches are also available for fluorescent lighting.

Dimmer switches are controlled by a solid-state device within the switch that alternately turns the current on and off as many as 120 times per second. By restricting the flow of current, the switch dims the light. The longer the current is off, the dimmer the light. Standard dimmer switches are rated for 600 watts.

Dimmer Types

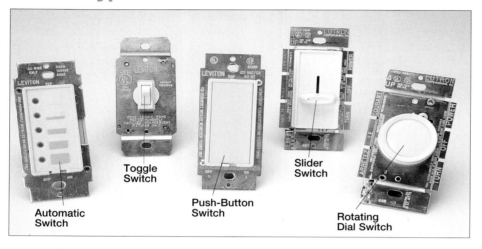

Automatic Switch · Toggle Switch · Push-Button Switch · Slider Switch · Rotating Dial Switch

Dimmers are available in a variety of types, from conventional toggle switches to rotating dials or switches that slide up and down. Automatic dimmers are operated by electronic sensors. Others have faces that light up in the dark.

Switch Types

End of run switch; the travelers are a red wire and a white wire taped black, or a black wire and a white wire taped black.

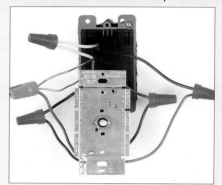

A middle-of-run dimmer switch has traveler wires that are either both black or one red and the other black.

Testing a Switch

When you flip on a switch and the switch circuit doesn't work, the problem may not be with the switch. It could be a blown fuse, a tripped circuit breaker, or a faulty fixture. First check the service panel; then test the switch. Begin by removing the fuse or setting the breaker on the switch circuit to the off position; then remove the switch coverplate. Apply the probes on a multitester to the black and white wire terminal screws to verify that the power is turned off; then turn on the switch. Next, touch the probe and clip of a battery-operated continuity tester to the wire terminals. If the switch is good, then the tester will either light up or buzz. Finally, turn off the switch. If it is good, then the tester should no longer light up or buzz. Replace the switch if it fails any of these tests. If the switch is good, the fault must be in the fixture.

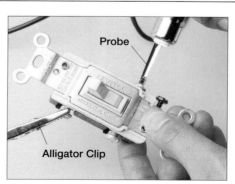

A continuity tester tests the integrity of an unwired or disconnected switch. The probe will light when you turn the switch on.

WIRING SWITCHES

Wiring a Single-Pole Dimmer Switch

Turn Off Power, and Disconnect the Circuit Wires. Turn off power to the existing switch circuit, remove the cover plate on the switch box, and pull out the switch. Using a neon circuit tester, check the circuit to be sure that the power is turned off, taking care not to touch the wire terminals while doing so. Disconnect the circuit wiring to the old single-pole switch; then discard the switch. Check for damage to the wire or box.

Strip the Circuit Wires, and Connect the Dimmer Switch. Clip off the stripped ends of the existing circuit wires; then strip them again—leaving about ½ inch of exposed wire at the ends. Using red wire connectors, splice the lead wires from the dimmer switch to the existing circuit wires in the switch box. Because they are interchangeable, either of the dimmer wires may be spliced to either of the hot circuit wires. However, be certain that the white circuit wire is taped black to label it as a hot wire. If there was a bare copper grounding wire connected to the old switch, then pigtail it to the grounding screw in the switch box if it is a metal box. Place the dimmer switch inside the switch box and replace the cover plate.

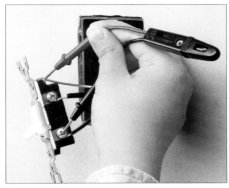
Before replacing a switch, check the circuit with a neon circuit tester to be sure that the power is off.

After disconnecting and discarding the old switch, examine the existing wiring and box for any damage.

To make a crisp, new connection, cut away damaged wire ends, and restrip the circuit wires.

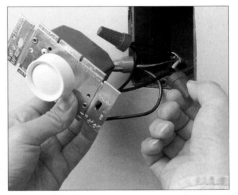
Splice the wires from the new switch to those from the circuit; either lead can connect to either circuit wire.

Splicing Grounding Wires

In existing wiring you're likely to come across the pigtail method of splicing grounding wires, near right, so that's the method demonstrated throughout this book. However, grounding wire connectors are manufactured with a hole at the top so that wires can be spliced as shown at far right. This is the method actually preferred by electricians these days.

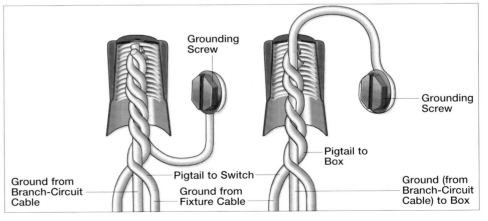

WIRING DIMMERS 131

7 Electrical

Typical Switch and Wiring Layouts

Dedicated Neutral for Switch Circuits

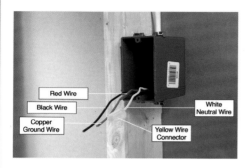

1 Switches at the end of a circuit require three wire cables with a ground. The neutral wire is dedicated for a room occupied sensor switch.

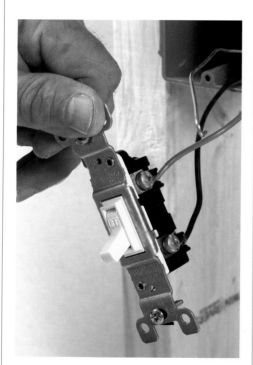

2 Installing a single pole switch at the end of a circuit. The black wire of the three-wired cable is is attached to the OFF side screw terminal. The red wire is attached to the ON side screw terminal.

Light Fixture to an End-of-Run Single-Pole Switch

Use two-wire cable to wire a light fixture where the switch comes at the end of the cable run. This configuration is known as a switch loop. Mark the white neutral wire with black tape to indicate that it is hot.

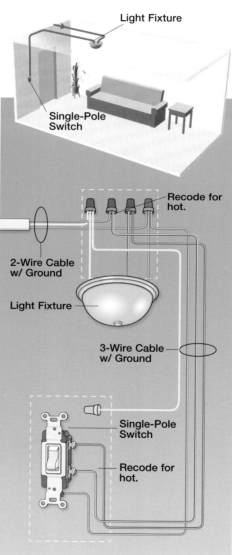

Double-Ganged Switches to End-of-Run Light Fixtures

In this setup, power is fed first through the switches and then to the light fixtures. Only two-wire cable is needed for the wiring connections. The switches occupy one double-ganged electrical box.

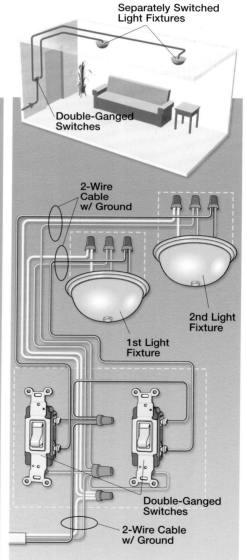

132 ELECTRICAL / WIRING SWITCHES

WIRING SWITCHES

Three-Way Switches with Fixture at End-of-Run
In this switch circuit, power goes from the first switch box through the second, and then to the light fixture. A three-wire cable with ground is run between the switches and a two-wire cable runs between the second switch and the fixture.

Three-Way Switches with Fixture at Start-of-Run
In this setup, power enters the light fixture on a two-wire grounded cable. It proceeds to the three-way switches and then returns to the fixture. Two-wire cable connects the fixture to the first switch and three-wire cable runs between the switches.

Three-Way Switches with Fixture at Middle-of-Run
Here, the light fixture is positioned between the two three-way switches. Power comes to the first switch on a two-wire grounded cable. It passes through the light fixture, proceeds to the second switch, and then returns to the fixture on three-wire cable.

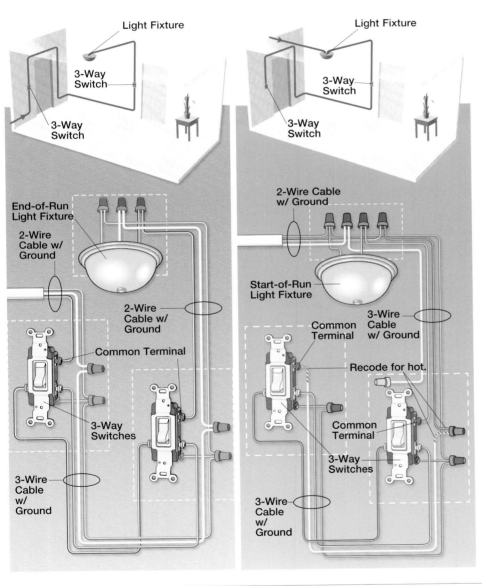

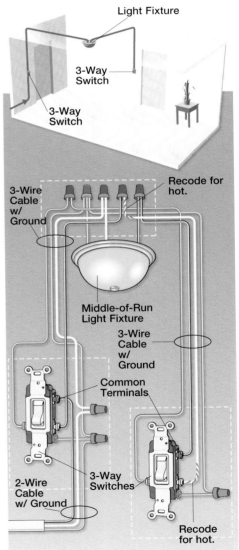

TYPICAL SWITCH AND WIRING LAYOUTS

7 Electrical

BULB TYPES

Lightbulbs are rated by lumens, which measure the amount of nondirectional light the bulb produces, and watts, which measure the rate at which the electrical energy is used. The ratio of lumens per watt is an indicator of a bulb's efficacy. Watts don't measure brightness: though a 100-watt incandescent bulb is brighter than a 40-watt one, a 13-watt compact fluorescent bulb may be brighter than the 40-watt as well.

Compared with an energy-guzzling 100-watt incandescent, compact fluorescents use 75 percent less electricity and last longer—but you can't dim them without installing special ballasts and wiring. The harsh, bluish light of a fluorescent is also not what most people want in the dining room or next to the sofa. For some uses, fluorescents are fine—for example, over the washer and dryer or tucked under upper kitchen cabinets for task lighting. If you are stuck with fluorescent fixtures, a lighting expert can help by choosing warmer bulbs or cooler tubes to suit the situation.

Halogen bulbs have a kind of clear-white quality, are about 25 percent brighter than standard incandescent bulbs of the same wattage, and can be dimmed without special wiring—but they do require special fixtures. They are also extremely hot and should be treated with caution. High-intensity-discharge (HID) bulbs, such as halide and high-pressure sodium, are also bright and efficacious, but require special fixtures. They are often used outdoors because of their brightness and long lifetime.

GREEN SOLUTION
Compact Fluorescents

In an era when saving energy has become a top priority, many experts now recommend switching to compact fluorescent lamps (CFLs). According to the Energy Star program, qualified CFLs use about 75 percent less energy and last as much as 10 times longer than standard incandescent bulbs.

The major difference between CFLs and other fluorescents is that CFLs are designed to be used in standard lamps and lighting fixtures, replacing incandescent bulbs.

When shopping, check the lumens or look for labels with information such as "60-Watt Replacement." This means that you are getting the same brightness as a 60-watt incandescent bulb while using less wattage.

CFLs contain mercury and should be recycled or treated as hazardous waste and not discarded in the trash.

Lamp Repair

Bulb lamps usually fail from heat or wear that breaks continuity between the hot terminal (brass or black screw) and the base of the socket, or between the neutral terminal (silver screw) and the socket. After unplugging the lamp, your first step in a repair is to test for continuity between the screw terminals and these points.

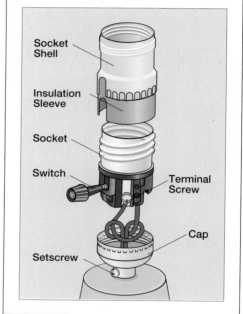

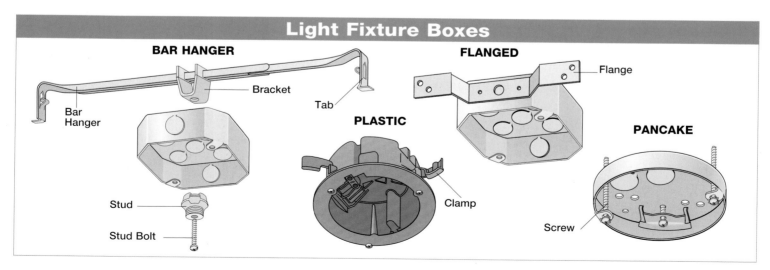

134 ELECTRICAL / LIGHTING FIXTURES

LIGHTING FIXTURES

Installing a Ceiling Box

USE: ▶ pencil • keyhole saw or saber saw • screwdriver ▶ ceiling fixture box • screws • wire connectors

1 Using the paper template a cut-in box lets you add lightweight fixtures without the support of joists or studs. Mark and cut the drywall opening.

2 Bring cables into the box, and press the box into the opening. Engage the holding tabs with a screwdriver.

3 Connect the circuit/switch-loop wires to the new fixture with wire nuts. Then, screw the fixture to its base.

Installing a Fluorescent Fixture

USE: ▶ neon tester • keyhole saw or saber saw • screwdriver • combination tool • fish tape (optional) ▶ fluorescent fixture & box • wire connectors • screws

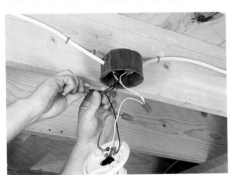

1 Take power from a nearby fixture box—in this case, in the floor below an installation of kitchen cabinet lights.

2 After securing connections with wire connectors and closing the box, mark and cut an opening in the wall for your switch.

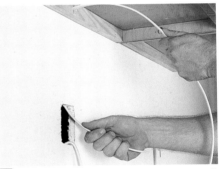

3 Fish the power leg up through the wall and the switch leg down from an opening in the back of the cabinets.

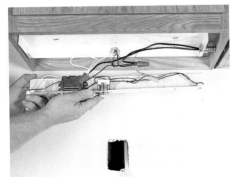

4 Screw the lamp fixtures to the cabinets, and wire them to each other through flexible conduit.

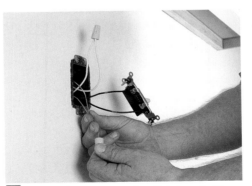

5 Clamp the power and switch legs in the wall box; strip the leads; and connect the black wires to the switch.

6 Finish by clipping diffusers on the lamp cases. They help fluorescents cast a more even light on the counters.

BULB TYPES

7 Electrical

WIRING CEILING FIXTURES

After turning off the power, test the circuit to confirm it has no power. Pull the cable from the switch into the electrical box; secure it in place; rip the cable sheathing; and strip the wires in the cable. If the box does not have a built-in hanger stud, attach a mounting strap to the box tabs. Screw a threaded nipple into the collar of the crossbar to support the weight of the light fixture. Make certain that it will extend through the fixture to engage the mounting nut.

Wire the Fixture

Using wire connectors, splice the hot black wire from the switch to the hot black lead wire from the fixture. Next, connect the neutral white wire from the switch to the neutral white wire from the fixture. Then, splice together the grounding wires, and pigtail them to the green terminal screw in the box or on the mounting strap. Push the completed wiring neatly into the box; install the fixture cover; tighten the mounting nut.

Surface-Mounted

Surface-mounted fixtures are usually installed on a ceiling or wall. They may use incandescent, fluorescent, or quartz halogen bulbs. Wall sconces, globe lights, above-vanity strip lighting, and ceiling fixtures are all examples of this type of lighting. Surface-mounted lights are generally attached to lighting outlet boxes.

Surface-Mounted

Surface-mounted fixtures come in a variety of styles appropriate for ceiling or wall installation.

Installing a Ceiling-Mounted Fixture

USE: ▸ cable ripper • wire stripper • cable clamps ▸ surface-mounted light fixture • mounting strap • electrical box • threaded nipple • wire connectors

1 Feed cable in through the back of the box. Use a screwdriver to help create the opening. Do not break the tab!

2 After making sure power is off, pull the cable into the box, and strip the insulated wires.

3 If needed, attach a mounting strap to the ceiling box. Use the screws supplied with the strap.

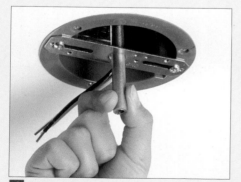

4 Screw a threaded nipple into the collar of the support. Do not run the wires through the center hole of the mounting strap.

5 Splice the fixture wires to the power wires, pigtailing the grounding wires to the grounding screw.

6 Place the fixture cover over the nipple, and tighten the mounting nut. Overtightening will damage lightweight covers.

LIGHTING FIXTURES

TRACK-MOUNTED FIXTURES

Track lighting is used primarily to focus light on a particular object or work surface. Because the light fixtures can be moved to any point along a modular power track, this type of lighting affords the homeowner a great deal of flexibility in lighting design. Noted for ease of installation, track lighting can be powered off an existing or new switch-controlled electrical box. The wiring remains concealed as the exposed lighting track runs along the wall or ceiling surface. Track sections are connected straight from the outlet (or electric) box and can branch off in different directions with the aid of special connectors that come with the light kit.

Turn off power to the electrical box where the track lighting will be connected. Use a neon circuit tester to verify that the power is off, and disconnect the existing fixture wiring. Measure and draw guidelines out from the ceiling box to center the new track lighting. Then, wire the track lighting power connector to the electrical box. Using wire connectors, splice the neutral white wire from the cable to the neutral white fixture wire and the hot black switch wire to the hot black fixture wire. Next, splice together the grounding wires, and pigtail them to the green terminal screw in the electrical box, if the box is metal. Secure all of the wires in the box. (Some nonmetallic boxes are provided with a grounding plate and green terminal screws.)

Fasten the power connector plate to the electrical box, and screw or bolt the first section of lighting track temporarily into place. Insert the power connector into the first section of track. Twist-lock the connector in place and attach the power-box cover.

Install additional lengths of track to the first section, using T-connectors and L-connectors, if needed. Cover all connections using connector covers. Mark the positions for screws or Molly bolts on the ceiling. Then take down the temporary track sections, and drill pilot holes as needed. Install the track and tighten all fasteners. Attach the lighting fixtures on the track.

Installing Track-Mounted Lighting

USE: ▶ measuring tape • multipurpose tool ▶ track lights • track connectors and cover • molly bolts or wood screws • power-box cover • wire connectors

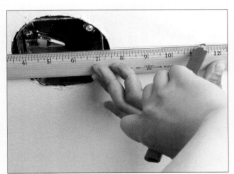

1 Measure and draw guidelines for the precise placement of your track lighting.

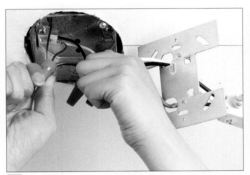

2 Wire the track-light power connector directly into a ceiling or wall-mounted electrical box.

3 Fasten the power-connector plate to the electrical box; then attach the first section of track.

4 Twist-lock the connector in place, and attach the power-box cover. Unlock the connector to reposition if needed.

5 Install additional connectors and track sections, as needed, using the appropriate connectors.

6 Slide the lighting fixtures into the track, and lock them into the desired position.

WIRING CEILING FIXTURES

7 Electrical

CHANDELIERS

Hanging a chandelier differs from installing a ceiling-mounted light fixture because of the added weight of the fixture. Special chandelier-hanging hardware is required, including a threaded stud and nipple, a hickey, and locknuts.

Prepare the Ceiling Box
After turning off the power, use a neon circuit tester to confirm that there is no power at the light circuit. Remove the fasteners holding the old light fixture to the ceiling box. Carefully support the fixture, or have someone else do it as you disconnect the lead wires from the box. Screw a short threaded stud into the center knockout at the top of the electrical box to support a hickey. Inside the box, screw a hickey onto the stud. Then screw a threaded nipple into the hole at the bottom of the hickey. Secure the hickey and threaded nipple using locknuts.

Wire and Install the Chandelier
Pull the chandelier wires through the nipple and into the electrical box. Using a wire connector, splice the black hot wire from the chandelier to the black hot wire in the ceiling box. Do the same with the white neutral wires. To complete the wiring, pigtail the ground wires to the green grounding screw in the electrical box. Tuck the wiring in the box. Then screw the chandelier support to the threaded nipple, and slide the escutcheon plate up to the ceiling box. Thread the chandelier's collar nut onto the nipple, and secure the escutcheon plate.

Heavy Chandeliers

The added weight of a chandelier must be supported by the ceiling box.

Installing a Chandelier

USE: ▶ standard tools ▶ stud • hickey • threaded nipple • chandelier

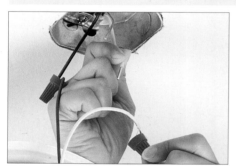

1 Turn off power at the main panel before disconnecting wires on the existing fixture.

2 Install a stud in the box. Then screw a hickey onto the stud and a threaded nipple into the hickey.

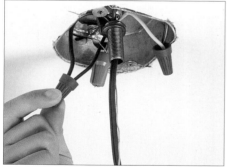

3 Pull the chandelier wires through the nipple, and wire the fixture. Install the chandelier and escutcheon plate.

4 Thread the chandelier's collar nut onto the nipple, and tighten the nut. Pliers can damage the nut.

Vanity Switch

Here's another wiring method common for vanities other than the method described opposite. In this configuration a GFCI-protected outlet is always hot and shares a box with a switch that controls the light on the vanity. The black wire from the power source is attached to the brass screw of the outlet. A jumper wire connects the outlet and the switch, bringing GFCI protection to the vanity light.

LIGHTING FIXTURES

VANITY LIGHTING

If there is no receptacle outlet within 30 inches of the washbasin, install a new 20-ampere circuit as shown here.

Bringing Power to the Bathroom

Install 12/2g cable from the panel box to the bathroom. In this sequence, power comes into a GFCI receptacle on the right. Run 12/2g cable from the receptacle to the light fixture. Run 12/3g cable from the light fixture to the outlet on the left. Finish by running 12/2g cable from the outlet box to the switch box.

Connect a pigtail to the two equipment grounding conductors that go to the receptacles. Connect the two white wires in the fixture outlet box to a pigtail. Connect the black wire from the load side off the GFCI receptacle outlet to the black wire from the standard receptacle (the one on the left in our example).

Fixture Connections

This type fixture has a mounting plate that covers the box as shown. Connect the white wires from the lighting fixture to the white pigtail and the grounding wire to the grounding pigtail. Connect the black wires to the red wire in the three-wire cable.

Outlet Connections

Attach the black wire from the three-wire cable to the brass terminal screw on the receptacle. Connect the white wire from the three-wire cable to the silver terminal screw. Attach the red wire from the three-wire cable to the white wire from the switch, and tape the white wire black. Attach the black wire from the switch to the other brass terminal.

Light Switch Box

Connect all of the equipment grounding conductors. Place black tape on the white wire located in the light switch box identifying it as black. Holding the switch so that it stays on, connect the taped wire to one of the terminal screws. Connect the black wire to the other terminal. Turn off the the switch, place it in the outlet box, and secure it with the screws.

Turn the circuit breaker on at the panel box. Plug a lamp into the GFCI receptacle. If it does not light, press the reset button on the GFCI.

Installing Vanity Lighting

USE: ▶ 20-amp GFCI receptacle • 15-amp standard receptacle • single-pole switch • lighting fixture(s) • switch box • 12/2g NM cable • 12/3g NM cable

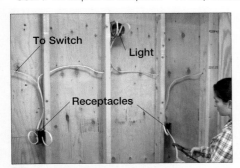

1 Run two-wire cable (yellow) to the receptacle and fixture. Three-wire cable connects the fixture to the other outlet.

2 Connect the power cable to the line side of the GFCI. Connect wires running from the outlet to the fixture to the load side.

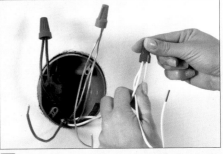

3 The black wire from the GFCI load side is connected to the black wire that runs to the other outlet. Pigtail the white wire.

4 Connect the white fixture wire and white pigtail together. Attach the red wire to the black fixture wire.

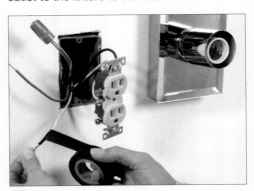

5 Second receptacle. Tape black the white wire running to the switch, and attach it to the red wire from the three-cable wire.

6 Attach the black wire and the white wire taped black to the terminals on the switch.

CHANDELIERS

7 Electrical

QUARTZ HALOGEN

Quartz halogen lamps have become very popular in recent years. However you must be extremely careful using any light fixture that employs a quartz halogen bulb. This type of bulb gets extremely hot and can easily cause a fire. Never allow anything flammable to come near the glass covering on the bulb. Even an extension cord placed too close to the lamp can heat up and ignite. Exposed skin will burn immediately if it comes into contact with the glass. Though advertised as having a long life span, quartz halogen bulbs sometimes last only for minutes and have been known to explode within a fixture. Use extreme caution if you decide to use this type of bulb.

Recessed Lamp Housing

Recessed lamp housings come in various sizes and shapes. They may or may not be designed for insulation contact. Be sure to check the labels. Installing the wrong kind of recessed lamp housing can present a fire hazard.

Replacing a Quartz Halogen Bulb

Remove the Burned-Out Bulb. Turn off the switch to the existing halogen light fixture. Allow the bulb to cool for several minutes before attempting to replace it. Remove the light diffuser and fixture cover. Be sure to keep the bulb glass absolutely clean, and never touch it with your fingers. Oil from the skin causes "hot spots" on the lamp when lit, which destroys the lamp. One side of the bulb holder is typically spring-loaded. Reach in with gloved hands, and push the bulb toward the spring-loaded end while lifting out the bulb.

Install the New Bulb. Install a new quartz halogen bulb by simply pushing it toward the spring-loaded end of the bulb holder and snapping it into place. Reattach the fixture cover and light diffuser, turn the power on again, and test the bulb. Using latex gloves will keep the bulb from contacting your skin. Be sure to the switch is off when installing new bulbs.

1 After letting it cool, push and lift the burned-out bulb toward its spring-loaded end.

2 Install new bulb by pushing it into the spring-loaded end and snapping the other end into place.

Recessed Fixture

A recessed ceiling fixture has its own prewired box attached. Just bring the cable into the box and splice the wires.

Quartz

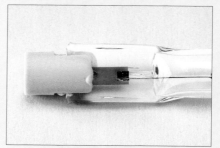

Quartz halogen bulbs are protected by a glass tube, and they generate intense heat.

LIGHTING FIXTURES

RECESSED LIGHTING

Recessed lights are characteristically used where spot lighting is needed and/or low-hanging fixtures are not desirable. Most commonly used in kitchens and living areas where they provide concealed lighting, recessed light can also illuminate a specific area. Some recessed lights rotate and focus at an angle to illuminate or bathe a particular object—a favorite painting, sculpture, or antique—in light.

Recessed light fixtures fall into two categories: insulated ceiling (IC) and non-insulated ceiling (NIC). It is best to opt for IC light fixtures so that you can put insulation right up against the metal fixture housing in an attic floor or cathedral ceiling. Non-insulated ceiling fixtures require, for fire safety, a minimum clearance of 3 inches between the housing and insulation and a clearance above the fixture so as not to entrap heat.

Most housings have two adjustable arms that are mounted to adjoining ceiling joists. The exposed surface of the fixture is extended below the joists just far enough to be flush with the finished ceiling. A decorative cover is snapped into place when the ceiling is finished. After the fixture has been mounted, the power branch-circuit cable is run into the electrical box attached to the housing.

Recessed lights don't draw much electrical current, so they can be wired into a standard light or receptacle circuit. They can be wired using either 14- or 12-gauge wire.

Wire the Electrical Box

The wiring is contained in an attached electrical box. Using wire connectors, splice the white switch-leg wire to the white wire from the fixture. Next, connect the black switch-leg wire to the black wire from the fixture. Pigtail the grounding wires from the switch-leg and fixture cables to the green grounding screw terminal in the electrical box, and attach the box cover. When the ceiling is finished, install the decorative housing cover.

Installing a Recessed Light Fixture

USE: ▶ insulated screwdrivers • power drill • cable ripper • long-nose pliers • multipurpose tool • cable clamps • wire connectors ▶ lamp housing

1 Install the bracket for the lamp housing, positioning it between two joists.

2 Adjust the housing, and fasten the extension bars to the adjoining joists. A cordless drill will speed the job.

3 Use a screwdriver to remove one of the knockouts to accommodate the power cable.

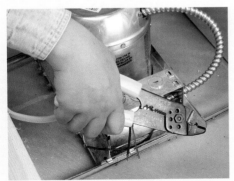

4 Pull the power cable coming from the switch into the junction box, and clamp it securely in place.

5 Splice the black hot wires and white neutral wires from the switch-leg cable and fixture.

6 Pigtail the ground wires to the green grounding screw in the electrical box. Attach the cover.

QUARTZ HALOGEN

7 Electrical

LOW-VOLT CIRCUITS

Doorbells and thermostats need less than 120 volts of power to operate. To install either one, you will also have to install a transformer to step down the house voltage to 12 or 24 volts.

A doorbell circuit must have three connections: a transformer to 120-volt house current and two low-voltage wires, one from the transformer to the bell and one from the bell to the button. The transformer may be wired directly into a junction box in a basement, attic, or above a dropped ceiling. Use transformer wire, 20-gauge or larger, or equivalent bell wire rated at 30 volts.

Remote Switching

In new wiring, you can save a great deal of time and money by installing a low-voltage remote control. Standard 12- or 14-gauge wire is used only between the main panel and the outlet or fixture boxes. Switches are wired with low-cost 18- or 16- gauge wire that does not have to run in conduit, and the switches do not have to be mounted in boxes (unless this is required by local codes). The lightweight wires are easy to run, strip, and connect.

You need three kinds of devices for a low-voltage remote-control system: a transformer, a switch, and a relay. A transformer takes the 120-volt primary power from the service panel and supplies 24 volts to the switching system. Wires between the transformer and the remote switches connect through a relay, an electromagnetic switch that acts as a go-between.

Phone Wiring

You can run phone wires in or on the wall. Partially conceal exposed wires on molding, secured with nonpinch staples.

If a phone works plugged into the company's exterior jack but not inside, the fault is in your interior wiring.

Installing a Transformer

USE: ▶ screwdriver • needlenose pliers • combination tool ▶ low-voltage transformer • transformer wire

1 Select a junction box near the bell or other transformer-fed appliance. Cut power to the box, and remove its cover.

2 Bend and remove a metal knockout in the side of the junction box where the transformer will be mounted.

Fixing a Doorbell

USE: ▶ screwdriver • soft-bristled brush • utility knife • continuity tester ▶ grommet

1 When the bell won't ring, check wiring connections behind the exterior button, which may corrode or come loose.

2 If the wiring is okay, try cleaning dust from the bell contacts. An unlikely tool (a soft-bristled toothbrush) works well.

3 If the bell tone sounds like a dull thud, cut out the grommet that helps to suspend the tone bar. It may be brittle or broken.

SPECIALTY WIRING

Replacing a Thermostat

USE: ▶ screwdriver • combination tool • pen ▶ thermostat • masking tape

1 Loosen the old thermostat from its wall mount, and mark the terminal location and color of each wire you disconnect.

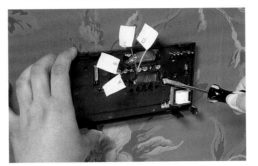

2 Mount the base of the new thermostat in the same location (away from heat sources), and pull through the tagged wires.

3 Remove the tags, and attach old wires to the new thermostat one at a time. Finish by installing the new faceplate.

• wire connectors

3 Mount the transformer to the box by inserting its threaded stub through the knockout and securing it with a nut.

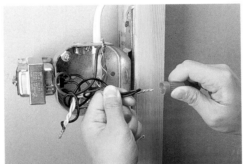

4 After removing the old wire connectors, join the transformer leads to their corresponding power leads in the junction box.

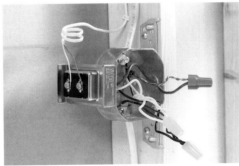

5 Run transformer-gauge wires from the bell, thermostat, or other appliance, and connect them to the transformer terminals.

4 Add a new rubber grommet if needed, so that the tone bar can vibrate freely and create sustained bell tones.

5 Use a continuity tester to check electrical circuits at the transformer that powers the low-voltage bell system.

6 If the transformer is not working, cut power to the junction box; remove the wire leads; and install a new transformer.

LOW-VOLT CIRCUITS

7 Electrical

TELECOMMUNICATIONS

Telecommunications wiring includes both conventional telephone and data transmission wiring for computers. In this chapter, discussion is limited to telephone wiring. National regulations, set by the Federal Communications Commission (FCC), also define your responsibilities with regard to system maintenance and hookup. For example, the telephone company may be responsible for central wiring to your home, while you are responsible for all of the wiring beyond the point of entrance, commonly called the point of demarcation. A modular wiring jack must also be provided that allows you to disconnect everything on your side of the demarcation point from everything on the other, public side of the telephone network.

Telephone wiring is commonly available as four-conductor line cord and telephone station cable. Line cord is the flat cord that connects your telephone equipment to a telephone jack. It should not be used for anything else. Telephone station cable for residential use typically consists of D-station wire that is color-coded for easy identification. Most home telephone systems require only four conducting wires (two-pair wire), with one pair for the phone and the remaining pair for a secondary line for a fax machine or modem. Color codes for telephone wire may consist of solid colors or two-color banding. The standard solid colors are red, green, yellow, and black. Banded colors are more varied. The telephone station cable variations shown in the phone-jack illustration below consist of alternating bands of green and white, orange and white, and blue and white.

You can route telephone wires from the network junction box, independently to each jack, in a straight line connecting the jacks in an open loop, or in a closed loop that returns to the junction box. A wire break to an independent jack will only cut service to the phone on that one line. A break in an open-loop system cuts service to any phones beyond the break. But a break in a closed-loop system won't stop a signal from traveling to the break point from either direction.

Wiring Telephones

A basic telephone system consists of a service entrance, a wire junction, telephone station cable, a surface- or flush-mounted wall jack, flat cable, and a telephone.

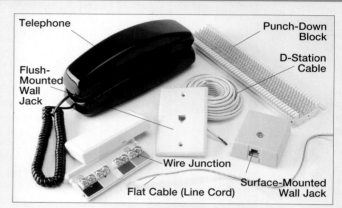

How Many Phones?

The power required to ring your phones is represented by a number, called a ringer equivalency number (REN). The total number of telephones that you can install on your line is determined by adding up the RENs on your phones. If the number is less than five, you will not have any problems. If the number exceeds five, your phones will still work, but not all phones on the line will ring.

Wiring Phone Plugs

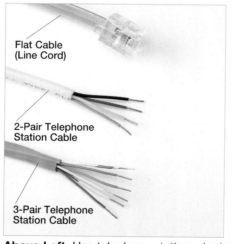
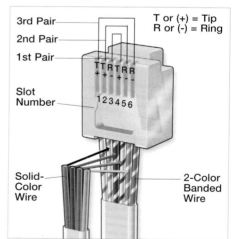

Above Left: Use telephone station wire between wire junctions and modular telephone jacks. Use flat wire or line cord to connect telephone or data transmission equipment to the jacks.

Above Right: This phone-jack diagram illustrates a standard 3-pair color scheme for telephone wiring. A 4-pair scheme would also have a white wire with brown banding and a brown wire with white banding. The slot numbers apply to 4- and 6-terminal telephone jacks only and would not apply to a fourth pair of wires. In a pair of telephone wires, each wire is identified by a tip (+) or a ring (−) polarity. Each pair of wires is further coded by a color scheme consisting of a primary color or a primary and secondary color.

PHONE WIRING

Wiring a Telephone Jack

Wire the Telephone Junction. Remove 2 inches of sheathing from the end of a length of D-station telephone cable, exposing the insulated wires inside. No jack should be more than 200 feet from the telephone service entrance. If your wire junction, or connecting block, has color-coded screw terminals, then strip the cable wires, and connect them to the appropriately colored terminals in the wire junction.

Run the Telephone Cable. Run the telephone cable from the wire junction to the location of your new telephone jack. Be careful not to run telephone wiring within 6 inches of parallel circuit wiring or within 5 feet of any bare wiring.

Wire and Mount the Jack. Fish the phone cable through a hole cut in the wall. Remove 2 inches of sheathing from the cable; then strip the telephone wires. Connect the telephone wires and jack leads to the matching color-coded screw terminals in the telephone jack; replace the cover; and mount the jack on the wall.

Test the Jack. Now the jack is ready for testing. Insert the plug end of a telephone line tester into the jack. If the jack is correctly wired, then the LED on the box will glow green; if not, it will glow red or not at all. The line tester will not light if no connection has been made.

Plug the telephone into the jack; then listen for a dial tone. If there is no tone, disassemble the jack and recheck the wiring. If there is a dial tone, then try dialing a number. The dial tone should cease after you begin to dial. If it does not, then the wires must be reversed.

Newer phone systems, instead of screw-terminal junction blocks, use punch-down, or connection, blocks. They are also known as insulation displacement connectors (IDCs). A standard M, or 66, block has connections for 25 pairs of wires. Additional blocks can be added if needed. A special punch-down tool presses the telephone wires into a 66 block, eliminating the need to strip the wires before connecting them to the block.

USE: ▶ fish tape (nonmetallic) • wire stripper • cable staples • cable ▶ telephone wire junction • telephone jack • telephone line tester • D-station telephone

1 Connect station wires to screw terminals or a punch-down block in a telephone wire junction.

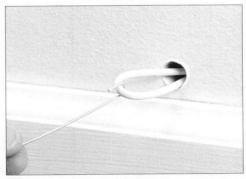
2 Fish the telephone cable through the wall to where the telephone jack will be mounted.

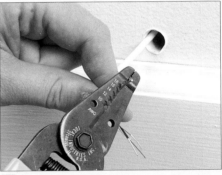
3 Pull telephone cable through a hole cut in the wall; rip the cable sheathing; and strip the telephone wires.

4 Mount the telephone jack; connect the lead (spade tip) wires to the jack; and attach the jack cover plate.

5 Using a telephone line tester, test the polarity of the telephone jack. A green light indicates correct wiring.

6 Plug a telephone receiver into the jack; then listen for a dial tone. If there's no tone, recheck the wiring.

TELECOMMUNICATIONS

7 Electrical

RANGE RECEPTACLES

High-voltage 220/240-volt outlets are most commonly used for air conditioners, clothes dryers, electric ranges, hot water heaters, and other major appliances that consume high voltage with motors or heating elements. The vast majority of U.S. households have circuit breaker panels fed with 240 volts. The 120 volt circuits in your home use just half of the available voltage. 120 volt circuits can safely provide the required wattage required for such things as lights and vacuum cleaners. But larger appliances need more power and a traditional 20-amp circuit can't carry enough of it. So, thicker wires are needed to carry higher wattage currents.

In the past, cable for kitchen ranges included two hot wires and a stranded ground/neutral. This type of cable is called service-entrance conductor (SEU) cable. The range receptacle for this kind of cable accommodated a three-prong plug configuration. The problem with this arrangement was that the current-carrying neutral was also used as the grounding conductor for the appliance frame. Today, cables for kitchen ranges are still required to carry two hot conductors, but the neutral wire must also be insulated, and the grounding conductor can be either bare or insulated. It is a four-conductor cable containing three insulated wires and one grounding conductor. Usually, there are two hot wires, a neutral, and a ground wire that is green or bare. This category of cable is called service entrance round (SER) cable. Type NM cable with three conductors and a ground is also allowed. The size used for a kitchen range is usually 6/3g SER. A range receptacle must accept a 4-prong plug configuration for this type of cable.

Of the four wires in SER cable, the two hot wires carry the 240 volts required to power the heating elements. The 120-volt power is carried across the neutral to either of the two hot wires—it doesn't matter which one. The 120-volt power is used to run the timer, clock, buzzer, etc. Drawing the neutral current away from the grounding conductor causes the return current to flow safely through an insulated conductor rather than through a bare copper grounding wire. The 4-slotted female receptacle into which the SER cable is wired can be surface- or flush-mounted. For the average homeowner, a surface-mounted receptacle is often preferred because it is easier to wire. When a range is not hardwired but has a cord and plug, the plug must have four prongs to match the receptacle, as mentioned earlier, so that the neutral and grounding conductor will remain separate.

4-Slot Outlet

Insert the wires under their appropriate clamps, tighten the screws, and mount the receptacle housing.

Direct-Wiring a Dishwasher

USE: ▶ insulated screwdrivers • long-nose pliers ▶ wire connectors • dishwasher • 12/2g NM cable

Pull the Cable. The internal wiring of a dishwasher is terminated in the power junction box, just behind the dishwasher's kick panel. The connecting wires are black, white, and green. A green grounding screw is sometimes substituted for the green wire. Using a knockout punch, remove a knockout on the dishwasher's junction box, and pull the 12/2g NM branch-circuit cable into the box. Be sure to leave extra wire (3 to 4 feet) looped behind the dishwasher so that the appliance can be pulled away from the wall for later servicing. Rip back the cable sheathing; cut away the excess; and secure the cable in the box using a cable clamp.

Wire the Box. After exposing them, strip the insulated wires from the branch-circuit cable, and wire the junction box. Using wire connectors, splice the black hot wire from the branch-circuit cable to the black hot wire from the dishwasher. Then, do the same with the white neutral wires. Lastly, splice the bare copper grounding wire to the green appliance wire, or connect it to the green grounding screw in the junction box to complete the appliance wiring.

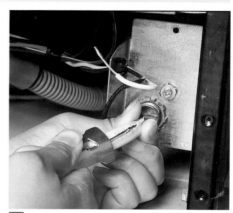

1 Pull the power cable into the dishwasher's junction box. Leave a wire loop behind the dishwasher.

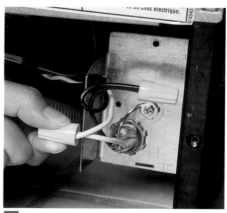

2 Splice the black to black and white to white wires. Then connect the ground wire to the grounding screw.

WIRING APPLIANCES

Waste-Disposal Unit

A kitchen waste-disposal unit is fairly easy to wire, but you will also need to install a switch to operate the appliance. Ordinarily, you should mount a disposal-unit switch on the wall off to one side of the kitchen sink. However, if the disposal unit is an add-on, you must choose between the expense of mounting the switch on the wall or simply placing it in the cabinet beneath the sink. Should you need to shut off the disposal unit in an emergency, having the switch readily accessible is safer and more convenient. To mount your switch on the wall, follow the instructions below.

Because disposals are meant to be run when the water is running, be sure to situate the electric switch where it can be reached by someone turning the water on or off. It is advisable to make this entire circuit a GFCI circuit.

A waste-disposal unit can be safely controlled by a switch mounted on the wall to the side of your kitchen sink or within the sink's base cabinet.

Wiring a Waste-Disposal Unit

USE: ▶ cable ripper • multipurpose tool ▶ waste-disposal unit • 12/2g NM cable • single-pole switch

1 Fish the disposal-unit cable from the sink cabinet to the switch box. Fish the power cable separately.

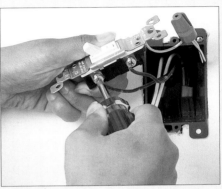

2 Wire the switch; tuck in the wires carefully; and then install the switch-box cover plate.

3 Remove the cover plate from the splice box at the bottom of the waste-disposal unit.

4 Pull the power cable into the splice box, and clamp it securely in place.

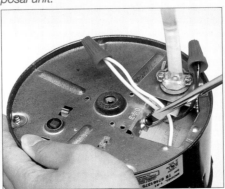

5 Splice the switch and appliance wires, and connect the grounding wire to the grounding screw.

6 Tuck the wires carefully into the box, and replace the splice box cover plate.

7 Electrical

CEILING FANS

When it comes to installing ceiling (paddle) fans, homeowners commonly assume that a fan can be suspended from any existing ceiling box. This is not the case, because a ceiling fan can weigh 35 pounds and the electrical box has to be approved for that weight. Make sure your ceiling box is secured to the structural framing.

Cut an opening in the wall for the light-switch box and another in the ceiling for the fan/light box. Fish 12 or 14/2g NM cable from the breaker panel to the switch-box opening and 12 or 14/3g NM cable from the switch box to the fan/light-box opening in the ceiling. Install the switch box, and pull both cables into the box. Rip open 10 inches of each cable, and remove the excess sheathing. Secure the cables in the box, and strip the wires. Do the same for the cable in the fan/light box. The three-wire cable from the switch box to the fan/light box allows you the option of controlling the light and fan independently. If you decide that you do not want to have independent control, run the three-wire cable anyway so you will preserve the option for later.

Wire the Electrical Boxes

Although there are two ways to run power to a ceiling fan/light—through the ceiling box or through the switch box—the preferred way is to bring power to the fan/light through the switch box. This makes trouble-shooting easier and minimizes crowding wires in the ceiling box. Using wire connectors, in the fan/light ceiling box, splice together the black hot wire from the fixture and the hot black lead wire from the fan switch. Connect the hot red wire from the fan/light to the hot black lead wire from the light switch. Next, connect the white neutral wire from the switch box and the white neutral wire from the fixture. Then splice the green grounding wire from the fixture to the bare copper grounding wire from the switch box. In the switch box, pigtail the hot black feeder wire to both switches, and connect the white neutral feeder to the white neutral from the fixture. Then connect the hot wires from the fixture to their respective switches. Push the switch wires into the box; screw the switches in place; and install the cover plate. Then install the mounting plate and ceiling fixture.

Ceiling Fan Support

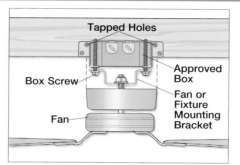
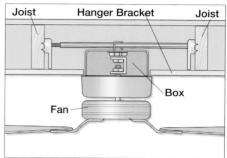

The NEC (Section 422.18) requires that ceiling fans up to 35 lbs. use an electrical box listed for that purpose. For fixtures over 35 lbs., the fixture must be supported independently of the box.

Ceiling Fan or Light — GREEN SOLUTION

USE: ▶ standard tools ▶ ceiling fan/light • ceiling-fixture • hanger bracket (for box fans over 35 lbs.)

1 Install the switch box, and fish the cables from the panel and fixture to the box.

2 Install the fan/light box, and secure and strip the three-wire cable from the switch box.

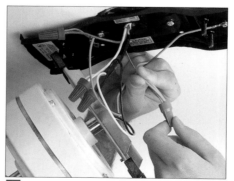

3 In the fan/light ceiling box, splice the fan and light switch wires to the fixture wires.

4 In the switch box, splice the fixture wires to the fan and light switch wires. Then energize and test the switches.

WIRING APPLIANCES

RANGE HOODS

Range hoods are essential for ridding a room (typically a kitchen) of greasy and smoky air before it escapes throughout the house. They also play a large role in removing excess moisture that accumulates during cooking, an important function in today's tightly constructed houses. Range hoods are not automatic (as a bath fan is when hard-wired to a light) so the occupant must exercise discipline when cooking to turn the fan on and off.

Range hoods come in two styles: ducted and ductless. A ducted hood removes heated air by exhausting it outside, while a ductless hood filters smoke and odor from the air and returns the air directly to the room. An average-size range hood circulates or removes 400 to 600 cubic feet of air per minute (CFM). The fan rating will determine the size duct you need for a ducted system. Because installation steps vary for each manufacturer, be sure to follow the specific instructions given by each for installing ductwork and wiring.

Install the Ductwork

Measure and mark the proposed location for the metal ductwork. The ductwork may exit the range hood from the rear and extend directly through an outside wall, or it may exit from the top and then either elbow out through the wall or extend to the roof and then out. Cut the opening as required, carefully following the manufacturer's written instructions, and install the ductwork and a roof or wall cap. Be sure to flash and caulk around any exterior openings to ensure a good weathertight and waterproof seal.

Fasten the hood assembly to the wall and/or cabinets using the screws or toggle bolts provided. Adhere carefully to the manufacturer's installation recommendations and requirements. Then, fasten the branch-circuit cable to the range-hood junction box using a cable clamp. Splice the white cable wire to the white range-hood wires and the black cable wire to the black range-hood wires. Next, connect the branch-circuit grounding wire to the grounding screw in the range-hood junction box and cover the box. Restore power, and test the unit.

Ducted Range Hood — GREEN SOLUTION

USE: ▶ ducted range hood • ducting • wall or roof cap • caulking • aluminum flashing

1 Mark a wall cutout, and then cut the opening. Extend the ductwork from the range hood to the exhaust outlet.

2 For through-wall ducts, install a wall cap. If the duct goes through the roof, cover it with a metal roof vent cap.

3 Install the range-hood unit by attaching it to the cabinet above using the appropriate fasteners.

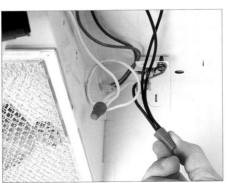

4 Run power to the range-hood box, and wire the fan. Then turn on power, and test the unit.

Duct Types

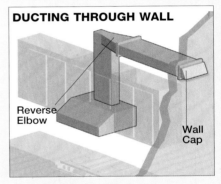

DUCTING THROUGH WALL
Reverse Elbow — Wall Cap

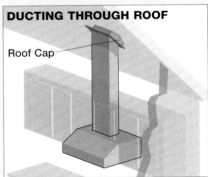

DUCTING THROUGH ROOF
Roof Cap

Ductwork can go directly through an exterior wall or up through the kitchen ceiling and out the roof. Either method requires capping the duct with a backdraft louver so exhaust cannot re-enter the home and wind can be kept out.

7 Electrical

ELECTRIC RADIANT FLOOR HEATING

Ceramic, slate, and tile floors are elegant and durable, although they often feel cold even at comfortable room temperatures. Radiant floor-heating systems are an effective and economical method of removing the chill from ceramic, marble, or stone-type floors. Installed between the subfloor and the finish flooring material, these cables will heat the floor to a comfortable warm temperature with a minimal use of electricity.

There are a number of these products on the market, but in general they consist of an insulated, flexible resistance-type heating element with attached nonheating leads. The product shown here has the conductors contained in a fabric material. The fabric keeps the conductors spaced properly.

Where the final flooring is tile or stone, the system should have a heat density of 10 or 15 watts per square foot. The NEC requires that these systems be provided with ground-fault circuit-interrupter protection.

Floor Preparation

Heating cable may be installed over wood flooring. Drive protruding nails flush or below the flooring, and sand uneven edges where floorboards come together. Nail down any loose flooring. For concrete floor installation, remove all debris, and grind down sharp edges of small cracks. Some manufacturers recommend installing a thermal barrier or layer of insulation under the heating cables. Secure the thermal barrier to the floor with a high-temperature adhesive.

Placing Cables

Plan the heating-cable layout. Remember that the nonheating conductors must be able to reach the control unit, which will be mounted on the wall. Don't overlap the heating cables, and do not allow the nonheating leads to overlap the heating area of the mat. Section 424.44(E) of the NEC requires nonheating conductors to be protected where they leave the floor by rigid metal conduit or by other approved means.

Spread a layer of thinset (thickness may vary by manufacturer) over the floor, and allow it to dry. Place the heating cables according to the planned layout. Secure to the floor.

The sensor connects to the control thermostat to regulate the temperature of the floor. Position the sensor between the heating cables. Secure it with duct tape or per manufacturer's directions. Route the sensor leads to the control box and thermostat. Keep the sensor wires separate from nonheating power leads.

Calculating the Power Supply

Base the size of the branch circuit supplying the heating cable on the size of the room, the amount of heating cable installed, and the watt density per square foot of the heating cable. The branch circuit must be sized 125 percent of the connected load. For example, a room consisting of 40 square feet of floor heating cable with a watt density of 15 watts per square foot results in 600 watts of demand, and 125 percent of 600 watts = 750 watts. If the heating cable is rated 120 volts, then 750 W divided by 120 V equals 6.25 amperes.

Install the Finished Flooring

Cover the heating cables with one or two layers of thinset cement or flooring adhesive. Tile and grout as with any tiling installation. It takes approximately 2 to 14 days for the thinset to cure before you can switch on the system.

Connect the conductors from the incoming power supply to the line-side leads or terminals of the control unit. Most control units contain GFCI protection. The sensor conductors are low-energy conductors; connect them to the terminals on the control unit.

Radiant Floor Heating — GREEN SOLUTION

USE: ▶ tiling tools and materials ▶ radiant heating kit and controls • digital multimeter

1 Draw up a plan for the heating cable. Install fabric and cable over a layer of cured thinset.

2 Secure the temperature sensor to the heating pad. Position the sensor between heating elements.

3 Spread a layer of thinset, and install tile and grout. Do not nick the cables with your trowel.

4 Attach the sensor conductors to the terminals indicated in the manufacturer's instructions.

WIRING APPLIANCES

SCHLUTER® UNDERLAYMENT SYSTEMS

Schluter Systems has designed two different types of underlayment mat system that will work extremely well with all tile products, including floors, wall, and showers, which can include an array of heat applications to keep you warm in the bathroom.

Every job requires careful planning and decision-making on form, function, and style of the complete tile job to be installed, including which Schluter underlayment mat to choose—either the Schluter-DITRA or the Schluter-DITRA-HEAT. Both products provide uncoupling to prevent cracked tile and grout as well as waterproofing; the HEAT product also provides selective heating. In addition, consideration must be given for the type of floor substrate the system will be installed over. Is it wood, concrete, or another building material? In any case, the design parameters of the system will accommodate just about every type of material.

Installing a Schluter® Underlayment System

USE (per square feet required): ▶ Schluter-DITRA-HEAT underlayment roll • thin-set adhesive • 5-inch Kerdi tape roll • desired floor tiles • sanded floor grout • Schluter programmable thermostat, heat sensor, and heat wire kit • Simpson steel stud guard screwdriver • notched floor trowel

1 *Begin over any floor substrate that is structurally sound and does not have surface defects. Use a wet mop to clean the floor prior to beginning layout of the underlayment. This will ensure the top surface is free of dust and will increase the bonding strength of the adhesive layer that must be added beneath the underlayment.*

2 *Burnish the adhesive layer into the top of the subfloor surface with the straight edge of a notched floor trowel, covering the entire floor.*

3 *Roll the underlayment out along the entire area to be tiled. Then lift the underlayment, starting at one end, so that half of the floor is exposed. Using the notched end of the notched trowel, spread more adhesive over the floor area, then place the underlayment material back down over the notched adhesive.*

4 *Using a rubber grout float or a flat wooden trowel, smooth the underlayment material from the center to the outside edges to eliminate any air pockets under it. Repeat the adhesive spreading step and this step at the opposite end of the section and on every section until the floor is covered with the underlayment.*

5 *Schluter-DITRA material will require that all the seams running the length of the material have adhesive applied over the joint with a flat trowel; usually, a 4-inch drywall taping knife works best for this.*

6 *After spreading the adhesive along a joint, apply the 5-inch-wide Kerdi tape over the seam. Then use the drywall knife to burnish the tape into the adhesive. Complete this for every seam between the rows of the underlayment.*

7 *You can also create a watertight perimeter around the room wherever the underlayment material meets with a finished wall surface. Just as each seam was covered and sealed previously, cover and seal at the bottom of all vertical walls that meet the floor system.*

8 *Using the drywall taping knife to embed the Kerdi tape into the floor and wall adhesive for a smooth application. When complete around the room, the Kerdi tape provides a strong watertight application, which is not achievable when using older-style, rigid underlayment boards.*

ELECTRIC RADIANT FLOOR HEATING

7 Electrical

Installing a Schluter® Underlayment System

USE (per square feet required): ▶ Schluter-DITRA-HEAT underlayment roll • thin-set adhesive • 5-inch Kerdi tape roll • desired floor tiles • sanded floor grout • Schluter programmable thermostat, heat sensor, and heat wire kit • Simpson steel stud guard screwdriver • notched floor trowel

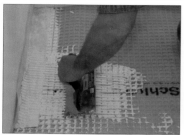

9 To install floor tiles over the underlayment, use a notched floor trowel to create an even, wide area larger than the tiles to form an adhesive bed for the tiles to set on.

10 Install the tiles according to the manufacturer's instructions. You will typically apply some adhesive to an area, install a tile or two, and then repeat, rather than applying adhesive to the entire area and installing all the tiles at once.

11 When using the Schluter-DITRA-HEAT underlayment, the project will require using a Schluter programmable thermostat, heat sensor, and heat wire. The floor heat sensor has to be placed over the underlayment, and the sensor's wires have to be run behind the wall to an approved junction box, typically a single gang size box. You will need to notch the underlayment so that the end of the sensor can lay unobstructed in the mat design.

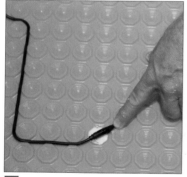
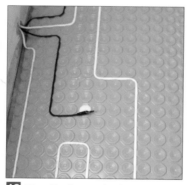
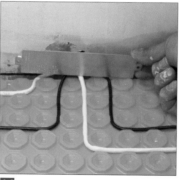

12 Cut the required notch and place the sensor flush with the mat design.

13 Run the heat wire behind a finished wall and apply it over the underlayment in a specified pattern, not crossing over the heat sensor or sensor wires.

14 Install an approved stud guard plate over the opening at the bottom of the wall, below the programmable thermostat, to protect the wires running behind the wall to the junction box.

15 Inside the thermostat junction box will be 120-volt line wires (black and white), heat wire ends that create the closed loop, and heat sensor wires.

16 Connect the heat sensor wires, heat wires, and line voltage wires to the programmable thermostat as detailed in the instruction booklet provided by the manufacturer.

17 Finish installing the programmable thermostat by mounting it to the junction box with the screws included. Follow the instruction booklet to program all of the digital functions of your new thermostat, and enjoy years of a stable, warm, watertight floor system.

WIRING APPLIANCES

DRYER RECEPTACLES

Like an electric range, an electric dryer uses both 120- and 240-volt power and requires a four-conductor cable. The greater voltage supplies only the heating element while the lesser voltage powers the motor timer and buzzer. Also like ranges, dryers were once powered by two-conductor cable with a ground. Such dryers were fed with 10/2g NM copper cable containing a black hot wire, a white neutral wire, and a bare copper grounding wire. 10-gauge NM cable is still used, but it must now have three conductors—red, black, and white wires—plus a bare copper grounding wire. This 10/3g NM cable must be connected to a dedicated female four-slot receptacle rated at 30 amps. The cord on the dryer must have a four-prong male plug and the neutral must not be connected to the dryer frame. The receptacle is usually surface-mounted but can be flush-mounted. When the dryer is shipped to a distributor, the plug and cord are not sent with it. The distributor must install the cord onto the dryer. Before you have a dryer delivered to your home, ask the distributor to install an extra-long rather than standard-length cord. The extra cord will permit you to pull the dryer away from the wall for servicing. It is also a good idea to install the plug high enough on the wall so that you will not have to bend far to unplug it.

Fish the Cable

Turn off the main breaker at the service-entrance panel. Locate a position on the wall for the dryer receptacle, and cut an opening in the drywall to fish the cable. Run 10/3g NM copper cable from the main panel to the dryer receptacle location. Fasten the cable to the framing at intervals not to exceed 54 inches and within 12 inches of each end or joint. Using fish tape, pull the NM cable up through the wall and out the cut opening. If you're mounting the receptacle on a basement masonry wall, run the cable in EMT conduit, and attach the conduit to the wall using conduit straps and masonry screws. Pull the cable into the receptacle housing, and secure it in place using a cable clamp. Rip the sheathing from the cable, remove the excess, and strip the inside wires.

Wire and Mount the Receptacle

Connect each of the hot red and black wires to either of the brass screw terminals on the receptacle. (They are interchangeable.) Next, connect the white neutral wire to the silver screw terminal and the green or bare copper grounding wire to the green grounding screw terminal on the receptacle housing. Mount the housing on the wall, and install the receptacle cover.

At the main service-entrance panel, wire the dryer cable to its own dedicated 30-amp double-pole circuit breaker. Connect the hot red and black wires to each of the two brass screw terminals on the breaker. Then connect the white neutral wire and the green or bare copper grounding wire to the neutral/grounding bus bar in the breaker panel. Turn on the power to the panel by flipping the switch at the main breaker, and test the dryer circuit.

Choosing a Dryer

Purchasing a clothes dryer often means choosing between a gas- or an electric-heated model. Gas dryers are slightly more expensive to purchase and maintain but less expensive to operate. For some homeowners, however, gas may not be an option. Regardless, you can consider other options. Older dryers, for example, rely on timers or thermostats to sense when clothing is dry. Newer models gauge humidity in a dryer, allowing precise heat control to protect clothing from damage. Another option is a wrinkle-guard. When heat turns off, this keeps clothing tumbling until it's removed from the dryer. If you want clothing dried in a hurry, consider a larger dryer. Extra air space between clothing exposes more fabric surface, resulting in faster drying times. For aesthetics, think about built-in units that blend with the room's decor. If you lack space, look at combination or stackable washer/dryers. Also, take into account ease of use. If lifting is a problem, choose a unit with angled or large doors, or down- or side-opening doors. Select electronic controls for more cycle options and custom programming.

Installing a Dryer Receptacle

USE: ▶ insulated screwdriver • fish tape • cable ripper • multipurpose tool ▶ 30-amp dryer receptacle • 10/3g NM cable • 30-amp double-pole breaker

1 Locate a box to a stud. Install 10-3 with ground wire. Strip conductors to length as shown.

2 Wire the dryer receptacle. Black and red wires connected to the brass screws. White wire connected to the silver screw. Ground wire connected to ground screw.

3 Mount the receptacle to the box and install cover plate.

4 At the main panel, connect the dryer cable wires to a 30-amp circuit breaker designated for the dryer only.

7 Electrical

TYPES OF SMOKE DETECTORS

Though battery-powered smoke detectors are widely used and readily available, most building codes now require hardwired smoke detectors (with battery backup) in a home. In addition, all hardwired residential smoke detectors must be interconnected so that when one alarm sounds, they all sound. If you purchase any smoke detectors, be certain that they have these capabilities. They should also include smoke and carbon monoxide detection for basements with gas- or oil-fired furnaces, and smoke and heat detection for attached garages and basements with electric heat pumps. **Be sure to read the manufacturer's description on the packaging to assure you have the correct type.** Continue to use your battery-powered detectors as a backup system, changeing the batteries at least twice a year.

When a battery runs low, the detector will give a short beep every minute or so. When replacing detector batteries, replace them all at the same time throughout the house or apartment. It is easier to keep track that way.

Smoke detectors have a service life of around 10 years. After that, they should be replaced.

Where to Install Detectors

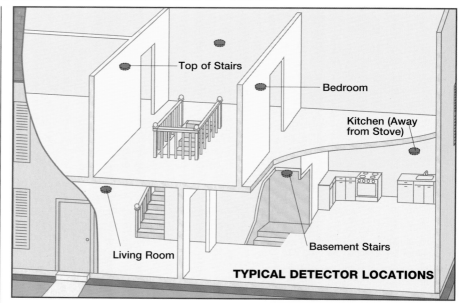

TYPICAL DETECTOR LOCATIONS

Place smoke detectors high on the walls where their mechanisms can sense rising smoke. Ideal locations include the top of stairs and bedrooms. Most codes call for at least one detector on every level of the house.

Wiring Smoke Detectors

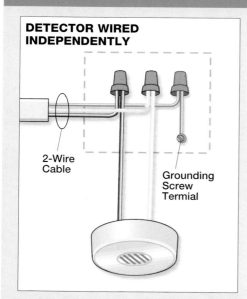

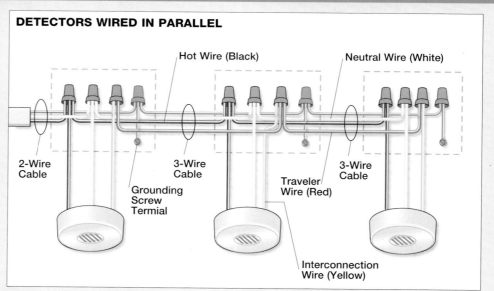

Smoke detectors can be wired independently or in parallel. To wire in parallel, use a three-wire smoke detector. When one alarm sounds, they will all sound.

154 ELECTRICAL / SAFETY DEVICES

SAFETY DEVICES

HARDWARE SMOKE DETECTORS

Install the Ceiling Boxes
Cut ceiling openings, and fish 14/2 or 12/2 NM branch-circuit cable to the first opening in the series. Run the 14/3 or 12/3 NM traveler cable between the remaining ceiling openings, and install ceiling boxes. Pull the cables into the boxes. If they are not self-clamping, use cable clamps to secure the cable in each box. Rip back about 10 inches of the cable sheathing, and cut away the excess to expose the insulated wires. Then strip ¾ inch of insulation off the ends of the wires.

Wire the First Detector
To hardwire the detectors in parallel, first splice the hot black branch-circuit wire to the black traveler wire and the black wire from the first detector's wiring module. Next, connect the white branch-circuit wire (neutral) to the white traveler wire and the white wire from the first detector's wiring module. Then splice the red traveler wire to the oddly-colored wire (in this case, the yellow wire) from the device.

The oddly-colored third wire connects the parallel-wired detectors so that when one alarm sounds, they will all sound. For this reason, be sure to use a three-conductor cable between alarms.

Changing Batteries

Most building codes now require homes to have hardwired smoke detectors. Use battery-powered detectors only as a backup. Hardwired detectors have battery backup for when the power fails, and these batteries should all be changed yearly, ideally on the same day. After around ten years, smoke detectors should be replaced.

Wire the Remaining Detectors
Wire the rest of the detectors by simply splicing together the like-color wires in each box. Then attach each of the detector mounting brackets to the ceiling. Plug the wiring modules into each of the smoke detectors, and mount the foam gaskets and detectors on the ceiling. Turn the power on at the main panel, and test the system by pushing the test button.

There are generally two types of smoke detectors: **photoelectric** smoke detectors and ionization smoke detectors. The photoelectric devices trigger an alarm when a beam of light is redirected to a sensor if it is blocked by the buildup of smoke. When the light hits the sensor, it triggers the audible alarm, which should trigger all the interconnected smoke alarms.

An **ionization** smoke detector monitors the current between two plates that have differential positive and negative electric charges. If the charges vary—as when they are altered or interrupted by smoke in the air—the audible alarm is set off, which should trigger all the interconnected smoke alarms.

Wiring Smoke Alarms in Series

USE: ▶ cable ripper • wire stripper • fish tape (nonmetallic) ▶ smoke detectors • ceiling boxes • 14/2 or 12/2 NM cable • cable clamps • wire connectors

1 In parallel, run two-wire cable into the first detector and three-wire cable between detectors.

2 Splice two-wire cable to the first smoke detector and three-wire cable to the next detector.

3 Wire the remaining detectors; plug in the wiring modules; and mount the detectors.

TYPES OF SMOKE DETECTORS

7 Electrical

WIRING CO DETECTORS

A carbon monoxide (CO) detector is as important to life safety as a smoke detector. Carbon monoxide can accumulate from any number of sources, including clogged chimneys, gas or wood-burning stoves, portable kerosene or gas heaters, or even a car left running in a garage. A carbon monoxide detector is absolutely essential in a home with a gas heating and hot-water system or other gas appliances. It must be relied upon to detect harmful levels of carbon monoxide gas, which is odorless and colorless, yet deadly. Like smoke detectors, carbon monoxide detectors can be battery operated. However, for greater protection, these should also be hardwired, leaving battery-operated detectors as a backup system. Wiring is similar to that for smoke detectors and can be done in parallel.

Underwriters Laboratories Inc. (UL) which tests and rates carbon monoxide detectors (among many other products), has issued UL Standard 2034. It says that home carbon monoxide detectors must sound a warning before carbon monoxide levels reach 100 parts per million over 90 minutes, 200 parts per million over 35 minutes, or 400 parts per million over 15 minutes. The standard requires that the alarm sound "before an average, healthy adult begins to experience symptoms of carbon monoxide poisoning." The warning provides time to evacuate the premises, according to UL.

CO and Gas Appliances

Every home with a gas furnace or other gas appliance should have a hardwired carbon monoxide (CO) detection system.

Hardwiring Carbon Monoxide (CO) Detectors

USE: ▶ cable ripper • long-nose pliers • fish tape (nonmetallic) ▶ carbon monoxide (CO) detectors • wall or ceiling boxes • wire connectors • 14/3 or 12/3 NM cable

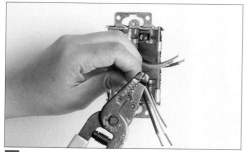

1 Locate the electrical boxes for carbon monoxide detectors on a wall or ceiling. The boxes are standard, non-specialty items.

2 Using 14/2 or 12/2 NM cable, bring power into the first detector box from the main panel.

3 Wire the remaining detectors in parallel, using 14/3 or 12/3 NM cable. One circuit can often serve an entire house.

Installing a Floodlight

USE: ▶ screwdriver • power drill/driver • caulking gun ▶ floodlight • weatherproof fixture box • wire connectors • cable • staples • screws • caulk

1 To bring power to your new floodlight, punch a knockout opening in an existing ceiling electrical box.

2 Extend power to the switch box, stapling the cable as you go, and then continue to the exterior light location.

3 A typical box fitting has a mounting stub for the light fixture. The box must be installed securely to support the lights.

OUTDOOR LIGHTING

OUTSIDE POWER

Outdoor wiring is needed for the same reasons you put electrical lights and outlets inside: to light living space, to make steps and walks safer at night, and to run fixtures and appliances (such as hot tubs). There are also motion-sensor lights that are installed to startle intruders. Installing outdoor wiring is similar to doing it indoors, but it's done with special weatherproof switches, outlets, and light fixtures.

Generally, local codes require that outdoor wiring be protected by rigid metal or intermediate metallic conduit (IMC) whenever it is installed aboveground. Most codes allow buried cable to be Type UF (underground feeder); some require that Type TW (thermoplastic—wet) wire and conduit be used. Always check your local codes before planning an outdoor wiring project.

Installing Outdoor Cable

The most difficult part of installing outdoor wiring is digging trenches for the cable; the longer your cables, the more time-consuming the project will be. Cables need to be buried 18 inches underground (12 inches if the circuit is less than 20 amps and is protected by a GFCI). Weatherproof receptacles can be attached to the side of your house, screwed to a deck post, or left freestanding. (Freestanding outlets must be supported by a footing, like a buried bucket filled with concrete). Outlet boxes are attached to the conduit with watertight compression fittings.

Outdoor Circuits

There are two basic types of outdoor systems. One includes outlets and floodlights and is like your standard house circuits. But the outlets must be waterproof and in many regions must include a built-in circuit breaker (GFCI). Also, wiring run outside the building must be shielded from exposure, typically in metal conduit. The other basic system is low-voltage lighting, typically to light a walkway. These are easy to install, and are run from a transformer plugged into the house.

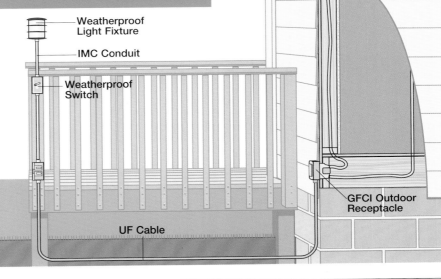

4 *Most fixtures have two lamp holders that swivel to cover a large area. Many also have motion sensors and timer switches.*

5 *Connect wire leads from the lamps to the power leads (and ground) in the box with wire connectors. Caulk the top of the box.*

6 *Mount the box cover, and adjust the lights to suit. Use bulbs rated for outdoor use, even under a roof overhang.*

7 Electrical

WEATHERPROOFING

Outdoor electrical materials and equipment, such as fixtures, electrical boxes, receptacles, connectors, and fittings must be manufactured not only to meet code requirements but to resist the elements. Outdoor electrical equipment must be weatherproof and, in some cases, watertight. For these reasons, you use different materials and equipment for outdoor electrical work than for indoor work.

Outdoor Electrical Boxes

Outdoor electrical boxes are either rain-tight or watertight. Rain-tight boxes typically have spring-loaded, self-closing covers, but they are not waterproof. This type of box has a gasket seal and is rated for wet locations as long as the cover is kept closed. It is best to mount a rain-tight box where it cannot be penetrated by driving rains or flooding. Watertight boxes, on the other hand, are sealed with a waterproof gasket and can withstand a soaking rain or temporary saturation. These boxes are rated for wet locations. Even though an electrical fixture is rated watertight or raintight, this doesn't mean the fixture is waterproof. The fixture or outlet should always be GFCI protected, and if the fixture ever becomes submerged in water, it should be de-energized, and taken apart for inspection when entirely dry.

Outdoor Power

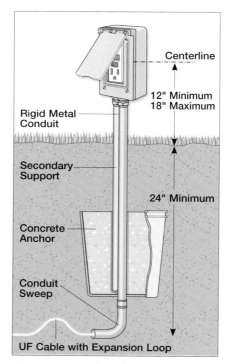

A concrete-anchored, rigid metal conduit may be used as a support on which to mount a weatherproof outdoor receptacle (with secondary support). Check local code for above-grade minimum and maximum height requirements.

Conduit & Boxes

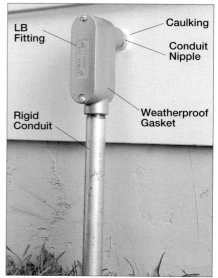

Outdoor cable run underground must be protected in rigid conduit where it enters or emerges from the trench.

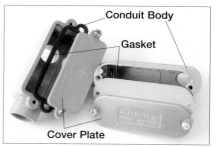

An LB fitting protects cable at the junction of interior NM cable and exterior UF cable.

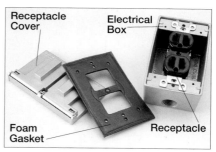

Watertight receptacle boxes are sealed with waterproof foam gaskets. Receptacle covers snap shut over each outlet. Switch covers have watertight levers.

Weatherproof Boxes

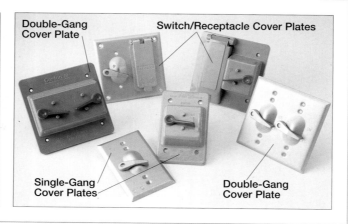

Weatherproof boxes and cover plates are available for single-pole, double-pole, and three-way switches. Covers also exist for switch/pilot lights or switch/receptacles.

OUTDOOR LIGHTING

RECEPTACLES AND SWITCHES

Any receptacles that provide outdoor power for a residential dwelling, even if they are in an outbuilding, must have ground-fault-circuit-interrupter protection [NEC Section 210.8(A)]. Although GFCI receptacles may be used, they tend to nuisance-trip when exposed to the weather. It is better to have your outdoor branch circuit powered by a cable connected directly to a GFCI circuit breaker.

Every residence must have at least one receptacle installed at the front and back of the house. These receptacles must be within 6½ feet of the finished grade [Section 210.52(E)]. In addition, any outdoor receptacle that will be in unattended use, such as one that supplies power to a pump motor, must have a weatherproof box and a cover that protects the box even when the plug is in the receptacle [Section 406.8(B)]. Receptacle covers are available for both vertical and horizontal installations and are either on the device in the box or attached to the box itself.

An outdoor receptacle may be mounted on a wall, post, or any secure location. If you choose to screw a receptacle box onto a wooden post, then be sure that the post is pressure-treated to inhibit rotting. You can also mount a weatherproof electrical box on the end of two ½-inch-diameter sections of galvanized rigid metal conduit that are threaded on one end and anchored in concrete at the other. A two-gallon bucket can be filled with concrete to form the anchor. Burial depths vary across the country.

Weatherproof boxes and covers are also required to protect outdoor switches from exposure to the elements. Covers to single-, double-, and triple-gang boxes operated by toggle levers are available for outdoor switches, and there is also a cover for a combination single-pole switch with a duplex receptacle.

GFCI regulations aren't the only ones that affect outdoor wiring. Conduit types, cable types, and even the depth of the cable are regulated by NEC code, and it applies whether your electrical contractor installs the equipment or you do it yourself.

Cover Plate Types

Receptacle cover-plates for vertical or horizontal boxes may be box- or device-mounted. The cover types can be snap-shut, screw-cap, or flip-top. In environments where rain or blown grit or sand are common, a tight cover can extend the life of the fixture. All of these types are types you may encounter in existing installations, but they are no longer code compliant—when replacing them, you must install the weatherproof type.

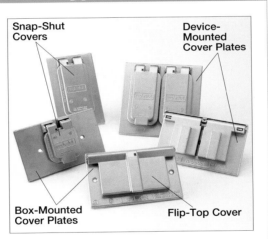

Snap-Shut Covers, Device-Mounted Cover Plates, Box-Mounted Cover Plates, Flip-Top Cover

Weatherproof Box

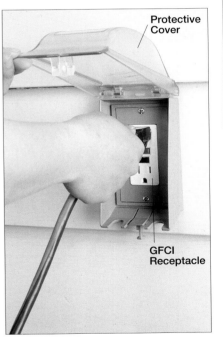

Protective Cover, GFCI Receptacle

Outdoor equipment in constant and unattended use must be connected to a weatherproof box. The cover must protect the box even when the plug is in use.

GFCI Outdoor Power

A freestanding receptacle box supported by rigid metal conduit must be mounted 12 to 18 inches above the ground and have secondary support.

WEATHERPROOFING

7 Electrical

Extending Power Outdoors

USE: ▶ measuring tape • drill with spade bit ▶ 14/2g NM cable • cable staples • mounting bracket • junction box • rigid conduit • LB fitting • conduit compression

1 Mark an access hole to extend power outdoors at least 3 in. from any structural framing. Check other side for obstructions.

2 Using a spade or masonry bit, drill an access hole through the exterior wall or header joist.

3 Mount the junction box over the access opening, and bring the branch-circuit cable into the box.

Installing UF Cable

USE: ▶ round-head shovel • wooden stakes • adjustable pliers • chalk-line box • 6-mil plastic sheeting • mason's string ▶ conduit compression • connectors

1 Stake out a trench, running from the LB fitting to wherever your outdoor box or fixture will be located.

2 Carefully set aside the sod as you dig the cable trench so that it can be replaced when the work is done.

3 Extend a conduit sweep into the trench. Then feed cable through the sweep bend to the junction box.

Installing an Outdoor Receptacle

USE: ▶ insulated screwdrivers • utility knife • multipurpose tool • adjustable wrench • long-nose pliers ▶ rigid conduit • UF direct-burial cable • conduit compression

1 Mount an outdoor receptacle on a wall or freestanding post, or attach it to a deck.

2 Another way to mount an outdoor receptacle is on 2 or more lengths of rigid metal conduit in concrete.

3 To prevent cable from chafing on sharp edges of conduit, attach plastic bushings on the conduit ends.

OUTDOOR LIGHTING

• conduit sweep bend connector (if metal) • conduit nipple • caulking compound • pipe straps • masonry anchors

4 Use a short length of conduit (conduit nipple) to make a connection between the LB fitting and junction box.

5 Connect a length of conduit from the LB fitting into the cable trench. Exposed sections should be in metal conduit.

6 A conduit sweep bend safeguards the cable as it goes underground. A bushing at the end prevents chafing.

and bushings • LB fitting • rigid conduit

4 Splice the NM cable and exterior UF cable inside the interior junction box you installed earlier.

5 When laying cable in a trench, provide an expansion loop. This prevents unnecessary stress on the cable.

6 After refilling a trench, carefully replace the sod that you removed, and gently tamp it back in place.

• receptacle (GFCI if needed) • mounting ears • weatherproof box • conduit sweep bend

4 Because UF cable has a thermoplastic coating applied over the wires, the wires are difficult to strip.

5 Join the hot and neutral wires to the receptacle terminal screws and grounding wire to the grounding screw.

6 Seal the outdoor receptacle box, using the waterproof foam gasket that comes with it.

EXTENDING POWER OUTDOORS 161

7 Electrical

LOW-VOLTAGE LIGHTING DESIGNS

Installing low-voltage lighting is one of the most popular do-it-yourself projects. Because the systems operate on only 12 volts of power as opposed to the 120 volts of standard line voltage, installing a low-voltage system is much safer than working on house wiring. Some manufacturers recommend turning on the power to connect the lights, so you can see the results right away. (Always follow the manufacturer's directions.)

Installation is easier, too. For most systems, plug in a step-down transformer to a standard GFCI-protected outdoor outlet, and run the wires to the light fixtures. Working with low voltage wiring means there is no need to bury wires in conduit or as deeply as standard wiring. Requirements vary, but most manufacturers call for direct burial of a few inches.

The quick installation also means it is easy to change the system by adding new fixtures or moving fixtures to new locations. That is a real plus because landscapes tend to change over time, and a low-voltage system can change with the environment.

Design Basics

Survey your property to identify the best locations for low-voltage lights. Note how moonlight affects your landscape, keeping in mind how you plan to use the property at night. Most homeowners use low-voltage lights to achieve a variety of design goals.

- *Highlighting Focal Points.* Use lights to uplight a distinctive tree or shrub. Or train a floodlight on garden statuary or a water feature. It is best to choose no more than one or two focal points. Choosing too many will make the yard look chaotic.
- *Safety Lighting.* Fixtures installed along paths and driveways and on stairs add a measure of safety to your yard and outdoor living areas.
- *Decks and Patios.* Low-voltage fixtures installed in deck railings or along the perimeter of patios provide good ambient lighting. They help extend the time these areas can be used.

Design Mistakes

Don't over-light. Many homeowners make the mistake of installing too many lights, especially along walkways and driveways. Over-lighting in these areas creates an airport runway look, which is something to be avoided. Place fixtures so that they do not produce glare or shine into your home's windows or the windows of your nearby neighbors.

Create a balanced lighting plan by varying the lighting techniques. For example, try teaming a dramatically uplighted tree with subtle walkway lighting that casts small pools of light.

Pick a System

You will find it helpful to sketch your lighting plan on paper. The perspective a scaled drawing affords you makes it easy to make changes quickly.

Once you have a plan, choose the fixtures you want to use. Home centers and lighting dealers usually have a selection of fixtures and low-voltage systems in stock. Lighting kits usually contain everything you will need. The disadvantage is that you are limited in fixture selection. Fixtures that will be visible should look as good during the day as they do when illuminated at night.

Wiring

When assembling components for low-voltage lighting, determine the length of the necessary wiring runs. The length will tell you what gauge wiring to buy. Unlike line-voltage lights, low-voltage lights experience a drop-off in power the farther away the light fixture is from the transformer, so follow manufacturer's recommendations when buying wiring.

Accessories

One of the most practical accessories for a low-voltage system is a mechanism for switching the lights on and off. Many systems are light sensitive, meaning that they turn on at dusk and off at dawn. An alternative is a system that switches on at dusk and then remains on for a predetermined number of hours.

The Right Wiring

Wire Gauge	Length of System	Maximum Wattage
16	100 feet	150 watts
14	150	200
12	200	250

Lighting Options

Light fixtures installed in steps make the deck safer. Direct lights so that they do not cause glare.

Create focal points with light. Use lighting to draw attention to specific areas of your yard.

OUTDOOR LIGHTING

LOW-VOLTAGE LIGHTING

Because of the low voltage needed to power this type of lighting, it is much safer to use outdoors than lighting powered by a conventional 120-volt line. It is so low, in fact, that a short-circuit in low-voltage underwater lighting would not even be felt by a swimmer. For this reason, it is the ideal type of lighting for in-ground pools. More often, though, low-voltage lighting is employed to light a drive or pathway or to accent landscaping. Lamps for low-voltage lighting commonly range between 25 to 50 watts. You need only install a transformer to step down 120-volt power to 12 volts. Lighting controlled from a transformer can be strung together and connected to fixtures that can then be spiked into the ground along the length of the low-voltage wiring.

Install the GFCI Receptacle

In the outdoor receptacle box, attach the white neutral wire from the NM cable to the GFCI receptacle. Connect the black hot wire from the house branch-circuit cable to the terminal screw on the GFCI receptacle. Then, pigtail the grounding wire from the house branch-circuit cable to the grounding screw on the GFCI receptacle and the box (if necessary). Secure the GFCI receptacle in the box. Install the weatherproof GFCI receptacle cover plate and gasket, drawing the mounting screws tight for a good seal.

Using the attached ground spikes, install your low-voltage lighting system by driving the spikes into the ground. Low-voltage lights can be positioned along a wall, drive, or patio. They may also be used to accent your garden or some other landscaping feature. Lay the low-voltage cable that connects the lights in a trench at least 6 inches deep. Clip the wire leads from each light fixture to the underground cable.

Leaving an expansion loop in the trench, run the low-voltage cable to the transformer box. Connect the low-voltage wiring to the 12-volt step-down transformer. Then mount the transformer below or alongside the GFCI receptacle box you installed previously. Plug the transformer into the GFCI receptacle, turn on the power, and test the system.

Installing Low-Voltage Lighting

USE: ▶ saber saw • screwdriver ▶ 14/2g cable • GFCI receptacle • wire connectors • watertight low-voltage transformer • low-voltage lighting fixtures

1 Make a cutout in an exterior wall for a retrofit switch/GFCI receptacle box. Pull cable into the box.

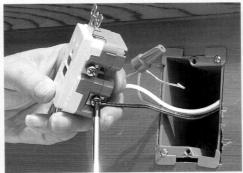
2 Attach hot and neutral wires to the line side in the GFCI receptacle. Pigtail the ground wires to the grounding screw.

3 Position your low-voltage lighting system along a walk, drive, or other landscaping feature.

4 Lay the cable in a trench at least 6 in. deep, and clip the wire fixture leads to the cable.

5 Following the manufacturer's directions, connect the low-voltage wires to the 12-volt step-down transformer.

6 Plug the transformer into the GFCI receptacle; turn on the power; and test the lights.

LOW-VOLTAGE LIGHTING DESIGNS

8 Plumbing

166 BASICS
- What Those Pipes Do ◆ Water Flow
- Anatomy of a Plumbing System
- Cutoff Valves ◆ Traps & Vents ◆ Types of Pipe
- Interior Water Supply

170 TOOLS
- Shopping for Tools ◆ Holding and Turning Tools
- Basic Tools ◆ Clearing Pipes
- Rental Tools ◆ Plumbing Supplies

172 WORKING WITH PIPE
- Supporting Pipe ◆ Cutting Pipe ◆ Materials
- Running Water ◆ Connecting Plastic
- Measuring ◆ Connecting Copper ◆ Making Flared Fittings ◆ PEX with Tubing ◆ Drains & Vents
- ■ LOW-VOLUME TOILETS ◆ Tying into Old Cast Iron ◆ Unblocking a Cleanout

180 SINKS & FAUCETS
- Sink Basics ◆ Fixing a Faucet
- Installing a Sink ◆ Aerators ◆ Sink Fittings

182 FAUCET REPAIRS
- ■ FIXING A SEAT WASHER ◆ Replacing an O-Ring ■ REPAIRING A PACKING WASHER
- ■ FIXING BALL-TYPE FAUCETS
- ■ REPAIRING A SINGLE-HANDLE KITCHEN FAUCET ■ REPAIRING CARTRIDGE FAUCETS
- ■ REPAIRING CERAMIC DISK FAUCETS
- ■ REPLACING A SPRAY ATTACHMENT
- ■ REPAIRING A TWO-HANDLE FAUCET SPOUT ■ REPAIRING A SINGLE-HANDLE FAUCET SPOUT ◆ Reaching Recessed Faucets ◆ Tub & Shower Faucets
- ◆ Scald-Control Faucets

194 INSTALLING SINKS & FAUCETS
- Bathroom Faucet & Drain ◆ Removing a Drain ◆ Water Connections for Faucets
- New Bathroom Faucets & Drains
- Sink-Connection Basics ◆ Vanity Sink Anatomy
- Wall-Hung Sink ◆ Bathroom Sink in Plywood Top
- Replacing a Kitchen Sink ◆ Attaching Drain Fittings to New Sink ◆ Installing a Metal-Rim Sink
- Attaching Faucets ◆ Installing Laundry Sinks
- Connecting the Water ◆ Replacing an S-Trap

206 TOILETS
- How Toilets Work ◆ Replacing a Wax Seal
- ■ UPGRADING A WATER CLOSET
- ■ WATER-SAVING TOILETS
- ■ FIXING A PRESSURE-ASSISTED TOILET
- Trouble-Shooting Gravity-Flow Toilets
- Cleaning a Bacteria-Clogged Toilet
- ■ FIXING A SLOW-FILLING TOILET
- Removing Dirt from the Diaphragm
- ■ REPLACING A TANK BALL
- Fixing a Running Toilet
- ■ REPLACING A FLAPPER
- ■ FIXING FILL VALVES
- ■ REPLACING A FLUSH VALVE ■ RETROFIT KITS
- ■ REPLACING A FILL VALVE ◆ Cleaning a Fouled Air Inducer ◆ Flanges & Gaskets ◆ Leaky Flange Gaskets
- Taking Up and Resetting a Toilet
- Installing a New Toilet

228 TUBS & SHOWERS
- Shower Options ■ FIXING A FAUCET
- Replacing a Tub Drain Assembly
- Replacing a Tub Shower Faucet
- Installing a Cast-Iron Tub
- Installing a Shower

238 WATER HEATERS
- Hot Water ■ WATER HEATER INSULATION
- ■ TANKLESS HEATERS ◆ Safety & Maintenance
- Hot-Water Dispensers ◆ Anode Rods
- Servicing Gas-Fired Heaters
- ■ MAINTAINING THE BURNER
- Servicing Electric Water Heaters

250 PUMPS
- Defying Gravity ◆ Sump Pumps
- Cellar Toilets ◆ Installing a Recirculating Pump
- ■ PUMP PIPING

252 APPLIANCES
- Washers & Dryers ◆ Dishwashers
- Water-Inlet Valves

256 WATER & WASTE TREATMENT
- Getting Better Water ■ SEDIMENT FILTERS
- ■ FAUCET FILTERS ■ REVERSE-OSMOSIS UNITS
- Septic Systems ◆ Waste-Disposal Units

262 SUMP PUMPS & WELLS
- Submersible Sump Pumps ◆ Water Softeners
- Wells ◆ Replacing a Leaking Pressure Tank

266 FIXTURE IMPROVEMENTS
- Replacing a Showerhead ◆ Tub Surrounds
- ■ REGLAZING A TUB ■ RELINING A TUB

268 SOLVING PROBLEMS
- Plumber's Helper ■ ANTI-FREEZE FAUCETS
- Temporary Repairs ◆ Quieting Noisy Pipes
- Drain Leaks ◆ Clearing a Waste-Line Clog
- Clearing

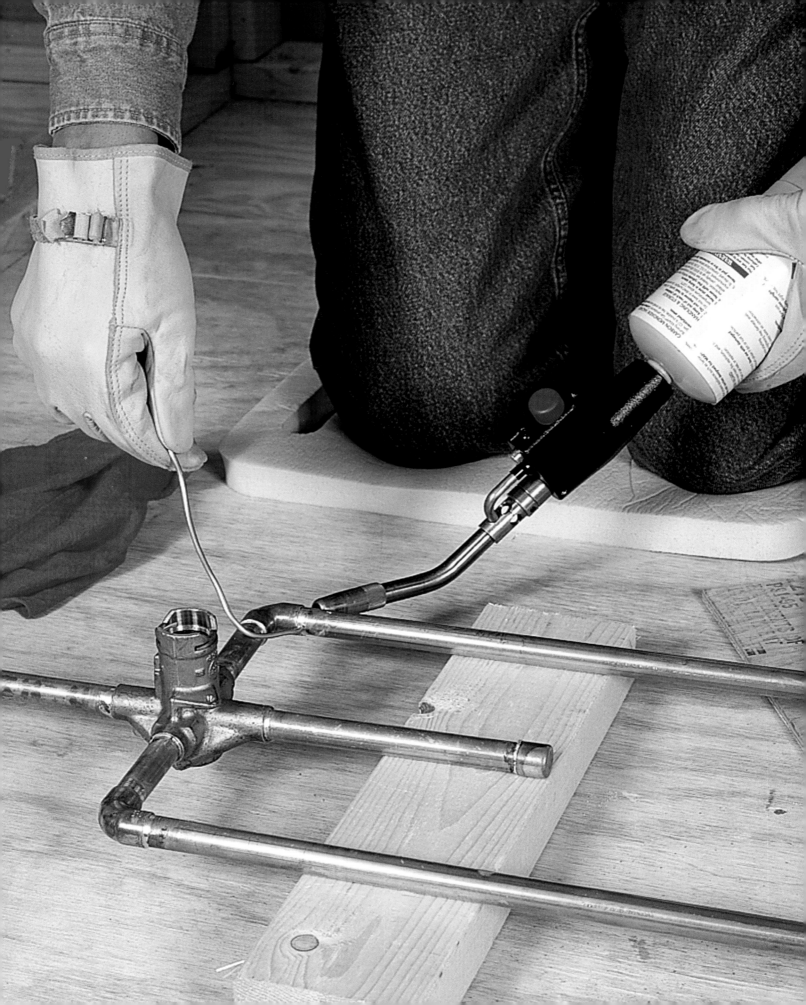

8 Plumbing

WHAT THOSE PIPES DO

If dealing with plumbing problems seems like a pain, remember that before indoor plumbing, any water used for cooking and washing had to be carried into the house by hand, and whatever wastewater remained went out the same way. And then there was the outhouse, where answering the call of nature could mean trekking outside in the rain or dark of night.

Parts of the Plumbing System

A modern household plumbing system has three basic parts: water supply pipes that distribute water (hot and cold) throughout the house; appliances, faucets, toilets, and various other plumbing fixtures that draw water; and the system that carries wastewater out of the house (the DWV, or drain-waste-vent system). Country houses draw water from underground wells, but most houses today are supplied (for a price) by city water pipes. A water meter keeps track of how much is used.

Inside the house, the main supply line splits, with one branch feeding all the cold-water pipes in the house and the other supplying water to the water heater. The hot-water line coming out of the water heater branches, paralleling cold-water lines throughout the house. The two lines supply hot and cold water to sinks, tubs, showers, dishwashers, and all other plumbing fixtures as needed. (Toilets and outside faucets don't need hot-water pipes.)

When water goes down the drain, it enters the DWV system. Part of the system actually channels the water down to the main house drain; the other part consists of pipes (called vents) that rise up out of the drainpipes to the roof. Vents allow in outside air to replace the air displaced by flowing water; otherwise, the negative pressure would suck the water out of your traps. (See "Traps & Vents," opposite.)

Besides supply lines and drainpipes, gas pipes and hot-water heating pipes may also run through your house. Gas pipes are usually steel or copper, and run between the gas meter and a stove or other appliance that uses gas. (Copper tubing may also be used for gas.) Other than shutting a gas cutoff valve in an emergency, you should always call professionals to work on gas pipes. The same goes for heating pipes: steam and scalding water can be very dangerous.

Water Flow

If you use water from a municipal system, the main shutoff valve and the water meter are located where the water-company supply line enters your house, generally along a foundation wall. The water meter keeps track of how many gallons flow into your plumbing system. Most companies meter water by CCFs. One CCF is 100 cubic feet of water, which equals 748 gallons. Most bills also list use in gallons used per day for the billing period.

Water meters are the property of the water utility, who should be called if yours is leaking or not working properly.

Anatomy of the Home Plumbing System

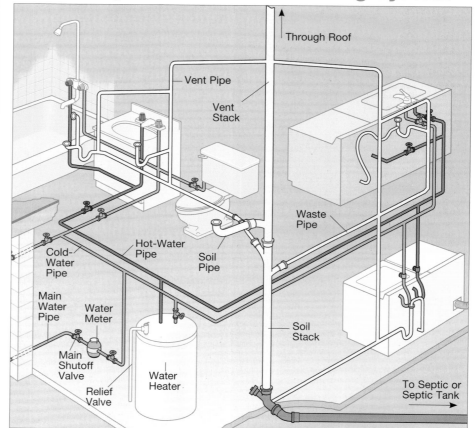

Residential plumbing has three basic components: supply pipes that deliver water (typically copper or plastic), drainpipes that carry away waste (generally plastic), and vent pipes (also plastic) that allow the system to drain freely.

BASICS

Cutoff Valves

The main cutoff valve for your water supply system will be near the spot where the municipal water pipe enters your home or, if you have well water, near the storage tank. Other cutoff valves farther along enable you to turn off just part of the water supply system, allowing you to close a valve to fix a leak in the bathroom, for example, and still have water in the rest of the house. Though sometimes hard to find, there should be valves on supply lines near toilets and water heaters; other fixtures depend on local regulations.

Every water-using appliance or fixture, like this toilet, should be isolated with its own easily accessible cutoff valve.

If you can't find or turn a fixture cutoff and need to stop an emergency leak, use the main valve located near the meter.

Traps & Vents

Most people know about the drainpipe running from each sink, bathtub, and other plumbing fixture in a house. It carries wastewater down to the main pipes of the drainage system. But the drainpipe for each one of those plumbing fixtures also has a trap—the curved section of pipe that you can usually see right below a sink.

The trap's main purpose is to prevent noxious (and potentially lethal) sewer gases from rising up through the drains. Because of their shape, traps always have water in them, which provides an airtight seal. The traps are always near a drain opening, although some may be hidden below the floor (such as those on tubs) or out of sight in the basement.

Each drain must be vented, as well. Vent pipes are usually located near the drain opening after the trap; sometimes multiple drains are served by one vent (called wet venting). Vents connect to a drainpipe and run either directly to the roof or horizontally to another vent. Without them, wastewater flowing down the drain would empty the trap due to siphoning action. Instead, vents allow outside air to flow into the pipe, breaking the siphon action.

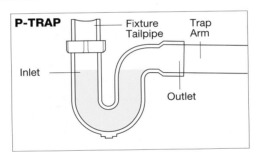

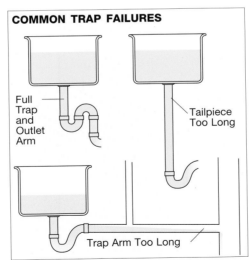

Traps are designed to keep a small amount of water in the drain line. The water effectively seals out sewer gas.

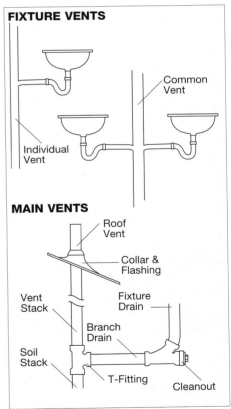

Vent pipes are designed to let water drain freely and to release sewer gas outside the house.

WHAT THOSE PIPES DO

8 Plumbing

Types of Pipe

CHARACTERISTICS	CUTTING TOOLS	JOINING METHOD	
PVC PIPE. Polyvinyl chloride plastic pipe is the preferred drain and vent piping for houses. It can't corrode, and it's easy to assemble.	Wheel cutter, hacksaw, scissor cutter	PVC solvent cement	Drain, Waste, Vent
ABS PIPE. Acrylonitrile butadiene styrene is a black plastic used in the same applications as PVC. It is not as rigid as PVC.	Wheel cutter, hacksaw	ABS or PVC solvent cement	
CAST-IRON PIPE. Once used in drain and vent systems, this durable but brittle metal pipe has been largely replaced by PVC and ABS plastic.	Snap cutter, chisel	Banded neoprene couplings or rubber gaskets	
RIGID COPPER TUBE. The dominant water piping material today, copper pipe is usually joined with soldered (sweat) fittings.	Wheel cutter, hacksaw	Sweat or compression fittings	Water Supply
SOFT COPPER TUBE. Used primarily for natural gas and propane but also for water, this pipe is allowed under concrete.	Wheel cutter, hacksaw	Compression, solder, or flare fittings	
CHROMED COPPER TUBING. This flexible piping is used as fixture water-supply tubes between fixtures and permanent piping.	Wheel cutter, hacksaw	Compression fittings	
FLEXIBLE BRAIDED-STEEL SUPPLY LINE. This flexible piping, often used as fixture supply tubing, is easier to use than chromed copper tubing.	Fixed length	Factory-installed fittings	
CHROMED RIBBED COPPER TUBE. Available only as fixture supply tubing, the ribbed section of this pipe makes it easy to bend.	Can't cut	Compression fittings	
CPVC PIPE. Chlorinated polyvinyl chloride plastic water piping was created to replace rigid copper. Does not meet all local codes.	Wheel cutter, hacksaw, or scissor cutter	PVC cement or compression or crimp-ring fittings	
PEX TUBE. Cross-linked polyethylene plastic pipe is a flexible piping material gaining acceptance for in-house water systems. It requires few fittings.	Scissor tool, hacksaw	Several brands of proprietary fittings	
GALVANIZED-STEEL PIPE. Once used for in-house water systems, steel pipe is now used mostly in repair situations.	Wheel cutter and threading dies	Threaded fittings	
BLACK STEEL PIPE. Steel pipe was once used for in-house gas piping, though it's fast losing ground to soft copper and CSST.	Wheel cutter and threading dies	Threaded fittings	Gas
CSST. Corrugated stainless-steel tubing is a flexible, plastic-coated pipe made of stainless steel for in-house natural gas and propane.	Hacksaw or wheel cutter	Proprietary compression fittings	

PLUMBING / BASICS

BASICS

INTERIOR WATER SUPPLY SYSTEMS

A home's in-house water system starts with the water meter or, in the case of a private well, a pressure tank. If your home has an outdoor meter pit, consider the first full-size shutoff valve in the house as the starting point. From here, a single supply line—the cold-water trunk line—travels to a central location where a T-fitting splits it into two lines. One of these enters the top of a water heater while the other continues to feed cold water to the house. A third line exits the heater and becomes the system's hot-water trunk line. The hot-and-cold-water trunk lines usually run side by side along the center beam of the house, branching to serve isolated fixtures or fixture groups along the way.

In most cases, the trunk lines are ¾ inch in diameter and run under the floor joists, near the beam. The branch lines are usually tucked up between the joists so that most of the basement ceiling can be finished. While trunk lines are typically ¾ inch in diameter, branch lines are usually reduced to ½ inch in diameter at some point. Most codes allow only two fixtures on a ½-inch line, so plumbers will run ¾-inch branch lines until they reach the third-to-last fixture on the run. They also reduce to ½ inch in diameter the dedicated fixture risers that extend directly from the trunk lines.

No Basement?

Where are the water pipes if you live in a house without a basement? It depends on the type of home. If your ground floor spans a crawl space, the piping is likely to resemble that of a basement installation, except that the water heater may be located on the main floor in a utility closet.

If yours is a slab-on-grade home with a concrete floor, expect to find soft-copper water piping buried under the concrete slab. The water service in this case will usually enter a utility room through the floor. In most cases, the fixture supply piping will also be run under the slab, surfacing in the utility room, near the meter and water heater, on one end and near each fixture or group of fixtures on the other. In the extreme southern reaches of the country, where hard freezes are unlikely, copper or plastic water lines may be run in the attic, with branch lines dropping into plumbing walls.

Water Supply Lines

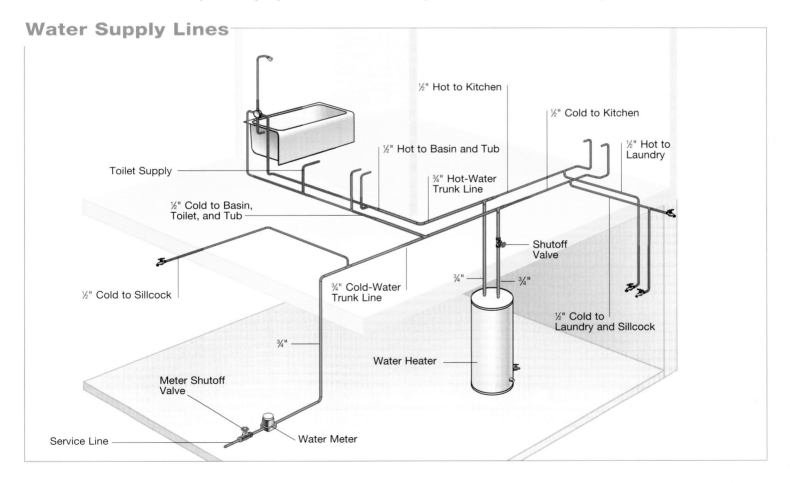

TYPES OF PIPE 169

8 Plumbing

TOOLS

It's easy to get carried away when buying tools, so focus on the basics first. A good rule is to hold off buying specialty tools until you need them. And if you can't imagine needing a large expensive tool more than once or twice, rent it if possible. You're likely to find many large tools, like chain wrenches and oversize pipe wrenches, in the local rental store. And finally, if you need to do special work on piping, many full-service hardware stores will cut, thread, or flare pipes for you in the store.

Of the specialty holding and turning tools shown here, the basin wrench and spud wrench are the real problem solvers. They will save you a lot of headaches, and they are not that expensive to buy.

Good tools can last a lifetime, so buy the best you can afford, especially when it comes to essentials you'll use often: pliers, screwdrivers, adjustable wrenches, and hammers.

Shopping for Tools

The right tools really do make the work easier

Holding and Turning Tools

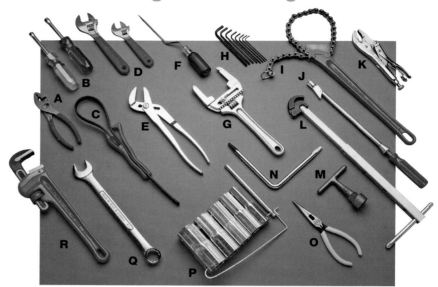

A—slip-joint pliers, **B**—nut drivers, **C**—strap wrench, **D**—adjustable wrenches, **E**—groove-joint pliers, **F**—offset screwdriver, **G**—spud wrench, **H**—Allen wrenches, **I**—chain wrench, **J**—sink-clip (Hootie) wrench, **K**—locking pliers, **L**—basin wrench, **M**—stop-box wrench, **N**—faucet-seat wrench, **O**—needle-nose pliers, **P**—deep-set faucet sockets, **Q**—combination wrench, **R**—pipe wrench

Drill Bits

All drill bits cut holes, but they do it differently, with differing results. The two self-feed bits shown here are for rough-in work, for chewing quick, crude holes through framing lumber. The hole saw, in contrast, cuts a neat, precise hole, so it's better suited to drilling countertops, plastic sinks, and tub-shower surrounds where chipping may be a problem. Speed bits are good starter bits. Spiral bits work best when you need to drill a lot of holes.

A—nut driver, **B**—1-in. self-feed bit, **C**—2-in. self-feed bit, **D**—hole saw, **E**—speed bit, **F**—auger bit, **G**—speed bit, **H**—high-speed bit

Cutting Tools

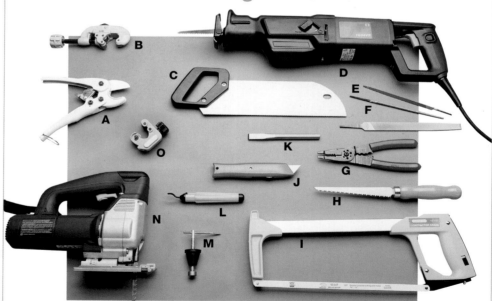

A—scissor-type tubing cutter, **B**—wheel cutter, **C**—plastic-pipe saw, **D**—reciprocating saw, **E**—rat-tail and slim tapered files, **F**—bastard-cut (flat) file, **G**—multitool, **H**—utility saw, **I**—hacksaw, **J**—utility knife, **K**—cold chisel, **L**—plastic-pipe reamer, **M**—faucet valve seat grinder, **N**—saber saw, **O**—miniature tubing cutter

TOOLS

Common Plumbing Tools

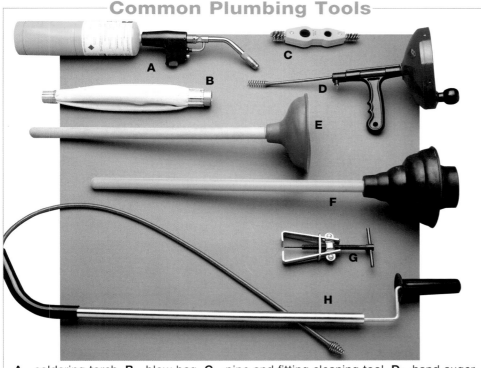

A—soldering torch, **B**—blow bag, **C**—pipe and fitting cleaning tool, **D**—hand auger, **E**—standard plunger, **F**—combination plunger, **G**—handle puller, **H**—closet auger

and, in the end, better. While you can certainly cut water and gas piping with a hacksaw, a wheel cutter is faster and leaves a cleaner pipe edge. This may seem insignificant, but a crooked or ragged pipe end can affect fit and create turbulence as water passes through it. Excess turbulence can erode the pipe wall, greatly shortening its service life.

Similarly, a high-quality saber saw with an auto-scroll feature cuts more accurately and is less damaging to countertop laminates than bargain saws, and a good heavy-duty reciprocating saw makes difficult jobs much easier. And finally, no one who remembers life before cordless drills will ever take them for granted.

Plumbing Tools

Most plumbing tools are very specialized, so they have to earn their keep on plumbing projects alone. Basic tools, like plungers, pipe wrenches, and drain augers, should be part of every homeowner's tool kit. You'll use these items many times over the years.

Rental Tools

A—power drain auger, **B**—steel-pipe cutter, **C**—steel-pipe threading ratchet handle, **D**—right-angle drill, **E**—snap-cutter for cast-iron pipe, **F**—snap-cutter drive wrench

Plumbing Supplies

Plumbing depends on caulks, tapes, and sealants. There are a wide variety of specialty products from which to choose, including the ones shown below, which range from latex tub and tile caulk and pipe-thread sealing tape.

A—latex tub and tile caulk, **B**—PVC solvent cement, **C**—plumber's putty, **D**—PVC primer, **E**—leak-detection fluid, **F**—ABS solvent cement, **G**—silicone caulk, **H**—pipe joint compound, **I**—solder, **J**—flux, **K**—emery cloth, **L**—abrasive pad, **M**—pipe-thread sealing tape (yellow spool: gas, blue spool: water)

8 Plumbing

SUPPORTING PIPE

When you run pipe through stud walls or through floor and ceiling joists, always drill the pipe holes slightly larger than the outside diameter of the pipe. Plastic pipe expands when warm water passes through it, and if the pipe fits too tightly, you'll hear a steady ticking sound when warm water is used. This annoying sound is the pipe rubbing against the wood as the plastic expands and contracts.

When you hang plastic drainpipes and vent pipes under floor joists, support the pipe with hole strapping or the appropriate pipe hangers. Support plastic waste piping at least every 4 feet.

Cutting Structural Timbers

You'll rarely need to cut into load-bearing timbers to install drainpipes. You can usually hang a horizontal fixture line (extending from a vertical stack) under the joists near the center beam of the house and box the area in later. When a drain line needs to travel with the joists toward an outside wall, you can tuck it up between the joists.

You may occasionally need to run pipes through a few joists to maintain an adequate ceiling height, however. This is usually not a problem, but drill only the center one-third of each joist, and where possible, stay within a few feet of a support wall. Never notch the bottom of a joist, because the bottom carries a disproportionate share of the load. Break these rules, and you could threaten the floor's load-bearing capacity.

The best approach is to use a right-angle drill equipped with a self-feed bit. Lacking these, drill four small holes to form a square (or circle), and cut between them using a reciprocating saw. Remember that each succeeding hole must be slightly higher (or lower) than the last to maintain an adequate slope. With 1½-inch pipe, you can often bend it enough to start it through the first two holes. After that, drive it through with a hammer, using a block of wood to protect the pipe. With larger pipes, you'll have to splice short lengths together using couplings.

When installing 1½-inch plastic pipe through stud walls, you can usually bend it between the first two studs and drive it the rest of the way. With an open wall, it's sometimes easier to remove one stud to get it started, and then toenail the stud back in place.

Installing In-Ground Piping

Every plumbing system has some portion of its drainage piping underground, even if it's only the sewer service line. Many houses with basements and all slab-on-grade homes will have soil pipes trenched in place before the concrete goes down. If you build an addition that requires below-grade piping or if you break out some portion of your existing concrete to add piping, you'll need to know the basics.

In-ground piping must be able to support the weight of the soil and concrete above it. This means that the trench must be uniform, with adequate slope, and without extreme high spots or voids. High spots can squeeze a pipe out of round, hindering flow. Voids beneath the pipe can cause sags and breaks.

Running Pipe

Wire hangers, which you just hammer into joists to support pipes, are quick and affordable.

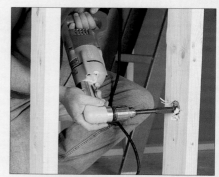

Use a right-angle drill and a large self-feed bit to bore pipe holes in studs.

Installing Pipe through Joists

Drill four holes if you don't have a large bit or hole saw.

Use a saber saw to cut out the lumber between the holes.

Fit short pieces of pipe between joists, and couple them together.

WORKING WITH PIPE

CUTTING PIPE

If you want to take care of plumbing repairs yourself, sooner or later you will have to cut pieces of pipe. For copper pipe, get a pipe cutter, which scores the pipe around its perimeter with a cutting wheel. It is possible to make clean-edged, accurate cuts using a hacksaw, but that can be difficult on existing pipes tucked between wall studs where there isn't much room. Wheel-type cutters work where hacksaws can't go because they are more compact.

Try out the cutter a few times on a scrap piece of pipe to get the idea. First, fit it around the pipe, and tighten the handle until you feel resistance. After you rotate the cutter around the pipe once or twice, you'll feel less resistance as the wheel deepens its cutting groove. Tighten the handle again to make the wheel bite in a little deeper; keep rotating and tightening until the wheel cuts all the way through the pipe. With some practice, you'll get the knack of tightening the handle gradually while you're rotating the cutter. If you find it difficult to rotate, the wheel is biting in too deeply, and you should back off on the handle a bit.

Sometimes even a relatively clean cut needs smoothing around the inner edges, which improves water flow and prevents buildup of mineral deposits on the pipe's inner wall. There are reaming tools for cleaning up burrs; many wheel cutters have fold-out reamers. You can also use carbide sandpaper or a round metal file.

Cutting Plastic & Iron

Cut rigid PVC plastic pipe with a coarse-toothed hacksaw or a fine-toothed carpenter's saw. To ensure a square cut, wedge the pipe firmly against a piece of wood screwed down to your workbench. Pipe already in place can be stabilized for cutting by taping it tightly to a nearby joist, stud, or pipe. The inside edge of cut plastic pipe must be cleaned and smoothed before installation. A razor knife works best.

Small steel pipe (½ to 2 inches) is best cut with a wheel cutter; a coarse-toothed hacksaw also works well. Larger steel pipe is cut with a tool called a chain cutter, which is available at rental stores. Alternatively, you can score a cut in the pipe with a hammer and a cold chisel.

Materials

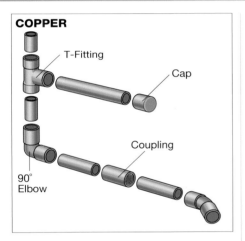
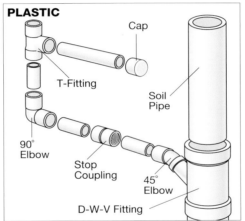
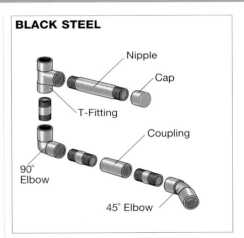

Soldered copper is the material of choice (and of many local codes) for carrying water to fixtures and appliances.

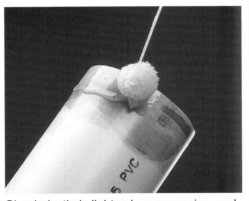

Glued plastic is lighter, less expensive, and easier to install than copper. It is widely used for water supply, drains, and vents.

Threaded black steel pipe is the standard for carrying natural gas but not for water or wastes.

8 Plumbing

RUNNING WATER

Water supply lines, which distribute hot and cold water to all the plumbing fixtures in your house, are much smaller than drainpipes—usually they range from ½ to 1 inch in diameter. The pipe walls have to be reasonably thick, though, because the water these pipes carry is under pressure. (The pressure makes the water flow when you turn on the tap.) Houses built before the 1960s may have galvanized iron water pipes. Newer homes have copper or plastic pipes. Local codes regulate what kinds of supply lines can be used in your area.

Pipes & Fittings

Copper water supply lines inside the house come in two grades: Type M or the thicker-walled Type L. Plumbing-supply stores sell this kind of hard copper pipe in ½-, ¾-, and 1-inch diameters and 10- and 20-foot lengths. Type K copper pipe has the thickest walls and is used for underground water lines.

Plastic supply lines are CPVC (chlorinated polyvinyl chloride), which is rigid and off-white, or PEX (cross-linked polyethylene), which is white and flexible. PE (polyethylene), which is black, is most often used for underground watering systems. Black Schedule 40 ABS (acrylonitrile butadiene styrene) and white Schedule 40 PVC are both used for drain and vent lines.

Fittings, such as elbows and reducers, come in the same materials as the piping. You can also buy fittings for transitions between iron and copper pipe or CPVC and copper.

Connecting Plastic

USE: ▶ fine-toothed saw or tubing cutter • utility knife

1 You can cut plastic pipes (supply lines, drains, and vents) with almost any saw, but a PVC saw makes a cleaner cut.

5 Primer coat the PVC pipe and fitting for the full socket depth for adequate cleaning and primering.

USE: ▶ two-piece flaring tool ▶ copper tube

Measuring

Until you get used to the system, you have to remember to measure the length of pipe between fittings, and add the amount of pipe that rests inside elbows, tees, and nipples. The safest approach is to hold a rough-cut length in place (with copper and plastic), and mark the overall distance from the barrel of one fitting to another. To be sure everything fits, test-fit several sections and fittings. Then, disassemble the pieces, make any adjustments, and solder or glue them together.

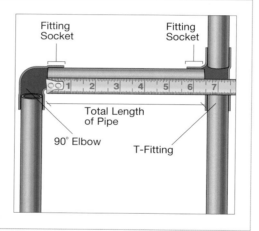

Connecting Copper

USE: ▶ hacksaw or tubing cutter • propane torch • spark lighter • wire brush or sandpaper • flux brush • work gloves • eye protection • rags ▶ copper tube and fittings • solder • flux

1 You can cut copper tube to length with a hacksaw, but a tubing cutter that rotates around the pipe makes a much cleaner cut.

2 Plumbers use a small tool with metal wires inside to brighten copper for the best solder bond. Use sandpaper in a pinch.

3 To brighten the interior surfaces of a connection, use a wire-brush tool, and rotate it several times in the fitting.

WORKING WITH PIPE

• felt pen (for marking) ▶ rigid plastic pipe & fittings • solvent glue • pipe primer

2 When you cut through plastic, even a fine-toothed saw can leave burrs and small shavings. Trim them off with a utility knife.

3 Measure the depth of the fitting socket for proper primer coating.

4 Mark the socket depth on the pipe for proper primer coating.

6 Apply PVC glue over primered area of pipe.

7 Apply PVC glue over the fitting socket. Be sure to read and follow all label cautions.

8 Plastic pipe adhesive softens mating surfaces. They become one when the surfaces harden. You need to work quickly.

Making Flared Fittings

• brass flare fittings

1 Use flared fittings where soldering would be unsafe. The base part of a flaring tool clamps around the pipe. Add flare nut to pipe coil before flaring.

2 The top of the flaring set forces the lip of the pipe against the clamp to create a bell-shaped flare.

3 A flaring nut on one side of the joint threads onto a flaring union on the other. Most codes don't allow flares in walls.

solder • flux

4 To draw solder completely into the joint (even uphill against gravity), coat mating surfaces with soldering paste, called flux.

5 Assemble the connection, and apply heat evenly to the entire joint. Wear gloves or use clamps to handle heated tubing.

6 When the copper is hot enough to melt solder, remove the flame and apply solder around the joint. The hot metal can scorch. Solder both sides of fitting together.

8 Plumbing

PEX WATER TUBING

Cross-linked polyethylene, or PEX, water tubing has been allowed in some locations for 20 years and pointedly disallowed in most others, but with the recent volatility in copper prices, resistance has begun to melt away.

Advantages

PEX has a lot going for it. It's easy to install using crimp and barbed fittings. Water flows through PEX silently, which is a big plus. PEX can take a freeze, and it's impervious to acidic soils, which destroy copper pipes. It will never rust or corrode, internal turbulence is less likely to erode tubing walls, and soft water won't leave traces of copper in your drinking glass.

Some codes allow water-supply installations made entirely of PEX, while others require copper in the initial stages of the system, at water heaters, and at fixture stub-outs.

As most household electrical systems are grounded, at least in part, through the home's metal water piping, an alternative ground needs to be established.

PEX tubing comes in 20-foot sticks and 100- and 300-foot coils. PEX is made in a variety of colors, plus clear, but most companies are gravitating toward red and blue, to signify hot and cold water.

Cutting and Making Connections

PEX tubing comes in a variety of colors and is easy to cut with a plastic tubing cutter.

PEX tubing comes with a variety of conversion fittings, which allow you to tap into existing plumbing materials.

Barbed PEX tubing fittings come in plastic and brass, with brass having greater code acceptance.

This brass conversion is designed to solder into a copper sweat fitting for a neat conversion to PEX.

Tools and Equipment for PEX Tubing

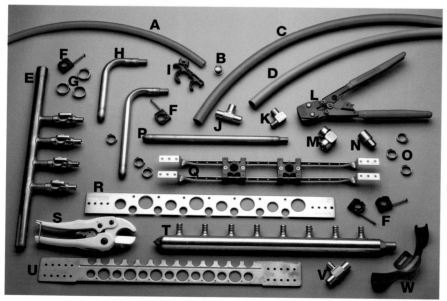

A—½-in. PEX, **B**—plug, **C**—¾-in. PEX hot, **D**—¾-in. PEX cold, **E**—copper manifold with shutoffs, **F**—tubing hangers, **G**—¾-in. crimp rings, **H**—copper stub-outs, **I**—dual tubing hanger, **J**—brass T-fitting, **K**—brass 90-deg. elbow, **L**—crimping tool, **M**—elbow with crimp collar, **N**—sweat x PEX coupling, **O**—½-in. crimp rings, **P**—straight stub-out, **Q**—carriage stub-out bracket, **R**—copper stub-out bracket, **S**—tubing cutter, **T**—manifold without shutoffs, **U**—alternate copper bracket, **V**—¾ x ¾ x ½-in. T-fitting, **W**—90-deg. tubing bracket.

WORKING WITH PIPE

Installing a PEX Manifold System

USE: ▶ crimping tool • hammer ▶ manifolds • copper tubing • PEX tubing • crimp rings • tubing brackets and anchors

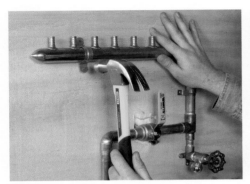

1 A popular approach with PEX tubing is to extend each line from a manufactured copper manifold.

2 Connect a new PEX line to each manifold nipple. In this way, each fixture has its own line.

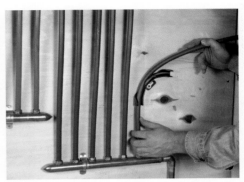

3 To change direction abruptly, use a plastic 90-deg. bracket. Screw the bracket in place, and snap the tubing into it.

4 Secure each feed line with a crimp ring. You will need a proprietary crimping tool for this function.

5 Use plastic tubing anchors to gather the various fixture feed lines. Anchors come with nails installed for quick attachment.

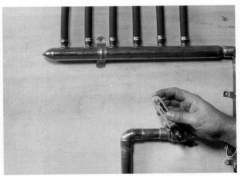

6 Install drain valves below the hot and cold manifolds, and test the system with water under pressure.

Installing PEX Stub-Outs

Use copper stub-outs mounted in a copper bracket. Just nail the bracket across two studs.

Slide two copper x PEX 90-deg. stub-outs into two of the bracket's holes, and solder them in place.

Secure each connection with a crimp ring; then anchor these risers every 3 ft. vertically.

PEX WATER TUBING

8 Plumbing

DRAINS & VENTS

Pipes in the drain-waste-vent (DWV) system rely on gravity to carry liquid and solid wastes downward to the main house drain and out to the underground sewer or a septic tank. Smaller drainpipes called waste pipes, usually at least 1½ inches in diameter, carry wastewater from sinks, showers, and appliances. Soil pipes, which handle solid wastes from toilets and serve as the main house drain, are larger: 3 or 4 inches around. Most drain lines lead to the house's main drain, called the soil stack. (The portion above the highest fixture that leads to the roof is called the vent stack.) The soil stack (and some fixture drains) lead to the house drain.

Both waste pipes and soil pipes have vent pipes that run upward to the roof or horizontal branch lines that tie into a vertical vent. The vents equalize air pressure inside pipes to keep the water from being siphoned out of the traps. Vent pipes tend to be slightly smaller than the drains they connect to—waste drains have, say, 1½-inch vents; soil pipes, 2-inch vents.

Soil pipes and their vents are made from cast iron or plastic. Waste drain lines and vents are usually iron, galvanized steel, or plastic.

Getting Good Drain Flow

If drainpipes always flowed straight downward they probably wouldn't back up as easily as they sometimes do. Instead, they must follow twisting paths inside walls and floors, and there will always be sections that run horizontally, not straight down. In these sections, the pipe still slopes downward, but the proper angle is critical—a ¼-inch downward slope for every foot of horizontal pipe is just right. Make the slope less than that, and water and solid waste flow too sluggishly and may back up. Make it steeper, and the extra slope causes the water to flow faster than the solids, which get left behind and build up in that section.

You can also improve drainage by avoiding 90-degree turns when laying out waste pipes; use two 45-degree elbows instead. If the space is tight, go with a long-sweep 90-degree elbow. Do not use tight-turning 90-degree vent pipe elbows in the waste portion of the drain system. And include Y cleanout fittings where pipes of the same size tie in and then make long horizontal runs to a larger waste pipe.

Low-Volume Toilets

In 1992, the Department of Energy mandated low-volume, 1.6-gallon toilets as a water-conservation measure. But a nationwide survey of builders and homeowners conducted by the National Association of Home Builders found that roughly four out of five experienced problems with low-flush units. Most builders surveyed said that they receive more callbacks on low-flush toilets than on anything else.

There are three common complaints: multiple flushes are needed to clear the bowl; residue remains even after multiple flushes; and they clog easily. New low-flush units work better than the first models, but many builders and owners still have to call in plumbers. And some service calls cost up to $500 due to damage from overflows. To deal with the problems, most people revert to double flushing, which defeats much of the water-conservation potential of the system.

If you have a choice, there are two basic types of low-flush units to choose

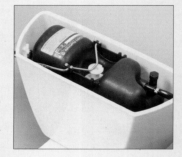

GREEN SOLUTION

from. Gravity-tank toilets, the most common and inexpensive, depend on the siphoning action of water in the tank to flush wastes.

Pressure-tank toilets (pictured) have a secondary container inside. It uses the pressure of water coming into the main tank to compress air and give each flush a pressure assist to push out wastes. This hybrid design is roughly twice the cost of gravity units, and making repairs may be unfamiliar to most homeowners. But many simple adjustments are easy with a little practice.

Low-volume toilets have the potential to save up to 12,000 gallons of water a year in a typical household. And they are required by all plumbing codes. When you buy a new toilet, it will be a low-flow model. And advances in toilet design make newer models much more efficient than those first models as engineers have been making refinements to the design of both types of low-flow toilets.

Tying into Old Cast Iron

USE: ▶ riser clamps (strap hangers for horizontal pipe) • chain cutter • adjustable wrenches • chalk

1 Support heavy cast iron on both sides of the cut with riser clamps. Mount the upper clamp on blocks before cutting.

2 To cut cast iron, rent a chain cutter. Wrap the chain and its cutting wheels around the pipe, tighten, and twist.

WORKING WITH PIPE

Pipe Slope

Proper pipe slope is critical to maintaining efficient drainage and clean, non-clogging waste pipes. As a general rule, use ¼ inch of vertical drop for every horizontal foot of the pipe. Use a taut mason's string to determine where pipe hangers should be positioned before beginning to assemble lengths of pipe. Hang pipes to minimize cutting into joists. Observe local codes covering both incoming water and outgoing wastes.

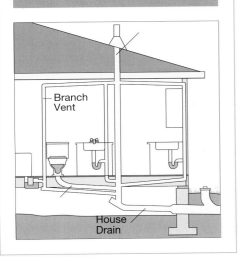

Unblocking a Cleanout

USE: ▶ Stilson wrench • hand or power auger ▶ Teflon tape or pipe dope

1 Use a Stilson wrench to unscrew the cap on a Y-shaped cleanout fitting and gain access for clearing blockages.

2 Feed a plumbing auger into the cleanout extension. Provide ventilation to carry away noxious sewer gas.

3 To clear blockages, most pros use a power auger that extends and turns the snake line. (You can rent one.)

4 To provide a positive seal against sewer gas, add tape or pipe dope to the cap before tightening it in place.

• socket wrench • measuring tape • screwdriver • work gloves ▶ plastic pipe and fittings • rubber gaskets • banded clamps

3 New plastic fittings tie into cast iron with heavy rubber gaskets and banded clamps. Slip the gaskets over each cut end.

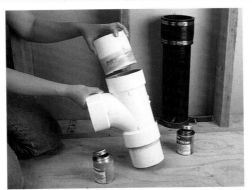

4 Fix short pipe stubs in the plastic fitting; set it in place; and slide the gaskets over the connecting joints.

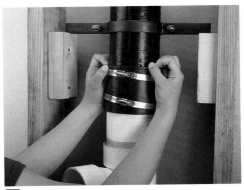

5 Leave the clamps loose so that you can maneuver the gaskets into position, and then tighten down the four clamp rings.

8 Plumbing

SINK BASICS

A sink is a sink, unless it's in the bathroom, in which case it's called a lavatory (from the Latin word meaning "to wash"). Sink or lavatory, the job it does is the same—to provide a basin to receive potable water, retain it as long as you want, and then drain it away when you're done.

If you ever have to buy a sink, you'll find out just how many different kinds there are. Sinks are made from enameled cast iron and steel, cultured marble, stainless steel, vitreous china, and even plastic. Bathroom lavs—round, oval, or square—can be mounted in the countertop of a vanity, hung from the wall, or supported on a pedestal attached to the bathroom floor. Kitchen sinks, often made of stainless steel or enameled iron, are usually mounted in kitchen countertops and may have one, two, or even three bowls.

Countertop-mounted sinks can be attached in three ways. Self-rimmed sinks have a molded lip that overhangs the edge of the hole in the counter and so holds the basin in place. Face-rimmed sinks are secured to the countertop by a metal strip around the basin edge. Unrimmed sinks are mounted to the underside of the counter with metal clips.

Bathroom sinks also have a few features in common. Most have pop-up stoppers controlled by lift rods to keep water in the sink when needed and an opening called an overflow that allows water to flow out of a stopped sink after the water reaches a certain height.

Fixing a Faucet

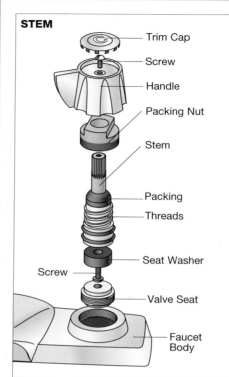

Many houses have stem faucets that screw up and down to open and close the water flow. The most common weak link is the washer at the base of the stem. You need to take the stem out and remove the holding screw to replace it.

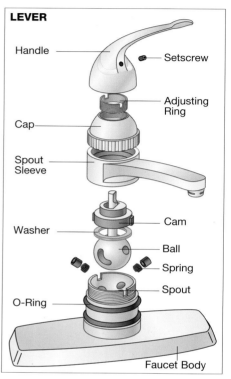

Lever faucets come in several types, such as disk- and ball-types that do not use threads. The common weak link is wear between disks. You have to remove a small setscrew at the base of the lever to remove it.

Installing a Sink

USE: ▶ adjustable wrenches • plumber's putty ▶ sink • faucet • drain kit • tailpiece • trap • drainpipe • supply risers

1 A widespread faucet has an 8-in. spread between valves. Join the control valves to the spout with flexible tubing.

2 Press putty around the drain flange, and thread the flange onto the drain tube inserted through the sink drain opening.

3 Tighten the nut on the drain tube from below, and attach the pop-up lever to the lift rod inserted through the faucet.

SINKS & FAUCETS

Aerators

An aerator screws onto a faucet nozzle to decrease the force and amount of the water flow. If the flow seems sluggish, the aerator may be clogged. If you can't unscrew the aerator by hand, wrap electrical tape around the jaws of pliers, lock them on the aerator, and turn counter-clockwise to loosen it; finish unscrewing it by hand. Flush out deposits from the screen and soak it in vinegar, or use a toothbrush to loosen residue. If that doesn't help, replace it.

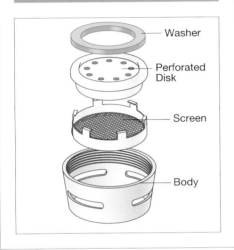

- Washer
- Perforated Disk
- Screen
- Body

Sink Fittings

To work in tight spaces under sinks, you need to use special tools. A spud wrench makes work on sink drains easier, while a basin wrench is indispensable for faucet work—its swiveling head can be wormed into spaces too confining for other wrenches. Plastic sink fittings are easier to work with than iron, copper, or brass, but be careful because plastic breaks more easily than metal. When working with threaded pieces, remember that they usually require only tightening by hand. If you have hard water, avoid mixing copper fittings with iron pipe.

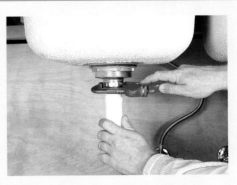

The nut joining the plastic tailpiece to the drain spud is metallic, so tighten it in place with a wrench.

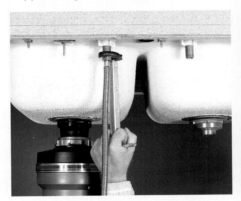

Use a basin wrench to reach between the sink and wall to tighten the supply risers to the faucet.

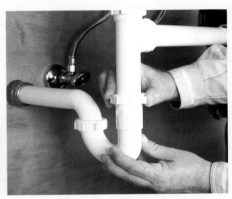

To connect a sink to the drain line, buy a plastic sink-waste kit. Tighten the fittings only hand tight.

4 The trap for a pedestal lavatory is hard to reach later, so attach it to the basin drain before setting the basin.

5 Set the basin on the pedestal, and join the faucet to the valves with chrome supply tubes. Tighten the compression nuts.

6 To connect the trap, slide a compression washer and nut onto the trap arm. Insert the arm into the drain and tighten.

8 Plumbing

REPLACING A SEAT WASHER TO FIX A LEAK

To service the faucet, remove the handles after you turn off the water supply. To gain access to the handle screws, pry under the index caps with a sharp knife, and set them aside. To keep from reversing the hot- and cold-side stems, work on only one side at a time.

Remove the screw, and lift the handle from the first stem.

Stuck Handle

When faced with really stuck handles, you'll have two choices. You can cut the handle with a hacksaw, slicing along one side to release its grip. After overhauling the faucet, install universal handles.

Another option is to buy an inexpensive faucet handle puller that consists of a threaded stem, a T-handle, two side clamps, and a sliding collar. To use the puller, insert the stem into the handle's screw hole until it bottoms out in the faucet stem. Press the side clamps under the handle, and slide the collar down to lock the clamps in place. Then twist the stem in a clockwise direction until you feel the handle break free.

Getting at the Washer

With the handle removed, look for the hex-head bonnet nut that locks the stem into the faucet. Loosen the bonnet nut. If it binds before you can remove the stem, rotate the stem up or down. This should free the nut, allowing you to lift the stem from the faucet.

Expect to find a worn or broken rubber seat washer attached to the stem. Put the handle back on the stem to make it easier to work on. To remove the washer, back the screw off the end of the stem. Carefully pry the washer from the stem using a sharp knife.

Examine the brass seat. (This is also a good time to check the seat for damage.) If the faucet's seat has a raised rim, approximately $1/16$ inch tall, use a flat washer. If the seat is concave, without a pronounced rim, use a beveled washer.

The washer you install needs to fit the stem perfectly, so it's best to take the stem with you to your local hardware store. You can also buy a washer assortment kit. These kits usually include a variety of washer screws as well. Look for a kit with a dozen or more washer sizes.

Installing the Washer

When you've located the right washer, press it into the stem's retainer, and tighten the screw through it. Before returning the stem to its faucet port, lubricate the washer with heatproof or food-grade plumber's grease, available at hardware and plumbing outlets. Grease can sometimes double the life of a washer, especially if the seat is a little rough. Apply a dab of grease to the stem threads and to the top of the stem as well. Lubricating the coarse stem threads keeps the stem operating smoothly, and lubricating the stem keeps the handle from sticking. Don't grease the threads of the washer screws, however. With the new seat washer installed on the stem and lubricated, thread the stem back into its port about halfway. Then thread the bonnet nut into the faucet or over the faucet port. At some point, the stem may cause the bonnet nut to bind. Thread the stem up or down a little to release the bind. Continue tightening the bonnet nut until it feels snug and the stem turns freely. Finally, replace the faucet handle, and repeat the process with the remaining stem.

Fixing a Seat Washer — MONEY SAVER

USE: ▶ utility knife • screwdriver • handle puller • adjustable wrench ▶ repair kit

1 Use an adjustable wrench to loosen the bonnet nut. Turn the stem counter-clockwise as far as it will go and try again.

2 Expose the seat washer and screw. Reattach the handle if necessary, and remove the washer screw.

3 Examine valve seat to determine the correct washer size, and press a new one into the stem retainer.

4 Tighten the washer screw, and coat the washer and stem threads with heatproof faucet grease. Reinstall the stem.

FAUCET REPAIRS

FIXING O-RING STEM LEAKS

Over the years, most faucet and valve companies have gone to O-ring packing seals. The O-ring is held against the stem by a bonnet or stem nut. This nut may have male threads turned into a recessed body port. A large, flat washer made of nylon or a composition material seals the joint between the nut and faucet body. These assemblies are less susceptible to wear than graphite or leather packing washers, but both the O-ring and flat washer can fail. When they do, water appears under the handle, just as it does with older packing assemblies.

Neither O-rings nor flat washers are hard to replace, but you'll need a close replacement match. Take the stem to a well-stocked plumbing outlet.

Replacing O-Ring Packing Seals

To deal with O-ring stem leaks, shut off the water and drain the faucet or valve. Then remove the handle, and loosen the stem nut. Lift the stem from the port, and pull the nut from the stem. (See the photo at right.) Roll or cut the old ring from the stem, and slide an exact replacement in place. If the old O-ring was seated in a groove in the stem, make sure the new ring seats as well. Lubricate the O-ring lightly with plumber's grease. Then press the stem nut back over the stem. As always with O-rings, try to find one that is made or at least recommended by the faucet manufacturer to ensure a good fit.

Although you won't usually need to, it's a good idea to replace the flat washer as a preventive measure, especially if one is included in the O-ring kit. If you find that the flat washer is leaking and don't have a replacement, you may be able to fortify the old washer with a thin layer of nonstick plumber's pipe-thread sealing tape. It works best if you stretch the tape around the washer, lapping it in the direction of the nut's rotation. Drop the washer into the recessed rim of the port, tighten the nut, replace the handle, and turn the water back on.

MONEY SAVER
Replacing an O-Ring

Expose the O-ring by lifting the bonnet nut from the faucet stem.

Roll a new O-ring onto the stem, and lubricate it using plumber's grease.

Turn Off the Water

The first step in servicing any faucet is to turn off the water, usually at the shutoff valves under the sink, and open the faucet to relieve any remaining water pressure in the supply lines. If you don't find valves under the sink, shut off the water at the meter (municipal water) or pressure tank (private well). And if you have plumbing fixtures on a floor or two above the sink on which you're working, open all upstairs faucets and drain the system.

Repairing a Packing Washer

USE: ▶ screwdriver • adjustable wrench • packing materials

MONEY SAVER

1 To repair a leak coming from the area of the packing washer, shut off the water, and remove the handle and bonnet nut.

2 Pry off the old graphite packing washer using a flat-blade screwdriver, and slide a new washer onto the stem.

3 Wrap packing string around the stem to finish the job. You can instead add packing string to the packing in an old faucet.

8 Plumbing

FIXING BALL-TYPE FAUCETS

To gain access to a ball-type sink faucet, turn off the water at the shutoff valves and tip up the handle. Loosen the Allen screw in the lower-front section of the handle, and lift off the handle. Where the handle had been, you'll find a large chrome cap with either wrenching surfaces or a knurled rim. If the nut you see has wrenching surfaces, loosen the nut with smooth-jaw pliers or an adjustable wrench. If the cap has a knurled rim, use either the Delta wrench that comes with each repair kit or large adjustable pliers padded with cloth or duct tape.

With the cap removed, lift the nylon and neoprene cam that covers the top of the ball. Then remove the ball. Set both aside. Reach into the faucet body, and using the Allen wrench or a small screwdriver, lift the cold-water rubber seal and its spring from the inlet port. Then lift the hot-water seal and spring. There's little noticeable difference between old and new seals and springs, so it's easy to get them mixed up. Throw out the old ones immediately. If you plan to replace the cam, discard it as well. If your faucet is more than 10 years old or has dripped for several months, replace the ball, too.

Reassembly

Assuming you'll be replacing everything in the faucet except the ball, press each rubber seal onto its spring. Slide the seal and spring onto an Allen wrench or screwdriver, with the seal facing up. With an index finger holding the assembly in place, insert the spring and seal into the inlet. Install the remaining seal and spring in the same way.

With the new seals installed, press the ball into the body. The ball will have a peglike key on one side that matches a slot in the body, so there's no chance you'll get it wrong. Press the new cam cover over the ball, and align its key with the keyway on the faucet body. Push it down until the key engages, and then thread the cap over it. Tighten the cap until it feels snug. Replace the handle, and test your work. If the faucet drips or water appears around the handle, remove the handle and tighten the cap a little more.

Repairing Ball-Type Faucets — MONEY SAVER

USE: ▶ screwdriver • Allen wrench • groove-joint pliers ▶ ball-type faucet repair kit

1 To reach the handle screw, tip back the handle. Insert an Allen wrench or faucet tool, and remove the screw.

2 Loosen the cam nut to gain access to the ball assembly. Delta faucets have slotted nuts.

3 Pearless units have wrenching surfaces. So a factory wrench fits into existing slots.

4 Lift the plastic cam to expose the ball assembly below. Plan at least to replace the cam and the faucet seals.

5 Lift the ball from the faucet body, and set it aside. Some kits come with replacement balls and some do not.

6 Use an Allen wrench or thin screwdriver to remove the rubber seals and springs. Replace them all.

FAUCET REPAIRS

REPAIRING A SINGLE-HANDLE KITCHEN FAUCET

Begin by turning off the water at the shutoff valves. The faucet handle may or may not have a chrome or plastic index cap. If it does, pry under the cap with a utility knife to gain access to the handle screw. If it doesn't, like the one shown in the photo, just pull off the cover.

Remove the screw from the handle, and tip the handle up and back. The handle's cam slot fits into a deep groove in the pivot nut, so expect to have to wiggle and coax it a bit. When the lever clears the pivot nut, lift it and its plastic hood from the faucet column. Loosen and remove the threaded pivot nut to reveal the top of the cartridge. Looking closely, you'll see that the cartridge is locked in place by a small U-shaped clip, positioned horizontally across the top of the cartridge. Use needle-nose pliers or a screwdriver to remove this clip. Then grasp the cartridge stem, and pull straight up (with pliers if need be): it will break free and come out.

Replacing the Cartridge and Handle

To make the repair, insert a new cartridge into the port, and press it down as far as it will go, aligning the flat notches in the stem with the brass-body slots. Insert the retainer clip. If it won't go in all the way, rotate the stem to correct a slight misalignment. Push the clip into its slot until it bottoms out. Then, thread the pivot nut back onto the column and replace the handle.

When replacing the handle, the cam opening must engage the groove of the pivot nut. If it doesn't, the handle won't operate through its full range. You'll barely be able to turn the water on. To avoid this, tip up the handle as high as it will go within its plastic hood. Carefully engage the back of the lever in the pivot nut's groove. When you feel it engage, press the handle down; install the stem screw; and replace the decorative cover. Turn the water back on. If you find that the hot water is now on the right side, remove the handle and rotate the stem 180 degrees.

Repairing a Kitchen Cartridge Faucet — MONEY SAVER

USE: ▶ screwdriver • needle-nose pliers • adjustable wrench ▶ new cartridge

1 To get access to the handle screw, lift the decorative cap from the column. If there's no lift-off cap, pry up the index cap.

2 Remove the handle screw using a Phillips-head screwdriver. The screw is threaded into the stem of the cartridge.

3 Lift the handle and hood from the faucet to reveal the pivot nut. The hood covers the top of the cartridge.

4 Use an adjustable wrench to remove the pivot nut. Rotate the nut counterclockwise to unscrew it from the faucet body.

5 Using needle-nose pliers, withdraw the retainer clip to remove the cartridge. Be careful to avoid scratching the finish.

6 Lift out the cartridge by the stem. Keep the old one as a reference for the replacement part.

FIXING BALL-TYPE FAUCETS

8 Plumbing

REPAIRING TWO-HANDLE CARTRIDGE FAUCETS

Many manufacturers offer two-handle cartridge faucets, usually as low-cost alternatives to their single-control models.

To fix a leaking faucet, shut off the water and pry the index cap from the handle. Remove the handle, and loosen the cartridge nut using an adjustable wrench, smooth-jaw pliers, or groove-joint pliers with the jaws wrapped in duct tape. Lift the original cartridge from the faucet body; throw it away; and stick a new one in its place. Restore the nut and handle. Be sure to work on only one side at a time to keep from accidentally reversing the cartridges. When you're finished, turn on the water to test your work.

Most dual-handle cartridge faucets have disposable cartridges similar to those just described. But some have spring-loaded rubber seals in the inlet ports like those you find in ball-type single-handle faucets. If upon lifting the cartridge you notice these accessible seals, keep the cartridge and replace only the seals and springs.

If you find that your faucet cartridges are hard to remove, hard-water calcification over the years may have stuck them in place. Some faucet manufacturers offer cartridge extraction tools, but it also helps to pour warm vinegar into the faucet port and around the cartridge. Give the vinegar a few minutes to work, and try pulling the cartridge again. Repeat if necessary, and give the vinegar time.

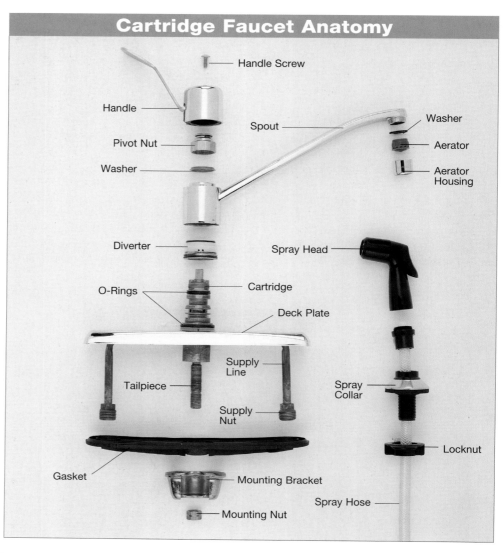

Cartridge Faucet Anatomy

Repairing a Two-Handle Cartridge Faucet

USE: ▶ utility knife • screwdriver • pliers • plumber's grease ▶ new cartridge

MONEY SAVER

1 Use a knife to pry the index cap from the handle. Then remove the retaining screw, and lift off the handle.

2 Using an adjustable wrench, turn the chrome retaining nut counterclockwise to remove it from the cartridge.

3 Lift out the old cartridge, and install an exact replacement. But first lubricate the O-rings with plumber's grease.

FAUCET REPAIRS

REPAIRING A SINGLE-HANDLE CARTRIDGE

To repair a cartridge faucet for the bathroom sink, turn off the water at the shutoff valves and tilt back the handle. Using a utility knife, pry the red-and-blue index cap from the underside of the handle. Loosen the screw under the cap with an Allen wrench, and set the handle aside. With the handle removed, you'll see a brass or cast-metal cover screwed in place. Remove the screw and cover to reveal a plastic (usually gray) retaining ring. Rotate the ring counter-clockwise until it lifts off, revealing a white nylon retaining nut. Pull the U-shaped (usually brass) retaining clip from the back of the nut, and remove the nut. Pull the cartridge from the faucet body using slip-joint pliers. Replace the entire cartridge, and assemble the remaining components in reverse order of removal.

If you can't get the cartridge out of the faucet, pour vinegar around the assembly (to loosen minerals). If that doesn't work, buy a cartridge-extraction tool from a plumbing-supply store.

Getting Rid of a Faucet Sprayer

If the faucet spray attachment doesn't work properly and you'd rather be rid of it entirely, you can remove it and in the process free up the sink deck hole for a soap dispenser or hot-water dispenser. You'll have to close off the faucet nipple with a threaded cap, however. (Some nipples are threaded inside as well, so you can buy a threaded plug to close it off instead.) To eliminate the hose spray, remove it; apply pipe joint compound to the faucet nipple threads; and tighten the plug or cap onto the nipple.

Repairing a Single-Handle Cartridge Faucet   MONEY SAVER

USE: ▶ utility knife • allen wrench • screwdriver • slip-joint pliers ▶ new cartridge

Many single-handle cartridge sink and tub-shower faucets require an extra step before you can remove the retaining clip. With the handle off the faucet, you may see a stainless-steel sleeve installed over the cartridge and column. This sleeve is decorative, but it also keeps the clip from backing out. Pull the sleeve from the column; remove the clip; and replace the cartridge. As is the case with many single-handle cartridge faucets, reversed hot and cold sides can be corrected by rotating the stem 180 degrees.

1 Pry out the decorative index plug using a utility knife, and remove the handle's Allen setscrew.

2 Loosen the fastening screw that secures the metal cover to the top of the faucet. Remove the screw and cover.

3 Rotate counterclockwise, and remove the ring to expose the nylon nut and cartridge retaining clip.

4 Pull out the U-shaped clip at the back of the assembly to free the nut and cartridge. Remove the nut, and set it aside.

5 Pull out the old cartridge. If it won't budge and minerals don't seem to be a problem, buy a cartridge-extraction tool.

8 Plumbing

REPAIRING CERAMIC-DISK FAUCETS

Ceramic-disk faucets are particularly vulnerable to sediment accumulations. For this reason, don't assume that a dripping faucet needs a complete overhaul. When a ceramic-disk faucet develops a steady drip, remove the aerator and move the handle through all positions several times. If sediment was the culprit, this should clear it. In general, a ceramic-disk faucet is not a good choice if you experience sediment problems with your water, especially if they are so severe that you require a filter.

Fixing a Leak

If you have a newer-style ceramic-disk faucet and you can't seem to clear the sediment by rotating the handle, you'll need to check the cartridge. Shut off the water; tip back the handle; and loosen the setscrew. Remove the handle, and lift off the decorative cartridge cap. Use a small flat-blade screwdriver to remove the retaining screws. Then lift the cartridge from the faucet. If you see sediment in the inlet ports, clean it out using tweezers. You can also remove the neoprene seals (visible in photo 3 below) to look for sediment. Clean them if you find sediment, but if you don't find any, the problem is likely in the cartridge. Most ceramic cartridges are not serviceable, so don't waste time looking for parts; simply replace the cartridge, and reinstall the handle.

Don't Shatter the Ceramic

Ceramic disks are extremely durable, but they have a weakness. When you drain a piping system for repairs and then turn the water back on, the air in the system escapes through the faucets in bursts and surges. These pressure shocks can shatter a ceramic disk, preventing the faucet from shutting off completely. To avoid ruining your ceramic-disk faucet, turn the water back on slowly after a plumbing repair. Allow the air to be pushed out gradually before moving to a full-open position.

Repairing Ceramic-Disk Faucets

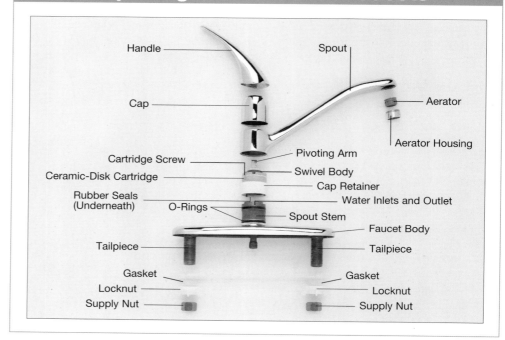

Fixing a Leak in a Ceramic-Disk Faucet

USE: ▶ Allen wrench • flat-blade screwdriver • groove-joint pliers • tweezers ▶ new cartridge

MONEY SAVER

1 To locate the handle screw, shut off the water and tip the handle back. Use an Allen wrench to remove the screw.

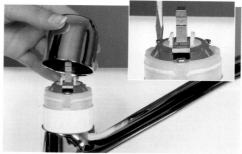

2 Lift off the decorative cap to expose the cartridge. Using a flat-blade screwdriver, loosen the screws at the top of the unit.

3 Look for sediment near either or both of the inlet ports. Clear the sediment and clean the seals, or replace the cartridge.

FAUCET REPAIRS

REPAIRING TWO-HANDLE CERAMIC-DISK FAUCETS

Turn off the water at the shutoff valves below the sink. Remove the first handle, and use a screwdriver to remove the cartridge cap. Use an adjustable wrench or smooth-jaw pliers to loosen the retaining nut. Lift out the ceramic-disk cylinder, and install an exact replacement.

How to Replace a Spray Attachment

When stopgap measures no longer work, it's time to replace the entire spray assembly. It's best to buy a replacement made by the manufacturer of your faucet, but universal kits will also work. Start by shutting off the water and draining the hose as much as possible. Then reach under the sink, and cut the old hose in two, catching the small amount of water that drains out. Pull the old hose through its deck fitting, and discard it. Then, from under the sink, unscrew the remaining length of hose from the faucet nipple. Remove the jamb nut from the deck fitting under the sink, and remove the fitting.

With the old assembly removed, install the new deck fitting, and then feed the hose through it from above. Apply a small amount of pipe joint compound to the male threads of the faucet nipple, and thread the new hose in place using an adjustable wrench or smooth-jaw pliers. Don't overtighten the fitting.

Removing Sediment

Soak a mineral-hardened spray nozzle in warm vinegar to dissolve calcified minerals and enable it to work properly.

Repairing a Two-Handle Ceramic-Disk Faucet — MONEY SAVER

USE: ▶ utility knife • screwdriver • adjustable wrench • slip-joint pliers ▶ new cartridges

1 Remove the first handle and use a screwdriver to undo the cartridge cap. Work on only one cartridge at a time.

2 Use an adjustable wrench to loosen the cartridge retaining nut. Rotate the nut counterclockwise to remove it.

3 Lift out the old cartridge using slip-joint or groove-joint pliers, and install an exact replacement. Replace both sides of the faucet.

Replacing a Spray Attachment — MONEY SAVER

USE: ▶ screwdriver • pliers • adjustable wrench • spray replacement kit

1 Slide the rubber gasket onto the new one's fitting shank, and feed the shank through the sink deck hole.

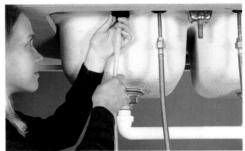

2 Slide the hose through the deck fitting from above, and install the jamb nut below. Tighten the nut until it's snug.

3 Connect the spray attachment's hose to the faucet's diverter nipple. Use pipe joint compound on the nipple.

8 Plumbing

FIXING FAUCET SPOUT LEAKS

Leaks frequently appear around the base of the faucet spout and from under the spout collar, and they should be fixed as soon as possible. If allowed to leak for long, the water that accumulates around the base of the faucet can work its way under the baseplate. From there, it can drop into the cabinet or seep into the countertop, where it can cause real damage.

Repairing Two-Handle Faucet Spouts

To start, turn off the water at the shutoff valves. Most modern faucets have O-ring spout seals. If your faucet has two handles, the spout is probably held in place by a threaded retaining nut, either exposed or hidden beneath a decorative cap. An exposed nut may be smooth or ridged around the perimeter. Remove this kind using a strap wrench or a pair of groove-joint pliers with thick padding on the jaws. A hidden nut may be a hex type. Remove it using an adjustable wrench or smooth-jaw pliers. With the retaining nut removed, lift and twist the spout until it slips off. With the spout out of the way, you will see one or more O-rings on the spout stem, which act as water seals. The O-rings are probably worn or broken. Cut them out using a utility knife, and install replacements, lubricating the new O-rings with heat-proof plumber's grease. Then reinstall the spout, and turn on the water.

Clearing Buildup

Mineral-encrusted aerators are easy to unscrew from the faucet spout for cleaning or replacement.

MONEY SAVER

Soak the scaled-over aerator parts in vinegar, and clear the screens with a straightened paper clip.

Repairing Single-Handle Faucet Spouts

If yours is a single-handle faucet, you'll need to remove the handle and a retaining cap nut. Shut off the water. If the faucet handle has an index cap or decorative cover, remove it, back out the screw holding the handle, and remove the handle. Many faucets have an Allen setscrew holding the control lever/handle in place. If yours does, tip the handle back and loosen this screw. If you find a knurled cap nut just under the lever, unscrew it as well. If you find a pull-off decorative cap, expect to find a retaining nut just under the cap. Loosen the nut, and pull the spout from its post.

With the spout removed, look for two or three rubber O-rings seated in grooves in the post. Pry or cut these seals off without scratching the post. Buy factory replacements from a local plumbing-supply retailer, and roll the new O-rings onto the post until they become seated. Finally, lubricate the O-rings with plumber's grease, and carefully press the spout collar over the post. Replace the fastening nut or cap and the handle, and turn the water back on.

Reaching Recessed Faucets

Full-skirt escutcheons or decorative chrome trim plates are indicators of deep-set faucets. You have to remove the escutcheons to reach

Repairing a Two-Handle Faucet Spout

USE: ▶ strap wrench • utility knife ▶ spout O-ring kit

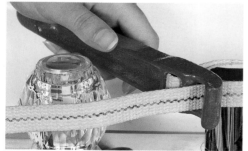

1 Use a strap wrench or smooth-jaw pliers, padded with cloth or duct tape, to remove the spout cap.

2 When the cap breaks free, finish unscrewing it by hand. If it's caked over with minerals, soak it in warm vinegar.

MONEY SAVER

3 Grip the spout near its base, and pull up on it. Cut off the old O-rings, and install and lubricate new ones.

FAUCET REPAIRS

the bonnet nuts, which hold the faucet stems in place. Escutcheons are easy to remove once you figure out the fastening mechanism. (See "Escutcheon Alternative," on page 192.)

To begin, remove the handles. Working on one side at a time, pry off the index cap to reach the handle screw. If the cap has knurled edges, it is most likely threaded. Unscrew it using needle-nose pliers. Once you have access to the handle screw, back it out and remove the handle. If you see a knurled stem nut, remove it using an adjustable wrench, smooth-jaw pliers, or groove-joint pliers padded with cloth fabric or duct tape. (If you don't see a nut, the escutcheon is attached to the faucet stem. See "Escutcheon Alternative," page 192.) If the escutcheon won't let go, check to see whether it's caulked to the wall tile. If it is, slice through the caulk with a utility knife to free it and then try to back it off.

The bonnet nut rests beneath the plane of the finished wall. Only a deep-socket faucet wrench will reach it. You can rent professional versions of these socket wrenches, but homeowner versions are also available at modest cost from hardware stores and home centers. Once you've got past the escutcheons and bonnet nuts, the mechanism is not different from other ordinary compression faucets. See "Fixing a Seat Washer," page 182, for more information about working with compression faucets.

Compression Faucet Repair

If you repair a faucet soon after it develops a drip, you can make a lasting repair with a 5¢ washer or two. But if you allow a faucet to drip for several months, the leak can do real damage. Pressurized water has incredible force. When allowed to seep past a defective washer, it actually cuts a channel across the rim of the brass seat. Stick a finger into the port of a relentlessly leaky faucet, and you'll feel the pits and voids in its damaged seat. If the seat feels smooth, leave it. If not, replace it. If the seat is not removable, grind it smooth with a dressing tool. A pitted seat can chew up a new washer in a matter of weeks.

Repairing a Single-Handle Faucet Spout — MONEY SAVER

USE: ▶ screwdriver • adjustable wrench • utility knife ▶ O-ring kit

1 Start by removing the handle and retaining nut or pivot nut under it. There's no need to remove the cartridge.

2 Grip the spout near its base; then twist and lift it off the faucet body. The spout should come free with steady pressure.

3 Carefully cut the old O-rings from the spout stem, and roll new ones in place. Lubricate the rings with plumber's grease.

Reaching Recessed Faucets

USE: ▶ needle-nose pliers • adjustable wrench • deep-socket faucet wrench • screwdriver ▶ faucet washers

1 Remove the handle index cover (inset) and screw; then remove the escutcheon nut from the stem or unscrew the escutcheon.

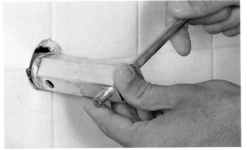
2 Press the deep socket wrench onto the recessed bonnet nut, and remove the nut and stem from the faucet body.

3 If you find a stem with a floating seat (an older design), unscrew the seat from the stem, and replace the washer.

8 Plumbing

TUB & SHOWER FAUCETS

Cartridge-type tub and shower faucets with single-handle controls have mechanisms similar to those used in their kitchen and bath counterparts, and the cartridges are held in place by similar U-shaped retaining clips. Gaining access is a little different, however.

Turn off the water at the shutoff valves. Pry the decorative insert from the plastic handle, and remove the handle screw and handle. In some models, a plastic cam fitting or bushing will be in place between the handle and stem. Remove it as well.

With the handle off, you'll see a metal sleeve installed over the cartridge and column.

Escutcheon Alternative

Deep-set escutcheons use two mounting designs. In one, a knurled nut under the faucet handle holds the escutcheon in place. In the other, the escutcheon has internal threads and is turned directly onto the faucet stem. If you don't see a nut, grip the escutcheon to remove it.

The sleeve is decorative, but it also keeps you from being able to remove the retaining clip. Pull the sleeve from the column. With the sleeve removed, you can pull the retaining clip using needle-nose pliers. Slide the cartridge out from the column, and replace it. Reassemble the faucet, and turn the water back on. As with other cartridge units, you can usually correct reversed hot and cold sides by rotating the stem cartridge 180 degrees.

Fixing Single-Handle Tub and Shower Faucets — MONEY SAVER

USE: ▶ utility knife • screwdriver • needle-nose pliers ▶ replacement cartridge

Some faucet handles have a separate bushing between the handle and stem. Be sure not to damage or lose it during the repair procedure. You may want to use masking tape to cover the polished plates. It takes just a minute to prevent what would be a lifelong unsightly scratch. An alternative to masking tape is a line of "paint-on" scratch protection systems. These paint-on systems provide a thick rubbery layer of scratch protection that you peel off later. They can be applied to nearly any working surface.

1 To gain access to the handle screw, start by removing the handle's decorative index cap. Pry under it with a knife.

2 Use a Phillips-head screwdriver to remove the handle. Be careful not to lose the plastic inset bushing.

3 Carefully pull the decorative stainless-steel inner sleeve from the valve body and trim plate.

4 Use needle-nose pliers to pull the U-shape retaining clip. Be careful not to bend it or you'll have difficulty replacing it.

5 Pull the old cartridge and install an exact replacement. Replace the retaining clip, inner sleeve, and handle.

FAUCET REPAIRS

WORKING WITH SCALD-CONTROL FAUCETS

Scald-control faucets have long been installed in hospitals and nursing homes, and in the past few years many codes have been updated to require scald control for residential bath and shower faucets. It takes only a couple of seconds of exposure to 140° F water to produce a third-degree burn—and only one second at 150° F. Many homeowners have their water heaters set that high. (A 125° F setting is safer, and your heater will give you longer service at that setting.) Small children, the elderly, and anyone with limited mobility are at greatest risk.

Every manufacturer now makes affordable scald-control faucets for residential use. Most are single-control faucets, but there are a few two-handle models.

Handle-Rotation Stop

Scald control is delivered in two ways. First is a temperature-limit adjustment, in the form of a handle-rotation stop. With a handle stop, you remove the handle and dial in a comfortable water temperature, with the water running, and then lock the setting and replace the handle. Thereafter, when you turn the handle to "Hot," it will rotate only to the stop position. You can reduce the temperature with cold water, but you can't exceed the hot limit. Because ground temperatures change with the seasons, affecting water temperatures, you may need to adjust these settings twice a year.

Pressure-Balance Spool

The second mechanism, a pressure-balance spool, is designed to accommodate a sudden drop in pressure on one side of the piping system. Pressure drops are common. The most familiar scenario: you're taking a shower when someone in an adjacent bathroom flushes the toilet. The toilet diverts half the line pressure from the cold side of the faucet, upsetting the ratio of hot-to-cold water. The result is a sudden blast of hot water.

Balance spools come in several forms, but the most common is a perforated cylinder with a similarly perforated internal slide. The slightest drop in line pressure on one side of the faucet moves the slide over a bit, realigning the perforations and reducing intake from the high-pressure side. When pressure is restored to the weak side, the slide returns to its original position. You'll find pressure-balance spools in two locations. Some are built into the faucet body—with front access—and some are built into the valve cartridge.

Cleaning Sediment from a Pressure Balance Spool

Pressure-balance spools are vulnerable sediment problems. When a spool clogs with sediment, the faucet will deliver only a trickle of water or only hot or cold water.

To deal with a gradual accumulation of sediment in a spool that's installed in the faucet body, remove the faucet handle and escutcheon. Close the integral stops by turning the exposed screwheads clockwise with a flat-blade screwdriver. (See the photo at top left.) You'll find the stops just behind the faucet's trim plate on each side of the faucet. Use a large flat-blade screwdriver to unscrew the spool from the faucet body, and remove it. (See the photo at top right.) Tap the spool several times on a hard surface, and then rinse it clean. The internal slide should easily slip back and forth when you tip the cylinder. If it doesn't, you'll need a new spool. Reinstall the spool; open the stops; and reassemble the faucet. Make sure both stops are fully open. It's easy to mistake a nearly closed stop for a clogged balance spool.

If the spool is built into the faucet cartridge, you may need a new cartridge.

Cleaning a Pressure-Balance Spool

Some scald-control faucets have integral water stops. Use a screwdriver to shut off the water.

To clear a balancing spool of sediment, pull it out and tap it on the counter. Flush the spool with water, and replace it.

Pressure-Balanced Cartridges

Scald-control measures have changed the way you work on familiar cartridge faucets. Many cartridges and faucets look like standard units, but they now contain a balancing spool, and they are installed differently. You can't remove them from the faucet without a special tool, so don't try pulling them out, the way you would a standard cartridge. Either buy a metal twist tool, like the one shown here, or use the little tool that comes with each replacement cartridge. Manufacturers normally provide toll-free numbers and websites with their instructions, so there's plenty of help available.

8 Plumbing

BATHROOM FAUCET & DRAIN

Bathroom faucets are sold with and without drain assemblies. Because the lift rod that operates the drain's pop-up plug is installed through the faucet-body cover, it's a good idea to replace both when changing out an old faucet. The procedure described here assumes that your sink is installed in a vanity cabinet.

First, remove the old faucet and drain assembly. Shut off the water, and drain the faucet lines. If you have to shut off the main valve and you have plumbing fixtures on the floor above this bath, open those fixtures as well. Otherwise, water from upstairs will drain onto you as you lie beneath your work.

Removing the Drain

With the lines bled, slide a pan or shallow bucket under the old sink trap, and loosen the trap nuts. Use a padded pipe wrench or groove-joint pliers on chrome traps. The nuts on plastic traps should be only hand tight. Drop the trap, and empty it into the pan or bucket. Next, loosen the nut at the wall fitting (or floor fitting if you have an old S-trap), and pull the trap arm from the drainpipe.

Move to the sink's pop-up assembly. You should see a horizontal pop-up lever extending from the back of the drain's tailpiece. The vertical lift rod will be connected to this lever by means of a clevis and tension clip. Squeeze the clip, and pull the clevis from the lever. Then loosen the nut holding the pop-up lever in the tailpiece, and pull the lever from the tailpiece to release the pop-up plug in the drain. You'll see a swivel ball in the end of the lever that allows it to pivot up and down, controlling the pop-up, and the pop-up should be free.

To remove the drain from the sink basin, loosen the large fastening nut that holds the drain in place. This hex-shaped nut is threaded over the drain extension at the bottom of the basin. Grip the nut with a wrench or a pair of groove-joint pliers, and back it down the threaded extension about 1 inch. Then push up on the drain to break the flange seal inside the sink basin. Reach into the sink, and unscrew the flange from the threaded extension. Pull the

Removing the Drain

USE: ▶ groove-joint pliers • bucket or shallow pan • putty knife

The flat washers used with chrome traps need to be lubricated. If you don't have pipe joint compound, liquid dish detergent is an acceptable substitute. Also, be careful when applying force with a wrench. These pipes can be deceptively fragile. When working on them, it is good to backhold the pipes (using two wrenches with opposing force) so that the force you apply to the pipe with a wrench is countered with an opposing force from another wrench.

1 To remove a sink-basin drain, loosen the two P-trap nuts and remove the trap. Keep a bucket under the trap to catch wastewater.

2 To remove the P-trap arm, disconnect the friction nut at the wall, and pull the arm straight out from the drainpipe.

3 To disconnect the sink drain's pop-up linkage, squeeze the tension clip, and slide the clevis from the lever.

4 Use groove-joint pliers to loosen the nut that holds the sink drain in place. Backhold with a second pair of pliers if it spins.

5 Unscrew the basin flange from the sink drain (working from above), and pull the drain out of the basin from below.

INSTALLING SINKS & FAUCETS

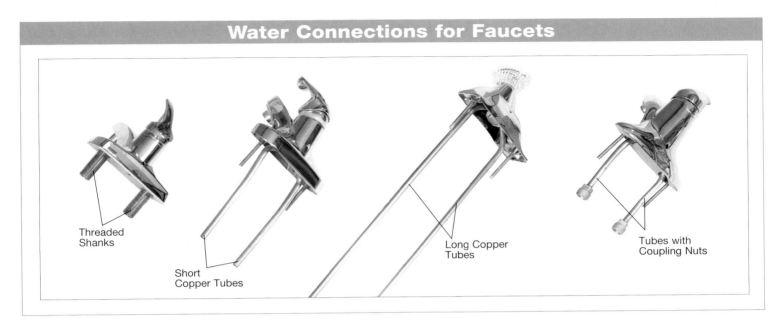

Water Connections for Faucets

Threaded Shanks

Short Copper Tubes

Long Copper Tubes

Tubes with Coupling Nuts

drain out, and scrape any old putty or pipe joint compound from the basin.

Removing Old Bathroom Faucets
How you remove the old faucet depends on whether it's a top-mounted or bottom-mounted faucet. Top-mounted faucets are the most common.

Top-Mounted Faucets
A top-mounted faucet is held in place from below by threaded shanks or fastening bolts, which fit through the basin's deck holes. If the faucet is a two-handle model, expect to find jamb nuts tightened onto the shanks from below. In this case, you'll connect the supply tubes to the ends of the shanks. If your faucet has copper tubes instead of brass shanks—usually the case with single-control faucets—the faucet will be held in place by threaded bolts. One look under the sink, and you'll know the type you have.

When working within the cramped spaces behind a sink, you'll find loosening and tightening nuts a lot easier with a basin wrench. A basin wrench is really just a horizontal wrench on a vertical handle. Lay its spring-loaded jaws to one side, and the wrench loosens; lay it to the other, and it tightens. The extended handle allows you to work high up under a basin or sink deck without having to reach.

Loosen the two coupling nuts that secure the supply tubes to the faucet. Then, using a standard adjustable wrench, disconnect the lower ends of the tubes from their compression fittings. Set the supply tubes aside, and use the basin wrench to remove the fastening nuts from the faucet shanks or fastening bolts. Finally, lift the old faucet from its deck holes.

Bottom-Mounted Faucets
If the old faucet is a bottom-mounted model, with the body of the faucet installed below the sink deck, the initial approach is from above. Leave the supply tubes in place until after you've completed your work on top. Start by prying the index caps from the faucet handles. Remove the handle screws, and lift the handles from their stems. This will give you access to the flanges (or escutcheons) threaded onto the stem columns. Thread these flanges off; disconnect the supply tubes; and drop the faucet from its deck. Some bottom-mounted faucets consist of three isolated components, including two stems and a spout. These components are normally joined with tubing and are easy to disassemble from below the sink.

Removing Faucets

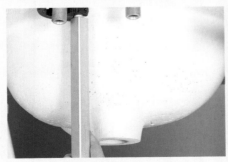

To remove a top-mounted faucet, use a basin wrench to unscrew the jamb nuts from the faucet shanks.

To remove a bottom-mounted faucet, undo the connecting tubing and remove the jamb nuts.

8 Plumbing

INSTALLING A NEW BATHROOM FAUCET

Installing a new faucet is not particularly difficult. If the faucet comes with a plastic or rubber base-plate gasket, put the gasket on it before setting it in place. If the new faucet has threaded brass shanks, insert the shanks through the deck holes; center the faucet; and align the back of the base plate with the wall. Then thread a jamb nut onto each shank from below, and tighten the nuts with a basin wrench. While you're tightening these nuts down, try to have a helper keep the faucet straight and centered. If you're working alone, tighten the nuts only until they begin to bind, and then recheck the alignment from above. If you find that the faucet has moved, make the necessary adjustments, and then tighten the jamb nuts until they feel snug.

If the faucet does not have threaded brass shanks (as most single-control faucets do not), expect to find two bolts alongside two copper tubes. When you have inserted the tubes and bolts through the deck and positioned the faucet, install an extended locking washer and a nut on each bolt, and tighten it all down.

Installing the New Drain

To install the drain assembly, first find the drain flange ring. Roll plumber's putty between your hands until you've made a soft, pliable rope approximately 5 inches long and ¼ inch in diameter. Stick the putty to the underside of the flange, and set the flange aside for the moment.

The drain tube should come with its hex nut, brass washer, and rubber gasket already in place. Apply a thin coating of pipe joint compound to the rubber gasket. Push the drain up through the bottom of the sink basin with one hand, and working from above with the other, thread the flange onto the drain.

With the two halves joined, orient the opening for the pop-up lever to the back of the sink, and tighten the hex nut, using large groove-joint pliers or a pipe wrench. Stop when the nut feels snug and the rubber washer is compressed against the basin outlet. Return topside, and trim the excess putty from around the flange using a knife. Then drop the lift rod through the opening

Installing a New Bathroom Faucet and Drain

USE: ▶ adjustable wrench • plumber's putty • pipe joint compound • groove-joint pliers ▶ new faucet

1 Set the base-plate gasket in place and insert the faucet's water lines through the deck holes.

2 Install the extended washers and the spacers (inset), and thread the nuts onto the faucet's fastening bolts.

3 With the faucet installed, press a roll of plumber's putty around the underside of the drain flange.

5 Insert the drain through the basin's drain opening from below, and thread the flange ring onto it from above.

6 Tighten the hex-shaped jamb nut with groove-joint pliers until it feels snug; then trim away the excess putty from above.

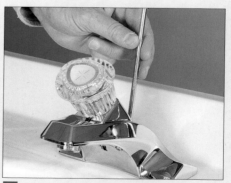

7 Lower the faucet's drain-plug lift rod through the hole in the top of the faucet until it bottoms out.

INSTALLING SINKS & FAUCETS

in the top of the faucet. Next, insert the pop-up plug into the drain so that its offset slot faces the back of the basin. It should rest on top of the lever, slightly above the closed position.

To engage the pop-up plug with the lift lever, loosen the nut that secures the lever in the drain tube. Withdraw the lever slightly, until you hear the pop-up plug drop, and then push the lever forward, into the plug's slot. Screw the lever's nut back into place; slide the clevis bracket onto the lift rod; and push the lever all the way down. Connect the lever to the second or third clevis hole from the bottom, and squeeze the clip onto the lever. Finally, with the lever and lift rod both in a lowered position, tighten the clevis screw against the rod.

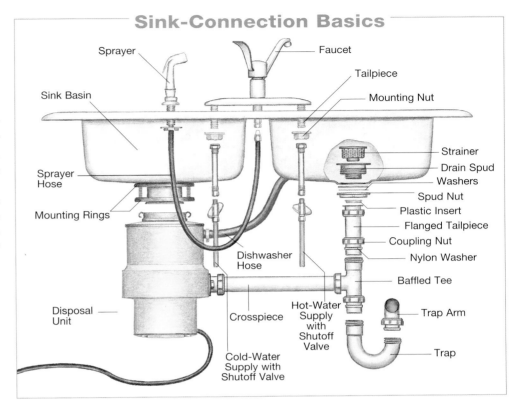

Sink-Connection Basics

4 If it's not already in place, slide the cone-shaped gasket onto the drain, and coat it with pipe joint compound.

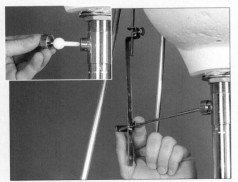

8 Insert the pop-up assembly (inset); connect the clevis through one of the lower holes; and install the clip.

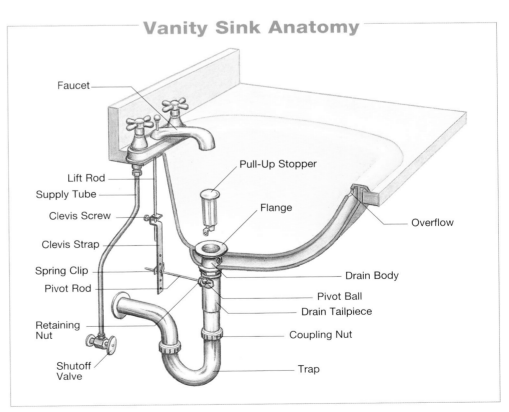

Vanity Sink Anatomy

INSTALLING A NEW BATHROOM FAUCET 197

8 Plumbing

WALL-HUNG SINK

Vanity cabinets come in a variety of standard widths and depths. They range from 18 to 60 inches wide and 16 to 24 inches deep. Unless your bathroom is unusually small or oddly shaped, you're sure to find one that will fit. You'll find everything from well-built hardwoods to unassembled particleboard. If all else fails, you might consider having one built, though that's an expensive option.

Sinks and Vanity Tops

When it comes to sinks and vanity tops, you'll find several choices in two basic categories. You can choose a molded top with an integral sink basin or a plywood top that is finished in tile or plastic laminate. In the latter case, you buy the basin separately and fit it into a hole cut into the top. (See "Installing a Sink in a Plywood Top," page 200.)

Removing the Wall-Hung Sink

Wall-hung bathroom sinks hang from a steel or cast-iron bracket that is screwed to the wall. The bracket lies just under the deck of the sink basin. The sink may also be screwed to the wall. These two screw holes can be found at the lower corners of the deck apron. When removing a sink, start by turning off the water at the shutoff valve and removing the trap. Then loosen the compression nuts that secure the supply tubes to the shutoff valves or adapters. With the plumbing disconnected, remove the two screws from the apron. If the sink is caulked to the wall along its deck, slice through the caulk with a utility knife. Grip the sink on each side, and carefully lift it from the hanger bracket. Remove the bracket, and scrape away any caulk on the wall.

Wall Repair

You may see some minor wall damage, ranging from torn drywall paper to screw holes to a gaping hole in the wall. To repair drywall, start by knocking down any high spots, including those around screw holes. Use a hammer or the end of the knife handle to batter high spots into slight depressions. Wipe on a thin coat of drywall joint compound with a 4-inch taping knife. This first application is just a base coat, so don't worry about the finish. Just put it on, and walk away. Better to apply three or four skim coats, which dry quickly, than one or two thick coats. After the first coat dries, knock down any high spots with a sanding block and give it at least one more skim coat, feathering the edges around the perimeter. When the final coat has dried, sand it lightly. Finally, paint the area with primer and a top coat.

At this point, you'll also need to remove the base trim from the wall where the new cabinet will go. Pry the base moldings loose using a pry bar.

Installing the Cabinet and Top

When walls and floors are plumb and level, vanity installations are easy. First, locate the wall studs behind where the cabinet will be. An electronic stud finder works best. Set the cabinet in place, and screw its back brace to the wall studs. Use 3-inch drywall screws, and predrill holes through the brace to clear the screw threads. You'll need to hit at least one stud. You may also want to trim out the toekick with baseboard.

Unfortunately, though, most floors and walls are a little out of square, either because the house has settled or because it was built that way. To check, move the cabinet into position and look for gaps along the wall. If you see a gap near the top, shim the bottom of the cabinet. Slide pine or cedar-shingle shims under the cabinet until you've closed the gap. Then mark the shims where they meet the cabinet, and pull them back out. Cut the excess from each one, and apply a spot of construction adhesive to the top side. Finally, slide the shims back in place, and screw the cabinet's back brace to the wall studs using 3-inch drywall screws. When you're finished, trim around the bottom of the vanity using baseboard or vinyl cove.

Attach the Top

Always install the faucet and drain assembly in the sink basin portion before setting a molded top. In fact, you can even install the supply tubes if you'd like. With the faucet and drain attached, lift the molded top over the cabinet, and set it in place. If the vanity sits out in the open, center the top on the cabinet and push it back against the plumbing wall. When it looks right, hook up the drain and the water supplies, and turn on the water for a test. If you don't find any leaks, secure the counter to the cabinet.

Removing the Wall-Hung Bathroom Sink

USE: ▶ screwdriver • groove-joint pliers • 4-to-6-in. taping knife • drywall joint compound • utility knife • sandpaper

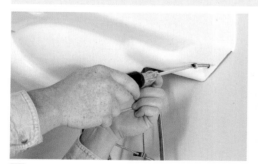

1 Begin by removing the fastening screws just under the sink deck; then cut the basin's caulk or paint seal.

2 Grip the basin at the sides, and carefully lift up off the hanger and out from the wall. Then unscrew the hanger from the wall.

3 With minor wall damage, cut out the caulk; knock down the high spots; and apply at least two coats of drywall joint compound.

INSTALLING SINKS & FAUCETS

Fixing Plaster

If a wall-hung sink is mounted on a plaster wall, the final coat of plaster may have been applied after the sink was installed. Expect to break a little plaster in the sink-removal process. You can fix minor plaster damage with standard drywall joint compound, but deeper damage, including missing chunks of plaster, require some perlite plaster or quick-setting drywall compound. Avoid filling deep holes with standard joint compound; it may take days to cure, shrinking as much as 30 percent.

Where plaster is loose or missing, carefully break out the loose material without enlarging the hole any more than necessary. Use a surface-forming plane or a rasp to knock down the perimeter of the hole. Mix plaster for the job in a plastic bucket, using enough cold water to bring the mixture to the consistency of toothpaste. Wait a few minutes for the plaster to stiffen; then stir and apply it directly over the wood or metal lath. Trowel on enough perlite plaster to bring the surface to within 1/8 inch of finish, and let it dry completely. Follow with several skim coats of drywall joint compound, allowing each to dry before applying the next; then prime and paint the new surface.

When finishing any plaster, it is human nature to rush the job and apply subsequent coats of plaster before the undercoats are fully dry. Be sure the plaster is fully dry before applying more. Also note that plaster is not as easily sanded as joint compound.

Do not expect damaged plaster to be as easy to fix when dry as joint compound. You can use joint compound for some repairs, but in general working with plaster means taking extra time to smooth the plaster as much as possible when it is workable and wet because it's hard to get a smooth finish later.

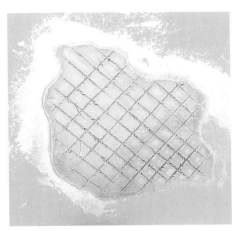
Water-spray the area before applying the rough plaster.

Follow the plaster with several coats of drywall mud.

Seal the area with primer and a coat or two of paint.

Installing the Cabinet and Top

USE: ▶ level • shims, pencil • saw or utility knife • variable-speed drill • belt sander ▶ vanity cabinet • cultured-marble top

1 Slide a shim under a high corner of the vanity cabinet to level it, and mark the shim with a pencil. Cut and glue the shim in place.

2 Locate the wall studs, and then use a variable-speed drill-driver and drywall screws to fasten the cabinet to the wall.

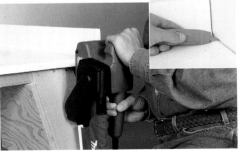
3 If an out-of-square corner makes for a bad fit, sand down the top or cut a recess in the drywall with a utility knife (inset).

WALL-HUNG SINK

8 Plumbing

INSTALLING A BATHROOM SINK IN A PLYWOOD TOP

Plywood or high-density-particleboard tops are finished with plastic laminate or tile. You can buy them prelaminated or install your own laminate or tile. Unlike a molded top, plywood and particleboard tops can hold screws, so it makes sense to screw them down to keep them from shifting. The standard approach is to screw from the bottom up, through the cabinet's corner brackets. These brackets may be wood, plastic, or metal.

Choose your screws carefully, however. (Don't use a screw long enough to pierce the laminate glued to the top of the counter.) Set the top in place; then measure from the bottom of the bracket to the bottom of the counter. Add the depth of the countertop (not the edge band) to this measurement, and subtract ⅜ inch to determine an overall length. If the brackets are made of plastic, remember that plastic can arch upward when you draw the screw tight. This lets the screw travel too far, which may put the laminate at risk. Stop when the screws begin to bind.

Cut the Sink Opening

If you have a new plastic-laminate top and would like to install a china, steel, cast-iron, or plastic-resin sink, don't be intimidated by the prospect of cutting an opening. (To cut an opening in a plywood countertop that you plan to tile, cut the opening before installing the tile.) You'll need a saber saw and a drill. If the new sink comes with a paper template, tape the template to the top, and cut along its dotted line. If you don't have a template, turn the sink upside down on the top; center it; and trace around it with a pencil. Remove the sink, and draw a second line roughly ½ inch inside the first. You'll cut along this interior line.

Drill a ⅜-inch hole through the top, just inside the line. Install a medium-course blade in a saber saw, and lower the blade into the hole. Cut along the line until you're within 6 inches of completing the circle. At this point, have a helper support the cutout. Just a little support from below will keep the cutout from falling abruptly and perhaps breaking the laminate in the process. (If the top's backsplash interferes with your saber saw, finish the cut using a utility saw from above or the saber saw from below.)

When you set the sink, adhere it with latex tub-and-tile caulk. In addition to having great adhesive qualities, latex caulk comes in a variety of colors.

Start by squeezing a liberal bead of caulk all along the joint. Draw a finger along the bead, smoothing it and forcing it into the joint. Finally, wipe away the excess caulk using a damp sponge or rag. Continue wiping until the joint appears uniform all the way around.

The phrase "measure twice and cut once" is never so applicable as now. After spending a good sum on a countertop and installing it, the last thing you want to do is cut a hole that is the wrong size. When in doubt, err to the side of a smaller hole. You can always enlarge a hole, but you can almost never make one smaller. Also, note the shape of the sink bowl you are installing. In bathrooms, the hole is rarely square.

Finally, be sure not to make the hole the shape of the top profile of the sink but of the insert portion instead. Some sinks have asymmetrical "lips." You may even want to create a template out of cardboard just to make sure for easy marking.

Installing a Sink in a Plywood Top

USE: ▶ pencil • saber saw • tub-and-tile caulk • sponge ▶ new sink • countertop

1 Set the basin upside down on the laminated countertop, and center it. Trace around it using a sharp pencil.

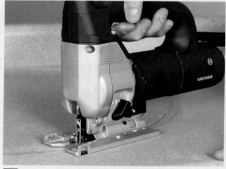

2 Draw a second line approximately ½ in. inside the traced basin outline. Set the basin into the cutout.

3 Install and connect the faucet and drain, and apply a bead of latex tub-and-tile caulk around the basin's rim.

4 Use a wet sponge to smooth the joint and remove excess caulk. Leave only a thin line of caulk behind.

INSTALLING SINKS & FAUCETS

HOW TO REPLACE A KITCHEN SINK

Kitchen sinks come in several rim styles: self-rimming (cast iron, porcelain, solid-surface resin, and stainless steel), metal-rim (enameled steel and cast iron), and rimless (stainless steel, cast iron, and solid-surface resin).

Self-Rimming Sinks

The most popular is the self-rimming variety. A self-rimming sink has a rolled lip, which rests directly on the counter. All that holds the cast-iron and solid-surface sinks steady is a bead of latex tub-and-tile caulk. Stainless-steel self-rimming sinks have a slightly different rim style. The sink rests on its rim, but special clips secure it from below. These clips slide into channels welded to the bottom of the sink rim. While this rim style is popular, a common complaint is that the rolled edge doesn't allow you to sweep food crumbs directly from the countertop into the sink.

Metal-Rim Sinks

This rim, which is usually made of stainless steel, clamps over the unfinished edge of the sink. You place the entire assembly into a counter opening and use a series of metal clips to bind the rim to the counter. A metal-rim sink requires a more precisely cut opening. Most of these sinks are inexpensive enameled-steel models.

Rimless Sinks

And finally, there are rimless cast-iron, stainless-steel, and solid-surface resin sinks made to be installed in tiled or solid-surface countertops. Cast-iron sinks are installed similarly to self-rimming sinks. The depth of the sink perimeter lip approximates the thickness of ceramic tiles. This makes for a more uniform appearance, especially when you match the tile and cast-iron sink colors. Under-mount stainless-steel sinks and solid-surface resin sinks are attached from below, with the finished joint all but invisible in solid-surface installations. This makes for an integral solid-surface look because sink and countertop are the same material. Note that, to maintain the manufacturer's warranty, a professional must install solid-surface countertops.

Stubborn Drain Fittings

When all goes well, it's easy to remove a kitchen sink drain. Just disconnect the tailpiece; grip the spud nut with large groove-joint pliers or a spud wrench; and unscrew the nut from the drain spud. But if the drain is old, the pot-metal spud nut may be fused to the brass with corrosion. If a spud wrench or large pliers won't do the trick, spray the nut with penetrating oil. Give the lubricant 10 minutes to work, and try it again. If lubrication doesn't help, try driving the nut loose with a hammer and cold chisel. Set the chisel against one of the tabs on the nut, and drive it in a counterclockwise direction. This will usually break the nut free, allowing you to finish with pliers or a wrench. If that doesn't work, switch to a hacksaw. Position the saw at a slight angle across the nut, and cut until you break through it.

The Sink Itself

If the old sink is a self-rimming cast-iron or porcelain model, undo the water and waste connections and slice through the caulk between the sink and countertop. A sharp utility knife works best. Then just lift out the sink. If the old sink is a self-rimming stainless-steel unit or a rim-style model, you'll first need to remove the rim clips. The easiest way to remove clips is with a special sink-clip wrench, called a Hootie wrench, though a long screwdriver or nut driver will also work. If the old sink is made of cast iron, take note of the sink rim's corner brackets. These brackets are all that keep an extremely heavy sink from falling into the cabinet space. Leave the brackets in place until you've removed the sink. In fact, it's good practice to prop up the sink with a short piece of lumber when undoing the clips from a cast-iron sink. Stainless-steel sinks are much lighter and don't present a problem.

To remove the rim clips, unscrew the hex-head bolt from each one. When lifting a cast-iron sink, you may have difficulty getting a starting grip. Try lifting by the faucet column until you can get a hand under the rim. If that feels too awkward, remove the drain fittings and reach through the openings.

Attaching Drain Fittings to the New Sink

USE: ▶ plumber's putty • spud wrench ▶ drain kit

1 Form plumber's putty into a ½-in. roll several inches long, and press it against the underside of the drain flange.

2 Insert the drain through the sink opening, and install the rubber gasket, paper gasket, and spud nut in that order.

3 Use a spud wrench or large pliers to tighten nut. Trim any excess putty from around the flange and tighten the nut again.

8 Plumbing

INSTALLING A METAL-RIM SINK

Sinks that require separate metal rims have the advantage of a neat, flush-fit appearance. The rim is T-shaped, with a horizontal support flange and a vertical extension. The bottom of the vertical extension is rolled over to form a lip. You need to cut a more precise opening in the countertop, and you'll have a dozen or so rim clips to install. As noted earlier, the easiest way to install clips is with a special clip wrench. Lacking such a wrench, a long, slotted screwdriver or nut driver will work.

Lay Out and Cut the Opening
When laying out the countertop opening, you won't need a paper template. Instead, the vertical extension of the rim serves as a template. Just position the rim on the countertop, with the flange on top, and trace around the outside of the vertical portion of the rim. Then cut carefully on this line using a saber saw.

Install the Sink Rim
Sink rims are made to work on cast-iron sinks and on enameled-steel sinks. (Stainless-steel sinks have built-in clip channels.) As you inspect the rim, you'll notice that it has two sets of knock-in tabs running around its vertical band. These tabs, when folded in, support the rim of the sink before the clips are installed. The row of tabs closest to the flange are for steel sinks, while those farther from the flange are for thicker cast-iron sinks.

When buying a rim for a cast-iron sink, choose one that has separate corner brackets. Unlike steel sinks, cast-iron sinks are too heavy to be supported safely by thin metal tabs. If you can't find corner brackets, make a homemade support.

To install a sink rim, set it over the sink and use a screwdriver to punch in all the perimeter tabs. Lift the sink into the opening, and settle it on its rim.

Attaching the Faucet
Faucets from different manufacturers differ slightly in the way you mount them, but the basic connections remain the same. You may have to vary some aspects of your installation, depending on your faucet design.

Place the faucet's base plate over the sink's deck holes, and insert the mounting shanks on each side. From underneath, slide a large washer onto each fastening shank, followed by a jamb nut. Turn the nuts until they're finger-tight. As you draw down the nuts, make sure the faucet's base plate is parallel with the back edge of the sink.

If the faucet is a single-handle type with central copper inlet tubes, straighten the tubes; position the decorative deck plate over the base plate; and insert the tubes through the center deck hole. Slide the mounting hardware (either a nut or a screw plate) onto the faucet shaft, and fasten the faucet in place. If the faucet has a pullout spout, connect the spout hose now. Attach the supplied adapter to the faucet nipple. Pull back the spring housing to reveal the male threads of the hose inlet end, and screw the hose into the adapter. Attach the outlet end of the hose to the faucet at the top of the spout housing to complete the connection.

Sinks come with either three or four deck holes. The fourth hole is for accessories like a separate hose sprayer, an instant hot-water dispenser, a soap dispenser, and the like. When codes require a backflow preventer in the dishwasher discharge hose, the fourth deck hole can also hold the backflow preventer. If you can't think of an add-on you'd care to own, plug the opening with a chrome or brass-plated sink-hole cover. Some snap in place, while others have a threaded fitting. The plug will need to be watertight, so caulk around it very lightly. The bead should be all but invisible.

Rim Clips

Rim clips like these hold metal-rim-style steel sinks in place. You must detach them before attempting to remove the sink from its opening.

Installing a Metal-Rim Sink

USE: ▶ pencil • saber saw • screwdriver • rim clips • sink-clip wrench ▶ sink and rim

1 Position the sink rim on the countertop, and trace around it. Use a saber saw to cut the sink opening (inset).

2 Use a screwdriver to bend the tabs inward on the sink rim. With a cast-iron sink, install supporting corner brackets as well.

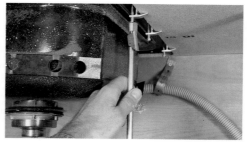

3 Use a sink-clip wrench (or a long nut driver or screwdriver) to drive the clip bolts against the underside of the counter.

INSTALLING SINKS & FAUCETS

Attaching the Faucet

USE: ▶ screwdriver • groove-joint pliers • latex tub-and-tile caulk ▶ new faucet

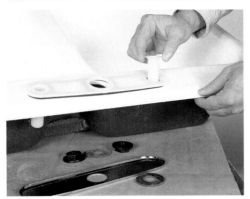

1 Center-column faucets work on sinks with one or three holes. For a multihole sink, install a base plate.

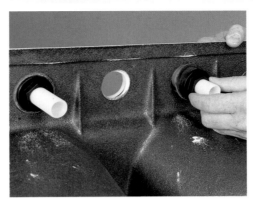

2 Fasten the plastic base-plate support/gasket from below with plastic jamb nuts. Make them finger-tight.

3 Snap the decorative base plate in place over the plastic support, and insert the faucet's column through the center hole.

4 Slide the mounting hardware onto the column and tighten it. The unit shown has a plastic spacer, steel washer, and brass nut.

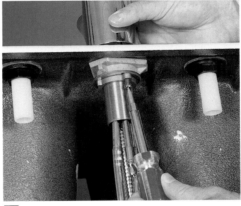

5 Use a screwdriver to drive the setscrews against the large washer. Stop when the screws feel snug.

6 Install the supply adapter on the faucet nipple. Most faucets use a slip fitting with an O-ring seal, as shown.

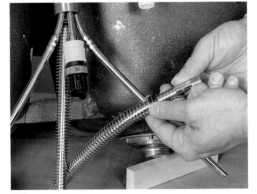

7 Pull back the spiral tension spring at the bottom end of the spray hose to expose the male attachment threads.

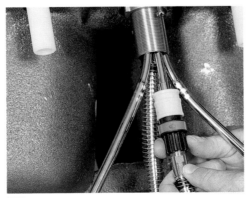

8 Tighten the male threads into the bottom of the faucet's supply adapter. Stop when it feels snug. Don't overtighten it.

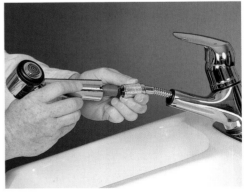

9 From above, thread the outlet end of the hose into the spray head. Pull the hose out several times to test it for ease of use.

INSTALLING A METAL-RIM SINK

8 Plumbing

HOW TO INSTALL A LAUNDRY SINK

Plastic and fiberglass laundry sinks come in several forms. Some are freestanding, some are wall-hung, and some are counter-mounted. Counter-mounted sinks are made to drop into cabinets, like self-rimming kitchen sinks. Fiberglass is sturdier than plastic.

Laundry sinks require lower piping connections than bathroom and kitchen sinks. While a kitchen sink drain is roughed-in 18 inches above the floor, a laundry sink's connection shouldn't be higher than 13 inches, measured from the center of the drainpipe. You can install the water-pipe stub-outs 15 to 16 inches off the floor.

Other differences involve the drain fittings and mounting methods. Drop-in models require that you install a basket-strainer drain, like those used on kitchen sinks, while wall-hung and freestanding models usually come with a drain fitting molded right into the bottom of the sink. These drains come with a rubber stopper, like those used in older bathtubs.

Installing a Freestanding Sink

The deck holes on laundry sinks have 4-inch center spreads, so kitchen faucets, which have 8-inch spreads, won't work here. Many people use lavatory faucets on laundry sinks because they're affordable, but special 4-inch-spread utility faucets are a better choice. They're often made of heavy brass and have spouts fitted with hose threads.

Install the faucet just as you would in a kitchen or bathroom sink. (See "Installing a New Bathroom Faucet and Drain," page 196.) Install the sink legs by snapping them into the molded slots on the bottom of the sink.

To hook up the trap and drainpipes, use a flanged tailpiece extension and the nylon-insert washer that comes with the sink. Set the washer on the drain; slip the tailpiece over the washer; and tighten the slip nut. Then trim the plastic tailpiece to length, and install a P-trap between the tailpiece and the permanent piping in the wall. Make the water connection with supply tubes, just as you would a kitchen-sink installation. (See "Connecting the Water," opposite.)

Each sink leg will have a hole in its base. Position the sink against the wall, and mark the floor through these openings. Then drill holes in the floor for screws. With a wooden floor, screw directly through the legs and into the floor with deck screws. On concrete, install plastic anchors in the floor and then screw into the anchors. Securing the legs is an important step, as it keeps the sink from getting bumped out of position.

If your freestanding sink has screw holes at the outer edges of its deck, screw the sink to the wall as well. This will make the fixture rock-steady.

Wall-Hung Laundry Sink

A wall-hung sink will come with a mounting bracket and two side covers. The trick is to screw the mounting bracket to a sturdy support feature. In new construction or a complete remodel, you'll be able to provide 2x6 lumber backing in the wall. Position the top of the backing 32 inches off the floor. In a retrofit installation, position the mounting bracket so that you can anchor firmly into at least one stud. For the other connection, use drywall anchors through the bracket on either side of the stud.

Standard sink height is between 32 and 34 inches off the floor. With the bracket mounted, hang the sink on the bracket and install the side covers. These covers, which may be plastic or steel, also provide support to keep the sink from tipping down in front. Finally, make the water and waste connections, and caulk the joint between the sink deck and wall.

"H" Is for Left

The hot-side faucet handle, usually marked "H" somewhere on the handle, should always be on the left as you face the faucet. If it's not, the faucet may have been installed backward or the water lines under the sink may be reversed. To see whether the faucet was installed backward, check the stems. They may look alike, but they don't rotate the same. With compression faucets, you turn the water on by rotating the hot-water handle counterclockwise and the cold-water handle clockwise. If your faucet doesn't work that way, someone inadvertently reversed the stems. A repair is your chance to correct them.

Installing a Freestanding Laundry Sink

USE: ▶ screwdriver • hacksaw • adjustable wrench ▶ laundry sink kit

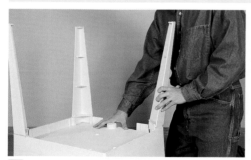

1 Turn the sink upside down to make it easier to install the legs. Just snap each leg into its molded slot.

2 Install a nylon insert washer in a flanged tailpiece, and connect the tailpiece with a slip nut. Attach the trap once the sink is in place.

3 Use wood or utility (deck) screws to anchor the legs to the floor. On concrete, use screws and plastic anchors.

INSTALLING SINKS & FAUCETS

Connecting the Water

USE: ▶ tubing cutter • pipe joint compound • adjustable wrench ▶ water supply tube

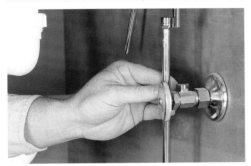

1 Hold the water supply tube between the faucet stub and the shut-off port, and mark it for length.

2 Connect the tube, and install the compression nut and ferrule. Coat the threads and ferrule with pipe joint compound.

3 Tighten the compression nuts at both ends of the supply tube. Backhold the upper fitting.

How To Replace an S-Trap

A sink that is drained through the floor via an S-trap is no longer legal, because the trap can't be vented. If your fixtures now drain through S-traps, you won't be required to change them, however, because they're covered by the grandfather clause. (And besides, an S-trap is just the most visible symptom of an outdated system. Ideally, the entire drainage system should be rebuilt to meet current standards, although that's an expensive option that most people can't afford.) On large projects, however, you may be required to make the change.

Still, if you're replacing a sink that has an S-trap, you may as well do all you can to improve the way that new fixture performs and replace the trap. The solution is surprisingly easy and costs just a few dollars as there is no need to replace the drain. It consists of installing an automatic vent device inside the sink cabinet.

Begin by attaching a banded coupling to the drain at floor level. (Or if possible, thread a 1½-inch PVC female adapter onto the drain's threads. These are the same threads used to connect the S-trap.) From the top of the coupling (or adapter), use two 45-degree PVC elbows to offset a riser to the back wall of the cabinet, about 4 inches left or right of center. Bring the riser up to trap level, about 18 inches off the floor, and install a sanitary T-fitting. Using a PVC ground-joint trap adapter, pipe the trap into the riser. Out of the top of the T, extend the riser up 6 inches, ending with another 1½-inch female adapter. Thread an automatic vent device into this adapter.

Automatic vent devices don't last indefinitely, so remove the vent every couple of years and check its operation. With the vent in hand, its spring-loaded diaphragm should be held firmly against its seat. If the diaphragm is down even slightly or if the rubber has deteriorated, install a whole new unit.

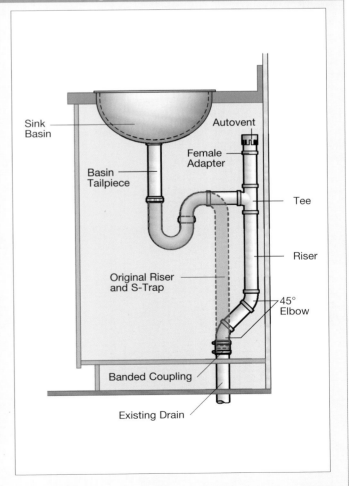

HOW TO INSTALL A LAUNDRY SINK

8 Plumbing

HOW TOILETS WORK

Though it may seem baffling when something goes wrong, there really isn't much to your toilet's inner workings. The two most basic parts are the toilet tank, which holds the water needed to flush, and the toilet bowl, which connects directly to the soil pipe by way of a built-in trap. (That's why there is always some water in the bowl.)

Pushing the toilet handle lifts a valve (often a rubber flapper type) and, thanks to gravity, water in the tank floods down into the bowl by way of the flush passages. As water rises in the bowl, it pushes the level up over the top of the internal trap. Water then flows down the drain, drawing with it any wastes in the bowl.

Meanwhile, back at the toilet tank, the flapper valve closes as the tank empties, and a float valve (or pressure sensor) turns on the water supply valve to refill the tank. The float rises with the water level and shuts off the supply valve when it reaches the preset level (or the water is deep enough to create the predetermined level of water pressure).

The tank also has an overflow tube that keeps water from running out over the top of the tank in the event the water supply valve fails to shut off. The small pipe running into the overflow tube is the bowl refill tube, which adds some water to the bowl after flushing ceases. That reseals the trap and prevents foul-smelling sewer gases from coming up through the toilet.

Replacing a Wax Seal

USE: ▶ adjustable wrench • hacksaw • putty knife or scraper • rags ▶ new wax ring • closet bolts

1 Before working on a toilet, turn off the fixture cutoff valve. Then disconnect the supply tube, and empty the holding tank.

2 Empty the bowl; pry the caps from the closet bolts; and undo the nuts. If the nuts are corroded, cut them off with a hacksaw.

Adjusting a Ball Float

You can adjust the ball float to change the height of the water in the reservoir. Adjust it too high, and the toilet runs constantly; too low, and flushing isn't efficient. Adjust the float by bending the brass float arm up or down, as needed. If the float is on a pull rod, adjust the spring clip on the lower end of the rod.

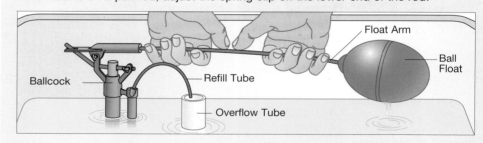

Upgrading a Water Closet

USE: ▶ adjustable wrench • spud wrench • needle-nose pliers • screwdriver • wire brush ▶ fill valve kit • pipe dope • new flapper valve • retrofit flush-valve

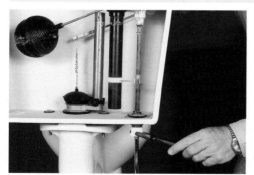

1 To replace a fill valve, drain the tank; loosen the supply riser fastening nut; and then loosen the jamb nut above it.

2 Clean the tank around the opening; coat the new fill valve's washer with pipe dope; and install the valve in the opening.

3 Attach the fill valve's small tube to the tank overflow; use the connecting clip packaged with the fill-valve kit.

TOILETS

- penetrating oil (optional)

3 To remove a toilet, grip it by the bowl next to the seat hinges. Set it on newspaper, and stuff a rag into the soil pipe.

4 Scrape away any old wax from the worn gasket with a putty knife or scraper. Remove the old closet bolts as well.

5 Install new closet bolts in the flange and a new wax ring. Then, remove the rag; reset the toilet; and refasten the connections. (See page 221 for types of toilet seals.)

Eliminating Tank Condensation

The cold water supply can make the bowl cool enough to cause moisture in warm interior air to condense. And flowing condensation can damage floors. To defeat this problem you can buy a toilet with factory-installed insulation, or use a retrofit insulation kit. Another option is a special mixing valve that adds hot water to temper the cold supply.

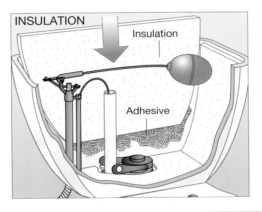

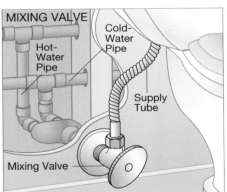

MONEY SAVER

seat kit • emery cloth

4 To upgrade a flapper (sealing the outlet), remove the old flapper, and hook the eyelets of a new flapper over the flush-valve hooks.

5 To install a retrofit flush-valve seat, use the kit's epoxy putty to stick the new seat over the old.

6 With the new seat fastened in place, feed the plastic lift chain through the hole in the flush lever.

HOW TOILETS WORK 207

8 Plumbing

WATER-SAVING TOILETS

Responding to legislation enacted several years ago, manufacturers reduced the volume of water needed to flush a toilet efficiently. Ironically, the very first toilets had small flush tanks, holding only about 3 gallons. These toilets worked because the tanks were mounted on the wall, high above the bowl. Elevating water in a column increases its downward force—an effect known as head pressure. The more pressure generated for the flush, the less water you need. The reverse is also true, so when tanks were moved down the wall and mounted on or just above the bowl, manufacturers increased the size of the tanks to as much as 8 gallons. This was clearly more volume than was needed for normal use, and the industry eventually settled on 5 gallons.

During the late 1960s and into the '70s, fresh water came to be recognized as a limited resource, and tank size was reduced again, to 3.5 gallons.

Water-Savers Are the Law

In the late 1970s, a number of Scandinavian countries began using—and mandating—super-low-flow toilets, which flushed with an amazingly skimpy 1.6 gallons of water. Before long, these toilets appeared at trade shows here in the United States, and manufacturers began experimenting with low-flow toilets, trying to improve performance. Eventually, the U.S. enacted a national standard that limits to 1.6 gallons per flush (gpf) the water used by residential toilets made in this country after January 1, 1994.

Do low-flow toilets work as well as 3.5-gpf ones? The early models certainly didn't. From the start, manufacturers offered two distinctly different low-flow toilets: a gravity-flow model and a pressure-assisted model. The gravity-flow models were like traditional toilets but with minor changes all around, including new fill valves and flush valves. Engineers reduced trap geometry, along with water spots—the surface area of the bowl water—and outlet diameters to cut down the flow of water. These early water misers were so sluggish that they often needed additional flushes to clear and clean the bowl, and clogs were common.

Fixing a Running Pressure-Assisted Toilet — MONEY SAVER

USE: ▶ screwdriver • groove-joint pliers • adjustable wrench ▶ new pressure-regulating valve

If you need a new toilet, which toilet should you buy: a pressure-assisted model or a gravity-flow unit? Price is one consideration. Pressure-assisted toilets are more expensive than conventional models, upwards of $150. Past experience and the condition of your existing plumbing is another consideration. If your old toilet clogged frequently or you have a system that is more than 50 years old, a pressure-assisted unit is probably your best bet. Plus, pay more now, because plumbers are expensive to hire for any repair.

1 Adjust the flush activator setscrew so that the flush button has at least ⅛ in. of clearance before it contacts the activator.

2 Adjust the flush lever or activator (or both) as needed so that there is ⅛ in. of clearance between lever and activator.

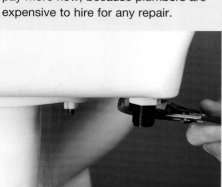

3 Shut off the water, and using large groove-joint pliers, loosen the jamb nut that holds the water supply group.

4 Twist the supply-group assembly apart to expose the pressure-regulating valve. Install a new valve.

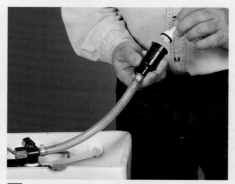

5 Assemble the supply group, and tighten all connections using an adjustable wrench.

TOILETS

Water Savers

1.6-GALLON GRAVITY-FLOW TOILET

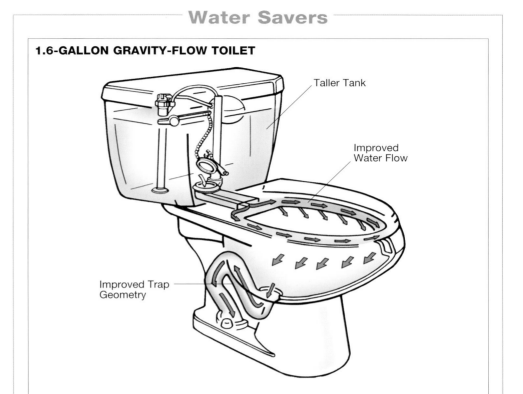

PRESSURE-ASSISTED TOILET

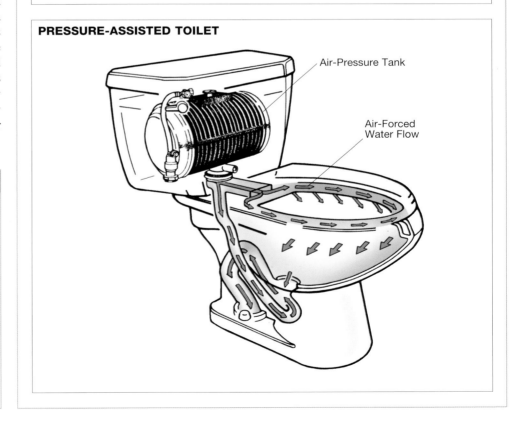

With steady engineering refinements, however, gravity-flow toilets now work reasonably well and are a good choice for most households.

Pressure-Assisted Toilets

In general, pressure-assisted toilets work better than gravity models. These toilets use incoming water to compress air in a chamber inside the tank. (A water pressure of at least 20 pounds per square inch is required.) Flushing releases this compressed air in a burst, forcing water to prime the trap almost instantly. Air assist allows the tank to operate with less water, making more water available for the bowl. More water in the bowl means a larger water spot and a cleaner bowl.

You'd think everyone would want a pressure-assisted toilet, but that hasn't been the case. The most common complaints are that they're too noisy and too complicated. Starting each flush with a burst of compressed air does make them noisy, and they're certainly less familiar. Most people would recognize the tank components in a traditional toilet, but lift the lid on a pressure-assisted unit, and all you'll see is a sealed plastic drum, a water-inlet mechanism, a hose, and a flush cartridge. Most manufacturers use almost identical tank components, all made by a single supplier. It's easy to repair and replace these parts, but not all hardware stores carry them. Plumbing wholesalers do, but they don't often sell to nonplumbers. Plumbers may sell the parts, but they'd rather that you pay for installation as well.

Checking for Clogs

Use a pocket mirror and Allen wrench to ream mineral-clogged holes around the underside of the rim.

WATER-SAVING TOILETS 209

8 Plumbing

TROUBLE-SHOOTING GRAVITY-FLOW TOILETS

Now that you know how traditional gravity-flow toilets are supposed to work, it's time to learn how and why they may not work and what to do when they don't. Keep in mind that poorly maintained toilets may display more than one symptom.

Slow Toilet

Your toilet seems sluggish. It once flushed vigorously, but now the water seems to move slowly through its cycle, often rising high in the bowl before passing through the trap. Large bubbles rise out of the trap during the flush. Sometimes the bowl even seems to double flush.

These are classic symptoms of a partially blocked trap. An obstruction, such as a toy, comb, cotton swab, and the like, has made its way to the top of the trap and lodged in the opening. Paper then begins to accumulate on the obstruction, further closing the opening. In many cases, enough of the trap remains open to keep the toilet working, but in time, partial clogs become complete clogs.

To clear a partial blockage, start with a toilet plunger, forcing the cup forward and pulling it back with equal pressure. If a plunger doesn't clear the clog, try a closet auger. If the closet auger doesn't do it, bail out the bowl with a paper cup or other small container, and place a pocket mirror in the outlet. Shine a flashlight onto the mirror, bouncing light to the top of the trap. The mirror should allow you to see the obstruction. When you know exactly what and where it is, you should be able to pull the obstruction into the bowl by using a piece of wire. In rare cases, you may need to remove the toilet and work from the other side.

Dirty Toilet

The toilet does not appear clogged, because water doesn't rise unusually high in the bowl, but it flushes sluggishly, and the bowl does not stay clean for long.

These symptoms suggest that the toilet bowl's rim holes—and possibly the siphon jet hole—are clogged with calcified minerals from hard water or with bacteria. To make sure, watch the water as it passes through the bowl. Open rim holes should send lots of water coursing diagonally across the sides of the bowl. If the water slides straight down, that may be a sign that the rim holes are partially clogged, either by bacteria or calcification. Dark, vertical stains beneath some of the holes suggest bacteria. Clogged siphon jets are almost always caused by bacteria. (See "Checking for Clogs," page 209.)

How will you know whether blockages are made of mineral deposits or bacteria? Bacteria accumulations are soft and dark, ranging from orange to black. Mineral deposits are hard, scaly, and usually light in color.

Bacteria

To remove bacteria, first kill as much of it as possible, not just in the bowl but in the bowl's rim and rim holes. Pour a mixture of 1 part household bleach and 10 parts water directly into the tank's overflow tube. Just lift the tank lid, and direct a cup or more of the bleach solution into the overflow. Allow the bleach to work for a few minutes; then flush the toilet, and carefully ream the rim holes with a pen knife or a piece of wire. You can't see the rim holes from above, so use a pocket mirror to check your progress. Scour any bacteria stains from the underside of the rim, using bowl cleaner and an abrasive pad. Add

Slow Toilet Fix

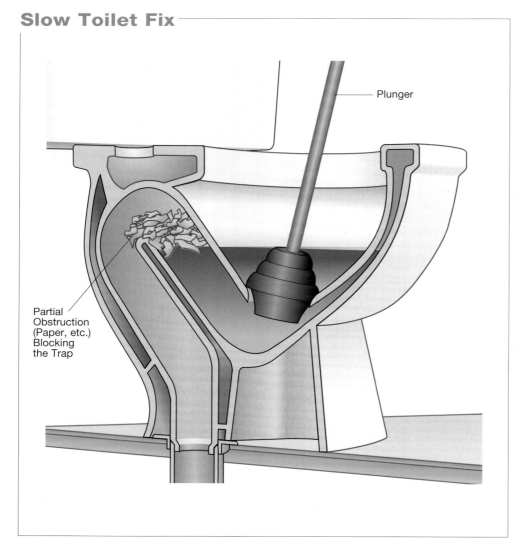

Plunger

Partial Obstruction (Paper, etc.) Blocking the Trap

TOILETS

a final dose of bleach through the overflow; wait a few minutes; and flush the toilet. To clear a clogged siphon jet, ream it thoroughly with a stiff wire. You'll probably have to do this all-out ream-and-scour cleaning only once or twice a year if you add one or two tablespoons of bleach to the overflow tube periodically.

Mineral Deposits

To remove calcified minerals left by hard water, you'll need slightly different tools. Instead of bleach, pour vinegar into the overflow tube, and let it stand for at least 30 minutes. Vinegar dissolves and loosens mineral deposits, allowing you to break and scrape thick accumulations that may have built up around the rim holes. Vinegar seems to work better when it's heated. Don't boil it, just heat it to shower temperature, about 104° F.

After letting the vinegar stand, ream each hole thoroughly. On heavily clogged holes, use Allen wrenches as reaming tools. Start with a small wrench, and use larger ones as you gradually unclog the hole. Remember that porcelain chips easily, so work carefully and use a pocket mirror to check your work. This problem is a good indication that you might need to install a water softener.

A water softener will remove calcium from the water and prevent buildup. But they are more costly than an occasional cleaning.

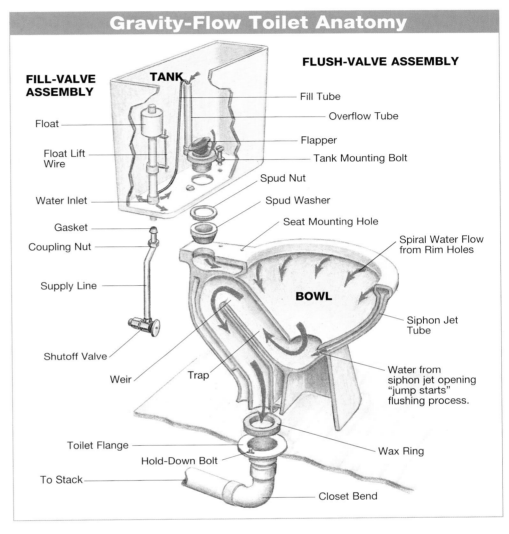

Gravity-Flow Toilet Anatomy

Cleaning a Bacteria-Clogged Toilet

USE: ▶ measuring cup • bleach solution • insulated wire • pocket mirror (if necessary)

1 To kill bacteria buildup in and under the toilet bowl's rim, pour a bleach solution directly into the overflow tube.

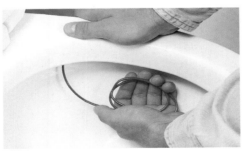

2 Clear bacteria from the rim holes using a short length of insulated electrical wire. Try approaching the hole from several angles.

3 Clear bacteria from the siphon jet again using a piece of wire. A dark opening may indicate the presence of bacteria.

TROUBLE-SHOOTING GRAVITY-FLOW TOILETS

8 Plumbing

Fixing a Slow-Filling Toilet

MONEY SAVER

The toilet flushes well enough, but the tank takes 10 to 15 minutes to fill. You also hear a slight hissing sound when the house is quiet.

This symptom indicates the presence of sediment in the fill-valve diaphragm of the toilet. (Significant sediment problems may occur after the installation of a new toilet, after work is done on a nearby water main, or after a well is put into service.)

The solution is to remove the diaphragm cover and pick the grit from the valve. This is usually a quick and easy fix. If your toilet has a conventional brass or plastic ballcock (a fill valve with a float ball), shut off the water, and remove the two or three screws that secure the cover. Lift the cover and float rod from the riser. You should be able to see sand or rust flakes scattered around the rim of the valve seat. Remove this grit using tweezers, and replace the cover.

Not all sediment in a line will work its way through at the same time, so you may need to clean the fill valve several times in the next few days or weeks. If sediment routinely plagues your water system, the best approach is to install a sediment filter in the cold-water trunk line.

FIXING A RUNNING TOILET

The toilet often keeps running until you wiggle the flush handle.

This symptom signals an adjustment problem, either in the flapper chain or tank-ball linkage. (It may also indicate a corroded flush lever, but in this case symptoms usually include the lever sticking in the up position.) If your toilet has a flapper, the lift chain may be too long. To correct the problem, remove the tank lid and lift the chain from the wire hook that secures it to the flush lever. Reconnect it so that it has less slack. With the chain hooked, top and bottom, press it to one side. You should see roughly 1 inch of sideways deflection. (See the photo opposite top.)

If your toilet has a tank ball, expect the lift-wire guide to be out of alignment. You'll find this guide clamped around the flush valve's overflow tube, secured by a set-screw. (See the photo opposite middle.) Begin by removing the tank lid; then flush the toilet several times until you see the tank ball fall off-center, showing you which way to move the guide. If the ball falls to the left, for example, move the guide to the right.

Shut off the water, and loosen the set-screw about two full turns. Rotate the guide about $1/16$ inch, and then reset the screw. Turn the water back on, and watch the tank ball fall through several flushes. It should hit dead center.

When making these adjustments, try not to apply too much pressure to the brass overflow tube. Brass gets brittle with age, and an old overflow tube might break off. If this happens, pry the remaining bit of tube from its valve threads and screw a new brass tube in place.

FIXING RIPPLING WATER

The toilet comes on by itself, runs for a few seconds, and then shuts off. You may also hear a steady trickle and see tiny ripples in the bowl water.

This is most likely a sign that your toilet's flapper or tank ball is worn out. It may also signal a bad flush valve, but check the flapper/ball first. In any case, a small stream of water is leaking through the valve. As the tank level drops, the float activates the fill valve and replenishes the water.

When a flapper or tank ball fails, it's usually because the rubber has broken down. A warning sign is a stubborn black slime on the rubber.

Replacing the parts is cheap and easy to do. Flappers and tank balls are universal, so just about any brand will work.

Removing Grit from the Diaphragm

USE: ▶ screwdriver • tweezers

1 To remove the ballcock's diaphragm cover, remove the three or four brass screws holding it down and lift the float mechanism.

2 With the cover removed, lift the valve's plunger and diaphragm gasket to look for sand grit or other sediment.

3 Use tweezers to lift out any sediment particles or rust flakes. You may need to repeat this sequence over time.

212 **PLUMBING** / TOILETS

TOILETS

One-Piece Silent-Flush Toilet

One-piece silent-flush toilets were the first alternatives to conventional two-piece toilets, hitting the market around the time Cadillacs grew fins and aimed at the same market. They were expensive, stylish, and discreetly quiet.

There are two basic types: one is a gravity-flow toilet with a conventional fill valve and flush valve. This type is easy to repair if you can find factory replacement parts. The other has a more complicated system of valves, which need to be calibrated to match a home's static water line pressure. One look at a repair kit, which resembles an automotive carburetor repair kit, and you'll get the picture: working on it is not easy. You should hire a good service plumber for this job. Ask up front whether the plumber is familiar with your particular make and model.

A silent-flush toilet is made in one piece. They are usually very quiet but may be difficult to work on.

Fixing a Running Toilet

Adjust the flapper chain so that it has about 1 in. of slack. This is easier to do when the tank is empty.

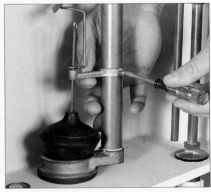

Adjust the lift-wire support if the tank ball falls off center. Dry flush to check.

When you replace an overflow tube that breaks off, coat the threads with pipe joint compound.

Replacing a Tank Ball — MONEY SAVER

USE: ▶ slip-joint pliers (if necessary) ▶ tank ball replacement

1 After you've drained the tank water, remove the tank ball from the lift wire. The tank ball may crumble if it is old.

2 Thread the lift wire into the new tank ball and flush it to check. This replacement type has a weighted bottom.

FIXING A SLOW-FILLING TOILET

8 Plumbing

FLAPPER REPLACEMENT

With a flapper, shut off the water; unhook the chain; and lift the flapper from its pegs at the base of the flush valve or from around the overflow tube if the flapper has a collar. Before installing the new flapper, clear the valve seat of any old rubber (slime) or mineral deposits. Wipe the seat rim with a paper towel; then sand it lightly with fine sandpaper or steel wool. An abrasive pad will also work. You may need to use a little vinegar to dissolve mineral deposits.

The new flapper will likely have two types of flush-valve attachments for a universal fit: side tabs that hook over the valve's side pegs and a rubber collar for use on valves without side pegs. In the latter case, you'd slide the collar over the overflow tube. If side pegs are present at the base of the flush valve, cut the rubber collar from the flapper and throw it away. Then hook the flapper tabs over the pegs. Once you've attached the flapper, reconnect the chain.

Tank-Ball Replacement

To replace a tank ball, first shut off the water and grip the top of the lift wire with pliers. With your remaining hand, thread the ball from the lift wire. If the ball crumbles, leaving only the brass insert attached to the wire, use a second pair of pliers to grip this fitting. Brass is soft, so you won't have any trouble backing the ball fitting from the lift wire with a good grip. In a few cases, the lift wire may break, but replacements are common hardware store items.

Again, dress the flush-valve seat with an abrasive and a paper towel; then insert the wire through its guide and into the new tank ball. Finally, make any needed adjustments in the lift-wire guide.

If the flapper (or tank ball) seems to be in good shape and creates a good seal, the problem lies with the flush valve, and you'll have to replace it.

Fixing a Hissing Toilet

The toilet doesn't shut off completely. You hear a hissing noise and see ripples in the bowl. This behavior starts intermittently but over time becomes constant.

These symptoms suggest a problem with the fill valve or ballcock. (Remember that all ballcocks are fill valves, but not all fill valves are ballcocks. The term "ballcock" applies only to traditional fill valves, which have ball floats on the end of a pivoting arm.) It may be that you need to remove sediment from the fill-valve diaphragm. Or you may be able to solve the problem by making a simple float adjustment. In most cases, however, the valve needs to be repaired or replaced. Don't put it off. A toilet that won't shut off completely wastes lots of water.

Do the simple things first. If the water

Spot Repairs

Use a chlorine-resistant flapper for a longer useful service life. This flapper is clear.

MONEY SAVER

Tighten the new seat bolt using a screwdriver, and snap the hinged cover in place.

Replacing a Flapper — MONEY SAVER

USE: ▶ abrasive pad, steel wool, or scouring pad • vinegar (if necessary) • scissors (if necessary) ▶ flapper replacement

1 Clean the flush-valve seat by wiping it with steel wool or a scouring pad. Feel for imperfections in the seat's surface.

2 Universal flush-valve flappers are made to fit either of two situations. If the flush valve has no side pegs at the bottom of the overflow tube, slide the collar over the tube (photo A). For side pegs, cut off the collar, and hook the eyelets over the pegs (photo B).

TOILETS

level is so high that it spills into the overflow tube before the float ball or cup can shut off the fill valve, adjust the water level. Aim for a level about 1 inch below the top of the overflow. With a ballcock assembly, tighten the adjustment screw on top of the fill-valve riser. If that doesn't work, bend the float-ball rod down slightly. (See the photos below.) Use both hands, and work carefully. Newer fill valves have other float adjustments. For example, a common type has a stainless-steel clip that you use to adjust the height of a float cup. (See "Replacing a Fill Valve," on page 218.) If adjustment doesn't solve the problem, a tiny amount of sediment may be in the diaphragm. See "Removing Grit from the Diaphragm," page 212, for how to clean the assembly.

Assuming that the float setting is fine and there is no sediment in the diaphragm, your next option is to replace the diaphragm and float-plunger seals. Begin by shutting off the water and removing the two or three screws that secure the combination diaphragm cover and float-arm assembly to the top of the riser. Lift the cover-and-float-arm assembly from the valve, and pull out the rubber seals. Expect a large rubber disk or stopper. Take these parts with you to a well-stocked plumbing outlet. If you can find replacement seals, install them in reverse order of removal, and test your work. In some cases, there are problems that are not visible and you may be better off replacing the entire fill valve, especially if it's old.

Easy-Fix Flush Valve

Here's a flush-valve assembly unlike any other. It consists of an inverted cone that rides up and down on a plastic tower.

A simple rubber gasket seals the valve. The design holds up well, but when it fails, all you'll need to do is replace the gasket.

Thread the cap from the tower, lift the cone, and press a new gasket in place.

(Address is: Mansfield Company, 150 First St., Perrysville, Ohio 44864; toll-free phone number: 1-877-850-3060.)

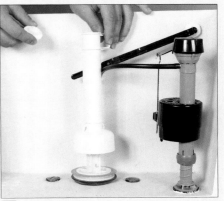

1 *Thread the plastic cap from the flush-valve tower.*

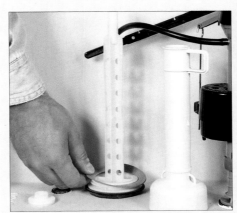

2 *Lift the tower to expose the sealing gasket.*

3 *Replace the gasket, and reinstall the tower.*

Fixing Fill Valves to Cure Hissing

MONEY SAVER

Adjust the float rod by bending it downward carefully.

Fine-tune the float by turning its adjustment screw.

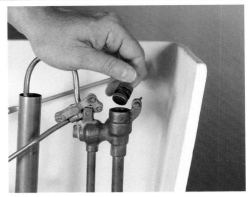

Replace the rubber seals on the plunger of a brass ballcock.

8 Plumbing

REPLACING A FLUSH VALVE

A flush valve fails when its valve seat can no longer form a seal to hold water. As water leaks past a defective flapper or tank ball, it can cut channels through the valve seat. When these voids appear, you have two repair options: install a retrofit flush-valve seat right over the damaged seat, or separate the tank from the bowl and replace the flush valve. Retrofit kits are easier, but a new flush valve makes a longer-lasting repair. (See "Retrofit Kits to the Rescue," right, for information on how to install a retrofit flush-valve seat.)

To replace a flush valve, start by shutting off the water, either at the main valve or at the shutoff valve beneath the toilet. Lift the lid from the tank, and flush the toilet to drain as much water as possible. Sponge out any remaining water.

Remove the Tank

With the tank empty, loosen the coupling nut that secures the water-supply tube to the fill-valve shank, and remove the tube. Reach under the tank, and using a socket wrench, remove the nuts from the two tank bolts. If the bolts spin, backhold them with a large screwdriver. In most cases, the nuts will turn free, but if your toilet is old and the nuts haven't been disturbed in many years, it's reasonable to expect them to resist. If the nuts seem really stuck, forgo the wrench and reach for a hacksaw blade. Brass bolts are relatively soft, and you should be able to cut through them quickly and easily. Use the blade from a standard hacksaw, and wrap one end with duct tape to serve as a makeshift handle. You'll use only the blade because you can't fit a hacksaw between the tank and the bowl.

With the tank free, lift it from the bowl, and lay it on its back. (The photos show a cut-away tank upside down for clarity.) If you're working on a finished floor or other surface that you don't want to mar, spread newspapers or a folded drop cloth. Pull the rubber spud washer from the shank to expose the spud nut as shown in photo 3. Back the spud nut from the shank with groove-joint adjustable pliers or a pipe wrench. You might have to grip the flush valve with one hand inside the tank or get a helper to lend a hand. Pull the old flush valve out, and scrape away any old putty from around the tank hole.

Install the New Flush Valve

Slide the rubber spud gasket over the new flush-valve unit's shank threads, tapered edge down. Insert the shank of the new valve body through the tank hole, and thread the spud nut onto the shank, finger-tight. Before tightening the nut further, rotate the flush valve so that the overflow tube is nearest the ballcock or fill valve and the seat and flapper face away from it. Then, while holding the valve in place, draw the spud nut down with pliers or a wrench until it feels snug. Slide the new spud washer over the spud nut, and coat its tapered edge with pipe joint compound.

Next, set the tank on the bowl. Slide washers onto the tank bolts; coat them with pipe joint compound; and install the bolts. Place a washer and nut on each bolt, and draw the nuts tight alternately, a little at a time. Keep in mind that the tank is not meant to rest directly on the bowl but should remain suspended on the flush-valve's spud washer. If you attempt to draw the tank down to meet the bowl, chances are you'll break one or the other. When the nuts begin to feel snug, stop and check the level and plumb alignment of the tank. The goal is a level tank that sits parallel with the back wall. Don't be concerned if the tank wobbles a bit on the spud washer. It will firm up when you fill the tank with water.

Connect the flapper. (See "Replacing a Flapper," page 214, for how to do this.) Then reinstall the water supply tube, and turn the water back on. Watch the flapper operate through several flushes, and make adjustments as needed. Check the bottom of the tank bolts for leaks, and continue to check them several times over the next few days. If no leaks appear in the first week or so, none should appear thereafter. If you feel dampness or see water at the ends of the tank bolts, tighten them just a little more and dry them. If they leak again, tighten them more, but only one-quarter to one-half turn or so at a time. Don't overdo it. The greater hazard is in over-tightening the bolts.

Retrofit Kits

If you're not up to separating the tank from the bowl to repair a damaged flush valve or if the toilet has its flush valve and over-flow cast into the china, then a retrofit valve-seat replacement kit is a good choice. These kits usually consist of a stainless-steel or plastic seat valve, an epoxy-putty ring, and a flapper. The super-tough epoxy putty ring adheres the entire unit onto the old seat. The kit can be used over brass, china, or plastic flush valves. Another type of kit consists of a tube of silicone adhesive, and a plastic replacement seat valve.

Replacing a Flush Valve

USE: ▶ socket wrench • flat-blade screwdriver

1 Loosen the tank bolts with a socket wrench while backholding the slotted bolt head using a flat-blade screwdriver.

5 Install the new flush valve through the tank hole, and position the overflow tube to fit the fill tube on the fill valve.

TOILETS

USE: ▶ replacement kit

1 Peel the protective paper from the epoxy ring.

2 Press the replacement seat over the old seat.

3 Connect the chain with about ½ in. of slack.

MONEY SAVER

• hacksaw blade • adjustable wrench • groove-joint pliers ▶ new flush valve • pipe joint compound

2 If the tank bolt won't break free, you'll have to use a hacksaw blade on it. Wrap the end of the blade you will be holding.

3 Lift the tank off of the bowl, and turn it upside down on the floor. Remove the old flush valve's spud-nut washer.

4 Loosen the spud nut using groove-joint pliers. The nut should break free easily.

6 Thread the large spud nut onto the flush valve at the bottom of the tank, and tighten the nut with pliers until it squeaks.

7 Fit the large rubber spud washer over the spud nut, and coat the washer with a generous layer of pipe joint compound.

8 Install washers on the tank bolts, and coat them with pipe joint compound. Insert the bolts through the tank and bowl.

REPLACING A FLUSH VALVE

8 Plumbing

REPLACING A FILL VALVE

If you're having trouble with a toilet's fill valve, it's usually best to replace the entire unit. You'll see several types on the market. The most familiar is probably the traditional ballcock, in brass or plastic, but you'll also see some that have floats that slide up and down on a vertical riser and some low-profile valves that are activated by head-water pressure.

Codes require fill valves to have built-in backflow protection. Backflow preventers keep tank water from back-siphoning into the water system. Although all codes require them, some manufacturers make both protected and nonprotected fill valves. Check the product labeling, and choose a valve that lists built-in backflow prevention as one of its features.

Remove the Old Ballcock

To remove the old fill valve (ballcock), start by shutting off the water and flushing the toilet. Sponge any remaining water from the bottom of the tank. Loosen the coupling nut that binds the supply tube to the ballcock shank. Then loosen the compression nut that binds the bottom of the supply tube to the shutoff valve. Finally, remove the jamb nut from the ballcock shank, and lift the old assembly from the tank. Be prepared to catch any water that remained in the tank.

Install the New Valve

Scrape away any old putty or pipe joint compound from the area around the tank opening. Slide the sealing washer onto the new fill valve's threaded shank, and coat the bottom of the washer with pipe joint compound. Insert the shank through the tank hole, and thread the jamb nut onto the shank threads. Before tightening the jamb nut, make sure the fill valve is aligned properly in the tank. If you're installing a traditional-style ballcock with a float ball, make sure the ball doesn't contact the tank wall or the overflow tube. Ideally, the ball should ride at least ½ inch away from the back of the tank. Grip the fill valve to keep it from rotating

---

Toilet Supply Lines

The traditional (but not necessarily the best) supply tube is made of chromium-plated copper, with a rubber cone washer or a plastic flat washer, fitted to meet the fill valve's shank. You cut the other end to length, and join it to the shutoff valve with a compression fitting. These tubes are supple, and if the offset is not too severe, you can bend them without a tubing bender. Make the bend as high or low on the tube as possible.

Hold the tube between the fill valve's shank and the shutoff valve. This should give you a rough idea as to the degree of offset and the length you'll need. Make the offsets so that each end of the tube will enter its fitting dead straight. Trim the tube, and then slide on the coupling nut, followed by a compression nut and ferrule. Lubricate both ends with pipe joint compound.

To gain enough vertical clearance to fit the tube between the fittings, press down on the shutoff valve. If the valve won't budge, bend the supply tube just enough to clear the fittings, and then straighten it again. Make the compression connection at the shutoff valve first. To keep from overtightening the compression nut, turn it finger-tight while wiggling the tube a little to keep it from binding. Then tighten it one full turn using a wrench. Finally, tighten the coupling nut at the tank, and turn on the water to test for leaks.

In addition to chromed-copper tubes and the stainless-steel mesh tubing mentioned earlier, you'll also see ribbed-copper tubes, which are easier to bend, polybutylene tubes, and polymer tubes reinforced with nylon webbing. Only the chromed copper, ribbed copper, and stainless-steel mesh tubes have wide code approval.

Traditional supply risers use compression fittings to make the seal at the shutoff valve.

Replacing a Fill Valve

USE: ▶ groove-joint pliers • adjustable wrench

While the tank is empty for replacing the fill valve or any other repair, it's a good time to take care of any condensation problems. Puddles of water around the toilet tank during hot weather is a sure sign that the cold water in the tank is causing the moisture in the outside air to condense on the side of the tank and drip onto the floor. Condensation could cause damage to the floor around the toilet. You can buy sections of insulation for the inside of the tank to prevent this problem.

4 While holding the unit steady, carefully tighten the fill valve's jamb nut with pliers until the nut feels snug.

TOILETS

against the fill tube, and then tighten the jamb nut until the sealing washer flattens out and the nut feels snug.

Next, install the fill tube between the nipple on the fill valve and the flush valve's overflow tube. Use the provided fitting to hold the fill tube on the overflow. You can adjust the float (a ball float if you're installing a ballcock) now by approximating where you want the water level. With a vertical fill valve, pinch the stainless-steel adjustment clip on the float rod, and move the float cup up or down. To adjust a ballcock float, you may need to bend the float rod or tighten the adjustment screw. See "Fixing a Hissing Toilet," page 214.

Hook Up the Supply Line

The last step is to reconnect the water supply. If the new fill valve's shank extends the same amount from the bottom of the tank as the old one's, you can just reconnect the old water supply tube. If the new shank is more than ⅛ inch longer or shorter than the old one, however, you probably need a new supply tube. There are a number of different supply tubes available. Some are mentioned in "Toilet Supply Lines," opposite. Perhaps the easiest to install is a pre-fitted tube made of polymer plastic encased in stainless-steel mesh. You just attach the couplings at each end of the tube, and you're done.

Flush Valve Test

To test for a leaky flush-valve cartridge, pour water around the activator stem, and flush. Bubbles indicate a leak.

MONEY SAVER

▶ new flush valve • pipe joint compound • new supply tube (if necessary)

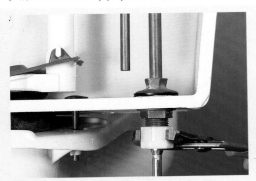

1 Using a pair of large groove-joint pliers, loosen the coupling nut that connects the supply riser to the ballcock shank.

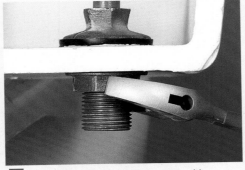

2 Loosen the ballcock jamb nut with an adjustable wrench while gripping the ballcock unit from above.

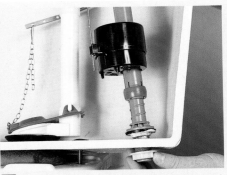

3 Insert the new fill valve through the tank hole, and tighten the jamb nut. Apply pipe joint compound to the washer.

5 Measure, cut, and install the fill tube. Don't allow it to kink. Connect one end to the fill valve and the other to the overflow tube.

6 Adjust the fill valve's float by compressing and sliding the clip up or down the lift wire. This will control the water level in the tank.

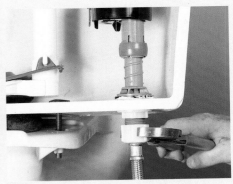

7 Using the pliers again, tighten the supply riser's coupling nut onto the fill valve's exposed shank threads.

8 Plumbing

PRESSURE-ASSISTED TOILETS

Pressure-assisted toilets function differently from gravity-flow toilets. They are based on a different operating principle: use of air pressure. The toilets offer limited repair options, but you can attempt a few remedies.

Running Toilet

The toilet keeps running between flushes or produces a loud buzzing noise after each flush.

This behavior suggests that something is keeping the flush-valve cartridge—the central, top-mounted fitting in the tank—from closing. Although an occasional trickle of water into the bowl between flushes is not unusual for these toilets, if the water actually seems to run while the toilet is idle, you'll need to investigate. (See "Fixing a Running Pressure-Assisted Toilet," p. 202.)

If the toilet has a flush button mounted in the tank lid, the button's trim collar may be riding on the activator. Remove the tank lid, and flush the toilet. If the toilet now flushes properly, the collar was the problem. Replace the tank lid, and remove the flush button. While sighting through the lid opening, move the collar until it is centered over the activator. When you replace the button, make sure it travels downward at least 1/8 inch before contacting the activator. If it doesn't, loosen the locking screw; adjust the activator; and reset the screw. If the flush button doesn't seem to be the problem or your toilet has a flush lever, shut off the water; drain the tank completely; and then turn the water back on. Hereafter, avoid pressing the flush button or lever before the tank fills completely. As with a flush button, the flush lever needs 1/8 inch of clearance.

If adjusting the activator doesn't help, suspect the pressure-regulating valve. (A faulty valve may also cause the toilet to take longer than normal to fill.) The regulating valve is sandwiched between the check valve and the relief valve in the water supply group. To replace it, you'll have to remove the entire supply group. Shut off the water; loosen the coupling nut beneath the tank; and remove the water supply tube. Then remove the jamb nut that holds the supply group in the tank. Lift the supply group out of the tank, and take it apart to expose the pressure-regulating valve. Remove the old regulating valve, and insert a new one. Reconnect the assembly, and tighten all connections with a wrench, backholding with a second wrench. Turn the water back on. If replacing the regulating valve doesn't correct the problem, suspect a faulty flush-valve cartridge. (See "Flush-Valve Cartridge," below.)

Inefficient Flushes

The toilet flushes sluggishly and doesn't clear the bowl.

This symptom may simply mean a clogged trap, so use a plunger or closet auger on the bowl before tearing into the tank. If the trap is clear, suspect a faulty flush-valve cartridge. Lift the tank lid, and check for water on top of the cartridge. If you see any water at all, replace the cartridge. If you don't, you'll need to conduct a little flush-valve test.

Turn off the water, and flush the toilet. Hold the flush lever down a full 60 seconds to drain the tank completely. Then pour a little water into the hollow around the activator stem. Turn the water back on, and allow the tank to fill completely. Then check for air bubbles around the activator. If no bubbles appear, assume that you have a fouled air inducer. Remove the inducer and clean it. If you do see bubbles around the activator, you'll need to replace the entire cartridge.

Cleaning a Fouled Air Inducer

USE: ▶ groove-joint pliers or adjustable wrench • vinegar (if necessary)

1 With the tank drained, remove the inducer's plastic nut. Place a hand under the inducer to catch the parts.

2 Hold the tiny brass or plastic poppit fitting under running water, and roll it between your fingers to clean it.

Flush-Valve Cartridge — MONEY SAVER

USE: ▶ needle-nose pliers • plumber's grease ▶ flush-valve cartridge

1 Insert the pointed ends of a large pair of needle-nose pliers into the fins of the cartridge's top nut.

2 Coat the top threads and adjacent O-ring of the cartridge with plumber's grease, and tighten the cartridge in place.

TOILETS

FLANGES & GASKETS

A toilet flange, or closet flange, is a slotted ring, usually connected to a vertical collar. The collar fits through the floor, while the slotted ring, or flange, rests on top of the floor. Toilets are bolted directly to this fitting. In the case of a cast-iron flange, the collar slides over a riser pipe, which extends to floor level. The gap between the pipe and collar is packed with lead and oakum. Beeswax rings, called bowl wax gaskets, eventually replaced putty as the preferred gasket material for toilets. (You'll also see the rings referred to as wax seals.)

These wax rings were able to accommodate the slight flexing that occurs between a toilet and floor, so for generations bowl wax gaskets have been and still are the standard. You can buy gaskets in 3- or 4-inch-diameter sizes by about 1 inch thick. They are inexpensive and durable—a hard combination to beat.

Still, improvements are always in the works. One improvement was to incorporate a plastic, funnel-like insert in the traditional wax gasket. The insert was designed to deliver the water well past the flange surface, thereby eliminating leaks between floor flange and the wax. These special seep-proof gaskets are often used when a toilet is installed on a concrete slab.

Rubber

Next in the progression came flexible-rubber gaskets, which when compressed, block the lateral migration of water. Rubber gaskets can also reseal themselves once disturbed, and they're reusable. They are sold in several thicknesses to accommodate a variety of flange heights relative to floor height, and you need to buy precisely the right thickness (unlike wax gaskets, which are more forgiving of small height differences). If a rubber gasket is even a little too thick, the toilet won't rest on the floor. If it's too thin, it won't seal.

The benefits of rubber and wax have recently been incorporated into a hybrid gasket, a neoprene rubber ring with a wax coating. (These gaskets are also available with seep-proof inserts.) The advantage is that the rubber will bounce back to reseal itself, while the layer of wax is more forgiving of sizing errors.

Measuring for Rubber

Wax gaskets come in standard thicknesses, and easily conform to slight job-site differences (flooring thickness and the like). In most cases, just knowing the horn length is good enough. To get the right thickness for a rubber flange gasket, you need to lay a straightedge across the base of the toilet and measure the length of the horn. Subtract the thickness of the toilet flange (minus any finish flooring such as tile), and add ⅛ to ¼ inch. The result is the ideal gasket thickness. Buy one as close to that thickness as possible. Measuring is not difficult, but it is bothersome. For that reason, you should use rubber gaskets only when the toilet is likely to be bumped frequently, as it might be when used by a physically handicapped person. A rubber gasket is better able to reseal itself after being disturbed.

Installing a Wax-Free Gasket

Fluidmaster has a new gasket assembly that fits inside 3-inch-diameter drainpipes (plastic, copper, or cast iron) and comes with a sleeve that fits 4-inch drainpipes, so you're ready for any situation. O-rings ensure a tight seal in the drainpipes.

Before you install the gasket assembly, insert the 3-inch-diameter gasket into the drainpipe. If it is too loose, you'll need to use the 4-inch sleeve. You will also find two O-rings, a thick one and a thin one, one of which goes at the bottom of the gasket or sleeve. Use whichever one gives you the best fit.

To install the gasket, just barely insert it into the drainpipe. Then wrap the square cardboard spacer that comes with the unit (not shown in the photo) around the gasket, and secure it at the two cut corners. Push the gasket down into the pipe until it touches the spacer or is 1 inch above the floor, whichever comes first.

Slip the supplied floor bolts into the toilet flange's mounting adjustment slots, and attach the supplied plastic retainer nuts to hold the bolts steady. Line up the holes in the toilet base with the bolts, and lower the toilet. Apply weight to the toilet to allow it to rest on the floor. The gasket will engage the toilet's horn, and the cardboard spacer will collapse and will not interfere with the gasket. Secure the toilet to the flange, and you're done.

Rubber Flange Gaskets
- Neoprene Gasket with Wax Coating
- Foam-Rubber Gasket
- Wax Free Toilet Seal
- Wax Gasket with Seep-Proof Insert
- Standard Wax Gasket

PRESSURE-ASSISTED TOILETS

8 Plumbing

LEAKY FLANGE GASKETS

If you see water on the floor around the toilet, first try to determine whether it is from leaking tank bolts or condensation. Often water that appears on the floor is not from the bowl but from the tank. If the tank has been bumped, water can leak past the bolts, dripping on the floor. Reach under the tank and feel for moisture clinging to the bolts. If they are wet, dry them and check back later. If they are still wet, try tightening the bolts a half-turn to a full turn to stop the leak.

If it isn't the bolts that are leaking or if more water appears with each use, the water is probably coming from the floor flange. Correct the situation as soon as possible. Water can delaminate plywood, blister underlayment, and rot the subfloor.

Quick Fix

If the toilet was installed within the past few months, then merely tightening the closet bolts on the base of the toilet may reseal the bowl's wax gasket. New wax gaskets almost always compress a little after installation, which can leave the bolts loose. Continued use can then cause the toilet to rock in place, breaking the seal. There's often enough wax to create a new seal—but only if you can draw the toilet and floor flange together.

Start by popping the caps from the closet bolts at the base of the toilet. Pry under them with a screwdriver or putty knife. With the caps removed, use a small wrench to test the tightness of the nuts. If they turn easily, tighten them only until they feel snug, and then watch the base of the toilet carefully over the next few days. If the floor stays dry, you've solved the problem.

Replacing the Gasket

If water reappears or if the bolts were snug in the first place, you'll need to take up the toilet and install a new wax gasket and closet bolts. (See "Taking Up and Resetting a Toilet," page 224.) If your toilet has been in place for years, don't expect a quick fix to work. Replace the wax gasket at the first sign of trouble.

Rubber Gaskets

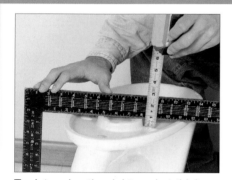

To determine the right gasket thickness use a straightedge and ruler to measure the toilet horn.

Flanges on Concrete

Drill anchor holes, and tap plastic anchors in place. Apply caulk to the bottom of the flange.

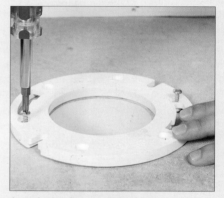

Attach the flange directly to the concrete using screws installed in the plastic anchors.

Installing Toilet Flanges on a Wood Subfloor

This plastic insert flange fits neatly inside a cast-iron riser.

Apply PVC cement to the flange and riser, and press in place.

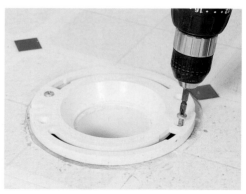

Attach the flange to the floor using panhead screws.

TOILETS

BROKEN FLANGES

Plastic toilet flanges are sturdy and seldom fail under normal conditions. Flanges made of cast iron and cast brass, on the other hand, are more vulnerable. The slotted portion of a flange is fairly narrow, so the slightest casting flaw will weaken it further. It's easy to break a flange by overtightening the closet bolts, but in time, even normal use can break a weak flange. When a flange breaks, the bolt on that side drifts outward, losing its grip on the flange. The toilet begins to feel loose, rocking in place when you sit on it. This movement soon breaks the gasket seal, and the toilet leaks with every flush.

If this happens, don't panic, because there's a fairly easy solution. Replacing the entire flange is the most professional repair, but doing so requires cutting and splicing drainpipe or replacing cast-iron fittings—a big job. The easier solution is a repair strap that works surprisingly well. The crescent-shaped strap mirrors the shape of a flange and has an opening for a closet bolt. To use this strap, first remove the toilet. (See "Taking Up and Resetting a Toilet," next page.) Insert a closet bolt through the repair strap; slide the strap under the old flange, next to the break; and install a wax gasket. If the strap won't slide under the flange, loosen the floor screws and lift the flange slightly with a pry bar. When you get the strap in place, retighten the screws.

Flange breaks involve small sections of metal, so there's usually enough flange left to support the repair strap. It's a neat trick, and one that saves hours of work.

To Caulk or Not To Caulk?

When it comes to caulking around the base of the toilet, opinions vary. Some local codes require it, some forbid it, and others don't address it at all. The advantages of caulking around the base of a toilet are that the caulk **1)** has good adhesive qualities, **2)** helps preserve the seal by keeping the toilet from moving, **3)** keeps the joint free of bacteria and other dirt, and **4)** looks better than a dark, soiled perimeter crack (as long as the joint is neatly done).

The disadvantage is that when a leak does occur, it can't be seen, so it does more damage.

There are generally no strong arguments one way or the other, except in two circumstances.

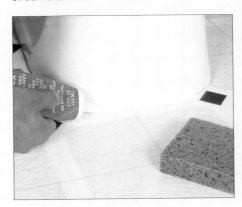

- Caulking is usually recommended for toilets that are set on concrete. Concrete is never level around a toilet flange, and the added support from the caulked joint helps keep the bowl from rocking.

- Caulking is usually not recommended in a bathroom with wood floors (a poor flooring choice). These floors buckle with the slightest penetration of water, and you'll want to see a leak as soon as it starts.

If you choose (and are permitted by local code) to caulk around the base of a toilet, use latex tub-and-tile caulk. Leave the back of the toilet base open for water to escape in case a leak occurs. Start by wetting the floor and toilet base; then apply a liberal bead of caulk in the seam on the three visible sides. Draw a finger all around the joint so that any voids are filled, and wipe away the excess using a damp rag or sponge. Keep in mind that latex caulk will shrink a bit when it cures, so leave a little more in the joint than seems necessary. The caulk will dry to the touch within the hour but will take several days to achieve adhesive strength. In the meantime, feel free to use the toilet.

Repairing a Cast-Iron Flange

USE: ▶ tools and materials for removing a toilet ▶ flange-repair kit • new wax gasket

1 After removing the toilet, lift the wax-and-plastic toilet flange insert from the drain, and scrape away any wax.

2 Insert a closet bolt into the repair strap, and slide the strap under the broken section of the toilet flange.

3 Install the other closet bolt, and press a new bowl gasket over the flange. A wax-over-neoprene gasket is shown.

LEAKY FLANGE GASKETS

8 Plumbing

TAKING UP & RESETTING A TOILET

Start by shutting off the water. Flush the toilet, holding the handle down to drain as much water as possible from the tank. Use a paper cup or other small container to scoop the water from the bowl, and use a sponge to soak up the remaining water in the bowl and tank.

Loosen the coupling nut that joins the water supply tube to the fill valve. Pry the caps from the closet bolts at the base of the bowl, and remove the nuts and washers from these bolts. If the bolts spin in the flange slot, keeping the nuts from threading off, try jamming them to one side so that they bind in the flange slot or against the toilet base. If this doesn't work, use needle-nose pliers to grip the top of the bolts while backing the nuts off. If all else fails, use a miniature hacksaw to cut through the bolts, just under the nuts.

With all connections undone, put down several layers of newspaper to protect the floor from the wax clinging to the outlet. Before moving the toilet to the newspapers, consider how you'll lift and carry it. A toilet is not terribly heavy, and one person can lift it with the right approach. Because the tank will make the toilet back-heavy when it clears the floor, the best place to grip the bowl is just in front of the tank. This deck area will have a lip, making it easy to grip. (Never lift a toilet by its tank.) Straddle the bowl, and with your feet just forward of the tank, tip the toilet up on one side. This will break the wax seal. Still straddling the bowl, carefully raise the toilet several inches and walk (duck-walk) it over to the waiting newspapers. (Rest your elbows on your knees when walking a toilet several feet.) Carefully lay the toilet on its side, and use a putty knife to scrape away any wax clinging to the bottom of the toilet.

Scrape the remaining wax from the surface of the flange, and slide the old closet bolts from their slots. Each slot has a large opening at one end where you can free the bolt head and remove the bolt. If you've cut off the old bolts, buy two bolts, nuts, and metal washers (usually as a kit), as well as a new wax gasket for the bowl. When shopping for a wax gasket, match your piping system's flange size, either 3 or 4 inches.

Set the Toilet

Slide the bolts into their slots, and center them across the outlet opening from each other, typically 12 inches from the wall. Then press the wax gasket onto the flange. Some sources suggest sticking the new gasket to the toilet, but when it's installed on the flange, the gasket keeps the bolts in place. In fact, you should make sure that each bolt is stuck to the wax gasket in an upright position. These bolts guide the toilet onto the flange, so it's important that they remain upright and centered.

To set the toilet, grip it near the hinge deck as before, and walk it over to the flange. Lift it just enough to clear the waiting closet bolts. Maneuver the toilet until you can see the closet bolts through the holes in the base of the bowl. When the bolts are visible, slowly lower the toilet onto the wax gasket. Check the alignment of the tank: the back of the tank should be parallel with the wall. If it isn't, rotate the toilet left or right until it is straight. Then press down on the rim of the bowl with all your weight. This will compress the wax enough for you to install the nuts on the closet bolts.

Secure the Toilet

If the decorative caps that covered the old closet bolts snapped over plastic retaining washers, install these washers under the metal washers that came with the closet-bolt kit. These plastic keeper-washers are marked "This Side Up." Place the plastic retainers over the bolts first, followed by the metal washers and the nuts. With the bolts ready, draw the nuts down with a 6-inch adjustable or combination wrench, alternating sides every few turns. Using a small wrench allows you to feel the resistance of the nut.

Continue drawing the nuts down until you feel resistance. Don't overtighten, or you'll break the toilet base. When the nuts feel snug, stop, and put all your weight on the bowl again. This will probably loosen the nuts enough to gain another turn with the wrench. Repeat this press-and-wrench sequence until the nuts no longer loosen under your weight. Then using a miniature hacksaw, cut the bolts just above the nuts. This may loosen the bolts, so retighten them as needed. Normally, a plumber would snap the decorative caps over the bolts at this point, but because you live in the house, leave them off for a few days. After some use, the nuts will probably loosen slightly. At this point, tighten them again, and then snap the caps in place. If your bolt caps don't have retainer washers, fill them with plumber's putty and press them over the nuts.

Hook Up the Water

You may be able to reuse the water supply tube. Just coat the washer with pipe joint compound, and tighten the nut over the fill-valve shank. If you've damaged the tube or just want to replace it, use a new stainless-steel mesh tube. Attach and tighten the valve end of the tube first. Use two wrenches, one to tighten the compression nut and one to backhold the shutoff valve. Then attach the tank end of the tube. Turn the coupling nut onto the fill-valve shank until it's finger-tight; then snug it using a pair of groove-joint pliers. Don't overtighten the nut.

Finally, turn the water back on, and test your work. As always, look for leaks at the tank bolts. If the bolts are wet, tighten them just a little.

Changing Floor Height

If you've taken up the toilet to install new flooring, you may need to alter your approach when resetting the toilet. If all you've done is lay new vinyl over old (not a bad idea considering that asbestos may lurk in old floor coverings and mastics), you'll be able to reset the toilet just as described. If you've installed ¼-inch plywood or cement-based backer board underlayment and/or glazed tile, however, you'll have added to the height of the floor relative to the flange. You'll need to compensate for this increased depth with a thicker bowl gasket.

If the added floor height is less than ¼ inch, you can simply knead and stretch a standard wax gasket to give it a taller profile. With an increase of ¼ inch up to about 1 inch, stack one standard wax gasket on top of another.

TOILETS

Taking Up and Resetting a Toilet

USE: ▶ adjustable wrenches • needle-nose pliers • newspaper • hacksaw • plumber's putty • putty knife • groove-joint pliers ▶ new wax-gasket kit • water supply tube

1 Unscrew the compression nut from the toilet shutoff valve. Then loosen the coupling nut at the top of the supply tube.

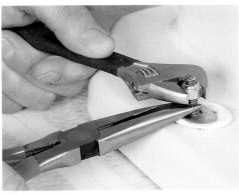

2 If the closet bolt spins in place, you probably have a broken flange. Hold the bolt with needle-nose pliers.

3 Grip the toilet just in front of the seat hinges; carefully lift using your legs; and carry the toilet to several layers of newspaper.

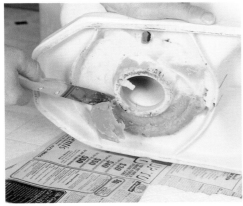

4 Gently tip the toilet on its side, being careful not to disturb the water tank, and use a putty knife to scrape away any old wax.

5 Install the new closet bolts and wax gasket. Carefully set the toilet over the bolts, and press the base into the wax.

6 Tighten the closet-bolt nuts with a small adjustable wrench (for a better feel), and stop when they feel moderately snug.

7 Using a hacksaw, cut off the bolt piece that extends above the nuts; retighten the nuts; snap on the decorative caps.

8 While backholding the shutoff valve with one wrench, tighten a new supply tube in place with another.

9 Finish by tightening the coupling nut using pliers. Stop when the nut feels snug and begins to squeak.

8 Plumbing

INSTALLING A NEW TOILET

A new two-piece toilet will come in two boxes, one containing only the bowl and another containing the tank, tank bolts, decorative bolt caps, retaining washers, and spud gasket.

Slide the new closet bolts into their flange slots so that they're centered across the opening and are the same distance from the back wall. Press a new bowl wax gasket onto the flange, and stretch the wax slightly so that the gasket holds the bolts upright.

Fasten the Toilet in Place

When carrying the new bowl, grip the seat hinge deck with one hand and the front rim with the other. Lift the bowl over the flange until you can see the bolts through the tank holes, and then settle the bowl onto the wax gasket. As when setting an assembled toilet, this is the time to step back and judge the straightness of the bowl. You won't be able to align the tank with the wall, so eyeball it as best you can. If you're not confident in sighting the unit, measure from the tank-bolt holes to the back wall. Each hole should be the same distance from the wall.

When you've aligned the toilet, unpack the cap-retaining washers and place one over each bolt. They'll be marked "This Side Up." Place a metal washer from the closet-bolt kit on top of each plastic retaining washer, and thread the nuts onto the bolts. Tighten the nuts.

Attach the Tank

With the bowl set, install the tank. The fill valve and flush valve are usually factory installed, but you should always check the tightness of the flush-valve spud nut.

Fit the large rubber spud gasket over the flush-valve threads and onto the tank bolts (along with the supplied rubber washers). Lubricate the tank side of the washers with pipe joint compound before installing them. Then lubricate the rest of the washers and the spud gasket with pipe joint compound. Insert the bolts through the tank holes, and set the tank on top of the bowl.

Start a washer and nut on each tank bolt. Tighten these nuts carefully with a small adjustable wrench, alternating from side to side.

Installing a New Toilet

USE: ▶ adjustable wrenches • closet bolts • pipe joint compound • groove-joint pliers ▶ new toilet • water supply tube • wax gasket

One of the mistakes many people make when installing a new toilet is that they forget about the accessories they must purchase separately. While picking out your new toilet at the home center or plumbing supply outlet, be sure to pick up a wax seal, a set of closet bolts, a water supply tube, and a toilet seat. Few things are more frustrating when attempting any home repair or improvement than to get home and not have all the parts you need.

1 Install new closet bolts in the toilet flange. If the bolts come with plastic washers, use the washers to hold the bolts in place.

2 Install the new wax gasket on the toilet flange. Press the bolts into the wax to hold them upright.

3 Using the closet bolts as guides, set the bowl onto the flange gasket. Site through the china bolt holes.

4 Insert the tank bolts through the bottom of the water tank, and install spacer washers if they are provided.

5 Set the water tank on the bowl deck, and tighten the tank bolts. Level the tank, but don't overtighten the bolt nuts.

TOILETS

Fixing a Rotted Floor

When you allow a toilet to leak long enough, the underlayment and possibly the subfloor around it swells and rots. The only option then, before replacing the flooring, is to cut out and replace the damaged underlayment area or cut the affected subfloor area back to the nearest joists to replace it. Remove the flooring, and probe the affected area using an awl or screw-driver to determine the extent of the damage and whether it extends past the underlayment to the subfloor.

REPLACING UNDERLAYMENT

When you replace just the underlayment around a toilet, you can leave the toilet flange in place and install the new plywood in two pieces. (The underlayment may be ½-inch or, more likely, ¼-inch plywood over ¾-inch subflooring.) Cut out the damaged area using a circular saw to cut just through the underlayment. Use a utility saw or reciprocating saw to cut through uncut areas in the corners or near walls. Cut the new plywood to size, and then cut it in half, with the cut intersecting the flange opening. Trim the plywood to accommodate the outside diameter of the flange collar (not the flange itself), and slide each half under the flange. (See top right.) Screw and glue the underlayment in place.

REPLACING SUBFLOORING

If the subfloor is damaged, you'll have to replace it. Cut the subfloor back to the nearest joist on each side where there is undamaged subflooring. A reciprocating saw is a good tool for this. Nail two-by blocking around the perimeter of the hole to provide nailing for the new plywood, sized and cut to match the existing subflooring. (See bottom right.)

There are two ways to proceed from here. One is to replace the old toilet flange and riser. If you have a cast-iron flange, convert the riser and flange to plastic. (See "Broken Flanges," p. 217.) If you have plastic pipes, cut off the toilet assembly at the horizontal pipe, and rebuild it exactly the same way using a coupling. Leave the flange off for now. Measure and cut a hole in the new plywood for the riser before installing it. Nail or screw down the subflooring with fasteners around the perimeter every 6 inches on center. Lay down new underlayment and flooring, and then install the new flange. From there you are ready to install the toilet.

The second way is to leave the flange in place. Nail two-by blocking between the joists on each side of the toilet riser. Then cut the plywood down the middle the same way as for underlayment, and install it, sliding it under the toilet flange. Then lay new underlayment and flooring.

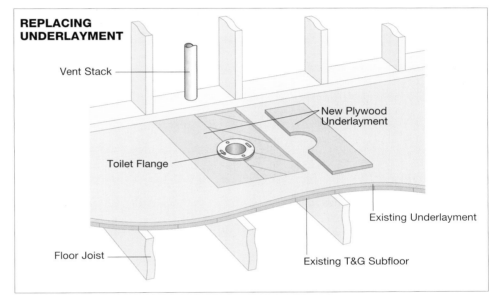

REPLACING UNDERLAYMENT

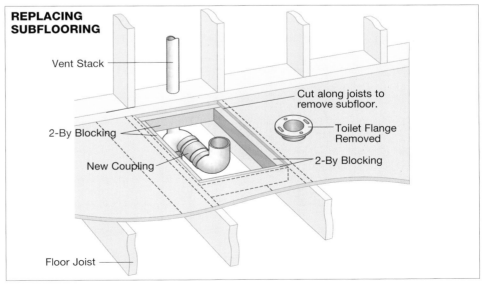

REPLACING SUBFLOORING

INSTALLING A NEW TOILET

8 Plumbing

SHOWER OPTIONS

Over anything but the most heavy-duty walls, using tile for showers is a case of good looks overwhelming good sense. You're building a watertight enclosure to resist leaks and to withstand the weight of people moving around inside but including weak links—namely, the leak-prone grout joints between tiles. The result is an attractive maintenance nightmare.

A popular alternative is to replace the tile with a seamless, one-piece fiberglass or solid-surface-material shower. Choices range from very narrow units that trigger claustrophobia to spacious combination tub-showers—some with luxurious whirlpool jets, extra spray nozzles, and steam generators.

The smallest sizes of one-piece units generally can be maneuvered through existing halls and stairwells, but before you buy, be sure to check that you can get the unit through the front door and into the bathroom. If you want to use more spacious models, there are two options. One is to take out a few extra studs during your bath remodeling job and to make sure that the shower is ordered in time for delivery while the walls are open. Another option is to order a special remodeling unit that comes in two or three pieces.

Cracking

Even well-made, rib-reinforced fiberglass can flex under body weight. That can lead to hairline cracks in the shower floor finish, which eventually lead to larger cracks that leak. To prevent flexing, mound a light masonry mix under the shower floor before the unit is installed and plumbed.

Consider this a mandatory part of the installation—even if it isn't included in the instructions and even if your contractor says that the fiberglass is strong enough without it. Making this supporting mud pie takes all of 5 minutes and could save the entire installation.

Sound Insulation

Packing the walls around the shower with insulation is not really necessary, but it's a nice touch that reduces the kind of tin-roof echoing of water against the thin shower walls. Some are even available with soundproofing insulation molded right into the unit.

Drains

Pop-up and plunger tub drains are notorious for getting plugged up, usually with some combination of hair and congealed soap. This is a particularly difficult blockage to move using only a plunger. So when you install a new tub enclosure, it pays to take three precautionary steps. First, install a wide-mouth plunger drain with both a removable, fine-mesh screen that is flush with the shower stall floor and another wide-mesh, or crosshair, screen directly underneath. These should catch most of the debris before it has a chance to move into the trap. Second, make sure that the drain linkages operate smoothly. Third, provide access to the trap, so you can take it apart and clear blockages.

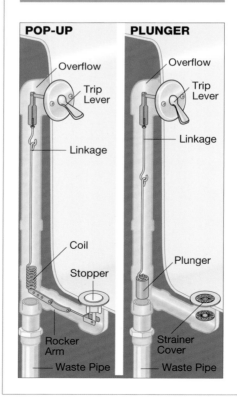

Fixing a Faucet

USE: ▶ slip-jaw pliers • socket wrench with deep

1 To repair a deep-set tub faucet, cut off the water supply; undo the handle; and back the retainer nut from the escutcheon.

Matching Faucet Types

When you replace a tub/shower faucet, the easiest option is to install a brass faucet that matches your old faucet's spread (its openings in the tile), especially if your piping is galvanized steel. In this case, you can use the new faucet's unions to connect the supply pipes. This approach is especially easy if the faucet has a union connecting the shower riser pipe. Of course, you may not want the same type of faucet. If not, you'll need to buy new tiles to patch the old openings, and you may need to make more complicated piping connections. When switching faucet types, it's usually better to convert to copper pipe.

A replacement faucet with the same center spread as the original is the easiest to install.

TUBS & SHOWERS

socket for faucets • screwdriver ▶ replacement parts (as needed)

MONEY SAVER

2 Use a deep socket kit designed to reach deep-set faucet fittings to back the faucet stem from the faucet.

3 Some older faucets have stems that seat inside replaceable sleeves. Stem parts and worn sleeves can be replaced.

4 Back the screw from the base of the faucet stem, and replace the washer. Make sure the replacement is the same size.

Drain Access

If you want to replace an inaccessible drain (that is, when the tub is installed on a concrete slab or above a finished ceiling below), you'll need to do the work through the back of the plumbing wall. If your tub has an access panel already in place behind the tub, you're in luck. If not, you'll need to create an access hole. To gain access, cut out the drywall between the two studs that straddle the center of the tub. When you've finished the new drain installation, you can either patch the drywall or install a permanent access cover.

You can make the cover out of plywood and door casing to create a frame around the access hole perimeter or purchase a ready-made plastic panel.

When the opening is in plain sight, it's best to repair the wall to avoid lookng at an unsightly access panel all the time. When the access hole is in a closet or concealed by furniture, you can install a permanently accessible panel.

Bathtub Installation Anatomy

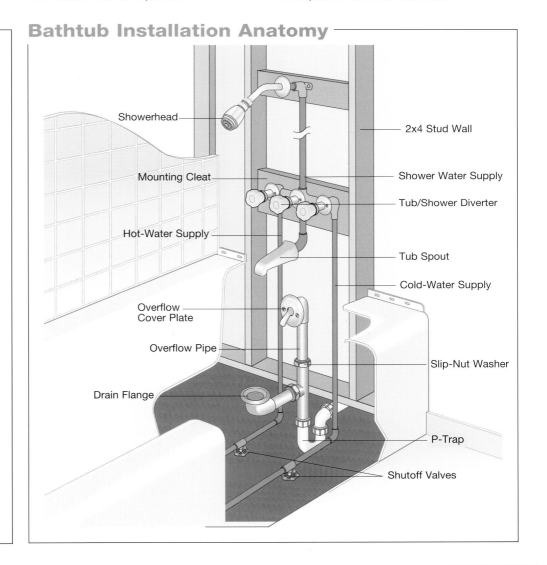

SHOWER OPTIONS 229

8 Plumbing

REPLACING A TUB DRAIN ASSEMBLY

At some point, every drain assembly needs to be replaced. Perhaps its finish has deteriorated, or the linkage is broken, or you'd like to switch from chrome to polished brass.

Start by removing the overflow plate and pulling the tripwaste linkage from the overflow tube. If your drain has a pop-up stopper in place, remove it as well. Then insert the handles of an old standard set of pliers into the drain opening, and engage the crosspiece in the bottom of the drain fitting. With a pop-up style drain, engage two brass tabs protruding from the inner rim of the fitting. (Special drain-spud wrenches are also made for this task.) Grip the jaws of the pliers with an adjustable wrench, and turn the pliers counterclockwise. You should be able to unscrew the fitting from the threaded drain shoe.

Disconnect the Drain

If the fitting won't budge, move to the underside of the drain. Between the drain shoe and the bottom of the tub, you'll see a rubber gasket. Place a hacksaw against the gasket, and saw straight through it and the drain spud. (You can also use a reciprocating saw.)

With the drain connection severed, the P-trap will hold the entire assembly in place. If you have room to maneuver, loosen the trap's friction nut and lift the entire assembly out in one piece. If not, loosen the friction nuts that hold the waste and overflow components together and take out the assembly in pieces. In most cases, the trap will be attached to the waste assembly by means of a friction nut and flat washer. If the trap is made of plastic, the joint usually comprises a ground-joint adapter and compression-style nylon washer.

Install the New Tub Drain

With a conventional brass waste-and-overflow assembly, begin by pressing plumber's putty around the underside of the drain flange, rolling the putty as you would for a sink drain. Then have a helper place the rubber gasket on the drain shoe, and hold the shoe up to the tub's drain outlet from below. Feed the drain fitting through the tub opening, and thread it into the shoe. Insert pliers handles into the drain opening so that the handles engage the side tabs or crosspiece. Grip the pliers with an adjustable wrench, and tighten the fitting until most of the putty squeezes from beneath the flange. Trim the excess putty, and continue to tighten the fitting until it feels snug.

Next, slide a nut and washer onto the overflow tube, and tighten the tube into the waste T-fitting. Press the large rubber gasket onto the overflow flange, and lift the tube into place. Place a second nut and washer over the drain shoe, and connect it to the waste T-fitting using groove-joint pliers. Apply a light coating of pipe joint compound to the fine threads of the brass tailpiece, and turn the tailpiece into the bottom of the waste T-fitting. Slide a third nut-and-washer set onto the tailpiece; insert the tailpiece into the trap riser (or ground joint adapter); and tighten the compression nut. If the trap riser is fitted with a ground joint adapter, you'll be able to reuse the original nylon compression washer. If the riser is made of metal pipe, buy a rubber friction washer to make the seal.

Connect the Tripwaste

With the tubing components assembled and connected, feed the tripwaste linkage into the overflow tube until the cover plate meets the tub. Thread the two bolts through the cover plate and into the overflow flange. Tighten these bolts until they feel snug.

Replacing a Tub Drain Assembly

USE: ▶ screwdriver • groove-joint pliers • hacksaw • pipe joint compound • plumber's putty

Unfortunately, most people do not have the access we show here when replacing a drain. If you are lucky the tub may sit over an unfinished basement where access, while difficult, is usually much easier that trying to work on a tub drain that rests over finished space. In those cases, the only course is to cut through the ceiling, make the repairs or improvements and then patch the area. It is also a good idea to have same-level access to the plumbing lines of a tub or shower. An access door in an adjoining room is the best bet.

1 Begin by unscrewing the tripwaste cover plate and pulling the tripwaste linkage from the overflow (inset).

5 Install the rubber gasket on the bathtub overflow tube, and slide the tube upward into the wall cavity from below.

6 Join the drain and overflow pipes in the drain T-fitting, and tighten all the nuts using large groove-joint pliers.

230 **PLUMBING** / TUBS & SHOWERS

TUBS & SHOWERS

If your tripwaste has a pop-up plug and lever, feed them into the drain shoe with the trip lever in the open position. If your tripwaste has an internal plunger, install the hair screen over the drain fitting. If your screen needs to be fastened with a screw, be careful not to overtighten the screw, or you'll bend the screen.

The only way to know whether your tripwaste was adjusted correctly at the factory—they're usually close—is to test it. If it drains too slowly, there is probably a clog in the line and you may have to clear the clog with a power auger.

When the line seems to drain correctly and is working well, test the overflow seal by filling the tub to overflowing. Be present during the filling process in case a problem develops. Look for water around the rubber gasket on the back side of the connection. If it leaks, remove the cover-plate screws and look for an obvious problem. Some types of gaskets come with a front flange designed to fit through the bottom half of the tub opening. Also, check to see that the overflow flange is centered in the tub opening. If it's not, loosen the nut securing the overflow tube to the waste T-fitting, and adjust the flange up or down as needed. If a minor leak persists, this usually isn't a problem—simply seal the gasket with silicone caulking.

Pop-Up Drain

A foot-operated drain plug just threads into its drain fitting. Press it to open and close the drain.

▶ new drain kit

2 If you can't free the old drain spud nut, cut through it using a hacksaw blade or a reciprocating saw.

3 Loosen the trap connection to free the overflow tube. Expect to find a friction washer or compression washer.

4 Stick plumber's putty to the drain flange (inset), and screw the spud into the drain shoe and gasket.

7 Pull the plastic trap down far enough to allow you to screw the tailpiece into the drain T-fitting.

8 Slip the compression nut and washer onto the tailpiece, and tighten the nut to secure the tailpiece in the trap riser.

9 From inside the tub, slip the tripwaste linkage into the overflow tube until the cover plate sits against the tube's gasket.

REPLACING A TUB DRAIN ASSEMBLY

8 Plumbing

REPLACING A TUB-SHOWER FAUCET

Begin by turning off the water at the shut-off valve. Pry the index caps from the handles, and remove the handle screws. Pull the handles off, and remove the stem escutcheons or trim plate. Escutcheons are threaded and screw on and off; trim plates are secured by two screws. Your faucet will have one type or the other.

Remove the Faucet

The tub spout will also be mounted in one of two ways; either threaded onto a ½-inch-diameter nipple or clamped onto a copper stub-out with a setscrew. Reach under the spout, and feel for an Allen screw recess. If you find a screw, loosen it and pull the spout off. If you don't find a screw, the spout is threaded. Grip the spout with a strap wrench (or padded pipe wrench or large pliers), and back it out. (You can also insert pliers handles or a large screwdriver into the spout opening and twist.)

For removing tile, it's best to buy an inexpensive grout-removal tool. Plan on removing an area of wall around the faucet measuring about 12 x 16 inches. Try to confine the opening to the space between studs. Draw the grout-removal tool along each seam repeatedly until you've gouged most of the grout from the joints. Then, using a sturdy putty knife, pry under one of the ceramic tiles until it pops loose. Pry at several points and in several directions. If the tile won't budge, try one of the tiles next to the faucet opening. Grip the edge of one of these tiles and pull. Remove the remaining tiles, working from the center outward. When you've pried all tiles from the removal area, cut the wallboard just inside the remaining tiles and pull this section out.

How you remove the old faucet will depend on the piping material in place. If you find galvanized steel or brass, slip a 10-inch pipe wrench into the opening and loosen the two union nuts at the bottom of the faucet. If the shower riser pipe is joined to the faucet with a union, loosen it as well. If the riser is threaded into the faucet, use a hacksaw or reciprocating saw to cut it just above the faucet. Finally, push the faucet back until the spout nipple clears the wall. If you find copper pipe and soldered joints, the easiest approach is to cut all pipes near their faucet connections and pull the faucet straight out.

Install the New Faucet

The faucet connection methods will be dictated by the kind of unit you buy: whether it's a single- or dual-control model; whether it has union fittings, threaded ports, or soldered joints; and so on. If your piping is made of galvanized steel, you'll also need to consider the problem of electrolytic corrosion, which results when you join copper and steel piping directly.

If direct copper-to-steel connections are not a problem in your area, begin by threading galvanized-steel couplings onto each riser, using pipe-thread sealing tape. (It's a good idea to use couplings, with their female ports, because copper stretches when threaded over steel, often enough to cause a leak.) As always, be sure to use a second wrench to backhold the supply pipes when tightening the new fittings. With the steel couplings in place, wrap three rounds of pipe-sealing tape counterclockwise over the threads of two copper male adapters, and tighten the adapters into the couplings.

If you had to cut the old steel shower riser to remove the faucet, you'll have to install

Replacing a Tub-Shower Faucet

USE: ▶ strap wrench • screwdriver • grout removal tool • pipe wrenches • solder, flux, torch ▶ new faucet

The installation shown here is typical of the situation in many bathroom: you must cut into the wall to upgrade the tub faucet. If your piping system is already made of copper, cut the existing copper risers or unscrew the faucet unions and solder new copper in place, using sweat couplings and 90-degree elbows. For old cast-iron pipe, unscrew the unions, disconnect the faucet, and attach galvanized couplings and copper male adapters to make the conversion to copper. Use pipe-thread sealing tape on the adapters. (See far right.)

1 If you'd like to save the old spout, use a strap wrench to unscrew it from the faucet and avoid damaging the finish.

4 Loosen the faucet unions with a pipe wrench, and lift out and discard the old faucet. Remove and discard the unions.

5 When adding piping for the shower-head, cut an opening in the wall to secure the drop-eared elbow to blocking.

TUBS & SHOWERS

a new one. There are several options. You can remove the old riser and fish Type L soft copper in its place through the faucet opening, uncoiling the pipe as you go. This requires cutting a second opening at showerhead level. If the basement ceiling beneath the tub is open, you can usually install rigid piping from below. In any case, solder a drop-eared elbow to the top of the pipe, and feed it into the wall. Screw the elbow to a backing board.

With a new shower riser ready, solder copper inlet stubs into the side ports of the valve. The easiest approach is to make them a little long; then hold the faucet in place and measure between the stubs and the copper adapters on the supply tubes. Cut these lengths, and test-fit the riser-to-faucet connections.

With the faucet temporarily connected to the supply tubes and shower riser, cut stubs for the tub spout. If your new spout is threaded, solder a ½-inch-diameter male adapter to the end of the horizontal pipe so that the threads of the adapter protrude through the finished tub wall about ½ inch. If the spout clamps onto the pipe with an Allen screw, bring the copper pipe through the wall about 3 inches.

When you have all of the pieces fitted, pull the assembly apart and flux the ends of the stubs. Before soldering, remove the faucet cartridge or stem to keep from warping the plastic components.

Solder as much of the assembly as possible outside the wall. Insert the stubs into the faucet ports; lay the faucet on the floor; and solder each joint carefully. After the valve cools, flux the remaining joints; connect the faucet to the in-wall piping; and solder these joints. Be careful when soldering next to combustible surfaces inside the wall. Use a flame-shield and wet the wood near the faucet. When the faucet cools, reassemble it and turn the water on to test your work. When you're sure you have no leaks, repair the wall and install the spout, showerhead, and faucet trim.

3 Start near the spout or faucet stems, and carefully pry each tile away from the wall. Set the tiles aside for reuse.

2 With the spout and escutcheons removed, use a grout saw to strip the grout from the tiles in the removal area.

7 Remove the nylon faucet cartridge from the faucet body, and sweat all copper fittings using lead-free solder.

6 Complete the new faucet piping in copper. Install the spout pipe so that the spout will be 6 in. below the faucet body.

Galvanized Pipe

1 Apply pipe-thread sealing tape, and then thread ½-in. galvanized couplings to the old faucet supply tubes.

2 Using pipe-thread sealing tape (not pipe joint compound), screw ½-in. copper male adapters into the couplings.

REPLACING A TUB-SHOWER FAUCET

8 Plumbing

INSTALLING A CAST-IRON BATHTUB & SHOWER

Whether working with new construction (in a new home or addition) or replacing an old tub, the process is similar. A new installation requires a few more steps, so this procedure assumes a new tub in a new bathroom. The time to install the tub is when the piping is roughed-in and before the drywall goes up.

First, you need to cut a drain opening in the floor. This should be an oversize rectangular hole because the drain shoe will extend below the tub about 10 inches back from the wall. The hole will need to be centered in the tub space, which is typically 30 inches, so mark a cutout that is centered 15 inches from the long wall. The actual size of the hole should be 8 x 12 inches, with the longer dimension extending away from the plumbing wall. Use a circular saw or reciprocating saw to make the cut.

Set the Tub

A cast-iron tub weighs more than 300 pounds, but if you're careful you can set one with little or no help. Uncrate the tub, and stand it up. The easiest way to move an upright tub is to "walk" it. Grip the upper end, tilt it toward you slightly, and then rock the apron from corner to corner. Each time a corner leaves the floor, pull the tub toward you a little. Walk it, corner to corner, into the bath until you have it standing upright in the tub area. Maneuver it into one corner, and carefully lay it over until the high end comes to rest against the opposite wall. In new construction, where the walls are open, use a 2x4 stud to pry under the low end of the tub, through the wall. Each time you lift, the high end will settle closer to the floor and the low end will slide closer toward you. When the tub drops to the floor, push it against the long wall.

Cast-iron tubs are not meant to be secured. They stay put because they are so heavy and are locked in place by underlayment and drywall. You should support the tub along the long wall, however. Begin by stepping into the tub and walking around. If the tub rocks corner to corner even a little, make an effort to find out why. The problem can be an out-of-level floor or a slightly warped tub. With the tub pushed firmly against the back wall, lay a spirit level across the ends of the tub. If you see a discrepancy, slide a wooden shim under one corner of the tub. If this corrects the level and the rocking, mark the exact position of the shim and remove it. Apply a dab of construction adhesive to the bottom of the shim and slide it back in place. Then measure between the bottom of the tub deck, along the wall, and the top of the wall's soleplate. Cut two pieces of 2x4 to this length, and wedge them between the tub and soleplate. Nail them to nearby studs. These supports help spread the weight.

Install the Faucet Plumbing

When assembling the faucet piping for a tub or tub-shower, it's important to keep all the piping square and straight. Most plumbers assemble and solder the piping to the faucet on a flat surface, like a floor, and install the resulting "shower tree" in one piece.

Begin by laying three 16-to-24-inch pieces of 2x4 lumber on the floor. One piece should be near the showerhead, one just below the valve and one near the bottom of the supply tubes. For a tub, cut two lengths of type M rigid copper 36 inches long, to be used for the supply tubes. Cut a second length for the shower riser, 44 to 48 inches long. Finally, cut two short stubs for the horizontal connections to the faucet inlets, and

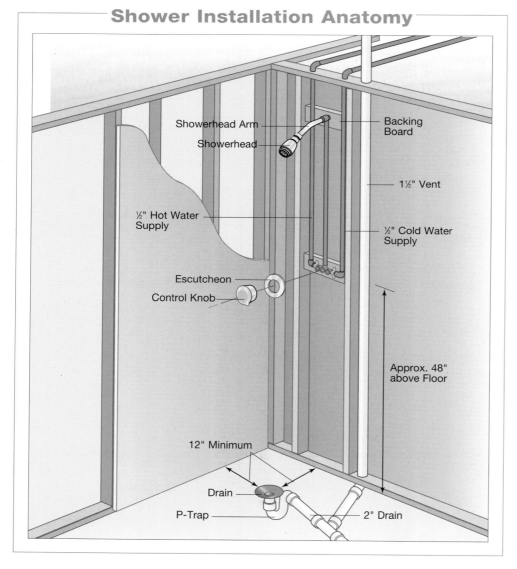

Shower Installation Anatomy

TUBS & SHOWERS

two 5-inch stubs for the spout piping. Clean and flux the pipe ends, and install a drop-eared elbow on the top of the shower riser. Insert the shower riser and pipe stubs into the remaining ports of the valve. Finish by joining the supply risers to the inlet stubs, using 90-degree elbows.

Lay this assembly on the three 2x4s, and check for straightness. When it all looks right, solder and wipe all joints and let the assembly cool. Then set the tree in place so that the front of the faucet's rough-in plate matches the approximate plane of the finished wall. Mark the wall's sole plate for the supply tube holes, and drill ¾-inch holes at each mark. Insert the supply tubes through the floor holes until the center of the valve is 28 inches off the floor. With a plastic tub/shower unit, insert the tubes through the tub's drain hole.

Let the tree stand by itself while you cut two 2x4 braces to fit between the studs. Nail these braces in place, positioned so that the valve's rough-in plate will be on the same plane as the finished wall, and that the showerhead's drop-eared elbow is held behind the face of the studs. Screw the ears of the elbow to the upper brace so that it is centered above the tub drain, and then center the valve above the tub drain and secure the supply risers to the lower brace, using two-hole copper straps. Loosely thread a ½ x 4-inch nipple into the drop-eared elbow. Connect the supply risers to the system water piping, and test your work.

Exterior Walls

If you plan to install a tub against an exterior wall, make sure that the wall is well insulated and that a plastic vapor barrier covers it. After you set the tub, and before installing drywall, stuff more fiberglass-batt insulation into the cavity between the wall and the tub. This simple step can greatly reduce heat loss. If the tub is made of porcelain steel, the insulation will also muffle the noise associated with steel tubs. It also helps to stuff insulation between the tub and the apron.

Installing a Cast-Iron Tub

USE: ▶ reciprocating saw • circular saw • groove-joint pliers • screwdrivers • hammer • solder, flux, torch ▶ new tub, faucet

1 *Cut the tub's drain opening (8 x 12 in.) with a reciprocating saw. Center the cutout 15 in. from the long wall.*

2 *An easy way to move a heavy cast-iron tub is to walk it, corner-to-corner, across the floor.*

3 *Use 2x4s, vertically, under the long-wall-side corners of the tub. Nail the blocking to the wall's corner studs.*

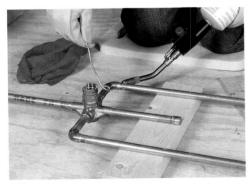

4 *Assemble the shower-tree on blocking laid on the floor, and when everything's straight, solder the copper tubing together.*

5 *Install backing boards behind the faucet and showerhead fittings for support, and secure the piping with straps.*

6 *Screw the showerhead's drop-eared elbow to the backing board, and install a temporary nipple in the fitting.*

8 Plumbing

INSTALLING A SHOWER STALL

Before installing the shower pan in its opening, measure off the back wall and cut a 6-inch circular opening in the floor. A good way to make sure you spot the drain in the right location is to set the pan in place and mark the floor though the pan opening.

Insert the rubber-gasketed drain spud through the opening in the pan. Then slide a rubber washer and paper washer over the spud from below. Thread the large plastic spud nut onto the drain, and tighten it. As with other spud-type drain fittings, tighten the nut until it feels snug. Don't overtighten it.

You'll need to plumb a 2-inch vented P-trap below the drain opening with a riser long enough to reach the pan drain. To determine the exact riser height, either set the pan in place or measure down from the floor. It helps to lay a straightedge across the opening. There are two possible connections. Some drains require that the trap riser be brought through the spud, stopping just below the drain screen. (These drains may be plastic or metal.) The seal is made with a rubber gasket, which you tamp into the gap between the pipe and spud with a hammer and packing iron or flattened piece of ½-inch copper pipe. Don't lubricate this gasket. Just set it in place and tamp it down. Other drain fittings require that you cement the riser pipe right into the hub. (These drains are always plastic.)

After you set the pan, install the water piping. (See "Shower Installation Anatomy," page 234.) Finish the enclosure with moisture-resistant drywall, or if you'll tile the walls, with a combination of moisture-resistant drywall and cement backer board. Cover the lower 24 to 30 inches of the walls with backer board, and drywall the rest. Attach the tile backer with construction adhesive and deck screws or galvanized roofing nails.

Install the Surround-Panel Kit

Next, you'll need to make the shower enclosure watertight. You can line the walls with ceramic tiles, but installing a plastic shower-surround kit made of PVC or ABS plastic or fiberglass is quicker and less skill intensive than installing ceramic tile.

Installing a Shower Stall

USE: ▶ measuring tape, pencil • saber saw • groove-joint pliers • cement backer board

Some surround kits come with three panels, others with five. A five-panel kit has separate corner panels, while a three-panel kit has the corners built into the end panels. If your walls are reasonably square and plumb, a three-panel kit is your best choice. (Use a 4-foot level to check the walls.) If not, a five-panel kit is more accommodating. Surround panels are held in place by two-sided foam-rubber tape and adhesive. You should seal the wall behind the panel with satin or semigloss latex or oil-based paint. If you use semigloss, scuff it with sandpaper before installing the panels.

1 Set the pan in place, and mark the floor through the drain. Remove the pan, and cut a 6-in. floor opening.

2 Insert the drain spud through the pan, and thread the spud nut over it from the other.

3 Tighten the spud nut with pliers or a spud wrench. Tighten until the nut feels snug.

4 Install a PVC drainpipe riser to meet the drain fitting. Use a straightedge to check the riser's length.

5 With drain installed over the riser, press the rubber gasket around the riser. Tamp it in using a hammer and packing tool.

TUBS & SHOWERS

• hammer, packing tool • level • drill, hole saw • adhesive • tub-and-tile caulk ▶ new shower stall

6 Nail cement backer board along the bottom 24 in. of the shower wall, and then continue with moisture-resistant wallboard.

7 Drill a faucet hole in the plumbing-wall panel using a drill and 3-in. hole saw. Measure carefully before cutting the hole.

8 Peel the paper from the foam tape around the panel edges and wherever else it appears; then apply panel adhesive.

9 Tape strips of cardboard to the shower pan; set the first panel on the strips; and press it against the wall from the bottom up.

11 Use silicone caulk in the joint between the shower pan and panels. Caulk the vertical seams, too.

10 Install the other side panel, then the back panel. Wipe all panels with a towel, pressing hard to spread the glue.

12 Caulk the shower faucet trim with a thin bead of silicone, and then wipe away all but an inconspicuous line.

INSTALLING A SHOWER STALL

8 Plumbing

HOT WATER

Showers just would not be the same without plenty of hot water, and it's the water heater's job to keep it coming. The heater's capacity does matter—a 40-gallon tank should be enough for a family of four—but if two showers are running and the dishwasher is going, it may not be able to keep up. Suddenly that soothing hot shower turns tepid, and then icy cold.

Elements of a Water Heater

The two most common water heaters are gas and electric. Both have a water tank (capacities range from 30 to 80 gallons), a cold-water inlet, and a hot-water outlet pipe. Turning on a hot-water faucet causes hot water to flow out of the heater's tank and draws cold water in through the dip tube, which carries it to the bottom of the tank. With cold water flowing in, the tank's water temperature will cool below the preset temperature (usually about 120° F). That activates the heater's thermostat, which turns on the gas burner (or the electrode in electric heaters) to raise the temperature.

The recovery rate measures how fast the cold water in the tank heats up. Gas heaters heat the water more quickly, having a better recovery rate, and because of this, electric models must have a larger tank capacity.

Water heaters have a temperature and pressure (T&P) relief valve that prevents the tank from exploding. Most also have a magnesium anode rod, which sheds electrons as it corrodes, helping to keep the tank from rusting. The size and number of anodes determine the life of the heater.

Capacity & Recovery Rates

Before installing a water heater, first estimate how much hot water you will need during the peak hour of use—for example, in the morning when three or four people need to shower, breakfast dishes need to be washed, and Dad needs to shave. (See the table below.) But a large-capacity tank is not the only way to avoid having that fourth shower be ice-cold. The heater's recovery rate is the number of gallons per hour that the heater can produce. A 30-gallon heater might be able to recover 55 gallons of hot water an hour, more than a 50-gallon model.

Connecting

Shown here is an example of a dielectric fitting (silver-colored) connecting the copper pipe to the top of the water heater. You need to ensure that you use one of these when connecting dissimilar metals. See more information about dielectric fittings on page 247.

Typical Hot-Water Use

ACTIVITY	GALLONS USED
Showering	3 gals./minute
Bathing	15–25 gals./bath
Shaving	1–3 gals.
Washing hands	1/2–2 gals.
Washing dishes	4–6 gals.
Running dishwasher	5–20 gals./load
Running clothes washer	25–40 gals./load
Cleaning house	5–12 gals.
Food preparation	1–6 gals.

WATER HEATERS

Gas Piping

The main supply line and valve may be black iron, although many building codes call for flexible copper tubing for gas lines, with flared compression fittings at the joints. To avoid kinks, bend tubing with a special springlike coil. Where abrupt turns are necessary, use 90-degree flared compression fittings. Once the gas is turned on, brush soapy water on fittings to test for leaks indicated by bubbles. Most leaks can be fixed by tightening the fitting.

To test the seal at a gas fitting, brush the joints with soapy water. If bubbles appear, turn off the supply to the pipe.

Water Heater Insulation

GREEN SOLUTION

Insulating your water heater, as well as the hot-water pipes, is the easiest way to improve hot-water efficiency. Where plumbing is exposed in a crawl space or cellar, cover hot-water lines with foam insulating tubes. Technically, heat radiating through pipe walls isn't lost because it helps to warm floors inside the house. But insulation that keeps water in the pipes warmer helps to deliver hotter water where you want it, and you waste less water running tepid water through the tap.

Cut partially across the top piece, and fit it carefully before taping it down. On a gas heater, be sure not to block the draft diverter.

Wrap the sides of the heater with the larger blanket, and tape the seams. Cut carefully around controls and the drain.

HOT WATER

8 Plumbing

Tankless Water Heaters

GREEN SOLUTION

These compact wall-hung units combine a coil of hot-water supply pipe and a heating element. The idea is to heat water on demand instead of storing it. For example, when you open the tap, the heating element on a gas unit envelops the coil in flame and heats the water passing through. A small unit may work in a remote bath far from the main heater but may not be able to handle the demand of baths that are used often.

Tankless water heaters have become part of improving a home's energy efficiency by only heating the water that is being used when the demand calls for it rather than heating a storage tank that holds 30, 50, or even 80 gallons of heated water at a time.

The primary fuel sources used to heat the water in a tankless water heater include natural gas, propane, and electricity. Of these three, the electric source is the easiest to install, as it does not require venting of any fuel exhaust, which both natural gas and propane require.

Installation of many types of water heaters may require a building permit and inspections when complete, depending on the individual municipality. In some instances, the code may also require a licensed plumber to complete the work. Check with your local building code department before beginning any work.

Tankless water heaters are rated by the gallons per minute (GPM) on the input side of the appliance. This is crucial to match, as many homes are equipped with private on-site wells that vary in the number of gallons per minute the pump can produce. Municipal water sources are more consistent and therefore regulated to provide a constant supply of water.

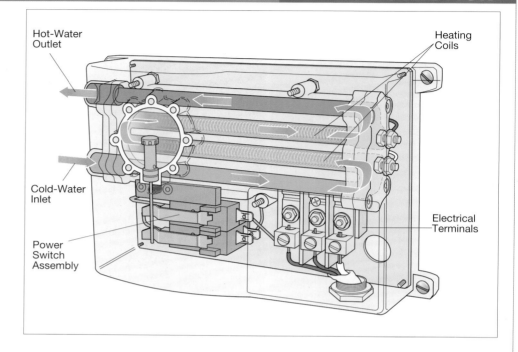

To determine the number of gallons your water system can provide, you can run a timing test. Collect up to 3 clean, empty 1-gallon containers, such as a milk container. Start the flow of water at maximum power and, after a few seconds, place the first empty jug under the faucet. At the same time, have someone start a timer and time until the water fills the container. Once it's filled, immediately stop the timer and write down the time to the second. Repeat until you have filled all 3 containers. This will give you a time interval for 3 gallons total and give you a better idea of the elapsed time per gallon.

For example, if it takes 20 seconds to fill a 1-gallon container, you have a 3-gallons-per-minute (3 GPM) flow of water as your input for the size of the tankless water heater. Be sure to read the manufacturer's website for the best model to suit your home's hot water needs and the number of bathrooms for which the device will supply the best performance.

An example of a tankless water heater. This one has a 3.8 GPM recovery rate.

WATER HEATERS

Safety & Maintenance

USE: ▶ socket wrench ▶ replacement valve • replacement anode rod • Teflon tape

All water heaters should be fitted with a pressure-relief valve. It's unlikely that the valve will be needed. But if problems in the heater developed to the point where the tank could explode, the relief valve would open and vent steam and overheated water. That's why the valve is connected to pipes that run to the edge of the heater and down toward the floor. You should test the valve periodically, and if it fails to operate, have a new one installed as soon as possible. Also periodically flush out sediments that collect in the bottom of the tank. Do this job once a year or more often if your water supply is hard and laden with minerals. Remember to turn off heating sources before draining the heater. Then attach a garden hose to the drain valve located near the bottom of the tank, open the valve, and let water flow out until accumulated sediment is cleared. To maximize the heater's life span, adjust the thermostat to maintain the water supply at approximately 120°F.

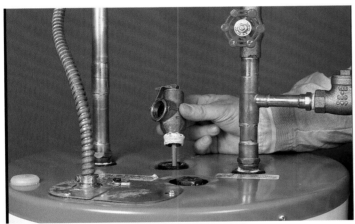

1 All water heaters require a valve to release pressure in an emergency. The fitting is threaded onto the top of the tank.

2 Connect the replacement valve with plumber's putty. A emergency discharge pipe from the valve leads to the floor.

1 To remove your water heater's anode rod, use a 11/16-in. socket wrench. If it's difficult to move, heat the tank fitting.

2 After removing the old rod, feed a new rod into the tank and tighten. This can double the life of your heater.

TANKLESS WATER HEATERS

8 Plumbing

HOT-WATER DISPENSERS

Instant-hot-water dispensers are custom made for busy lives. At 190° F, the water that these pint-sized appliances serve up is most likely at least 40 to 50 degrees hotter than that delivered by your water heater. (Water heaters are dangerous when used above 140° F.) Water at a temperature of 190° F is just right for blanching vegetables, making instant soups, and brewing real coffee, one cup at a time. Most units deliver 40 to 60 cups of hot water a day. The operating costs are about the same as a 40-watt lightbulb. Dispensers range in price from $90 to $250. The more expensive ones produce more hot water.

Basic Considerations

Hot-water dispensers are easy to install and don't require a dedicated water line. If you have copper pipes, you can steal water from the cold-water riser under the sink through a self-piercing saddle valve. This is the same kind of valve used to supply icemakers. (If your piping is made of brass or galvanized iron, you can still use a saddle valve, but you'll need to shut the system down and drill a tap hole into the riser.) If your cold-water riser ends at floor level or the back wall, keeping you from installing a saddle tap, you might consider splicing into the supply tube using a T-fitting and a ⅜-to-¼-inch reducing coupling.

Most units come with three-prong plugs, so a grounded receptacle inside the cabinet will do. Although a shared small-appliance circuit can power the dispenser, a dedicated circuit is a good idea. Expect the unit to consume between 4 and 5 amps of power. That's roughly one-third the capacity of a 15-amp waste-disposal-unit circuit or one-fifth of a 20-amp kitchen circuit. If you decide to run a new circuit, follow the same procedure and guidelines discussed in running circuits for waste-disposal units. (See "Installing a Waste-Disposal Unit," p. 252.)

Placing the Dispenser

Hot water dispensers are designed to be placed in the fourth deck hole of a kitchen sink. If you

Installing an Instant Hot-Water Dispenser

USE: ▶ screwdrivers • groove-joint pliers • open-end wrenches • tubing cutter ▶ dispenser, fittings

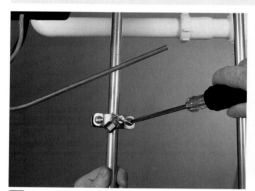

1 Make the water connection using the supplied saddle valve. All you have to do is clamp the valve onto the cold-water line.

2 Feed the unit through the sink hole from underneath the sink, and install the mounting nut from above.

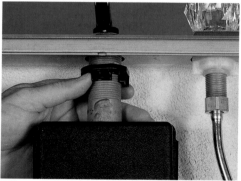

3 From below the sink, finger tighten the jamb nut on the threaded shank of the unit to hold it in place.

4 Install the spout in the sink-deck fitting, and tighten the faucet's setscrew to secure it. The spout seals with an O-ring.

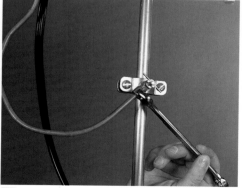

5 Connect the ¼-in. heater water line to the saddle valve using the supplied compression nut and ferrule.

6 If the heating unit delivers water above or below 190° F, adjust the temperature setting using a flat-blade screwdriver.

WATER HEATERS

have a deck hole that is covered with a plug, remove the cover. If you have a stainless-steel sink without a fourth hole, you can drill one with a knockout punch. You can also drill through a solid-surface sink, but not through a cast-iron or enameled-steel sink.

Installation

Begin the installation of an integral-tank-and-faucet instant hot-water dispenser by installing the self-piercing saddle valve on the cold-water riser pipe. Back the tapping pin out as far as it will go, and with the rubber tapping seal in place, bolt the two halves of the assembly together over the riser. Draw the two bolts down alternately, a little at a time, until they feel snug.

Next, install the tank. Feed its threaded shank up through the fourth sink-deck hole, and screw the mounting nut onto it from above. This will leave the unit hanging loosely from the deck. To secure it firmly, turn the jamb nut clockwise until it contacts the deck, and then tighten it with a basin wrench. With the unit in place, insert the chromed spout into the port atop the shank, and tighten the setscrew.

To make the water connection, carefully bend the unit's ¼-inch copper supply tube down to meet the saddle valve, and trim it to length. Then join the tube to the valve's compression fitting.

Make sure all the fittings are tight, plug in the unit, and test it for leaks. If the water doesn't seem hot enough, hold a thermometer under the running water. It should read 190° F.

ANODE RODS AND WATER HEATERS

Even though steel water-heater tanks are lined with a vitrified porcelain finish, the coating process is far from perfect. In fact, every glass lining has dozens of tiny pinholes where rust can start. The tank could rust through were it not for its sacrificial anode rod.

An anode rod works by sacrificing itself to corrosion, thereby preventing the tank from corroding. Most metals corrode, but at different rates. Water-heater anode rods are typically made of magnesium because it corrodes at a faster rate than does iron. As the magnesium corrodes, it sheds electrons, which migrate to the pinholes in the lining. This electron-rich environment keeps the tank from rusting. Under normal conditions, it takes 4 to 5 years for an anode to fail; thus, the five-year tank warranty. Ten-year heaters have larger anodes or two anodes. It's a good idea to replace the anode rod at the end of the warranty period.

Changing an Anode Rod

Anode rods are installed in heaters in one of two locations. The easiest to reach are those screwed into a female fitting in the top of the heater with a $^{11}/_{16}$-inch nut. These are usually visible through a hole in the top of the cabinet but are sometimes hidden under the top. In that case, you'll need to lift the sheet-metal top to reach the anode. The other type is threaded into the hot-water outlet. Called in-line anode rods, they are fused to a plastic-lined steel nipple.

To remove a top-nut anode, turn the water off and use a socket and breaker-bar wrench to remove the nut. If the nut won't budge, drain some water, peel the insulation away from the nut, and heat the tank fitting with a torch. This will expand the threads enough to allow you to wrench the nut free. Then lift out the old anode. To install the new anode, coat its threads lightly with pipe joint compound, and tighten it into the water heater. In order for the anode to work, it has to be in direct contact with the tank's iron fitting, so don't use tape and don't use much pipe joint compound.

The Anode Rod

The anode rod in some water heaters is attached to the hot-water outlet fitting.

Changing an Anode Rod — MONEY SAVER

USE: ▶ breaker bar with $^{11}/_{16}$-in. socket • torch (as necessary) • pipe joint compound ▶ new anode rod

1 Use a breaker bar and $^{11}/_{16}$-in. socket wrench to break the top nut of the anode rod free. Unscrew it, and lift the rod out.

2 This 5-year anode rod has exceeded its service life. It looks like a corrosion sculpture.

3 Use just a little pipe joint compound on the threads of the top nut, and tighten the new anode rod in place.

8 Plumbing

SERVICING GAS-FIRED WATER HEATERS

Gas-fired water heaters have service needs that are different from those for electric units. You may experience problems with the combustion process, the vent pipe, the burner itself, or the thermocouple and pilot.

Combustion-Air Problems

If your gas-fired water heater makes a puffing sound at the burner, it's not getting enough air. Air shortages are especially common in utility rooms, where a water heater, furnace, and clothes dryer all compete for air. Allow the water heater to run for a few minutes with the utility-room door closed. When it starts puffing, open the door and a nearby window. If the flame settles into smooth operation, inadequate combustion air was the culprit. Either cut a large vent or add a louvered door to the room.

Vent-Pipe Problems

When a gas-fired water heater runs, it creates a thermal draft, in which air from the room is also drawn up the flue. If there is insufficient secondary air or if the flue is partially clogged, deadly carbon monoxide gas may spill into your living space. Because carbon monoxide is odorless and invisible, it pays to test the efficiency of the flue from time to time. You can call your gas company for a sophisticated test, but the following method works. If the utility room has a door, close it. Wait a few minutes for the heater to develop a good draft, and then hold a smoking match, incense stick, or candle about 1 inch from the flue hat at the top of the heater (photo above, right). If the flue draws in the smoke, all is well. If the smoke is not drawn in or is pushed away from the flue, there's a problem. To determine whether the problem is an air-starved room or a clogged flue, repeat the procedure with a door and window open. If the smoke is now drawn into the flue, make the corrections noted in "Combustion-Air Problems," above. If not, call in a professional to inspect the flue.

Maintaining the Thermocouple and Burner

The job of the thermocouple is to hold the gas valve open so that gas can flow to the burner and pilot. If the pilot on your gas-fired water heater goes out and the heater goes out again after you re-light the pilot, the thermocouple is probably misaligned or defective.

Before buying a new thermocouple, perform this simple test, check to see whether the existing thermocouple's sensor is positioned directly in the path of the pilot flame. If not, bend the fastener clip so that the pilot flame surrounds the sensor. Then light the pilot to see whether it keeps burning. If not, a new thermocouple is in order.

Clean the Burner

As combustion gases degrade the inner surface of the water heater, flakes of rusty metal may fall onto the burner. The rust can cover gas jets around the burner's perimeter. With some jets blocked, the rest will flame orange and high, signaling a loss in efficiency.

Check the burner for rust as part of your routine maintenance. With the heater turned to pilot, remove the outer and inner access panels at the base of the heater. Shine a flashlight onto the burner. If you see rust on top of the burner, vacuum it off. Then turn on the water heater. If some of the gas openings still appear clogged, you'll have to remove the burner to clear them out. To do so, turn off the gas and loosen the nuts securing the three burner tubes to the gas control valve. Slide out the burner, and poke through the openings with a piece of wire.

Replace the Thermocouple

If you need to replace the thermocouple, pull the sensor from its clip. (Some are held in place by a screw.) Take the old thermocouple to a hardware store, and buy a matching one. Connect the new thermocouple, making sure that the sensor will catch the pilot flame. Reinstall the burner, and reconnect the burner tubes to the control valve.

To re-light the water heater, press down on the pilot button and hold it down for 30 seconds before lighting a match held with needle-nose pliers. Feed the flame into the heater. If the pilot goes out when you let up on the button, repeat the procedure. You may need to do this several times for the gas to push all the air from the pilot feed line. When the pilot stays on, replace the access panels and turn the control knob to "On."

Testing the Flue

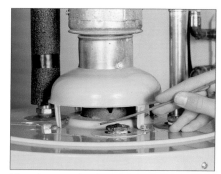

Use the smoke from a match, incense stick, or cigarette to test the flue.

MONEY SAVER
Replacing a Drain

Remove the leaky old drain valve using groove-joint pliers. The valve should simply twist off.

Install a new valve, turning it clockwise. Apply pipe-thread sealing tape to the threads.

WATER HEATERS

Water Heater Anatomy

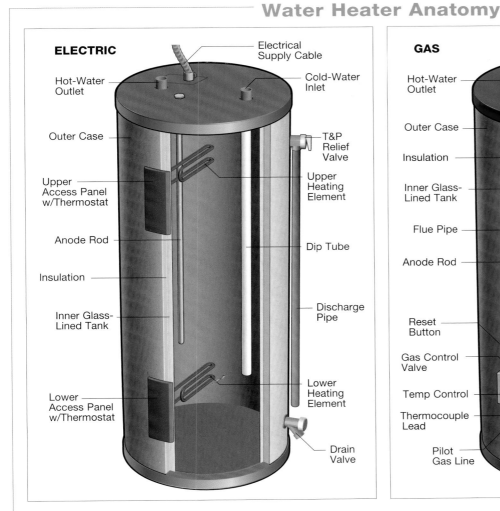
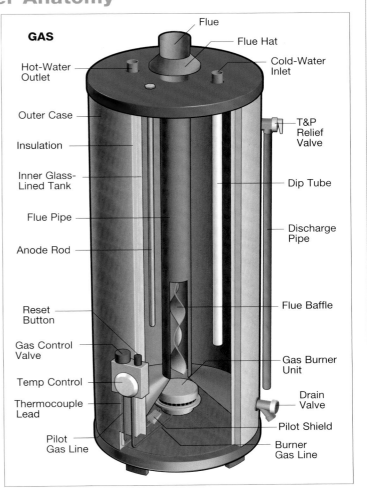

ELECTRIC — Electrical Supply Cable, Hot-Water Outlet, Cold-Water Inlet, Outer Case, T&P Relief Valve, Upper Access Panel w/Thermostat, Upper Heating Element, Anode Rod, Dip Tube, Insulation, Inner Glass-Lined Tank, Discharge Pipe, Lower Access Panel w/Thermostat, Lower Heating Element, Drain Valve

GAS — Flue, Flue Hat, Hot-Water Outlet, Cold-Water Inlet, Outer Case, T&P Relief Valve, Insulation, Inner Glass-Lined Tank, Dip Tube, Flue Pipe, Discharge Pipe, Anode Rod, Flue Baffle, Reset Button, Gas Control Valve, Gas Burner Unit, Temp Control, Drain Valve, Thermocouple Lead, Pilot Shield, Pilot Gas Line, Burner Gas Line

Maintaining the Burner and Thermocouple

MONEY SAVER

USE: ▶ adjustable wrench • wire • vacuum cleaner

1 Loosen the three connecting nuts, and remove the burner assembly. Tip the assembly down and out.

2 While you have the burner out, ream its jets with a piece of wire. Also, vacuum any rust from the floor of the water heater.

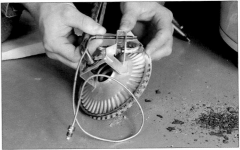

3 Snap the new thermocouple into the burner, and secure the clip. If the thermocouple wire is too long, coil it a bit.

SERVICING GAS-FIRED WATER HEATERS 245

8 Plumbing

SERVICING ELECTRIC WATER HEATERS

Electric water heaters are simple appliances. Diagnosis is easy using a voltohmmeter, which most plumbers carry. However, few homeowners own testing equipment, so a symptomatic is best.

Troubleshooting the Wiring and Thermostat

If the water heater stops working, the trouble may be as simple as a tripped circuit breaker, blown fuse, or loose wire. Or it could be the thermostat. Do the easy things first. Check the electrical service panel and the thermostat reset button.

Wiring

While tripped breakers often signal a defective heater component or a loose wire, the problem may be as simple as a momentary voltage spike.

Turn off the power, and remove the upper and lower access panels. Peel the insulation back, and look closely at all wire connections. If you see a wire that looks charred, loosen the binding screw holding it in place and trim the wire beyond the char mark and follow the directions below.

Replace the Thermostat

If you find a defective thermostat, you can buy an inexpensive, universal replacement at most hardware stores. Just make sure that it has the same voltage and wattage rating as the old one. *With the power off,* undo the wiring connections and push the wires aside. Note which wire goes to each terminal, however. Snap the thermostat into the clip on the face of the heater, and reconnect the wires. Press the reset button. Then, using a screwdriver, adjust the temperature setting to 130° F.

The thermostat doesn't have an in-tank sensing probe and instead senses the heat of the tank. Therefore, it's critically important that you replace the insulation and cover the thermostat completely.

Replacing a Heating Element

Faulty thermostats and heating elements can display some of the same symptoms (only testing will tell for sure), but elements are subject to more stress and generally fail more often than do thermostats. But short of testing, how will you know which element has failed? If the heater produces plenty of warm water, but no hot water, the upper element has probably failed. If you get a few gallons of hot water, followed almost immediately by cold water, it's most likely the lower element.

To remove a defective element, *shut off the power* and drain the tank. If you need to replace an upper element, just drain the tank to that level. Remove the access panel, and loosen the two terminal screws securing the wires to the element. If your heater has a bolted flange, remove all four bolts, and pull the element out. If it's threaded-in, grip the wrenching surface with a pipe wrench and back it out. If the sheet-metal cabinet keeps a pipe wrench from reaching the element, as it often does, use an element wrench.

To install the new element, slide the rubber gasket in place, and coat both sides with pipe joint compound. Then thread or bolt the element into the tank. Trim the stressed ends of each wire, and strip the insulation back about inch. Insert the wires under the binding clips or around the screws. Tighten both screws, making sure the wires don't drift outward. Press the reset buttons, and replace the insulation and access panel. Fill the tank and bleed all air through an upstairs faucet before switching on the tank. It should take 45 minutes to recover.

Troubleshooting   MONEY SAVER

USE: ▶ insulated screwdriver • electrician's pliers

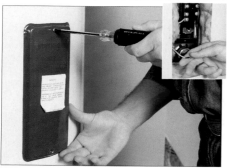

1 With the power shut off, remove the water-heater access panel and insulation. Check for charred wires (inset).

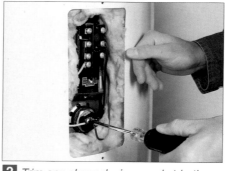

2 Trim any charred wires, and strip the insulation back ⅝ in. Insert the wire back into its slot, and tighten the terminal.

3 With the power still turned off, loosen the wiring connections and replace the thermostat.

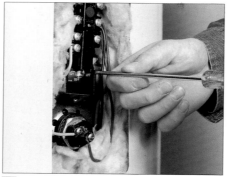

4 Snap it into its clip, it should fit right into its place easily. When secure, tighten the screws.

WATER HEATERS

REPLACING A GAS-FIRED WATER HEATER

To remove the old water heater, shut off the water; attach a hose to the heater's drain valve; and open the drain. Open some faucets to relieve air lock. Shut off the gas going to the water heater, and disconnect the gas line. If the gas piping is made of black iron, break the line apart at the union. Grip the union collar with one pipe wrench and backhold the upper nut. Disassemble the nipples and fittings between the union and water-heater connection. If the gas line is soft copper with flare fittings, loosen and remove all fittings below the shutoff valve. If a flexible connector is present, loosen the end nuts and remove the connector.

Remove the sheet-metal screws that join the vent pipe to the hat, and lift the vent pipe. If need be, wire the vent to the overhead pipes to keep it from falling. If the vent pipe enters a brick chimney, don't disturb that joint. If the vent pipe or its connection to the house is rusted, contact a professional to replace defective components.

With much of the water drained from the tank, disconnect the hot- and cold-water piping. If yours is an older home, with galvanized-steel or brass piping, loosen the union fittings just above the heater. Save the unions and nipples for now. The same approach applies to copper piping that joins the heater with dielectric unions. If you find a direct connection with soldered copper or threaded female fittings, cut the copper 6 to 12 inches above the heater or at least 2 inches below the cold-water shutoff. Walk the unit out from under the pipes.

Connect the New Unit

Walk the new unit under the water pipes. (See the sequence on the next page.) The water-piping alignment is the most critical, so make these connections first. Sometimes you can make the connection with existing unions and nipples. In most cases you'll need to adapt the piping. If your piping is threaded galvanized or brass, you'll be able to buy threaded nipples to extend the pipes. If the pipes are too long, you'll need to dismantle the pipes back to the first joint. If the local hardware store has a selection of longer nipples, in the 18-to-24-inch range, you can build back with those. If not, you'll have to cut and re-thread the existing pipes. Coat all the connections with pipe joint compound, and make the final connection with unions. (You might also convert, codes allowing, to plastic, flexible copper, or stainless-steel connectors.)

If the piping is made of copper and electrolysis is not a problem, adapt the lines with sweat couplings and male or female adapters. The heater connection is usually a threaded female fitting, so presolder ¾-inch copper pipe stubs into male adapters. Then thread the adapters into the heater fittings using pipe-thread sealing tape. Trim the stubs to length, and prepare them for soldering. Slide the unit into place, and check it for level. If the floor slopes toward a floor drain, shim the water heater's legs. Join the stubs to the existing pipes with repair couplings, and solder the couplings using lead-free solder. Coat the threads of the T&P relief valve with pipe joint compound, and tighten the valve into the tank fitting. With the valve in place, solder a copper male adapter to ¾-inch copper pipe that's long enough to extend from the valve to within 3 inches of the floor. Thread the pipe into the valve port to serve as an overflow tube. In earthquake zones, secure the water heater to the wall with straps.

Connect the Vent Pipe

In most cases, you'll be able to adapt the existing vent pipe to the new heater. If the heater comes with a 3-inch flue hat and the vent pipe is 4 inches in diameter, you'll need to screw a 3 x 4-inch sheet-metal increaser to the flue hat. Most codes require a 4-inch vent pipe. All joints, including the hat-to-heater and the hat-to-vent-pipe connections, should be screwed together. Use self-tapping, hex-head screws.

How you approach the gas-line hookup will depend on the existing material. Many codes now require soft copper or CSST in new construction. Existing black-steel piping need not be replaced, however, so continue the line to the heater with the same material. Install a drip-leg near the gas-control valve to keep dirt and condensation away from the valve. A drip-leg is made with black steel or brass fittings.

If your codes allow flexible copper or require CSST connectors, the gas hookup is quick and easy. Even with a CSST connector, however, a shutoff valve needs to be installed within 36 inches of the gas-control valve. Make all connections with pipe joint compound, but use it sparingly. A light coating on the male threads will do. When you're finished, light the burner and replace the access panel.

New Heating Element

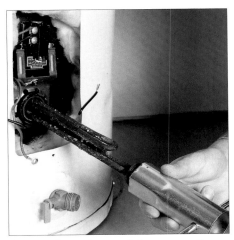

Use an element wrench to remove the old element from the heater. Some are held in place by four bolts.

MONEY SAVER

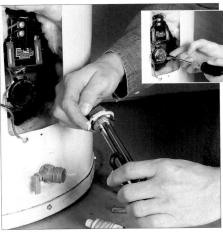

Coat the element's threads and rubber washer with pipe joint compound, and screw the element in place.

SERVICING ELECTRIC WATER HEATERS

8 Plumbing

Replacing a Gas-Fired Water Heater

USE: ▶ backsaw • pipe wrenches • nut driver • tubing and cutter ▶ self-tapping screws • pipe joint compound • solder, flux, torch ▶ new heater • T&P valve

1 Shut off the gas going to the water heater (inset), and break the gas line apart at the union fitting.

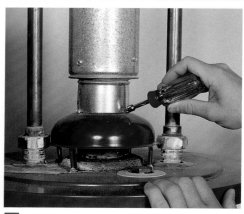

2 Use a nut driver to remove the screws from the flue hat, and then cut the water lines with a tubing cutter.

3 Presolder stubs into male adapters, and then tighten the adapters into the new water heater (inset).

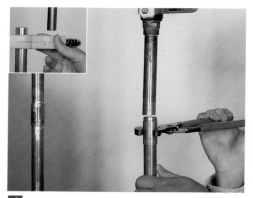

4 Cut and clean the pipe ends (inset). Install slip couplings with flux; slide the couplings up to meet the water piping.

5 Lightly coat the threads of a new T&P valve with pipe joint compound, and tighten the valve into the water heater (inset).

6 Thread a copper or plastic discharge pipe into the T&P valve. Extend the pipe to within 6 in. of the floor.

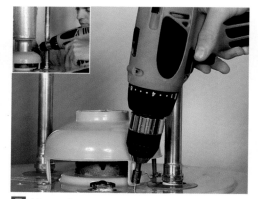

7 Use self-tapping hex-head screws to fasten the flue hat to the heater, and then screw the flue to the hat (inset).

8 Reconnect the gas piping. Be sure to include drip-leg piping (inset), and use a union as the final fitting.

9 Light the pilot for the water heater. A gas-grill lighter works best. Then install the access panel and cover, and turn on the water.

WATER HEATERS

INSTALLING AN ELECTRIC WATER HEATER

An electric water heater is easier to install than a gas-fired one. There are no flue or gas connections to make, and electric heaters usually come without legs, so any minor unevenness in the floor can be ignored. However, the unit does not to be reasonably level. If you floor slopes significantly, you will need to shim it. The water and T&P relief valve connections are the same as those for a gas-fired heater shown on the opposite page.

Electrical Connections

The big difference, of course, is the electrical connection. The heater will come with its own built-in electrical box, but you'll need to connect electricity to the box using a box connector and conduit. The conduit needs to run from the water heater to a disconnect box on the wall or into a joist space overhead. (Check with your building department.) Once in the joist space, the cable usually doesn't need to be in conduit. If your old heater was installed without conduit, it's a good idea to install it now. Use rigid, thin-wall conduit from the ceiling or flex conduit from a disconnect box on the wall.

Making the Connections

Begin by punching the knockout plug from the heater's box, and then install a conduit-to-box connector (photo bottom left). Bring the existing 240-volt cable through conduit and through the box connector. (If running new cable, make it a 3-wire with ground sized to meet the heater and breaker requirements, usually 10-gauge wire with a 30-amp breaker.) Tighten the conduit in the connector. Then, using approved twist connectors, join the black and red circuit wires to the black and red (or black and black) lead wires in the water-heater box (photo bottom right). Join the ground wire to the ground screw near the box. If you used rigid conduit, secure its upper end to the joists or a brace nailed between joists. Before turning on the power at the service panel, fill the tank with water and bleed all air from the tank through the faucets.

Water-Hammer

If you hear a pounding noise in your water system and see an occasional spill of water near your water heater, you have a water-hammer problem. Water-hammer arrestors have internal rubber bladders that act as shock absorbers for high-pressure back-shocks. You typically install them between the water piping and fixture.

For a clothes washer, screw the arrestors onto the stop valves and the hoses onto the arrestors.

To stop T&P valve leaks, install water-hammer arrestors on the washing machine shutoffs.

Saving Fuel Costs

Today's water heaters are better insulated than were those of even a few years ago. Still, you can boost efficiency by installing an aftermarket insulation blanket. The more hours you spend away from the house, the more benefit you'll get from this add-on. **Caution:** be careful not to cover the access cover, T&P relief valve, or control valve. You'll also need to hold the insulation away from the flue hat by several inches on gas-fired heaters.

An even better investment is to insulate all hot-water pipes. These pipes shed a good deal of heat, and the more you can do to slow heat loss, the lower your energy bills. You'll find several kinds of pipe insulation on the market. The best is pre-slit foam rubber. You can purchase the insulation at any home center or hardware store. The material comes in sizes to cover ½-, ¾-, and 1-inch pipe in 4-, 6-, and 8-foot lengths. Split the insulation along the precut slot and place it over the pipe. You can trim the insulation to cover short sections of pipe or to cover valves.

Installing an Electric Water Heater

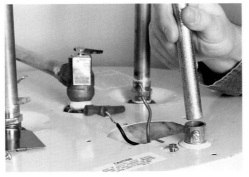

Thread a conduit connector into the water heater, and run the wires in ½-in. EMT conduit.

Bind the grounding wire under the grounding screw, and join the like-colored wires in connectors.

8 Plumbing

DEFYING GRAVITY

Pumps help us defy gravity by making water and other liquids flow uphill instead of down. Anyone who has a well, for example, relies on a pump to draw water from a hundred or more feet underground. Even municipal water systems use pumps—those water storage towers you see in towns and cities are kept full by huge pumps. But there are also pumps for the special situations covered here—sump pumps, wastewater pumps, and recirculating pumps.

Two basic types of pumps are reciprocating pumps and centrifugal pumps. A reciprocating pump uses a piston to draw water into a chamber. Then, after the inlet valve closes and an outlet valve opens, the piston pushes the water out of the chamber under pressure. Centrifugal pumps use a rotating impeller—a wheel with many blades—that draws water in at the center and forces it outward (by centrifugal force) along the spinning blades. This increases both the water's pressure and velocity of flow.

Well Pumps & Pressure Tanks

Because wells are located far belowground and the average person uses about 95 gallons of water a day, you need a powerful pump to supply a whole house. Submersible centrifugal pumps are commonly used—they have multiple impellers to increase lifting capacity and are powered by electric motors sealed in waterproof housings. The pump is attached to a drop pipe, and the whole unit is lowered into the well casing below the maximum draw-down level so that the pump's intake will always be submerged.

The pump's ability to send enough water up the pipe is one important concern, but it isn't the only one. For example, you also have to consider the well itself, which will only supply water at a given rate. Another concern is peak demand, which can outstrip the yield of even the best wells. A pressure tank helps solve that problem. With storage capacities of 20–80 gallons or more, it provides a reserve for those times when the washer is going, someone is filling a tub, and you are watering the lawn all at once. An air bladder inside the tank expands as the water is drawn out, keeping the pressure at about 40 psi. Meanwhile, the pump begins refilling the tank as water is being used and continues pumping until the tank is full again.

Sump Pumps

Sump pumps are the last resort against flooded basements. Typically, the pump is mounted in a small pit so water leaking into the basement will run into it. Because you may not be around to turn on the pump when the pit fills with water, there is an automatic switch triggered by a float mounted on a vertical rod set in the pit. When water fills the pit, the float rises, triggers the motor, and the pump runs until the water level recedes. It's wise to test your pump by dumping a pail of water into the pit.

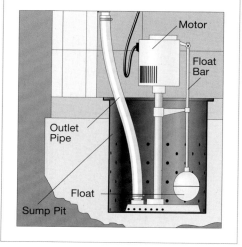

Cellar Toilets

Cellar toilets installed below the main waste pipe leaving the house require special plumbing. These units, called up-flush toilets, need an electrical connection and a pump to raise wastes up to the level where they flow by gravity to a municipal sewer line or private septic system. The expensive toilets require water supply and vent piping like standard installations, and a special valve, called a check valve, that prevents sewage from flowing back into the system if a blockage occurs anywhere in the house waste lines.

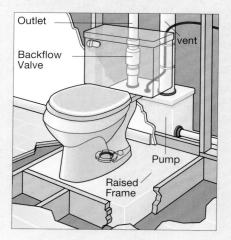

Installing a Recirculating System

This improvement can save 35–40 gallons a week of tepid water that pours down the drain while you wait for hot water to reach the tap. The system includes an extra pipe run from the heater and a small pump that keeps hot water slowly flowing through the loop. Hot water is always there when you want it because the pump recirculates water that normally would lose heat standing in the pipe. Installing the pump, electrical line, and extra pipe loop costs more up front, but the systems pays for itself long-term.

USE: ▶ adjustable wrench • propane torch • tubing

1 *Cut a tee into the cold-water supply line, and install a check valve. This allows water to flow only one way, toward the heater.*

PUMPS

GREEN SOLUTION
Pump Piping

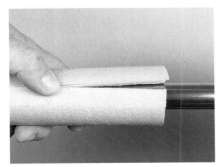

Preserve temperature with pipe insulation. Some foam tubes have zip-lock seams for easy installation.

Most pipe insulation comes partially split for retrofits. Trim to fit with a sharp knife, and tape over seams.

Well Systems

Most wells move water from a submersible pump at the bottom of a well casing into a pressurized tank in the house. When you open a tap, the pressure forces water through the pipes. This system conserves the well pump, which is costly to replace, by running the pump only periodically to refill the tank, not every time you call for water. If the power fails, the pump won't deliver more water to the tank, but you'll be able to use most of the water already stored under pressure.

The pressure in many holding tanks can be increased or decreased by pumping in or bleeding off air.

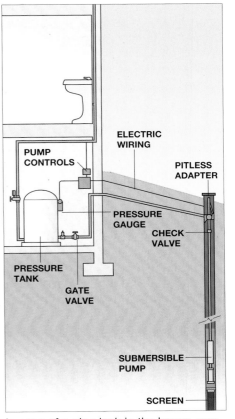

In case of a pipe leak in the house, you need to shut off the pump motor, and close the cutoff valve at the tank supply.

---

cutter or hacksaw • work gloves ▶ check valve • in-line pump • pipe and fittings • solder • flux • pipe insulation

2 Install the inline pump on the return loop near the heater and a power outlet. Pumps with timers are most efficient.

3 To make the return loop, cut a new pipe into the existing hot water line near the fixture farthest from the heater.

4 The loop should be insulated. (See above left.) The line should also have a shutoff and a drain valve to allow service.

DEFYING GRAVITY

8 Plumbing

WASHERS & DRYERS

The most obvious yet essential advice about installing washers and dryers is to follow the manufacturer's instructions. Although all plug into standard electrical outlets and take stock plumbing fittings, there is enough design variation so that no one set of instructions will do. But some pitfalls are common to most installations.

Hooking Them Up

Where you're hooking up to existing wiring and plumbing lines, installation is definitely a do-it-yourself job. The hardest part is likely to be maneuvering the cumbersome machines into position without battering door frames and scratching floors. But with enough extra muscle or a dolly, it's certainly possible.

New all-electric models are already wired, so you can simply plug them in to existing washer-dryer outlets. With gas dryers you have to be more careful, making sure that the gas cutoff valve is closed before disconnecting an existing gas line. (The top bar of the valve should be perpendicular to the supply pipe.) Also, take care after making a gas connection that you don't crimp the line as you nudge the dryer into its final position. If new wiring and plumbing lines are needed, you should call a licensed plumber and an electrician. After they provide the necessary electrical outlets, drainpipe, and water supply lines, you can continue on your own.

The new machines' height should not pose a problem unless you install cabinets over the new machines, a common setup. In that case, leave enough room to comfortably lift the doors of top-loaders (about 16 inches for most models) without banging into the bottom of the cabinet.

In a bone-dry climate, you may be tempted to skip installing an exhaust line for the dryer—although, despite lint filters, your comfortably moist environment would eventually become caked with little gobs of T-shirts and khakis. In the long run, you'll be better off choosing one of several venting options most manufacturers offer.

Back in 1908, washing machines weren't very automatic—you had to supply the hot water and hand-crank the tub. But this "Superba" model cost only $6.38.

Connections

The drain hose from a clothes washer is connected to a pump at the base of the machine. To prevent siphoning and backflows, the hose typically rises above the level of the washer drum and fits into an open drainpipe. To fix or replace a leaking hose, you're likely to find factory-installed spring clips on the connections. Remove them by squeezing the ends with a pliers. Install banded clamps on the new hose or to apply more pressure on the connections.

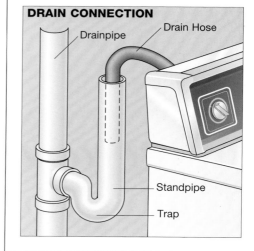

DRAIN CONNECTION — Drainpipe, Drain Hose, Standpipe, Trap

Removing an Old Dishwasher

USE: ▶ screwdriver • nut driver • adjustable wrench • bowl or pan

1 Begin by removing the old dishwasher's access panel. Look for several hex-head or slotted screws holding it.

2 Have a pan or bowl handy when you disconnect the water and discharge lines. Drain the lines into the bowl.

3 Disconnect the brackets screwed to the underside of the countertop, and carefully pull the dishwasher out by the door.

APPLIANCES

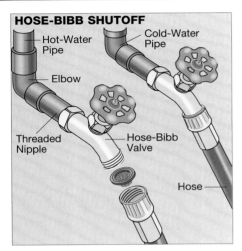

HOSE-BIBB SHUTOFF

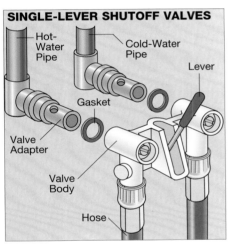

SINGLE-LEVER SHUTOFF VALVES

Water Inlet Valves

To protect automatic solenoid valves in the machine that control fill cycles, inlet valves have filter screens.

To clean an inlet screen, pry it out of the hose, rinse, carefully remove debris as necessary, and reinstall.

Leveling

Even on the sloping floors of older houses, you can adjust the legs of washers and dryers to make them level front to back and side to side. Adding a few drops of liquid detergent to the legs beforehand can make this job a lot easier. It also helps to unload the weight of the machine—for example, by raising it slightly with a crowbar. Once level, adjust the leg locknuts to hold the leg adjustments in position. If the washer or dryer wobbles or rocks during operation, you need to readjust the legs.

Use a level to check positioning, and adjust each leg up or down. Tighten locknuts to secure the position.

Installing a New Dishwasher

The best thing you can do for a dishwasher is to use it. If you don't for weeks on end, the water held in the pump may evaporate, allowing the seals to dry out and leak the next time you use the machine.

Check for a slimy dirt buildup on the lower section of the door seal at least once every few months. It's hard to see this accumulation from above, so use a pocket mirror. If you see signs of a buildup, clean the seal with detergent. While you're at it, lift the float from the bottom of the unit to check for dirt. A dirty float can increase the water level enough to cause a leak. And finally, check the spray arms for bits of plastic and other debris. If you see any debris in these holes, pick it out with tweezers.

A dishwasher requires hot water and electricity in order to function. Most often, the unit is located as close as possible to the kitchen sink for access to water. Location doesn't have much of an effect on electrical needs.

To wire the dishwasher, your minimum electrical requirements will be a dedicated 15-amp circuit run in 14/2g NM-B cable (14-gauge, two-wire-with-ground nonmetallic cable). If the dishwasher has a preheater, which boosts the temperature of the water, you may need a 20-amp circuit with heavier 12/2g cable. Check the manufacturer's specifications. Some local codes allow a direct connection, in which the cable is brought into the opening through the back wall or the floor and is connected directly to the dishwasher's electrical box.

WASHERS & DRYERS

8 Plumbing

Installing a New Dishwasher

USE: ▶ nut- and screwdriver • needle-nose pliers • adjustable wrench • wire strippers • wire connectors • pipe sealing tape ▶ dishwasher, fittings

Dishwasher leaks are not always visible, so they can do a lot of damage. Install a battery-operated water alarm in the cabinet. Place the sensor under the dishwasher. Another good tip is to be sure to install a separate stop valve for the dishwasher rather than the dual stop valve shown here. The separate valve allows you to stop the flow of water to the dishwasher and remove it for service without interrupting the water service to the rest of the kitchen sink. Install the valve under the sink for easy access.

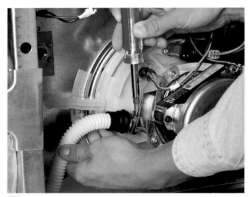

1 Attach the discharge hose to the dishwasher's pump, and lock it firmly in place using the supplied hose clamp or grip ring.

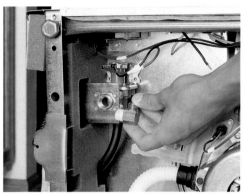

2 Thread a dishwasher elbow into the solenoid valve. Use pipe-thread sealing tape, and wrench it until it's snug.

6 Slide the dishwasher into the space. Stop several times to pull more cable and pipe into the cabinet as you proceed.

7 With the dishwasher in place under the countertop, thread the leveling legs down with an adjustable wrench and level the unit.

8 With the unit positioned and leveled, screw the fastening brackets to the bottom of the countertop's edge band.

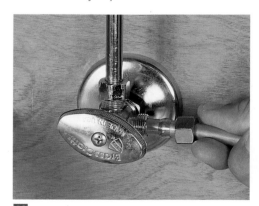

12 Install the dishwasher water supply line in one port of the dual stop valve, and connect the faucet supply tube to the other.

13 The discharge hose must connect to a backflow preventer or loop up to the top of the cabinet, secured with hole strap.

14 Attach the discharge hose to the waste-disposal-unit nipple (if applicable). Be sure to punch the plug from the nipple.

254 PLUMBING / APPLIANCES

APPLIANCES

3 Install a cable or conduit box connector in the unit's electrical box. It is required by the National Electrical Code.

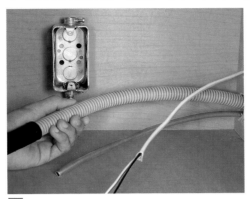

4 Attach a box in the cabinet (if needed), drill a hole in the cabinet side for the water lines and cable.

5 Remove the existing hot-water compression valve (or adapter) from under the sink, and install a new dual stop valve.

9 Slide a compression nut and ferrule onto the 3/8-in. copper water supply line, and connect the line to the dishwasher elbow.

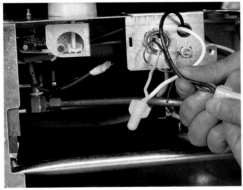

10 Join like-colored wires in twist connectors. Bond the grounding wire under the green ground screw.

11 Attach the black wires to the switch (if you're using one). Join the white wires in a connector, and bond the ground to the box.

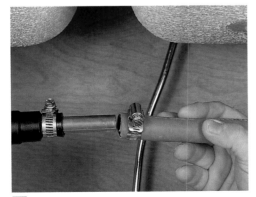

15 If you are using heater hose, you will need to connect to the disposal unit using an adapter.

16 If you don't have a disposal unit, use a waste T-fitting to connect the dishwasher discharge hose to the drain.

Test your work. With all of the final connections made, run the dishwasher through an entire cycle. Watch for leaks at the discharge and water lines. If a drip appears around a compression fitting, tighten it just enough to stop the leak. If the unit vibrates, look for a leveling leg that doesn't quite touch the floor and adjust it. If everything checks out, install the access panel.

A new dishwasher may break up years of accumulation in the pipes that can gather at a choke point and clog a drain. To fix, cable the entire line.

INSTALLING A NEW DISHWASHER 255

8 Plumbing

GETTING BETTER WATER

No matter what kind of contaminants get into your water supply, and no matter how sensitive you are to slight alterations of taste and odor, there is some form of conditioning equipment to fix the trouble. The first step, of course, is to have your water tested by a reputable agency.

The test for bacterial content costs only a few dollars. Having a full test done, which includes an examination for organic compounds, pesticides, dissolved gases, and solid particles, can cost several hundred dollars. Before deciding which tests you want, consider where your home is located. Are there gas stations in the vicinity? Have there ever been manufacturing facilities or a dump nearby? Is your house on (or downhill from) land that used to be a farm? Local officials can also be of help in deciding which tests should be done.

Selecting a Water-Treatment System

A single water-treatment system cannot remove all possible contaminating agents, but you have eight different systems to choose from. The most commonly used are activated-carbon filters, reverse-osmosis (RO) filters, and distillation units. Each of these eliminates more than one type of impurity. The remaining systems—ultraviolet radiation, chemical treatment, ion-exchange, sediment filtration, and aeration—have more narrowly defined roles.

UV radiation and chemical treatment are effective against bacteria. An ion-exchange system (or water softener) is effective if your water has a heavy concentration of calcium, magnesium, and/or iron. This kind of water, usually described as "hard," is not considered to be contaminated. A sediment filter screens out particles that make water look cloudy. It is also effective if asbestos fibers are present in the water. Often this type of filter is used in combination with an activated-carbon filter.

An aeration unit and activated-carbon filter are effective against radon gas and the odor caused by dissolved gases (such as sulphur) in the water. Activated-carbon filters also eliminate organic chemicals and pesticides. These filters are available in a wide variety of sizes and prices. Smaller units are attached to the spout of a faucet, while larger ones are tanks that are connected to the water supply line at its entry point into the house. An activated-carbon filter must be replaced periodically.

Both reverse-osmosis and distillation units remove a variety of heavy metals, such as lead, arsenic, and mercury. An RO unit will often be combined with an activated-carbon filter to weed out organic chemicals, pesticides, odors, and radon as well. Both units can cost from hundreds to thousands of dollars.

A distillation system doesn't filter the water but boils it and captures the steam in a condensing coil. This impurity-free water collects in a tank, where it is drawn off by the faucet. Distilled water, however, is flat and tasteless.

GREEN SOLUTION: Water Treatments

Problem	Solution
Bacteria	Chlorine feeder with activated-carbon filter; distiller; RO unit; UV unit
Low suds	Water softener; RO filter
Rusty stains	Water softener; oxidizing or activated-carbon filter
Green stains	Limestone neutralizer; neutralizing filter
"Rotten egg" smell	Oxidizing filter; chlorine feeder with sand filter; activated-carbon filter
Yellow/brown tinge	Water softener; activated-carbon filter; distiller; RO unit
Chlorine odor	Activated-carbon filter; RO unit
Pesticides, VOCs, benzene	Activated-carbon filter; distiller; RO unit
Lead, mercury	Distiller; RO unit
Nitrates, sulfates	Anion-exchange unit; distiller; RO unit

Installing a Sediment Filter

Clearing up cloudy water is a job for a sediment filter, which removes the inorganic particles causing the problem. It's not unusual to install a carbon filter as well. Both types can be installed as in-line fixtures, making the installation easy to set up. You'll have to do some soldering once you cut the water pipe, and you'll need a couple of cutoff valves to isolate the filters. Be sure to locate the filters in an accessible area so that you can get at them easily and clean them out on a regular basis.

USE: ▶ tubing cutter • propane torch • adjustable wrench • hose (for backflush) • sandpaper or wire brush

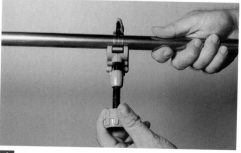

1 Remove a section of the supply line using a tubing cutter to accommodate the filter and threaded fittings.

2 To help new solder fittings bond on existing copper pipes, use a wire brush or sandpaper to brighten the old metal.

WATER & WASTE TREATMENT

GREEN SOLUTION
Faucet Filters

Modern filters can be built into faucets. This unit provides an aerated stream, spray, and filtered stream.

To keep the supply clean, regularly change the filter. On this unit, the cartridge tucks into the handle.

GREEN SOLUTION
Reverse-Osmosis Units

These units typically combine a reverse-osmosis membrane to filter out heavy metals, a carbon filter, a sediment filter, and a large tank to store the filtered water. Filtering membranes must be changed at least once a year or impurities will build up again. Although removing impurities is important, the taste of the water sometimes remains unaffected. Waterborne minerals such as manganese, iron, and sodium may need further treatment.

The combination of equipment required for reverse osmosis generally fills the cabinet space under a sink.

Softeners

Softening water is a process of reducing the amount of minerals that are picked up as water filters through the ground. Very soft water tastes bad but produces fewer deposits on fixtures, causes less corrosion in pipes, and creates more suds and cleaning action with soap. Effective systems split the supply between water for drinking, lawn watering, and such, and the softened supply for washing and cleaning. Installations such as these can cost between $1,000 and $3,000.

Water softeners need to be located in a convenient spot for maintenance and drainage.

GREEN SOLUTION

▶ sediment filter • male-adapter fittings • solder • flux • shutoff valves

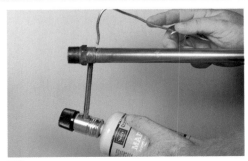

3 Solder a ¾-in. copper male-adapter fitting onto each open pipe end. Be sure to use lead-free solder.

4 Install the filter body with shutoff valves on each side. The shutoffs should be full-flow gate or ball valves.

5 To backflush a sediment filter, attach a hose, and drain it through the bottom. The element can also be replaced.

GETTING BETTER WATER

8 Plumbing

TREATING WASTE

Every ounce of water that leaves your house must be separated from solid wastes and treated before it can be returned to the environment. Homes that don't have septic tanks, as well as commercial properties and apartment houses, feed into enormous waste-treatment plants, some of which handle millions of gallons of sewage a day. These plants use the same chemical and biological processes that septic tanks do—and, like septic systems, they can be compromised by chemicals that are flushed into them. Pesticides, for example, not only interfere with the microorganisms that break sewage down but can find their way back into the area's fresh water supply.

Maintaining Septic Systems

A septic system has three main sections: the septic tank, the distribution box(es), and the leachfield (also called the drainfield). As sewage enters the tank, it is poured into a hot mix of waste and anaerobic bacteria, churning in endless loops inside the tank. Most of the solids are quickly digested, and the liquid effluent leaves the tank and enters the distribution box, from which it flows into the leachfield's perforated pipes and leaches into the ground. What remains behind in the tank are nondigestible solids called sludge (such as cigarette filters, apple seeds, and plastics), which sink to the bottom; and grease, which rises to the top. Accumulated sludge at the bottom of the tank won't fill the tank to overflowing and stop up your drains, but it will reduce the tank's capacity. If not cleaned out, eventually new solids in the tank won't have enough room to settle properly and can infiltrate the leachfield, clogging its pipes. The grease layer, if also left to accumulate, can flow past the tank's outlet baffle and out into the leachfield, where it coats the pipes and limits absorption, evaporation, and microbial activity in the soil. If the sludge or grease gets this far, the entire drainfield must be dug up and replaced, at a cost of thousands of dollars.

To keep this from happening, you should have a septic tank professionally pumped out periodically. How often depends on the tank

Removing a Waste-Disposal Unit

USE: ▶ screwdriver • groove-joint pliers

1 Unscrew the hose clamp on the dishwasher discharge hose. Pull the hose from the disposal unit's inlet nipple.

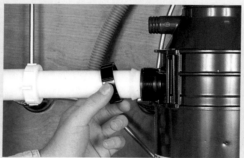

2 Remove the waste connection at the side of the disposal unit. Some are bolted in, and some have compression nuts.

3 Insert a screwdriver into one of the tabs of the retaining ring as shown, and rotate the ring counterclockwise.

5 With the power shut off at the main panel, reach into the box and pull out the wires. Remove the twist connectors.

6 Disconnect the ground wire, and loosen the box connector. Pull the conduit and wires from the connector.

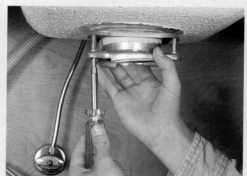

7 To remove the waste-disposal-unit drain fitting from under the sink, loosen the three bolts in the retaining ring.

WATER & WASTE TREATMENT

capacity and how many people live in your house. Two people using a 1,000-gallon tank only need to pump it out every six years or so, but six people using a 2,000-gallon tank should have it pumped out every three years. To help keep your tank working efficiently between cleanings, you should avoid flushing down a drain any chemicals that may kill the bacteria, such as paint thinner or photographic chemicals; instead, dispose of these at a local dump.

Major problems in a septic system will probably first come to your attention as a telltale sewage odor or permanently wet soil over the leachfield. Poor flow in every drain (not just one toilet) may also indicate a serious backup in the system.

Septic & Graywater Systems

If your home isn't connected to a municipal sewer system, you need a septic system on the property. A typical system has a tank for solids, which must be pumped out periodically, and a leachfield, where a series of perforated pipes gradually filters liquid wastes into the ground. Local codes for private septic systems are increasingly strict in most communities, and in some cases can cost up to $10,000 or more to install. The main concern is that the wastes do not contaminate underground supplies of water used for drinking, cooking, and washing. Expect to conduct a test of the soil, called a percolation (perc) test, and to excavate a substantial portion of the backyard. In some areas, you can install a complementary system to handle wastes from baths, showers, and washing machines called graywater. It may contain some soap and dirt but not sewage. With minimal treatment in a sand filter and a holding tank to eliminate contaminants, hundreds of gallons can be recycled for uses such as watering landscaping. Local codes are also strict about graywater; you can't simply dump soapy water on the grass.

4 Once you have disconnected the disposal unit, lower it and turn it over. Remove the electrical box cover.

8 Slide the retaining ring up, and use a screwdriver to pry off the snap ring to release the components.

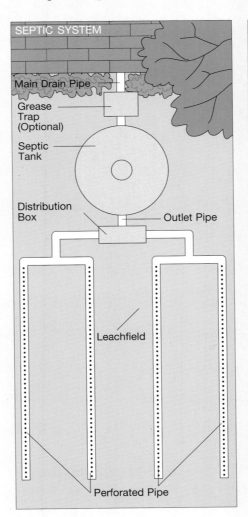

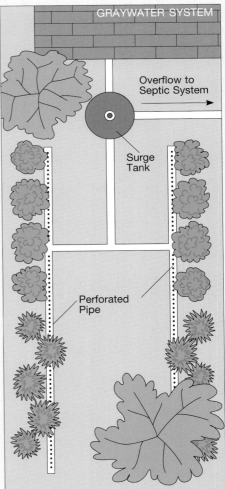

TREATING WASTE 259

8 Plumbing

WASTE-DISPOSAL UNITS

The first thing to know about waste-disposal units is that they're not substitutes for trashcans. Despite the claims of some manufacturers, the list of things that a disposal unit can safely handle is fairly short. Soft food items like boiled potatoes and oatmeal or crispy vegetables such as lettuce, carrot or potato peels, and the like are easily ground into a pulp that can be flushed away with enough water. Hard or stringy food items, on the other hand, are troublesome. Celery, egg shells, coffee grounds, and even apple seeds are common sources of trouble. And of course, you should keep trash such as paper, plastic, twist-ties, and so on from making its way into a waste-disposal unit. When in doubt, read the owner's manual for the unit.

General Repairs and Maintenance

When a waste-disposal unit won't work, the problem is most likely either a jammed drum or a burned-out motor. It's easy to clear a jammed drum, but unless the motor is still under warranty, it may not be worth repairing because the cost to fix a unit is often as much as a new unit.

Restarting a Jammed Unit

Manufacturers expect their units to stop once in a while, so they build in two useful features. One is a wrench slot in the unit's motor shaft; the other is an electric restart button. Both are located on the underside of the motor housing.

If you can't see or feel the obstruction from above, find the wrench that came with your waste-disposal unit or a large Allen wrench, and move to the bottom of the unit. Unplug the disposal and insert the wrench into the shaft at the bottom-center of the unit, and crank the motor back and forth. This will almost always clear the obstruction. You'll know you've made progress when the motor spins freely, without continuous resistance. Plug in the unit and press the reset button to allow it to run again. Once the unit starts up

Installing a Waste-Disposal Unit

Waste-disposal units can be convenient appliances, although they have some limitations. There are two basic types: batch-feed units that you load before running and the more common continuous-feed units that can run as you feed in wastes. To avoid problems with disposal units—mainly jamming—always run cold water as you feed in wastes, and to clear the drain afterward. Don't grind wastes with metal or glass, and never use chemical drain cleaners: they can damage inner seals.

USE: ▶ socket wrench waste-disposal unit (with drain fittings/waste kit) • supply tubes • shutoff valves

1 To install the sink drain, press plumber's putty around the drain flange, and insert the drain into the sink opening.

2 Disposal-unit drain fittings come with several types of fasteners. In this case, the lower component is drawn up with a bolt.

4 Lift the disposal unit up to the drain fitting, and rotate the metal collar to engage the connection at the top of the unit.

5 Use a plastic disposal waste kit to drain both sides of a double sink into the fixture trap. Connect the trap to the drain.

6 If your disposal unit has a cord with a household plug, provide a grounded, 15-amp switched outlet in the cabinet.

WATER & WASTE TREATMENT

again, test it. But make sure you use plenty of running water.

Routine Maintenance

Use cold water when grinding food scraps. To sharpen impeller blades, fill the waste-disposal unit with ice cubes and turn it on. Do this every couple of months. To keep a unit from developing a bad odor, use it often and with lots of running water. If your unit already smells, pour lemon juice into the drum and let it stand for a few minutes; then flush it. Run the unit with plenty of water thereafter. To clean the inner workings, quarter a potato, toss it in, and run the unit with cold water. When the drum is empty, run the unit with lots of hot water. And finally, avoid pouring your leftover sodas into the unit. They contain corrosive carbonic acid.

Removing a Waste-Disposal Unit

Unlike drain fittings, waste-disposal units don't become hopelessly stuck to sinks. The reason has to do with the mounting mechanisms, which range from simple hose-clamp fasteners to threaded-plastic collars to triple-layer bolt-on assemblies. The triple-layer mechanism described here is the most common—and the most complicated. (Also, see pages 258–259 for a photo sequence.)

To remove an old waste-disposal unit, start by shutting off the electrical power to the unit, either within the sink cabinet or at the main service panel. If your disposal unit also drains a dishwasher, loosen the hose clamp that secures the dishwasher discharge hose and pull the adapter from the waste-disposal-unit nipple. Next, loosen the horizontal waste tube's slip nut at the waste T-fitting, and undo the bolt or compression nut that secures the tube to the side of the disposal unit. Remove this tube.

To release the waste-disposal unit, look for three rolled-edge slots on the mounting ring. The ring is mounted at the top of the unit and has three such slots. Insert a screwdriver into one of the slots, and rotate the ring counterclockwise. If it won't budge, tap it with a hammer. As soon as the unit breaks free, support its bottom with one hand and rotate the nut about 2 inches until the unit falls away. This will leave only the bolted drain fitting in place.

With the disposal unit out, loosen the screw that holds the cover plate to the unit's electrical box. Pull the wires from the box, and undo the twist connectors and grounding screw. Then remove the fastening nut from the threaded box connector. This nut is located just inside the box, and you can turn it with your fingers as soon as you knock it loose with the screwdriver. Pull the connector and wires from the unit.

To undo the drain assembly, use a slotted screwdriver to loosen all three bolts separating the layers of the drain. With the bolts unscrewed about ½ inch, push the mounting flange up to reveal the locking ring. Pry this ring from its groove, and all the under-sink components will fall away. Lift the drain from the sink, and scrape away any old putty you find clinging to the basin under the flange. To install a new unit, see the photo sequence opposite.

As you shop, you'll notice a great range of prices. What's the difference? Materials and features. The inexpensive models may have only one impeller inside a steel drum, driven by a single-direction ⅓-hp motor. The higher-dollar models will likely have corrosion-proof stainless-steel upper bodies and ½-hp motors that are capable of driving in both directions. (A reversing motor alternates directions each time it is turned on.) These waste-disposal units are not as prone to the sort of one-way bind that an apple seed or fruit stem can cause. A reversing motor does automatically what you might have to do with a wrench.

- outlet box and wiring • plumber's putty

3 This disposal unit's drain fitting has a threaded collar with wings to help thread it over the drain extension, called a spud.

7 When connecting ¼-in. fixture supply tubes to ½-in. supply lines, use shutoff valves with compression fittings.

Restarting a Jammed Waste-Disposal Unit

USE: ▶ hex wrench

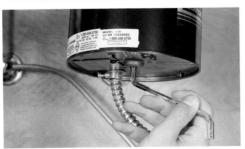

1 Insert a hex wrench (either a regular Allen wrench or one supplied with the unit) into the motor shaft and spin the motor.

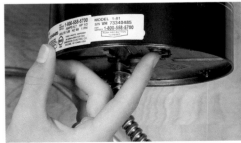

2 When the motor seems to spin freely, press the reset button to restart the stalled motor. Do this several times, if needed.

8 Plumbing

INSTALLING SUBMERSIBLE SUMP PUMPS

You'll find two types of sump-pit liners on the market. One has perforations around its upper half; the other doesn't. Use the perforated liner for groundwater sumps and the non-perforated type for gray-water pits. Dig a hole for the pit liner so that its top rim is flush with the top of the concrete floor.

Set and Pipe the Sump Pump

To install a submersible pump, begin by threading a 1½-inch plastic male adapter into the outlet fitting. Glue a 2- to 3-foot length of PVC pipe into the adapter to bring the riser up to check-valve level, preferably just above the liner lid. Lift the pump into the pit. If your check valve has threaded ports, make the connection with male adapters. If it has banded rubber connectors, tighten its lower end over the riser. Make sure the arrow on the side of the valve points up.

Before extending the riser, determine where you'll take it through the wall. The easiest spot is through the rim joist overhead. To keep from damaging the siding at the discharge spot outside the house, drill from the outside. Measure up the basement wall to determine the best exit point, and mark the rim joist several inches above the sill. Drill a ⅛-inch hole at the centerpoint to transfer the mark to the outside of the house. Drill a 1⅞-inch hole through the siding and rim joist from outside, and slide a short length of 1½-inch pipe through the hole.

Join this pipe to the drain line with a coupling. Replace any insulation you've removed to expose the rim joist, and secure the piping with hangers. The pump can cause the drain line to surge at start up, so you'll need good support to keep it from moving and making noise. Pipe movement can also break the exterior caulk seal.

On the outside of the house, trim the horizontal pipe stub ½ inch away from the siding, and cement a 90-degree elbow to this stub, with the open end pointing down. Extend the line down to ground level, and use another 90-degree elbow to direct the flow onto a splash block.

Installing a Sump Pump

USE: ▶ drill • saber saw • pliers, nut driver • hacksaw ▶ sump-pump kit • pvc pipe • primer cement • pipe hangers

1 Cut a 3-in. hole in the side of the pit liner corresponding with the location of the drainage pipe.

2 Connect the pipe to the liner, and place the liner in the pit. Backfill the pit below the pipe with soil, use gravel above.

3 Thread a 1½-in. PVC male adapter into the sump pump using pipe-thread sealing tape on the threads.

4 Cement a length of PVC pipe into the adapter, and lower the sump pump and pipe into the pit liner.

5 Install a check valve above the lid, and secure it with a nut driver (inset). Offset the riser against the wall using 45-deg. elbows.

6 Run the discharge pipe outdoors through the rim joist, using couplings for connections.

SUMP PUMPS & WELLS

INSTALLING WATER SOFTENERS

Shut off and drain the water at the main valve. When using copper male adapters, solder them to pipe stubs before threading them into the head of the softener. A softener piping loop needs a bypass just above the softener so that the water system can be kept in use when the softener is shut down for servicing. Factory-made bypasses are available, but you can also make your own using three ball valves.

Install the Bypass

If the softener comes with a built-in bypass, you won't need one in the piping loop. In this case, start by sliding the factory union nuts onto the provided copper stubs, and solder the stubs to ¾-inch copper risers using coupling fittings. (You should have previously measured and cut the risers to the right length.) Follow by installing the plastic purge line on the unit's head, locking it in place with a hose clamp. Attach the same type of hose to the brine tank's overflow fitting.

The bypass valve comes with lubricated O-rings, so just push it into the head until it locks in place. Slide rubber washers onto the copper stubs, and connect the copper to the bypass. Tighten the unions to secure the pipes. To preserve the piping system's electrical path to ground, install the copper clip that bonds the two pipes together.

To plumb the unit, slide the softener against the wall. Cut into the cold water line. Use 90-degree elbows to route the hard water line toward the softener and install a second line for the return. Cut a T-fitting into the supply line, and run a ¾-inch branch to serve any faucets and sillcocks that you want to get hard water. You can pipe one sillcock with soft water for washing the car, but try to pipe one with hard water for the lawn. If you can't isolate at least one, shut down the softener to water the lawn.

Secure the two lines to the wall, and make the softener connections. Solder all piping connections with lead-free solder. Finally, turn the water back on, purify the softener with bleach, rinse thoroughly, and add salt to the tank.

Installing a Water Softener

USE: ▶ pipe cutter • torch • groove-joint pliers • flame shield ▶ water-softener kit • copper pipe, fittings • flux, solder

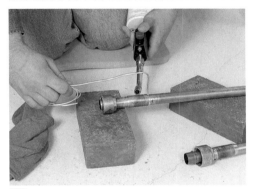

1 Slide the factory-supplied union nuts onto the prefitted stubs (also supplied), and solder the stubs to the copper riser pipes.

2 A softener needs a purge line and overflow tube. Slide and clamp these plastic tubes onto their barbed fittings.

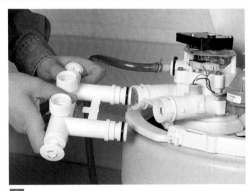

3 Mount the factory-supplied bypass valve on the softener head, and then attach the copper risers.

4 Use large groove-joint pliers to tighten the union nuts over the riser pipes. Connect the copper electrical-ground bonding wire.

5 Solder all water-pipe connections with lead-free solder, and protect combustible surfaces with a flame shield.

6 After disinfecting the unit, pour in several bags of special water-softener salt (sodium chloride or potassium chloride).

8 Plumbing

PRIVATE WATER WELLS

Well types range from the tireless windmill wells of generations past to shallow-well jet pumps to deep-well submersible pumps (most common). Most deliver water to a pressure tank located in a well pit or basement. Most wells these days are drilled by professionals and can be a couple of hundred feet deep. Well casings are 5 to 6 inches in diameter—large enough to fit a submersible pump.

The Pressure Tank

Pressure tanks maintain a steady line pressure and keep the pump from kicking on with every small draw of water. A pressure switch installed in the tank piping controls the electric well pump. The switch turns the pump on at a given low pressure and off it at a given high pressure. Different switches are used for different pressure combinations, but some adjustment is possible. Common "On-and-Off" settings are 20 and 40 psi for smaller tanks and 30 and 50 psi for larger tanks.

The most common pressure-tank problems are waterlogged tanks and pressure switches that have drifted off their settings or have simply worn out. Some pressure tanks (especially older ones) are simply hollow, galvanized containers that have a top-mounted snifter, or air, valve. The snifter, sometimes called a Schrader valve, looks and acts like a valve stem in a car tire. In these tanks, a quantity of air is held in the top of the tank as a pressure buffer and is in direct contact with the water in the tank. It's not uncommon for them to have waterlogging problems. Modern tanks have an air-filled rubber bladder inside. It has a snifter valve as well, but it's charged at the factory and usually needs no adjustment. With the air contained in the bladder, the water can't absorb it, so these tanks are nearly maintenance free.

Minor Well Problems and Solutions

If your well stops working abruptly, look first for tripped breakers and blown fuses, either in the main service panel or at a fused disconnect switch. Pump motors require roughly ten times the power to start than to keep running. Breakers and time-delay fuses usually accommodate this momentary overcurrent, but occasionally they will trip. Reset the breakers, and install new time-delay fuses. Don't use standard fuses.

The tank pressure must be drawn down for the pump to kick in when you turn the system back on, so open a few faucets to test your work. If the breaker trips or the fuse blows when the pump comes on, call a professional.

If the pump runs for an extended length of time before tripping the breaker or shuts off normally but seems to run much longer than it used to, you may have a pressure-switch problem.

Fixing a Pressure Switch

To gain access to the pressure switch, shut off the power, loosen the captive nut on the top of the switch cover, and remove the cover. You'll see several wire connections and two spring-loaded pressure sensors: one long and one short. The short spring will control the cut-out pressure (the range), which is usually the culprit, while the larger spring will control the cut-in (low) pressure while maintaining the cut-in/cut-out pressure differential. Both springs can be adjusted with the range nuts on top of the springs. Adjust the larger spring nut first. For a higher cut-in pressure, rotate the nut downward. For a lower cut-in pressure, rotate the nut upward. Follow with similar adjustments for the smaller cut-out spring. If the contacts are fouled, clean them according to the manu-

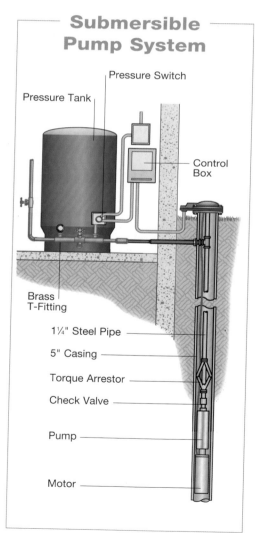

Submersible Pump System

Pressure Switches

To set the cut-in pressure while maintaining the pressure differential, adjust the nut on the long spring.

To set the cut-out pressure only, adjust the nut on the short spring using a wrench.

SUMP PUMPS & WELLS

facturer's instructions. Then restore power, and test your work. If these corrections don't help, or if the pump cycles erratically, have the pressure switch replaced.

If your pump seems to kick on every time you use a little water, expect a tank with too little air. While this problem is mostly limited to bladderless tanks, it is possible for a bladder style tank to leak as well, usually through the snifter valve. To make sure, test the tank pressure with a tire tester. If the high pressure is not as high as the tank specifications require (usually about 28 pounds), add some air.

If these simple procedures don't put your well equipment back in working order, call a professional.

HOW TO REPLACE A LEAKING PRESSURE TANK

Shut off the power and water, and drain the system through the boiler drain. Disconnect the power supply at the pressure switch, Then undo the feed line from the well and the union on the house side of the tank T-fitting. Slide the old tank out, and remove the tank fitting.

Fit the T-fitting with new components. Be sure to install a pressure-relief valve in one of the openings. Follow with a new pressure switch on a ⅜-inch nipple. Thread the nipple into the tank T-fitting with pipe-thread sealing tape and make connections as shown below.

Adjust the Pressure Switch

Have a helper open a faucet until the pump kicks in. With a 20-40 switch, check that the pump comes on when the gauge nears 20 pounds and shuts off near 40. If it shuts off before 40 pounds, adjust the switch's high-pressure spring by following manufacturer's directions. Next, wait for the pump to shut off and then have a helper open a faucet. The pressure should drop slowly; the pump should not come on until the gauge reads 20 pounds. If the pressure drops immediately, bleed the system of all water and take an air pressure reading. The tanks's air pressure should be within two pounds of the cut-in pressure. Add air through the snifter valve if needed.

Replacing a Leaking Pressure Tank

USE: ▶ pipe wrenches • adjustable wrench • screwdriver • wire-cutting tool • thread-sealing tape ▶ pressure-relief valve • pressure switch

1 With the power and water turned off, disconnect the wires and undo the box connector.

2 Then loosen the water-pipe unions on the house side of the tank T-fitting. Slide the old tank out of the way.

3 Tighten a new pressure-relief valve into one of the openings in the tank T-fitting. Use pipe-thread sealing tape on the valve's threads.

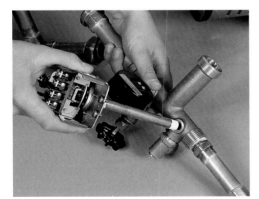

4 Mount a new pressure switch on a ⅜-in.-dia. threaded nipple, and thread the nipple into the tank T-fitting.

5 With the T-fitting assembled, install it on the pressure tank. Use the half union mounted on the tank for attachment.

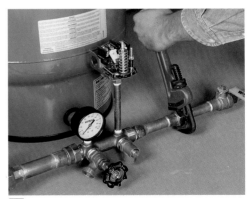

6 Attach the tank to the piping using unions with thread-sealing tape; reconnect the wiring and recharge.

PRIVATE WATER WELLS

8 Plumbing

CLEANUPS

The improvements shown here are some of the upgrades you can make—for example, to install a new showerhead and resurface an old tub or shower. Another way to spruce up plumbing fixtures is to get rid of mold, stains, and soap deposits with some heavy-duty cleaning.

To remove a buildup of soap residue on molded tubs and showers, spray the surfaces with an all-purpose nonabrasive cleaner, and let it soak in for a few minutes before rinsing. If some deposits remain, try a specialized product for soap scum or a liquid laundry detergent that also helps with mineral deposits left by hard water.

Try lightening porcelain stains by scouring with a proprietary cleaner and stain remover or a mixture of lemon juice and salt. For deep stains, add baking soda to the mixture to make a wet paste that you can leave on the stain overnight.

To remove mineral deposits on metal fixtures, use a softening solution of one cup of white vinegar in a quart of water or one of the specialized products on the market. You may need to wash the area several times and scrub with a nylon sponge to dislodge multi-layer deposits.

Remove mold stains from tile grout with a household scouring powder, or add enough household bleach to an abrasive cleanser to make a paste, scrub it on, and then rinse. Don't add bleach to a cleaner containing ammonia; the combination produces dangerous fumes.

Reglazing a Tub

1 A worn bathtub can be refinished. A contractor first chemically etches the surface and then sprays on a new finish.

Replacing a Showerhead

USE: ▶ pipe wrenches ▶ new showerhead • pipe dope or Teflon tape • scrap cloth

1 To remove the old head without marring the stem, wrap a cloth or thin towel around the fittings.

2 Use two wrenches to remove the old head: one to hold the stem in place and one to twist off the old head.

3 To seal the connection between the stem and head, wrap the threads with Teflon tape or add pipe dope.

4 Screw the new showerhead onto the threaded stem. New heads will come with flow restrictors to save water.

Tub Surrounds

USE: ▶ saber saw • caulking gun • measuring tape

1 Measure from the tub and sidewall to fix the faucet and spout locations. Cut the openings with a saber saw.

3 Mark the center of the tub and the center of the final panel. Stick the bottom of the panel first, and then press upward.

FIXTURE IMPROVEMENTS

MONEY SAVER

2 When the finish cures, the final step is to buff it to a high shine. High-quality refinishing will last 10 years or more.

Relining a Tub

1 Some contractors offer custom relining of old tubs. This alternative to reglazing takes only a few hours to install.

MONEY SAVER

2 This before-and-after photo shows half of the old tub and enclosure, and half relined with high-impact acrylic.

▶ tub surround • adhesive • caulk

2 If the panels are adhesive-backed, peel back the paper; otherwise, apply panel adhesive to the edges and center of the panel.

4 With all panels in place, caulk the vertical seams and the joint between the tub and surround.

Installing an Anti-Scald Faucet

USE: ▶ adjustable wrench • keyhole saw • torch • wheel cutter ▶ faucet • fittings • solder/flux

1 Start by removing the spout and faucet trim; the spout may thread or pull off. Look for an underside Allen screw.

2 Cut the drywall from the back of the plumbing wall, and cut out the old faucet with a close-quarters wheel cutter.

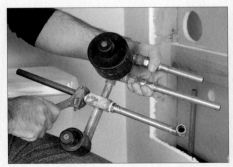

3 Solder male adapters to copper pipe stubs, and preassemble as much of the faucet piping as possible.

4 With the faucet assembly installed and secured, solder the remaining fittings with lead-free solder.

CLEANUPS

8 Plumbing

PLUMBER'S HELPER?

Sometimes learning from experience can be expensive. You're looking over the plumber's shoulder, acutely aware that the meter is running, thinking that you could have done what he's doing. Everyone needs professional help on some repairs, but there are a few good candidates for do-it-yourselfers. If in doubt about your ability to tackle them, pay for the education and watch a contractor do it. Soak up the details; then you'll know whether you can handle the job yourself next time.

DIY Pipe Repairs

Here are some general rules to observe whether you're trying to fix copper or plastic pipe. Start by turning off a valve to stop feeding the leak. If, in an emergency, you can't locate this cutoff valve, shut off the main meter valve until you find it.

If you try to heat up a pipe with a torch—to resolder a joint—first drain the line. You can't get copper hot enough to make solder flow when it's filled with water. Also, open a faucet just beyond the repair spot so that any steam that develops can escape.

Most plastic pipe is a snap to repair compared with copper. It can be easily cut and cemented to new fittings. Plastic fittings do require a lighter touch when threading them onto nonplastic fittings—it's easy to over-tighten a threaded plastic fitting and break it. Cross-threading a plastic fitting is also a problem at least until you get the feel for it.

Temporary Repairs

USE: ▶ metal file • screwdriver ▶ rubber pipe insulation • banded clamps

If you don't have the time or tools to make a permanent repair when a pipe springs a leak, temporarily plug a tiny pinhole by jamming in a sharpened pencil. To plug a split pipe, wrap the area with a piece of thick rubber, and tighten it down with a banded clamp. If you can't find a local cutoff, turn off the main valve while you search. Remember that in houses with wells and pressure tanks, leaks continue to flow even if you turn off the well pump.

1 To make a temporary repair that won't leak, first use a file to flatten out any ragged edges around the split.

2 Slit a thick piece of rubber or a short section of garden hose, and slip it over the damaged pipe.

3 Attach banded clamps over the rubber sleeve at each end of the split, and tighten the clamp screws.

Installing an Anti-Freeze Faucet

To protect an outside faucet against freezing, replace it with a special anti-freeze valve. You can still turn the water on and off outside. But the long stem of the faucet extends through the wall and controls a valve inside the house. Water doesn't stand in the portion of the pipe or faucet outside the wall where it could freeze. These faucets don't need a cutoff valve. You should take the standard installation steps of caulking and insulating the hole through the siding.

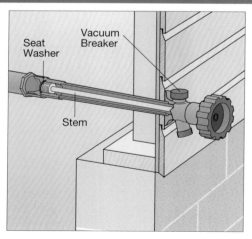

USE: ▶ tubing cutter • reciprocating saw (optional)

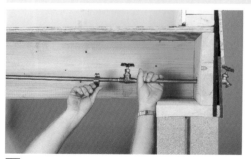

1 Shut off the water supply, and drain the system. Then, cut out the old faucet piping, including the shutoff valve.

268 **PLUMBING** / SOLVING PROBLEMS

SOLVING PROBLEMS

Heat Cables

Heat cables look like extension cords, but they are designed to convert electricity into heat so that you can wrap them around pipes that might freeze. Sealing air leaks and insulating pipes should be your first step. But if you install heat cables, use only UL-approved products, and follow manufacturer's installation instructions. Models with a built-in thermostat can be left plugged in. But take care not to wrap the cable on top of itself. It produces enough heat to melt the wire insulation and could start a fire. Do not use old, cracked, or damaged cables.

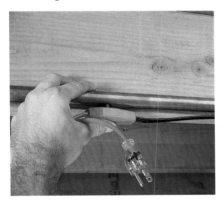

Modern heat cables have a built-in thermostat. Some do not have to be spiral wrapped around the pipe.

Quieting Noisy Pipes

To reduce the noise from supply and drainpipes, wrap them with foam tubes, and pack the wall cavities with insulation. (Cast iron drains are quieter than plastic ones.) Severe pipe banging, called water hammer, is caused by excessive water pressure or the abrupt shutoff produced by the solenoid valves on dishwashers and clothes washers. To fix water hammer, anchor the pipes to the framing with hangers or install a shock absorber. You can make a simple one by cutting in a T-fitting and a capped stem of pipe. Gas- and oil-filled shock absorber fittings also provide damping action.

To support water pipes and prevent rattling and banging against framing, use clip-on pipe hangers.

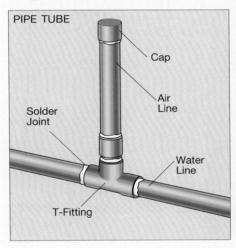

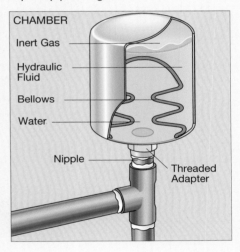

MONEY SAVER

• propane torch • caulking gun • work gloves/eye protection ▶ anti-freeze faucet • screws • solder/flux • braces • caulk • insulation

2 Enlarge the wall opening if needed, and insert the freeze-proof faucet through the siding and band joist. Secure it with screws.

3 Solder the copper supply line to the faucet with lead-free solder. Brace the line to ensure adequate drainage.

4 Although the valve is inside, it pays to protect water pipes at outside walls with foam tubing or batts of insulation.

8 Plumbing

DRAIN TROUBLES

What could cause gurgling in a rarely used wet-bar sink when water drains in the kitchen at one end of the house or in a bath at the other end? Chances are that the wet bar isn't vented or that its vent stack is blocked. You could check by looking for a plumbing vent pipe protruding through the roof above the area. It will look like the ones above the kitchen and bath.

Sluggish drainage is one symptom of this problem. Sewer smell is another. When water drains in the kitchen or bath, it siphons water out of the wet bar's drain trap, which produces a gurgling noise. Without the seal provided by water in the drain trap, sewer gas can rise up through the sink and into the house.

If you don't use the wet-bar sink because of these problems, call a plumber to install an auto-vent. With this in place, the other drains won't siphon water out of the trap. If you don't use it and don't really want it, have the plumber remove it and close the drain connection. In the meantime, you can reduce the sewer gas smell by periodically pouring some water in the wet bar drain to keep the trap full.

Big Backups

Even with proper venting, trap placement, and pipe sizing, drains still may be slow. This indicates a big blockage farther down the line. The best indicator that a blockage is in the waste line is sewage gurgling up through floor drains and basement fixtures. Solving this problem depends on your septic system type.

If you have a septic system with a leachfield, your septic tank may need to be pumped out. This is a job for a professional, but it's important to be around when the pumping concludes. Ask the pump operator if the tank was filled with enough greasy scum to cause the drain problems. If not, the pipes in the leachfield itself may be the problem. Digging them up may be the only solution.

If you're tied into a municipal sewage system, check to see whether any work has been done on your branch line lately. If so, it could have caused problems with your drainage. If not, tree roots may have gotten into the pipes. Hire a professional to auger the line and pull out the tree roots; thereafter, flush copper sulfate root treatment through the line twice a year.

Drain Leaks

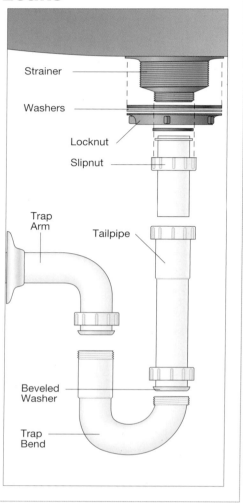

To find out whether the sink drainpipe or the drain flange in the sink is leaking, pour water directly into the drain. If water leaks below, the problem is in the piping. If not, the problem is likely in the flange seal. You may need to unscrew the flange, clean off old caulk or plumber's putty, install a fresh bead of caulk, and retighten the flange. Leaks in the drainpipes generally occur in the trap. Some traps have a cleanout nut that makes it easier to clean blockages.

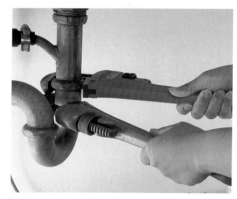

To disassemble a standard trap, use two pipe wrenches: one to hold the sink stem in place and one to turn the fitting.

Clearing a Waste-Line Clog

To clear a stopped drain, you should start by using a plunger. For best results, block the overflow and second-drain outlets; spread petroleum jelly on the rim of the plumber's helper; and (with 2 or 3 inches of water in the basin or tub) use steady, forceful downward strokes to clear the clog. If this doesn't work, you can try working a piece of wire through the cleanout plug and removing the blockage. However, snaking with a hand auger or power auger is much more effective. Disassemble the traps; insert the auger; and clear out any blockages.

To get the most force, plug the overflow fitting with a wet towel when plunging a bathtub. Standing water also helps.

SOLVING PROBLEMS

Cleaners

A plumber's auger is the safest way to clear stubborn jams in household pipes. It won't damage metal pipes, which is a risk with some caustic chemical drain cleaners, such as those that contain acid or lye. An auger is simply a flexible wound-wire cable that you turn into the pipe either by hand or with a special drill. Another option is to use water and air pressure to dislodge a jam. These pneumatic devices typically fit over the stopped drain and release a charge of air into the standing water.

Position a pneumatic cleaner over the stopped drain; seal the connection; and release the air charge.

Clearing a Tub Clog

Tubs fitted with an internal drain stopper, called a tripwaste, require extra attention when they become clogged. Removing the linkage helps you diagnose the problem. Check the drainage flow by running water down the drain. If the drain fails to empty properly, the clog is farther down the line. If the drain flows freely, the tripwaste is stretched and the stopper is too low in the drain tee, blocking flow even when open. Shorten the linkage by ¼ inch; tighten the locknut; and replace the linkage—the drain should work fine.

1 *To access a tripwaste linkage for cleaning, remove both screws; grip the overflow plate; and lift out the linkage.*

2 *Sluggish tub drainage may be caused by a stretched tripwaste linkage. Remove the linkage, and shorten it about ¼ in.*

3 *Pop-up drain plugs often clog with hair. Lift out the plug and linkage; remove the hair; and replace the assembly.*

When snaking a kitchen sink line, remove the trap, and bore directly into the line with a hand-held drain auger.

To snake out a bath drain, remove the overflow plate; pull out the tripwaste linkage; and bore through the overflow.

When plunging won't clear a clogged floor drain, it's best to remove the plug and auger the line through the cleanout.

DRAIN TROUBLES

9 Insulation

274 THERMAL PROTECTION
- Insulation Basics
- Insulation Terms
- Thermal Envelope
- R-Value Zone Map
- Insulation Performance

276 TYPES & APPLICATIONS
- Choosing Insulation
- Common Types of Insulation
- Foil Insulators
- Tools & Special Handling Equipment

278 INSTALLATION
- Foundations
  - INSULATING FOUNDATION EXTERIORS
  - INSULATING FOUNDATION INTERIORS
  - INSULATING CRAWL SPACES
- Diagnosing Wall Insulation
- Adding Insulation
  - INSULATING WALLS
  - INSULATING CEILINGS
  - INSULATING ROOFS

282 BLOWN-IN
- Blowing-in Insulation
  - INSULATING EXISTING WALLS
  - INSULATING ATTICS

284 COLD SPOTS
- Preserving Heat
  - EIGHT WAYS TO CONSERVE TEMPERATURE
  - VAPOR BARRIERS

9 Insulation

INSULATION BASICS

Like clothing, insulation comes in all shapes and sizes, but each has an R-value, the measure of its resistance to heat flow. In general, lightweight, air-filled materials such as fiberglass insulation batts have high R-values per inch of thickness and are good insulators. Standard fiberglass is rated at about R-3.5 per inch. Heavy, dense materials such as brick (R-0.2) and gypsum plaster (R-0.2) have R-values so low it's difficult even to think of them as insulators. R-values are stamped on the insulation itself and displayed in all insulation advertising. It is the only reliable way to determine how effective the insulation will be, and the only way to compare one type with another.

Comparing R-Values

Thicker is not necessarily better when it comes to insulation. That sometimes confusing subject becomes clear when you compare materials. For example, these three alternatives are rated at R-11 and offer the same thermal protection: 1½-inch-thick polyurethane board, 3½-inch-thick fiberglass batts, or 4 inches of loose-fill vermiculite. That means you can't pick one insulation over another based only on thickness. A full 5 inches of a traditional, poured-in insulation such as perlite rated at about R-13 would provide only about 60 percent of the thermal protection offered by an inch less (4 inches) of polyurethane board rated at about R-23. So in order to reach an insulation rating of R-11 in walls, R-19 in floors, and R-30 in ceilings, you would need different thicknesses of commonly used insulation materials.

Comparing R-values without splitting hairs should show which one of several alternatives will provide the most resistance to heat loss. But some insulation materials are better suited to certain kinds of installations. That means you must consider other insulation characteristics in addition to the R-value. (See "Choosing Insulation," p. 268.)

Also bear in mind the law of diminishing returns as it applies to insulation. This means that the first inch in an uninsulated wall offers the greatest benefit, while the second inch offers a bit less, and so on, even though the last inch costs as much as the first.

Insulation Terms

- **BATT OR BLANKET** insulation is usually made from fiberglass cut to fit into framing cavities. Batts are provided in sheets, while blankets come in rolls.

- **BRITISH THERMAL UNIT (BTU)** is a measurement of heat—one Btu is the heat required to raise the temperature of 1 pound of water 1 degree F.

- **RIGID BOARD INSULATION** is manufactured in panels that make it easy to clad a large area. Different types provide a range of R-values, although rigid foam generally is highly rated. It also resists water and rot in locations near the ground.

- **R-VALUE** is the standard measure of resistance to heat flow. Every type of insulation has an R-value per inch. The higher the value, the more resistance.

- **THERMAL ENVELOPE** describes the sum total of a home's insulation systems: walls, ceilings, foundation, floors, windows, and doors.

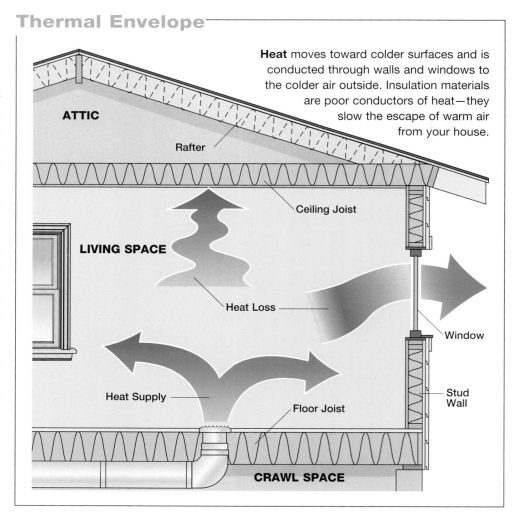

Thermal Envelope

Heat moves toward colder surfaces and is conducted through walls and windows to the colder air outside. Insulation materials are poor conductors of heat—they slow the escape of warm air from your house.

THERMAL PROTECTION

R-Value Zone Map

RECOMMENDED R-VALUES

ZONE	A	B	C	D
1	30–49	25–30	13	See note.
2	30–60	25–38	13–19	
3	30–60	25–38	19–25	See note, and add R-5 insulated sheating under new siding.
4	38–60	38	25–30	
5	49–60	38–49	25–30	See note, and add R-5 to R-6 insulated sheating under new siding.
6	49–60			
7	49–60			
8	49–60			

A=Uninsulated Attics
B=Existing 3–4 in. of Insulation in Attic
C=Uninsulated Floors
D=Uninsulated Wood-Frame Walls

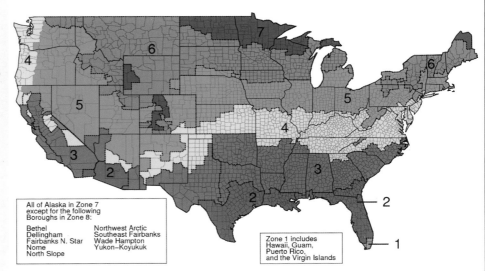

R-VALUE RECOMMENDATIONS FOR U.S. CLIMATE ZONES

All of Alaska in Zone 7 except for the following Boroughs in Zone 8:
Bethel, Dellingham, Fairbanks N. Star, Nome, North Slope, Northwest Arctic, Southeast Fairbanks, Wade Hampton, Yukon-Koyukuk

Zone 1 includes Hawaii, Guam, Puerto Rico, and the Virgin Islands

NOTE: When adding new siding on an uninsulated wall, drill holes in the sheathing and blow insulation into the empty wall cavity. For insulated walls in Zones 4–8, add R-5 insulated sheathing when adding new siding.

Insulation Performance

The most common type of insulation is fiberglass, which has R-values ranging from R-11 to R-38, depending on thickness. Other types of materials include blown cellulose, several varieties of rigid boards, sprayed foams, and many less common materials, such as cotton fiber and aluminum foil bonded to plastic bubble pack.

The table at right compares approximate R-values for different types of common insulation. Each material is evaluated on how much resistance to heat change it offers. Values can be slightly different for very similar materials. To be certain of the value (and be sure that you meet local energy codes), check product labels. All are required to list the R-value per inch.

Generally, extruded plastic boards like polystyrene and polyisocyanurate offer the highest R-values.

R-VALUE COMPARISONS

Fiberglass Batts

3½"	R-11
6½"	R-19
7"	R-22
9"	R-30
13"	R-38

Loose Fill (per inch)

Cellulose	R-3
Perlite	R-3
Vermiculite	R-2

Rigid Board (per inch)

Expanded Polystyrene	R-4
Dense Polystyrene	R-4
Extruded Polystyrene	R-5
Polyurethane	R-6
Polyisocyanurate	R-6–7

Sprayed or Foamed Fill (per inch)

Cellulose	R-3–4.0
Polyurethane	R-5.5–6.5

INSULATION BASICS

9 Insulation

CHOOSING INSULATION

Two main factors affect your choice of insulation: the configuration (for instance, loose fill or rigid foam board) and the R-value. There are many types available, so you can pick the most efficient product for the job. Some do-it-yourselfers also may consider ease of application, how the material is packaged, and potential drawbacks such as possible skin irritation. But many lumberyards and home centers stock only fiberglass and a few types of foam boards.

Basic Configurations

Insulation is commonly available in five forms: batts to fit between 16- or 24-inch-wide framing, either paper or foil-faced; loose fill to blow or pour into structural cavities; and foam boards, used mainly on roofs and on the outside of walls and foundations. The two other types, sprayed-in-place cellulose foams and foamed-in-place urethanes, are more expensive and not used as widely.

Some materials come in only one configuration, others in several. You can get by using scraps of one material in spaces where another product would offer more protection. But to do a thorough job of creating a thermal envelope around your living space, particularly in the framing compartments of an existing building, you may want to use more than one type of insulator.

Materials

Cellulose fiber is a paper-based product and has roughly the same R-value as fiberglass, about R-3.5 per inch. Typically, it is made from shredded recycled paper combined with a fire retardant. Loose fill can be blown in using a pressurized air hose. Several newer insulating materials use a mix of about 75 percent recycled cotton fiber (even scraps of old blue jeans) with 25 percent polyester to bind the fibers together. The material comes in batts and as loose fill. Its R-value is generally about the same as cellulose.

Polyurethane foam can be sprayed by contractors into open framing cavities where it provides a thorough seal against air leaks and a thermal rating of about R-6.0 per inch. It also comes in pint-sized quantities—in a can with a nozzle so you can use foam to fill small openings in the building envelope.

Common Types of Insulation

Fiberglass

♦ **FIBERGLASS**
The most common of wall and ceiling insulation materials, fiberglass insulation is installed in 80 percent of new homes. R-values available in a variety of different thicknesses range from R-11 to R-38. Unfaced batts can be laid on top of themselves to create super-insulated attics. Most residential applications use either rolls or precut batts.

Mineral Wool

♦ **MINERAL WOOL**
Like fiberglass, mineral wool is made from a hard mineral slag and spun into a soft material. Mineral wool gets clumpy when wet and will lose R-value. When dry, mineral wool has the same R-value as fiberglass.

Cellulose Loose Fill

♦ **CELLULOSE LOOSE FILL**
Cellulose is made from shredded newspapers that have been chemically treated with a fire retardant. It is sold in large bags and can be easily poured in between attic floor joists or professionally blown into wall cavities. When it is blown into walls some settling can occur, creating under-insulated slices along the ceiling line.

Extruded Polystyrene

♦ **EXTRUDED POLYSTYRENE**
This form of rigid board insulation has an R-value of about 5.0 per inch. The extruding production process creates a denser layer of polystyrene than expanded polystyrene. Boards are usually pink or blue in color. A similar material, called expanded polystyrene (EPS), has many tiny foam beads pressed together, like a styrene foam coffee cup or cooler. EPS is commonly called "beadboard" and has an R-value of about 3.5 per inch of thickness.

Polyurethane

♦ **POLYURETHANE**
This versatile type of foam has a white or yellowish color, and an R-value of about 6.0 per inch. Rigid panels can be faced with foil for radiant heat deflection. Used on a large scale on exposed framing, the material also can be mixed on site and sprayed into place as a dense liquid that fills both large areas and small spaces in irregular framing bays. The material bubbles up after application, and is later trimmed flush with framing.

Polyisocyanurate

♦ **POLYISOCYANURATE**
This plastic has an R-value of approximately 6.0 per inch. It has a white or yellowish appearance and is usually backed with foil for radiant heat reflection.

TYPES & APPLICATIONS

Cementitious foam is made from magnesium oxide compounds extracted from sea-water—a natural alternative to synthetic foams. This material (rated at about R-2–3 per inch) will not burn and does not shrink, but like polyurethane, is relatively expensive.

Expanded polystyrene board (beadboard) is similar to the material used in disposable coffee cups and is the first in a line of foam boards. Each offers a step up in quality, R-value, and, of course, cost. All are highly resistant to moisture and water damage. Expanded board is crumbly but can add R-4.0 per inch under new roofing and siding or over foundation walls when covered with an exterior finish. **Extruded polystyrene board** is a more expensive, somewhat denser board that offers R-5 per inch for more insulating value. **Polyisocyanurate board** is a more rigid foam board that carries a very high rating of about R-6.3 per inch. In one inch of space you get almost double the thermal resistance provided by fiberglass.

Bubble pack is a flexible, foil-backed sheet of plastic with air-filled bubbles. (It looks like packing material.) Use it where there isn't room for batts or foam boards. You can wrap the ¼- or ⁵⁄₁₆-inch-thick sheeting around ducts and even use it under new drywall, say, between the house and garage.

Foil Insulators

Foil-covered bubble pack sheets are thinner and more flexible than rigid boards, which makes them more versatile in their insulating applications. Foils are also impermeable to vapor, which allows them to double as a vapor barrier. Cold, sweaty pipes and air-conditioning vents benefit most from vapor-resistant insulation. Foil's flexibility makes it very useful in wrapping pipes and ductwork, and no safety precautions against airborne fiberglass particles or skin irritation are needed.

Foil also can be draped underneath plywood roof sheathing to improve energy efficiency. Laboratory testing of foil has shown that over 90 percent of radiant heat can be reflected from a foil's surface. When used with other insulations, reflective foils can boost energy efficiency while serving as an effective vapor barrier.

Foil sheeting installed under the roof deck between rafters can cut cooling bills by reflecting radiant heat.

New wall spaces can also be covered with reflective foil to reduce radiant heat flow into the living area.

Tools & Special Handling Equipment

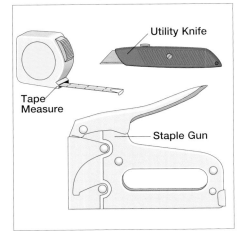

Most insulation is installed with simple tools that are probably already in your garage or workshop.

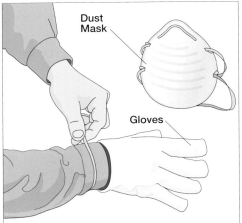

Protective gear for installing insulation includes gloves, dust mask, and a shirt with long sleeves secured with rubber bands.

An advantage rigid board and foil bubble pack have over fiberglass is that they require no protection against airborne particles. You can easily cut rigid boards using a utility knife, and you don't need to wear gloves and a dust mask. With fiberglass insulation, however, care must be taken to protect your eyes and lungs from glass fibers that can become lodged in your skin or breathed into your lungs.

WARNING: If you encounter asbestos insulation on old pipes, do not attempt to remove it yourself—call in a professional.

CHOOSING INSULATION

9 Insulation

FOUNDATIONS

Foundations can be insulated in two ways: from the inside and from the outside. Attaching insulation on the inside is much easier, but over the long term, insulating outside is more effective. Ideally, foundations should be insulated and waterproofed from the outside before they are backfilled during construction. But if your foundation is not insulated, and you are thinking about fixing basement leaks—which will require digging out around the foundation anyway—this may be the time to consider it.

From the Outside

An outside foundation insulation project requires digging out the backfill from the entire perimeter of your house—not an easy job. But if you go that route, remember to repair cracks with cement and check the overall foundation for signs of deterioration. Once the foundation is exposed, trowel a liberal layer of liquid asphalt on the outer surface for waterproofing; then attach rigid insulating board, and backfill the foundation. The backfill should consist of mainly gravel for the below-grade level and at least 1½ feet of topsoil for plantings and shrubbery.

If your house is built on a perimeter foundation with a shallow crawl space instead of a full basement, digging from the outside to expose the wall surface may be easier than trying to apply insulation from the confined space inside the foundation.

From the Inside

First, clean and paint the walls with a masonry paint. If efflorescence (white, powdery spots) occurs anywhere, then water is coming from the outside into the basement and should be addressed as a separate problem. If your basement is dry year-round and you want to add insulation, you can apply rigid foam board or batts and a modular system of furring strips or studs to hold a finishing layer of paneling or drywall.

You can stuff fiberglass insulation around openings in the foundation, for instance, where plumbing pipes exit the house, and then seal the opening with caulk or cement. Another option is to spray foam around the hole, and trim the expanded material flush with the wall.

Insulating Foundation Exteriors

USE: ▶ shovel • hose • chalk-line box with plumb bob • hammer • straightedge • trowel ▶ rigid insulation

1 *Excavate the dirt around the foundation. Check for cracks and holes; clean with a garden hose; and let dry.*

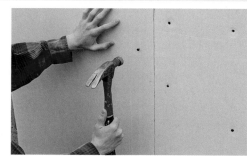

2 *Nail rigid foam board insulation directly into the concrete or block foundation using masonry nails.*

Insulating Foundation Interiors

USE: ▶ 4-ft. level • hammer • work gloves ▶ furring strips • wood shims • masonry nails • rigid insulation

1 *Install furring strips to support rigid foam panels. Use wood shims and a spirit level to plumb the strips.*

2 *Nail the strips in place with concrete nails once you have shimmed them the correct distance out from the wall.*

Insulating Crawl Spaces

USE: ▶ dust or respirator mask • work gloves • staple gun ▶ batt or blanket insulation • heavy-duty staples

1 *Batts of fiberglass insulation should be installed in the bays between the floor joists.*

2 *Staple wire mesh onto the joists to keep out animals and prevent the insulation from sagging.*

INSTALLATION

GREEN SOLUTION
- masonry nails • stucco

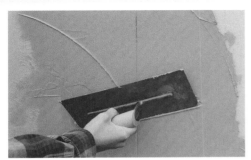

3 Apply cement plaster to protect the aboveground insulation from the weather and present a finished surface.

GREEN SOLUTION
- 6-mil polyethylene vapor barrier • drywall

3 Attach the foam between or on top of the furring, or both to increase the R-rating. Then cover with plastic and drywall to finish.

GREEN SOLUTION
- wire mesh

3 Staple fiberglass insulation batts onto knee walls above foundations. You can let the batts drape down over the masonry.

Diagnosing Wall Insulation

To rate the energy efficiency of your house, conduct an energy audit. The most thorough version, handled by professional testing firms, checks air leaks, insulation values, the efficiency of glazing, and more. Many utility companies provide this service free. Some also recommend specific improvements and estimate the expected return on your investment in the form of lower utility bills.

One easy way to discover if your home needs an audit is to conduct an insulation test with two thermometers. Tape one on an exterior wall; set the other in the middle of the room—off the floor and away from direct sunlight or heat registers that could skew the results. If the wall surface is within 5° of the ambient room temperature, the wall is adequately insulated.

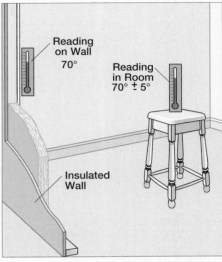

If a thermometer on an outside wall registers 5° less than one in a heated room, you could use more insulation.

To reduce drafts and cut energy loss around windows, fill narrow gaps around the frame with loose-fill insulation.

To stop air and water infiltration on the outside apply a bead of flexible caulking, such as silicone, to exterior seams.

A thermographic picture of your house can reveal where energy is escaping. This temperature-sensitive photograph highlights heat leaks in bright colors, typically at windows and doors. Also notice heat leaks under the double garage doors.

FOUNDATIONS

9 Insulation

ADDING INSULATION

If adding some insulation is a good idea, more must be better—but only up to a point. With insulation there is a law of diminishing returns: the first inch offers the greatest benefit and makes the most noticeable difference in comfort. But layer upon layer provides less and less benefit—even though the last one costs as much as the first.

So when is enough insulation enough? One sensible guideline is to fill the space between framing members. For instance, you can add 3½ inches of insulation to wall cavities framed with 2x4s, but trying to cram in 7 inches of insulation is counterproductive.

Interior Coverage

If you do decide to add to existing insulation, be sure to install only insulation, not a vapor barrier. A layer of plastic or foil buried between layers of insulation can cause condensation problems, decay surrounding wood, and greatly reduce the insulation's effectiveness.

For example, if you have only a few inches of cellulose between floor joists in the attic, you could add unfaced batts, rolls, or loose fill. You could build up the insulation depth between joists, and if you don't need the storage space, spread rolls of insulation above the joists at right angles to the framing.

This over-layer insulates the attic floor framing as well as the spaces between joists, which are in direct contact with the ceiling of the living space below and act as heat conductors. Because heat rises, extra ceiling insulation is often a cost-effective improvement.

Exterior Coverage

You can gain the same frame-covering benefits outside the house, too, and eliminate thermal weak links (such as the seam between foundation and framing) by cladding walls with a layer of rigid foam board. It can run from underneath the siding, down the exposed foundation wall, and into the ground because foam board won't rot.

Even with a layer of foam, it's important to fill spaces that don't match the size of standard insulation, such as openings between unevenly spaced framing members and gaps around window and door frames.

Insulating Walls

USE: ▶ dust or respirator mask • work gloves • knife • staple gun ▶ batt or blanket insulation • heavy-duty

1 Wear gloves and a long-sleeved shirt when you unroll fiberglass. Cut it to length using a sharp bread or paring knife.

2 Fit the batt between framing members by hand. You should wear a dust or respirator mask and gloves for protection.

Insulating Ceilings

USE: ▶ utility knife ▶ batt, blanket, or loose-fill insulation • plywood baffles

1 To avoid condensation due to trapped moisture, slit the facing of new batts or use unfaced blankets over existing insulation.

2 For maximum thermal effectiveness, run new insulation over both the old insulation and the framing members.

Insulating Roofs

USE: ▶ dust or respirator mask • work gloves • knife • staple gun ▶ air baffles • batt or blanket insulation

1 Staple air-chamber baffles to the underside of the roof sheathing before installing batts between rafters.

2 Because air can circulate through the baffles, you can fill the remaining space between rafters with batts.

INSTALLATION

staples • 6-mil polyethylene vapor barrier

GREEN SOLUTION

Insulated Headers

3 Run the insulation behind pipes, outlet boxes, and other obstacles to reduce thermal loss and prevent pipe freezing.

4 Flatten out the flanges extending from each side of the batt, and then staple them to the wall studs with plastic.

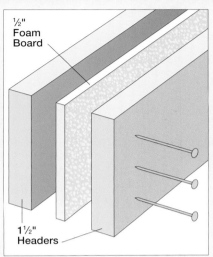

Headers that bridge openings over windows and doors normally have two timbers, a plywood core, and low insulation value. Where codes permit, raise the value by replacing the plywood with ½-inch foam board.

GREEN SOLUTION

3 Loose fill is poured into the spaces between joists. Keep it a few inches away from obstructions such as recessed lights.

4 Install plywood baffles above exterior walls to prevent loose fill from blocking vents in the roof overhang.

Air-Space Baffles

GREEN SOLUTION

• heavy-duty staples • 6-mil polyethylene vapor barrier

3 Position and flatten the insulation the same way you would between wall studs. Batts should not compress the air baffles.

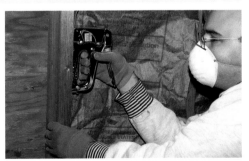

4 Trim and fit insulation in irregular openings and around obstructions to create a complete thermal barrier.

Air baffles stapled to the roof decking moderate roof temperature and protect against damage due to condensation problems. Air can travel beneath insulating batts from vents in the overhang to the ridge.

ADDING INSULATION 281

9 Insulation

BLOWING-IN INSULATION

Blowing-in insulation makes the most sense when the framing cavity is empty. New material won't be blocked by old batts and thermal improvement will be dramatic, even though the dead air trapped in an empty wall cavity does provide some insulation. A contractor can fill the empty space by cutting small holes through the drywall, inserting a hose, and pumping insulation into the bay between each pair of framing members. You do wind up with a row of little cutouts, but they can be patched, sanded, and painted.

Access from the Outside

It may be easier for the contractor to gain access from the outside, by removing a course of clapboards and cutting a channel in the sheathing over the studs. It depends on which way into the wall cavity causes the least damage while providing the best access. In most cases, it's simpler to remove and replace exterior siding than it is to patch and repaint dozens of small holes in an interior wall. But even working blind with a hose through a hole, experienced contractors should be able to gauge how much insulation the cavity should take and know when the flow of loose fill has been blocked—say, by a construction brace or plumbing. In those cases, they may have to make a second hole to be sure that the bay is completely filled.

You might want to make a thermographic or thermometer test when the job is done to confirm results—although to be completely fair, you would have to duplicate weather conditions of the first test. (See "Diagnosing Wall Insulation," p. 271.)

Insulating the Attic

Unlike wall cavities, you can't just fill all the empty spaces in the attic with blown-in insulation. You might find that there is room for a lot of new insulation—under the roof edges, for example—but you have to leave at least 1½ inches of air space under the roof for ventilation. Why? Because even when the ceiling insulation is protected with a vapor barrier, some moisture from the living space below gets through. Water vapor simply seeps through the insulation, rises against the cold roof, condenses, and drips back into the fiberglass or cellulose. This reduces the insulation's effectiveness and causes mildew, wood rot, and water "leaks" that, in the rooms below, will seem to be substantial enough to have come through the roof, not just from condensation beneath it.

Forming Dams

You need insulation directly over the exterior wall frame, but loose fill won't stay in neat piles along the wall until you close up the roof. And even if it did, over time the loose fill would spill down onto the soffit (the plywood running parallel with the ground on the underside of the overhang) and block the vents.

One way to solve this problem is by stapling foot-long batts of foil-backed fiberglass on their edges between the joists to form a dam running from the exterior wall back up and into the bays. Air coming up through the soffit vents can flow freely over the short batts and across the loose fill, which can't spill out onto the soffit.

Because of a lack of wood, early homesteaders in Nebraska built well-insulated houses out of sod—the 3-foot-thick walls (seen in the doorway) kept them warm in winter, cool in summer.

Obstructions

Blown-in insulation works best on walls where the wall cavities are empty. You won't gain much insulating value if existing insulation already fills most of the cavity. Another potential problem is that even if there is no insulation in the wall, some cavities will always be partly blocked by pipes, wires, and built-in obstructions such as horizontal fire-stops. Sometimes, the blown-in insulation will fill around the blockage. Usually, however, a new hole will have to be drilled higher on the wall to feed insulation into the blocked section.

Insulating Existing Walls

USE: ▶ zip tool (for vinyl siding) • pry bar (for wood siding) • stud finder • saber saw • drill/driver with hole saw

1 Remove vinyl siding using a zip tool. Once the seam is separated, slide the zip tool along to free the panels.

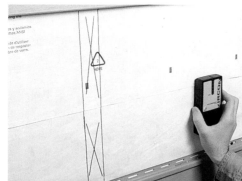

2 Use a stud finder to locate the studs beneath the sheathing. A hole will be drilled into each stud bay.

BLOWN-IN

Obstructions

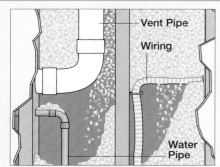

Pipes and wires can block the flow of loose insulation blown in through a hole in the top of the wall.

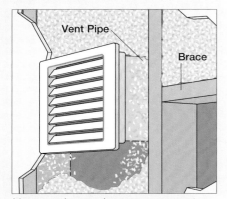

Vents and cross braces may create pockets without insulation. Another hole must be drilled to fill them.

Insulating Attics

You should keep loose fill away from recessed light fixtures in ceilings, which need air flow to prevent overheating. Also keep loose fill blown into attics away from eaves vents. It's pointless to continue insulating out onto the roof overhang. Install some form of dam above the exterior stud wall to hold back the loose fill, preserve its loft, and prevent it from retarding ventilation by spilling onto the soffit vents.

Spraying loose fill is a great way to add insulation in your attic and cut your heating bills.

GREEN SOLUTION

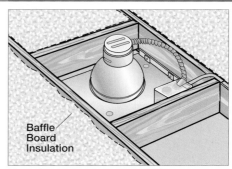

Keep loose insulation away from a recessed ceiling light fixture unless the fixture is rated for insulation contact.

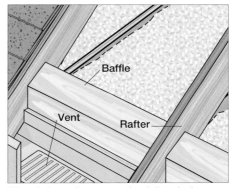

Wood dams between the joists help keep loose fill insulation from spilling onto soffit vents and blocking air flow.

GREEN SOLUTION

attachment (optional) • blower hose with nozzle • blower equipment ▶ loose-fill insulation • yellow wood glue • cork or plastic plug

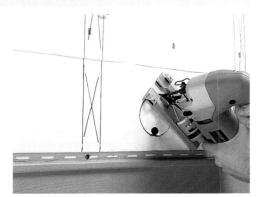

3 After you've shut off the power to the circuits in that wall, drill a hole with a saber saw or hole saw chucked into a power drill.

4 Place the nozzle in the hole, and load the hopper, operating the blowing machinery per the manufacturer's instructions.

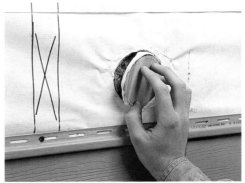

5 After a bay is filled, plug the hole with a cork or plastic plug; replace the siding; and move about 4 ft. up the wall; then repeat.

9 Insulation

PRESERVING HEAT

Insulation's main job is to keep heat from escaping from inside the house. So in many homes, heating ducts that run through the cellar or a crawl space are not insulated. They're not outside, after all. And any heat that radiates from the ducts is still inside the building. That's good and that's bad—good because the heat isn't lost; bad because it's not going where you need it, down the duct to a chilly bedroom far from the furnace.

In any case, the solution is simple enough: Wrap the ducts in insulation. The exact configuration (batts or rolls) or type (cellulose or fiberglass, for instance) hardly matters. And you don't have to hermetically seal each seam. Improvement in heat delivery from the ducts should be noticeable, particularly if the ducts travel through a chilly, vented crawl space.

The major sources of heat loss in a home, which obviously can't be fixed with insulation, are windows and doors. A popular method of sealing windows involves taping a thin sheet of shrink-wrap polyethylene over the window frame and blow-drying it for a shrink fit. On old double-hung windows with single glazing, this is a much cheaper plan than replacing the windows, although it is unsightly. Drafts are contained, and a layer of dead air is trapped between window and plastic, adding extra insulation.

Doors can be upgraded by attaching thermal stripping along the jambs and nailing a flexible threshold to the door's bottom edge. When installing weatherstripping, make sure you don't prevent the door from seating.

Preserving Hot Water

What's good for ducts is good for water pipes, too. Like a tea cozy on a teapot, insulation helps keep the hot-water pipes hot and will keep cold-water pipes from sweating in summer. You could wrap the pipe with insulation rolls or batts, or use widely available foam tubes made for the job. They are neater than a do-it-yourself spiral wrapping of batts but not any more effective.

If you are replacing or adding insulation in a wall and discover water pipes, be sure to place the insulation between the pipes and the outside wall. Don't bury pipes under insulation.

Eight Ways to Conserve Temperature

SEAL SILLS
In new construction, the space between the top of the foundation and the sill can be sealed with a narrow strip of foam insulation. Foam will expand to fill cracks that open as a house settles.

WRAP DUCTS
Ducts can be wrapped in paper-backed fiberglass insulation or foil-backed bubble wrap. Where ducts enter and exit through walls, ceilings, and roofs, seal the edges with foam insulation.

SEAL HOLES
Foam expands like shaving cream out of its can, and can be messy to work with if it's not contained in a hole or crevice. But foam is a good choice for where pipes go into walls.

CLOSE GAPS
Windows are often a major source of thermal loss. If you can feel a draft, remove the casing and stuff pieces of fiberglass insulation in any cracks between the window jambs and the framing.

COLD SPOTS

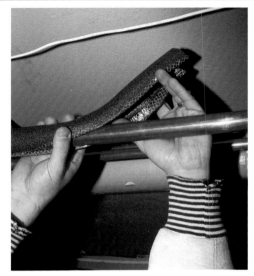

WRAP PIPES
To insulate pipes, buy preformed pipe sleeves that fit over the pipe, or wrap the pipe in thin fiberglass strips and secure them with duct tape. Both will prevent pipe sweating in the summer.

GREEN SOLUTION

ENCASE WATER HEATER
A water heater can be wrapped in a fiberglass thermal blanket to cut down heat loss. Water-heater blankets are sold in kits that include tape and a thermal blanket encased in a plastic sleeve.

GREEN SOLUTION
Vapor Barriers

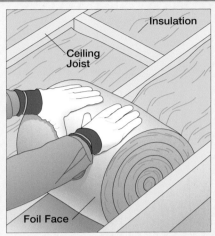

Vapor barriers block both air and water vapor. The only impervious barrier is foil. But you need to install foil-faced batts carefully, and tape the seams for maximum effect. Plastic sheeting is second best, and often used over paper-faced batts to improve moisture resistance. Clear polyethylene sheets 6 mils thick are standard. All vapor barriers are rated by permeance (the ability of air to penetrate). To be reasonably effective, the perm rating should be less than 1. Polyethylene sheets have a perm rating of 0.04 to 0.08.

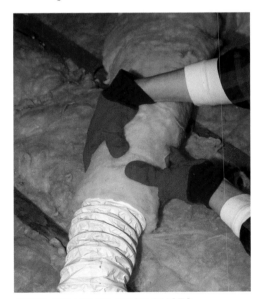

INSULATE VENTILATION DUCTS
Wrap attic ventilation ducts with thin batts of fiberglass insulation. This insulation prevents condensation from forming—and then leaking down through the ceiling—where hot vented vapor meets cold attic air.

SEAL UTILITY BOXES
An insulating pad inserted between a switch or receptacle and its cover will stop airflow. You can also inject silicone caulk around the box and around the drywall or plaster.

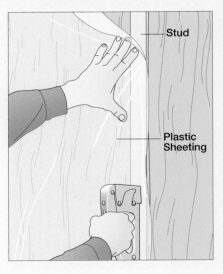

PRESERVING HEAT 285

10 Heating

288 BASIC SYSTEMS
- Comparing Systems
- Heat Savings
- Fuel Supply
- Hot-Air Systems
- Hot-Water Systems

290 BASIC INSTALLATION
- Expanding the System
- Baseboard Convectors

292 BASIC MAINTENANCE
- Filter and Flue Checks
- RELINING FLUES
- FURNACE MAINTENANCE
- IMPROVING EFFICIENCY

294 SPACE HEATERS
- Supplemental Heat
- Installing Toe-Space Heaters
- Gas Heaters
- Installing Wall Heaters
- Portables

296 STOVES & FIREPLACES
- Heating with Wood
- Stoves
- Gas Fireplaces
- METAL FLUES
- Installing a Masonry Fireplace

298 HUMIDIFYING
- Moisture & Air Quality
- Humidifier Types
- INSTALLING A HUMIDIFIER

10 Heating

COMPARING SYSTEMS

You can number-crunch comparisons among gas, oil, or electric furnaces, or even wood stoves, but when thinking about your home's heating system, don't forget to factor in basic, practical considerations. Switching to gas could mean that you'll have to pay to run a supply line into the house. Sticking with oil could mean replacing a rusting storage tank. Electric heat would cost more every month but save thousands on installation costs of alternative systems.

Ask heating contractors for installation estimates and fuel suppliers for approximate operating costs. Because all heat output is measured in British thermal units (Btu), you'll have a common denominator to make comparisons.

If you're thinking about replacing or upgrading an existing system, it's important to find out how efficient the new system will be compared to the old one. For a well-maintained existing system, you could subtract half the unit's age from the original efficiency rating—for example, rate a 20-year-old oil-burning furnace that was 65 percent efficient at 55 percent. Of course, if you pay a contractor to make combustion efficiency tests, you'll get a more accurate rating. Once you know the increase in efficiency with a new system, you can estimate how much less fuel you'll use every year, how much money this will save each year, and how many years of savings it will take to recover your investment in new equipment.

Fuel Supply

GAS-FIRED

Gas-fired furnaces burn natural gas or liquefied petroleum (LP) gas to heat either air that is blown through a system of ducts or water that is circulated to radiators or baseboard convectors through pipes. Older gas-fired appliances have pilot lights that are burning all the time. Improved modern systems have electronic igniters to light the flame as the gas starts to flow. Gas furnaces burn cleanly and convert up to 95 percent of the fuel into usable heat.

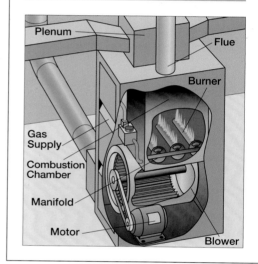

Heat Savings

HEATING SAVINGS

Savings for One Eight-hour Decrease of 10–15°F (5°C) Per Day

- 6–8%
- 9–11%
- 12–13%
- 14–15%
- 16–18%

Savings on your energy bill depend on home size and actual heat loss or gain, geographic location, frequency of temperature changes, and range in degrees of change.

Source: Honeywell Inc.

Hot-Air Systems

Hot-air systems use a blower to force heated air through a large supply plenum and into a system of ducts. The ducts lead to registers in the floor and walls of your living spaces. Cold-air registers and return ducts take cooled air back to the furnace for reheating. These systems require dust filters. Because they provide dry, hot air, a furnace-mounted humidifier often is needed to maintain indoor comfort.

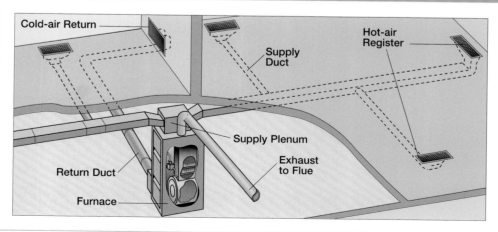

BASIC SYSTEMS

OIL-FIRED
A high-pressure oil burner, mounted either outside or inside the furnace, pumps a fine mist of oil and air into a combustion chamber, where it is ignited by an electric spark. This in turn heats a heat exchanger that passes the temperature onto air or water that circulates throughout the house. Unlike natural gas, which is fed from a gas main to provide a constant supply of fuel for gas systems, oil must be delivered. Oil burns less efficiently than gas, with furnace efficiencies under 90 percent.

ELECTRIC
Electric furnaces can heat air or water by passing current through heavy-duty heating coils. Electric heat elements also are included in heat pumps that provide both heating and cooling. Electricity can power central systems and is used to heat individual baseboard convectors. Electric systems require almost no maintenance because they generate heat without combustion. They are 100 percent efficient in your house, but not at the utility plant where power is generated.

HEAT PUMPS
In summer, heat pumps run like air conditioners. In winter, the system reverses, extracting heat energy from the air (or from the ground, in the case of ground-source heat pumps) to warm air in the house. But when the outdoor temperature drops to about 35°F, a back-up system of electric coils takes over, and the system loses fuel efficiency. The units are most cost-effective in regions with roughly equal heating and cooling demand.

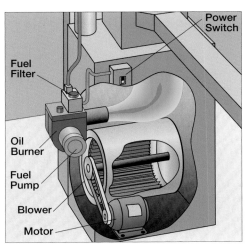
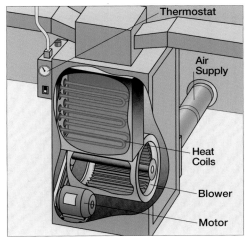
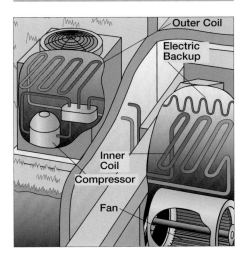

Hot-Water Systems

Hot-water (or hydronic) systems use a pump, called a circulator, to force heated water from a boiler through a network of pipes. Heat transfers from the pipes to the air at radiators or baseboard convectors and continues back to the boiler for reheating. Older homes have one large pipe loop. Newer homes have two or more loops, each with its own thermostat, to heat different zones of the house more efficiently.

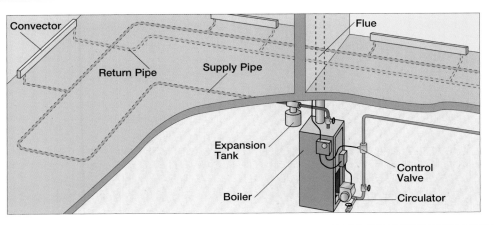

COMPARING SYSTEMS

10 Heating

EXPANDING THE SYSTEM

Installing a furnace or an entire network of HVAC ducts or heating pipes are not DIY projects—they must be put in by professional contractors. However, when you've added new living space to your home—whether by finishing an attic or garage, enclosing a porch, or building an addition—you will need to provide heat for the space. You have the option of extending your home's existing system into those spaces or installing individual electronic heaters or portable space heaters. (For more on space heaters, see pp. 286–287.)

Hot-Air Systems

New runs of ducts can be extended from the furnace's plenum or from a main duct in an extended-plenum system. Cut a hole in the plenum or main duct using sheet-metal snips; the hole should exactly fit a metal collar, either straight (to run sideways from the plenum) or takeoff (to run upward from a main duct).

Round metal ducts are then snapped or hammered together (depending on the type of duct) and attached to the system. There are T- and Y-fittings to make branches; for turns, use 45- and 90-degree-angle pieces or sections of flexible duct. Each new run must also have a damper, to shut off heat to that duct run and balance the system if needed. The final joint in a run of ducts should be attached with a drawband—a steel collar that is tightened with bolts (like a band clamp).

Use flexible metal straps, called hangers, to attach new ducts to basement ceiling joists. You can easily heat the first floor by placing heat registers (grilles with movable vents) in holes cut into the floor and running the ducts to the registers with transition fittings called boots. To bring heat to the second floor, you'll need to run the ducts up the wall or through closets and box them in with studs and drywall.

Hot-Water Systems

Hot-water heating pipes run in a circuit around the house. Usually, you won't need to add new pipes to the system to install a new convector but instead can tap into existing lines. Most systems have the excess capacity to handle one or two additional convectors.

There are three common layouts for hot-water systems. A series loop has the convectors as part of the circuit; hot water enters each unit through a supply riser and exits through a return riser, then moves on to the next convector in the loop. One-pipe systems have supply and return branch lines that feed each convector from a main supply loop. Two-pipe systems have entirely separate circuits for supply and return. You need to know what kind of system you have before you start cutting pipe. The main line may run around the perimeter of the basement or along a center beam. The steps for adding a hot-water convector are shown at right. (For information on cutting and soldering copper and cast-iron pipe, see "Plumbing," beginning on pp. 158.)

In 1777–1778, Washington's Continental Army wintered in Valley Forge, PA, in these primitive log huts with chimneys made from log sections chinked with mud.

Electric Systems

It may not be practical to extend your home's heating system into a finished garage or attic. Your other heating options include installing a wood stove, space heater, or electric convector. Although electric heaters aren't as efficient as gas or oil systems, a single room unit is still far cheaper and easier to install than extending ducts or pipes. Wall-mounted baseboard convectors run along the bottom of the wall like hot-water convectors; recessed models are installed through the exterior wall. Smaller 120-volt heaters will plug right into wall outlets; 240-volt models will be more efficient, but you will need to have an electrician install a new circuit at the service panel to operate them.

Baseboard Convectors

USE: ▶ screwdriver • power drill/driver • work gloves

1 *Locate the new baseboard unit over a supply pipe in the floor below. Start by installing the reflector panel on the wall.*

5 *After drilling a hole for the return pipe at the other end of the convector, solder on a bleeder valve. Protect the wall from flame.*

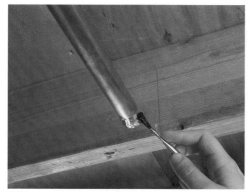

9 *To draw solder fully into the joint and prevent pinhole leaks, paint the connecting pipe parts with flux.*

BASIC INSTALLATION

• propane torch • pipe cutter or hacksaw • clamps ▶ baseboard convector • pipe & fittings • bleeder valve • solder • flux • scrap Type X drywall & foil

2 Position the convector element on brackets attached to the reflector panel. Be careful not to bend the heat-dispersing fins.

3 Temporarily fit the cutoff valve onto the end of the convector pipe, and mark the floor below where the supply pipe will rise.

4 Remove the valve, and drill a hole through the floor. The hole and valve will be hidden by the convector end cap.

6 Test-fit pipes and fittings to reach from the convector valves to the hot-water supply pipe below the floor.

7 Cut off the water supply, and use a pipe cutter to cut away a section of the supply pipe to install a T-fitting.

8 Use a wire-brush tool or sandpaper to brighten the mating edges (inside and out) of the old pipe and the new fitting.

10 Take care to protect surrounding wood from flames. A clamp holds this piece of nonburning drywall covered with foil.

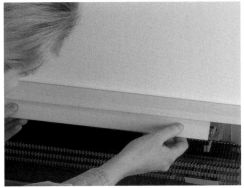

11 Once the pipe joints are soldered, slide on the adjustable heat-control flap and the front cover of the convector.

12 To finish the job, clip an end cap onto each end of the unit. A front flap lifts to provide access to the cutoff valve.

EXPANDING THE SYSTEM

10 Heating

FILTER & FLUE CHECKS

You'll probably remember to replace the big air filter in a forced hot-air system. If you don't, you'll eventually notice the reduced air flow—blocked by a thick mat of collected dust. But there may be several other potential clogs to remove. Before you work on them or on any electrical appliance, make sure that the system is turned off and its power supply is interrupted. Remember, a furnace that appears to be off can be suddenly triggered into operation by the thermostat.

On oil-fired systems, there is another filter to replace. Just like a car engine, the furnace has a filter in the oil line designed to trap sludge and other impurities that can clog the spray nozzle. They are very helpful if the furnace kicks in soon after an oil delivery, which stirs up sludge from the bottom of the tank.

A typical oil filter looks like a small canister attached to the oil feed line. The body of the canister unscrews to provide access to the removable filter cartridge inside. Because oil spills are smelly and difficult to clean up, put a pan beneath the filter (and wear rubber gloves) to make the change.

Some hot-air systems may have a filter on the return air grille—the oversized, centrally located, usually wall-mounted grille that returns cool air to the furnace for reheating. If an electronic air cleaner is added to the system, it's likely to have two filters—a wire-mesh grille for trapping larger particles of airborne debris, and electrostatic dust-collecting canisters. On most systems, the canisters are removable; you can pull them out of the cleaner and fit them in a dishwasher.

In addition to the seasonal maintenance on your furnace, normally performed by a contractor, it pays to check the exhaust flue. Aside from electric furnaces, which do not produce heat by combustion and don't need a flue, other systems require a clean, completely sealed escape route for exhaust. Even a small leak from a flue pipe inside the house can release carbon monoxide, which can be lethal.

In many houses, a metal exhaust pipe from the furnace leads into a masonry flue. It should have a lining, and be separated from other flues, such as a chimney flue, even when the two flues share one chimney. In older houses where brick chimneys deteriorate from exposure to the weather outside and exhaust gases inside, you may be able to reline the chimney instead of building a new one. There are two basic relining systems offered by specialty contractors. One is to insert a vibrating tube in the chimney, pour a fireproof cement mix around it, and gradually draw up the form, which vibrates to compact the mix. The other system relies on an inflatable form, which is centered in the old chimney while the fireproof mix is poured around it. The mix seeps into cracks and crevices, sealing and strengthening the chimney walls. When the form is removed, it leaves a newly formed flue.

MONEY SAVER: Relining Flues

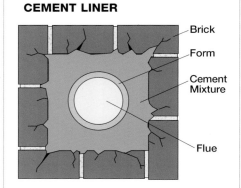

CEMENT LINER — Brick, Form, Cement Mixture, Flue

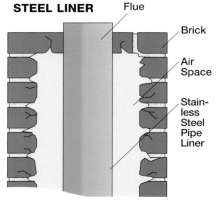

STEEL LINER — Flue, Brick, Air Space, Stainless Steel Pipe Liner

Contractors can reline damaged flues by pouring cement around a removable form or with flexible pipe.

Furnace Maintenance

If you are restarting a furnace, and it hasn't been serviced recently, begin the cold season by paying a pro to clean and tune the system. It's wise to check older systems every year, particularly to be sure that combustion furnaces (both gas- and oil-fired) are properly vented to safely exhaust potentially lethal gases. Look for nests, twigs and leaves, ash and soot—anything that could block escaping gas. If you can't get on the roof to look down the flue and don't want to disassemble the exhaust pipe, hire someone to do it.

USE: ▶ screwdriver • vacuum • bucket or watering can ▶ replacement dust filter • lubricating oil

1 Furnaces have air-intake grilles that are easily removed. Turn off the power supply first, and follow manufacturer's directions.

2 The most basic job, and one of the easiest, is to replace the dust filter. You may need to do this several times a year.

STOVES & FIREPLACES

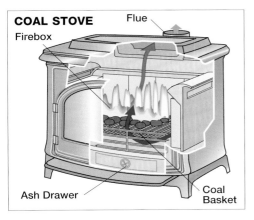

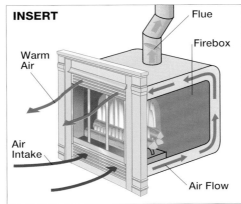

Firewood

SPECIES	BTU/CORD
Hickory	26 million
White oak	23 million
Sugar maple	21 million
Red oak	21 million
Spruce/hemlock	15 million
White pine	14 million
Aspen	13 million

Installing a Masonry Fireplace

USE: ▶ trowels • 4-ft. level • work gloves ▶ fireplace parts (throat, smoke chamber, damper, air intake) • concrete block • firebrick • mortar • flue liner • flue cap

1 This shallow but efficient masonry fireplace, called a Rumford, combines traditional hand work and preformed parts.

2 Once the opening is prepared, a team of contractors raises the firebrick firebox inside the concrete-block chimney.

3 A one-piece throat rests on top of the firebox. At the back of the firebox is an optional air-intake grille.

4 A two-piece smoke chamber sits over the damper, which must be protected from excess mortar that could foul its operation.

5 In about 4 hrs., this crew has completed the firebox. They continue to raise the block and interior flue.

6 On the way to finishing in one day (less the stucco and trim), the crew builds the chimney up around a clay flue liner.

HEATING WITH WOOD

10 Heating

MOISTURE & AIR QUALITY

To add moisture to dry winter air and create an indoor environment that's comfortable for you and good for your house as well, you can use a variety of portable or central humidifier systems. Here is a look at some of the options.

Portable & Central Systems

Portable, or console, humidifiers are concealed in small cabinets. They are helpful if one room is particularly dry or if you have a heating system without an air-distribution system, such as electric baseboards, that isn't suited to a central humidifier system. The drawback is that you have to add water to console storage tanks periodically. Also, they require maintenance much more often than central systems.

Central humidifiers are attached to the home heating system, normally at the plenum, where heated air is distributed to the ducts. The advantage is that the appliance is part of the house; you don't have to plug it in or add water. But an automatically replenished water supply can become a breeding ground for pollutants that are spread through the ducts and into living areas. Treating the water and doing seasonal maintenance can reduce this problem.

Types of Appliances

If you are shopping for a humidifier, bear in mind that two of the four basic types, just by their design, are more likely to disperse microorganisms. Ultrasonic humidifiers, which use high-frequency sound waves to generate a cool mist, and impeller humidifiers, which make a mist with a high-speed rotating disk, produce the greatest dispersions of microorganisms and minerals. Breathing misted air containing microscopic dust mites, mold, bacteria, and other pollutants can cause respiratory problems and allergic reactions.

The other two types of humidifiers generally disperse fewer pollutants. Evaporative units pick up water from a holding tank with a belt, sponge pad, or wick that is exposed to the airflow from the furnace. Warm-mist or similar steam-vaporizer humidifiers can completely eliminate pollution problems. A heating element boils the standing water before it is dispersed as mist into the air flow, which distills the minerals and kills bacteria and mold.

Humidifier Types

Heat from your furnace warms the air in your house, and dries it out, too. Forced-hot-air systems in particular can lower indoor humidity to the point at which people feel uncomfortable. You can add moisture to the air with portable humidifiers, although the most economical systems connect to the furnace. These have a moisture control, called a humidistat, and feed moisture directly into the warm air flow.

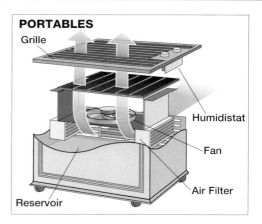

PORTABLES — Grille, Humidistat, Fan, Air Filter, Reservoir

Installing a Humidifier

USE: ▶ level • tape • marker • work gloves • metal shears • screwdriver • adjustable wrench • needle-nose

1 In a typical installation, you mount a paper template for the humidifier on the main return plenum above the furnace.

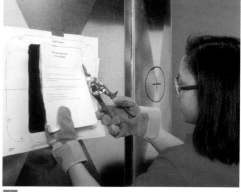

2 After marking the main supply plenum for the humidifier duct, cut the template through the sheet metal with a metal shears.

6 A typical humidifier has a solenoid valve to control water flow. This small pipe runs from the valve to the distribution tray.

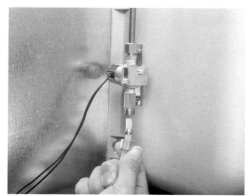

7 To bring water from your supply piping to the unit, most humidifiers supply either flexible copper pipe or hard plastic tubing.

HEATING / HUMIDIFYING

HUMIDIFYING

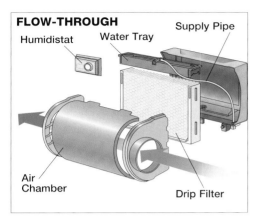
FLOW-THROUGH — Humidistat, Water Tray, Supply Pipe, Air Chamber, Drip Filter

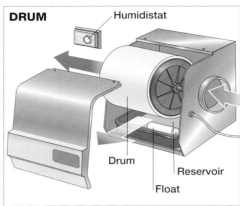
DRUM — Humidistat, Drum, Reservoir, Float

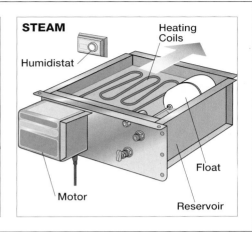
STEAM — Humidistat, Heating Coils, Float, Motor, Reservoir

GREEN SOLUTION

pliers • pipe clamp ▶ humidifier • humidistat • duct (flexible or metal) • mounting collar • saddle valve

3 It's important to level the humidifier for even water distribution. This unit has a small bubble level built into the water tray.

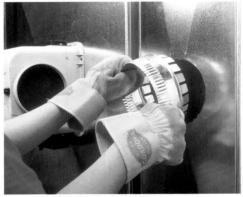
4 Cut through the main supply plenum to make a hole for the humidifier supply duct. This kit comes with a mounting collar.

5 Use flexible duct or a length of standard metal duct and an elbow fitting to connect the humidifier to the supply plenum.

8 Install a saddle valve (if permitted by code) on the supply pipe. Clamp it to the line, and turn the handle to pierce the pipe.

9 Central-system humidifiers typically have a catch basin that recirculates water or an overflow drainpipe like this one.

10 Install the humidistat, which allows you to regulate indoor humidity, on the plenum or near the existing thermostat.

11 Cooling

302 SYSTEMS
- Choosing a System
- Types of Units
- System Schematic
- Efficiency

304 WINDOW UNITS
- Picking a Window Unit
- Window Fittings
- Installing a Window Unit

306 IN-WALL UNITS
- In-Wall Unit Basics
- Installing an In-Wall Unit

308 MAINTENANCE
- Basic Maintenance
- Recharging
- BASIC CLEANING
- SAVING ENERGY

310 INDOOR AIR QUALITY
- Cleaner Air
- Portable Cleaners
- AIR EXCHANGERS
- Cleaning Ducts
- Media Cleaners
- Electronic Cleaners
- Electronic Air-Cleaning Installation

312 DEHUMIDIFIERS & THERMOSTATS
- Dry Air
- Dehumidifier Capacity
- Servicing a Dehumidifier
- Thermostats

11 Cooling

CHOOSING A SYSTEM

The time to collect your thoughts about keeping cool this summer is before the weather gets too hot and humid. By planning early, you'll avoid making a rash decision or being stuck with what's left at the home center. Before buying an air conditioner, ask some basic questions: will it fit in the window? Will it keep the room cool? Is it so noisy that you won't sleep?

Capacity

The cooling power of an air conditioner is measured is units of heat energy called British thermal unit (Btu). An air conditioner's Btu per hour (Btuh) rating indicates how much heat energy it can remove from the air in an hour. Some larger units are rated in tons, which measure the energy it takes to melt one ton of ice in a day. A ton is equal to 12,000 Btuh. As a general rule, 5,000 Btuh are needed to cool a 150-square-foot room. Add 1,000 Btu for every additional 50 square feet.

Central Air vs. Room Units

Central air conditioning is an attractive feature in the resale market, but it's costly and difficult to install in many homes. A contractor may be able to set up machinery and ducts in an unused attic or use forced-air heating ducts, but in a two-story house, you may have to give up some cabinet or closet space to install ducts on the first floor. In most homes, one or two window or in-wall units can keep crucial rooms comfortable and spill out enough cool, dry air to reduce heat and humidity in adjacent areas.

Types of Units

CENTRAL AC
Central ACs consist of an outside unit with a compressor, condenser coil and fan, and an interior evaporator coil installed in the supply duct of a warm-air furnace. Indoor heat is picked up and carried through pipes by a refrigerant to the condenser coil outside. Central air is expensive to install if your home lacks heating ducts but may still be cheaper (and quieter) than an array of room units. Modern, high-pressure lines can work with small-diameter hoses that are easy to install in existing spaces.

IN-WALL UNITS
Individual room units can be installed through the wall to avoid blocking the view through a window or having to remove the unit when it's cold. Like window units, in-wall units have two coils made of copper tubing and aluminum fins, one facing inside and one facing outside. These machines work like central systems, but all the components are built into one box. Most room units can be plugged into a standard 120-volt outlet but some require 240 volts. You should be sure that the unit does not overload the circuit.

System Schematic

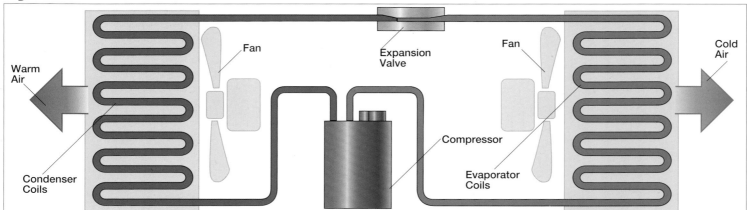

SYSTEMS

WINDOW UNITS
If you need to keep one room or area cool in the summertime, the easiest solution is to install a window unit. You don't have to make a hole in your house or install extensive ductwork, and most units can be plugged in and working an hour after you open the box. The best location is a double-hung window with a wall outlet nearby. The weight of the unit is carried on the sill and held in position with brackets at a slight downward slope for proper drainage of condensation. Extensions on each side of the unit slide out to seal the opening.

HEAT PUMPS
Heat pumps can heat and cool your home. They have an outdoor coil and compressor and an indoor coil and fan. (There are also self-contained through-the-wall units.) In hot weather, the heat pump acts like a conventional air conditioner. In cold weather, the cooling cycle of refrigerant is reversed to create a heat gain inside the house. But as the temperature outside drops, heat pumps lose efficiency, and an electric back-up heater kicks in. The units are most economical in areas with roughly equal heating and cooling demand.

CHILLERS
Evaporative chillers (sometimes called evaporative coolers) are typically used to cool the air in commercial buildings or large homes in the Southwest. Chiller units can deliver from 10 to 500 tons of cooling. Modern chillers with heat exchangers and high-efficiency motors can use as little energy as 0.5 kilowatt per ton of cooling. Chillers usually flow water through evaporator and condenser tubes surrounded by refrigerant. Hot refrigerant is then condensed back into liquid in a cooling tower.

Efficiency

You can compare air conditioners by checking their Energy Guide labels. These stickers explain annual electrical costs, compare efficiency among several units, and list an Energy Efficiency Rating (EER) number—a Seasonal Energy Efficiency Rating (SEER) on central air systems. This rating is the ratio of Btu used per hour of cooling to the watts used to produce those cooling Btu—fewer watts per Btu means greater efficiency.

SEER Rating	Recommendation
less than 9.7	old unit; replace with newer model
13	nat'l min. standard for single package/split-systems
14.5	recommended min. for Energy Star units
22	most efficient unit available

EER Rating	Recommendation
less than 8	old unit; replace
8.5–9.8	recommended min. for all units
9.4–10.8	recommended min. for Energy Star units (based on size)
10–11.7	most efficient available

CHOOSING A SYSTEM

11 Cooling

PICKING A WINDOW UNIT

If you don't have central air conditioning, you may find yourself lingering in front of the refrigerator this summer—unless you install at least one room air conditioning unit. And if you're only air conditioning one room, make it the bedroom. This way, you'll create an island of cool air where you can retreat on sweltering days and get a good night's sleep even if most of the house is hot.

There are many installation options for an individual room unit, but the basic choice is between a removable unit installed in a window or a permanent one built into a wall.

Site Requirements

Basic site requirements are similar for window and in-wall units. Look for a spot where the outside heat-dispersing coils won't broil under direct sunlight, which can decrease the unit's efficiency. You need a location near an outlet with sufficient capacity for the appliance. It's wise to pick a location away from a main entry or deck area where you spend time—so you won't be bothered by the humming and condensation dripping. An in-window machine may work if you have a convenient window. There are more design options with a through-the-wall machine because you can install it almost anywhere on the building.

Window units can be heavy—some big machines weigh over 100 pounds—but they are easy to install. Manufacturers include adjustable panels and foam weatherstripping that surrounds the machine and fills the gaps between the metal case and the window frame.

In-window machines look clunky and block some of your view, but they are portable. You could take the machine with you to another room (if it's not too heavy) or to another house. In-wall units are considered part of the building, like a furnace.

How Much Do You Need?

There are three capacity formulas you can use to help you decide what size unit to buy. The most general rule of thumb would be to buy one ton of cooling (12,000 Btu) per 500 square feet of floor space.

The more complex WHILE formula takes into account several characteristics of a building. In this formula, each letter in the word WHILE stands for a building characteristic for which you substitute a numerical value as follows. W stands for width of the room in feet. H stands for room height in feet. I stands for the amount of insulation. (Substitute 10 if the room is covered by an insulated, ventilated attic or another cool room, 18 for a top-floor room under an uninsulated attic.) L stands for length of the room in feet. And E stands for exposure factor. (Substitute 16 if the longest wall faces north, 17 if it faces east, 18 if it faces south, and 20 if it faces west.) Multiply the numbers and divide by 60 to estimate required Btu capacity. Here's how the formula works for a 15x20-foot room with 8-foot ceilings that is insulated and vented above with a southern long wall.

W x H x I x L x E / 60 = Btu needed.
15 x 8 x 10 x 20 x 18 = 432,000 / 60 = 7,200.

The third capacity formula is found in the Cooling Load Estimate Form available from the Association of Home Appliance Manufacturers (cooloff.org). While considerably longer and more complicated than the other two formulas, it is the most precise.

Invented in 1902, air conditioners were not common until the 1920s when they were installed in movie theaters. Stylish household units were not widely used until the 1940s.

Window Fittings

If you install a window unit improperly, it may fall from the window when you raise the sash, damaging the unit and whatever happens to be underneath it. Install all brackets with the hardware provided; if the wood of your sills seems soft and partly rotten, use another window or replace the sill. If your windows have metal sashes instead of wood, use sheet-metal screws to install them.

Installing a Window Unit

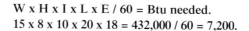
USE: ▶ pencil • measuring tape • level

1 Older units (and some very large machines) rest on external brackets, but modern ACs rest on a sill-mounted support.

5 Extensions on both sides of the unit slide out to make a snug fit in the opening. Screw each extension to the sash.

WINDOW UNITS

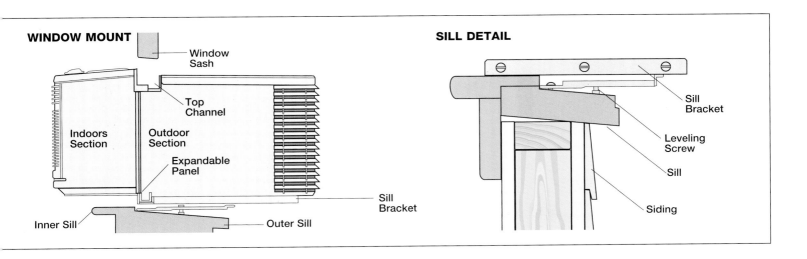

WINDOW MOUNT — Window Sash, Top Channel, Indoors Section, Outdoor Section, Expandable Panel, Sill Bracket, Inner Sill, Outer Sill

SILL DETAIL — Sill Bracket, Leveling Screw, Sill, Siding

- screwdriver or power drill/driver • caulking gun ▶ window AC unit • foam insulation

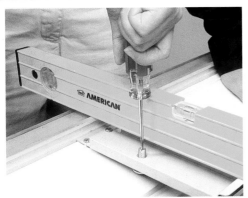

2 One end of this bracket is screwed to the sill. You adjust a center screw to level the unit and provide condensation drainage.

3 This self-contained AC unit has integral handles that make it easier to set in position on the sill over the mounting bracket.

4 As you slide the machine into the window opening, a pocket underneath the machine locks in place over the bracket.

6 Use the angle bracket provided with most machines to secure the two window sashes to each other over the AC unit.

7 To seal the installation inside, use a foam strip (provided with most machines) to seal the air gap where the sash overlaps.

8 Check the manufacturer's instructions for operation and maintenance; and plug in the unit.

PICKING A WINDOW UNIT

11 Cooling

IN-WALL UNIT BASICS

To install an in-wall air-conditioning unit, you will need to make a hole through the exterior wall of your house—a daunting prospect for some DIYers. But if you select a section that is free of pipes and wires, basic carpentry skills and tools will get you through the job.

If you're placing the unit in a solid wall, you will have to build a header—a horizontal beam that picks up loads from studs that are cut short and carries them to the sides of the opening. In most cases, you can avoid this step by installing an in-wall unit beneath a window. The space already has a header and should have double studs running down each side all the way to the floor. You may have to make the air-conditioner space smaller than the window, which is easy, but you won't have to worry about supporting loads from above. An in-wall unit also looks better if it's installed under a window. A metal box poking through a clean wall of siding will grab your eye. It's less noticeable under a windowsill, particularly if you trim the exterior of the conditioner the same way the window is trimmed.

Making the Opening

When cutting and building an opening for an in-wall unit, always follow the manufacturer's directions. Unless you have some experience cutting through concrete, brick, or stone, leave installations through solid masonry walls to professional contractors.

Select a location above the height of wall outlets in order to avoid any buried cables. The best choice often is directly under the framing that supports the windowsill. Wiring may run in the wall under a window, but it's generally at the same height or lower than the wall outlets. If you install the air conditioner in a wall of full-length studs rather than under a window, check with your local building inspector for the size of the header you'll need to install.

Always start work on the inside of the opening before the outside. That way, if you

Installing an In-Wall Unit

USE: ▶ drywall saw & crosscut saw (optional) • reciprocating saw, crosscut saw, circular saw, & saber saw (optional) • pry bar • hammer • pencil • level • screwdriver

Before beginning your project, plan a convenient installation, such as a section of wall beneath a window where you won't have to reroute plumbing pipes or electrical wiring. The air conditioner's instructions should explain what additional framing, if any, will be required to support your unit. Additional trimmer studs and cripple studs may be required to safely hold the unit in place. It's also important to follow the manufacturer's instructions for setting the unit slightly out of level to drain condensation.

1 Cut away a section of drywall to expose wall framing underneath. Remove the insulation, and lay out the opening.

2 Remove short studs in the opening by cutting them in half and prying out each section. Pull or cut exposed nails.

6 For a typical installation, you remove the machine from its chassis, and mount the chassis in the opening.

7 Use a level to match the manufacturer's specifications for sloping the chassis slightly downward for proper drainage.

8 Once the chassis is screwed to the frame, slide the machine into position. For heavy units, you may need a helper.

IN-WALL UNITS

uncover an unexpected obstruction—a gas pipe, for example—you'll only have to replace a section of drywall, and repaint the damaged wall. A pain, perhaps, but a much easier job than adding nailers to support the pieces of siding you've cut.

Don't cut blindly into a wall cavity with power tools such as a reciprocating saw or saber saw. Measure and mark the opening carefully, and then trim through surface gypsum by hand with a drywall knife or utility knife. This will produce clean edges that are easily trimmed and will help prevent surprise encounters with mechanical lines or hidden braces buried inside the wall.

Framing the Unit

To frame the unit, you can cut through studs under a window, and add cross timbers top and bottom and new side pieces. But for many people, it is easier to build a box—for example, from ¾-inch-thick plywood—to the manufacturer's specifications, and then use it to mark and cut away sections of studs. Allow an extra ½ inch of length and width so that you can plumb and level the box in the opening. Make sure the unit does not slope even the slightest amount to the inside. Pitch the casing slightly toward the exterior so that condensation will drain to the back of the unit and outside the house, not inside or into the wall cavity.

Transfer the perimeter of the opening to the outside wall by drilling small holes precisely at each corner or by driving nails through the corners. Then go outside, connect the dots, and cut through the siding and sheathing along the outline with a circular saw. For greater accuracy, nail a straightedge to the siding to serve as a saw guide.

Before installing the air conditioning unit in the box frame, reuse pieces of the old insulation to fill in any gaps between framing members. Then, before trimming the opening inside and out, seal small surface gaps with silicone caulk that is flexible enough to maintain a seal next to a vibrating machine.

or power drill/driver • drywall knives • caulking gun ▶ 2x4s as needed • framing & drywall nails • in-wall AC unit • drywall • joint compound • sandpaper • caulk

3 Add framing as specified by the manufacturer to build the rough opening and create support for the unit.

4 When the frame is complete, drive a long nail at each corner to mark the AC opening on the exterior wall.

5 Mark the opening on your siding; remove the nails; and cut through siding and sheathing with a saber saw or circular saw.

9 With the unit secured in the chassis, which should be flush with the finished wall, piece in drywall around the opening.

10 After finishing, sanding, and painting to conceal seams, trim the interior to suit, and reinstall your baseboard.

11 Use a flexible caulk around the exterior of the unit to seal the seams between the chassis and the siding.

IN-WALL UNIT BASICS

11 Cooling

BASIC MAINTENANCE

Air conditioners don't need a full seasonal tune-up the way most furnaces do. Some basic maintenance, however, will maximize cooling output. The well-maintained unit runs more efficiently, lasts longer, and makes rooms feel cooler at a lower, money-saving setting on the thermostat.

Before attempting anything more than superficial cleaning, unplug room units (or trip the breaker on central systems), and follow the manufacturer's directions for discharging the capacitor, an electrical storage device that can deliver a shock even when a unit is unplugged.

Room-Unit Maintenance

To clean the inside of a room unit, remove the access panel and the filter and, following the manufacturer's directions, either wash the filter or replace it. Clean the inside coil fins with a vacuum cleaner or a soft brush, taking care not to bend the fins. On the outside, remove the grille, and repeat the cleaning process on the exterior coils.

Even in a clean unit, the meeting of warm, wet air and cool, dry air produces humidity. In most room units, it collects in a pan at the base of the machine, which can stagnate. To keep the surrounding airflow clean, rinse the tray with a 50:50 solution of chlorine bleach and water. Be sure the tray drains when you rinse it with water so you won't get overflows.

To keep the fan from pulling warm air through to the inside, seal any leaks between the wall and the metal housing of the air conditioner. If you see moisture around the frame, warm air is probably seeping in, and you should caulk the seams inside and out.

Central-System Maintenance

The procedure for cleaning a central air system, illustrated below, is similar to that for a room unit, except that the machines are bigger, and the parts may be harder to reach—even though on most systems they are split into two sections. To clean the condenser fan and oil the fan motor (both in the outside unit), you will probably have to remove a cover grille, loosen a setscrew holding the fan on the motor shaft, and then remove the fan to gain access to oil ports on the motor. Typically, fan motors get two or three drops of non-detergent motor oil in each port, but follow the manufacturer's oiling guidelines.

Use a garden hose to clean the outside condenser coils, but only after removing the coil guard so you can spray from inside the unit. Otherwise, water-soaked debris will lodge in the fins. If the fins are bent against each other, which is more likely on exterior coils set away from the protection of the building, it's best to use a specialized tool called a fin comb to clean and straighten them. One of several sets of small teeth arranged around the tool head will fit between undamaged fins above the bent area. As you pull the tool downward, its teeth will separate the compressed fins.

Recharging

The refrigerant in older refrigerators, freezers, and air conditioners is typ-ically an ozone-damaging hydro-chlorofluorocarbon (HCFC) called Freon. Eventually, the closed refrigerant loop can develop a slow leak and cause the compressor to fail prematurely. As it's illegal to release HCFCs into the atmosphere, when an older machine needs repair, the service contractor is required to capture and recycle the refrigerant. Recycled HCFCs are used to recharge older machines. Most new models use less-damaging refrigerants.

Homeowners can take care of basic maintenance, but you need a contractor to recharge the refrigerant.

Basic Cleaning

USE: ▶ screwdriver • soft brush • fin comb ▶ lubricating oil • new air filter

Service on an AC compressor normally is left to professionals. But there are several steps you can take to improve the efficiency of your central AC system. Be sure to follow the manufacturer's instructions for basic cleaning, and always shut off power to the unit before working on it. Also take care not to compress the delicate rows of metal fins, which must be separated to transfer tem-perature efficiently.

1 The first step is to shut off power to the unit. Most systems have a cutoff box mounted outside near the fan unit.

2 Remove the access panel, and use a garden hose and brush to clear any debris or grass clippings from cabinet grilles.

MAINTENANCE

Saving Energy

GREEN SOLUTION

There are many ways to save on your air-conditioning bill—with high-efficiency windows, extra insulation, shades, and other devices. But the easiest way to save is simply to raise the thermostat a few degrees. Your savings will vary depending on where you live and how your house is constructed. But you can estimate a savings of 2 to 3 percent on your cooling bill for every degree of cooling you do without over 24 hours. You can also increase savings by installing a setback thermostat that can raise the setting automatically when you leave for the day, and lower it before you get home. (See "Thermostats," p. 305.)

You can also save money by improving the efficiency of the duct delivery system. Many systems lose 20 to 40 percent of the heating or cooling energy they carry, according to studies by the U.S. Department of Energy. The most obvious losses are at loose joints where you can stop leaks by reconnecting the ducts, securing them to each other with sheet metal screws, and wrapping the joints with duct tape. Insulating ducts helps as well. You can use plastic-wrapped sleeves or spiral-wrapped batts of wall insulation secured with tape.

Slide on insulating sleeves, or spiral-wrap ducts with conventional batts to reduce temperature loss in the ductwork.

Secure insulating sleeves with twist ties, or use duct tape to close the seams in spiral-wrapped batts.

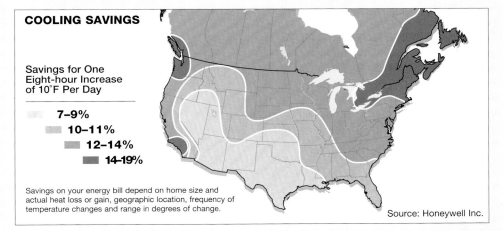

COOLING SAVINGS

Savings for One Eight-hour Increase of 10°F Per Day

- 7–9%
- 10–11%
- 12–14%
- 14–19%

Savings on your energy bill depend on home size and actual heat loss or gain, geographic location, frequency of temperature changes and range in degrees of change.

Source: Honeywell Inc.

MONEY SAVER

3 Clean the fan blades and fan motor housing, and lubricate the motor with oil as required by the manufacturer.

4 Use a soft brush to clear dust from coil fins. If some are bent, use a straightening tool or a fin comb to align them.

5 One of the easiest steps, but one of the most important, is to change air filters that keep dust out of the duct system.

11 Cooling

CLEANER AIR

Modern houses and apartments are built to be airtight for greater heating and cooling efficiency. They're so airtight, however, that new air circulates back into the house very slowly—it may take hours for the air in a new house to recycle itself. This creates not only stale air but a buildup of indoor pollutants. These pollutants include irritants such as dust, smoke, mold spores, pollen, and animal dander—which not only bother your lungs but clog up heating and cooling systems, computers, and other electronic equipment—but also more serious environmental hazards such as the outgassing of formaldehyde from construction materials.

Several devices are available to clean the air of a tightly-sealed house. Forced-air cooling and heating systems can have electronic filters installed right inside the HVAC ducts. If you don't have these systems, buy a portable air cleaner with a HEPA filter. Developed by the U.S. Atomic Energy Commission to remove almost all airborne particles, HEPA (high-efficiency particulate-arresting) filters are widely used in hospitals. A true HEPA filter removes 99.97 percent of particles as small as 0.3 micron. (A micron is one-millionth of a meter; the period at the end of this sentence is several hundred microns across.) An ultra HEPA, or ULPA, filter removes particles as small as 0.1 micron. HEPA filters need to be replaced periodically, typically every one to three years.

Portable Cleaners

A portable air cleaner may have several types of filters. The most basic filters are similar to the ones used in hot-air furnaces to trap dirt. Some have to be replaced, while others can be washed and reused. Some machines also include an ion generator to force particles against surfaces. The most effective filter is a HEPA filter, which removes nearly 100 percent of airborne particles. Manufacturer's literature, mainly the material data sheet, should list filter capacity and effectiveness against different pollutants.

This portable room air cleaner (about 18 in. wide and 12 in. high) will provide 6 air changes per hour in a 15 x 17-ft. room.

Air Exchangers

An air exchanger is often added to the ductwork of tightly built houses that have only minimal ventilation. Most are designed around the basic plan of two fan-assisted pipes: one to exhaust stale air and one to bring in fresh air. Where the pipes pass each other, a common wall or other media transfers up to 75 percent of the outgoing temperature to the incoming air supply. This way, you can exchange stale air year round without heating or cooling the fresh supply from scratch.

GREEN SOLUTION

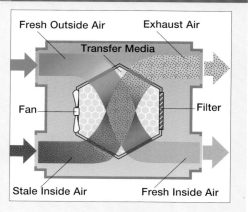

Cleaning Ducts

USE: ▶ screwdriver • glove • vacuum

1 If your ducts have never been cleaned (even if you use furnace filters), remove a register and check the duct walls inside.

2 A household vacuum may reach several feet down most ducts. A professional cleaning covers the entire duct system.

3 Dust that gets past your filters may eventually be trapped at the return-air grille, which also should be vacuumed.

INDOOR AIR QUALITY

Media Cleaners

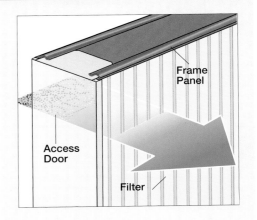

A media air filter performs the same tasks as a standard fiberglass mesh furnace filter. However, it does the job more thoroughly because instead of using a loosely woven, flat mesh to trap particles, these filters are made of a tightly interlocking network of microscopic fibers folded like an accordion. The configuration creates a large surface area in a small space. The increased density helps to capture more particles in the airflow than a standard filter. The increased area makes some media filters ten times more effective than a typical disposable filter found in most heating and cooling systems. Like an electronic unit, most media filters are mounted in the ductwork near your furnace. They are contained in a metal box frame that has an access door so you can remove the media filter for cleaning or replacement. These filters are passive compared to electronic systems. You need to make the same kind of alterations to existing ductwork to install them, but do not need to install wiring.

Electronic Cleaners

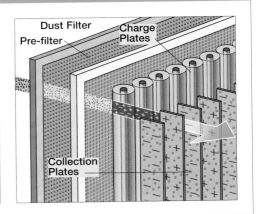

Electronic air filters are built into the ducts near your furnace. (A unit can be added to existing systems by modifying the ducts.) A typical electronic cleaner has a prefilter that is similar to a standard furnace filter. It traps large dust particles and can be removed for cleaning. Next in line is one or two metal boxes (also removable for cleaning) containing thin metal plates. Particles in the airflow, typically in the return duct, are positively charged on their way to the plates. The plates themselves are negatively charged to attract the particles, which are driven against the plate walls as they pass through the system. These appliances can remove over 90 percent of most airborne pollutants, including pollen and smoke particles. The drawback is that larger particles hitting the plates can make an annoying sound—like an outdoor bug-zapper.

Electronic Air-Cleaning Installation

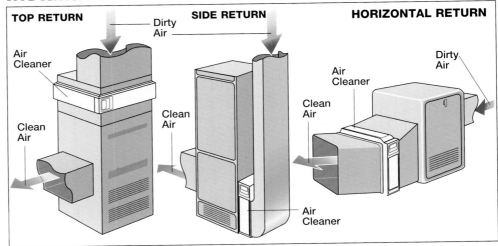

An electronic air cleaner can be mounted anywhere in the ductwork of a forced-air system. Generally, the most practical location is near the furnace at the end of the return plenum. Before dusty air from the house is sent back into the furnace for reheating, it passes through a prefilter (similar to a standard dust filter) and a row of removable cartridges that contain electronically charged plates. Aside from basic sheet-metal connections, you'll need to provide power for the plates and controls, including a sail switch, located in the return plenum. It activates the cleaner when the furnace blower runs and air flows through the ducts.

11 Cooling

DRY AIR

Moisture won't get a toehold in spaces covered by central air-conditioning or in the immediate vicinity of a room unit you run regularly. But many homes have a laundry, home shop, or storage room that doesn't really need cooler air, just drier air—that's where a dehumidifier comes in. These portable, plug-in appliances use a compressor to pump moisture out of the air, typically into a collection pan underneath the unit that has to be emptied periodically. To some extent, the same thing happens with a refrigerator, which collects condensation in a pan down by the compressor. You don't have to continually empty that pan because a fan blows air over the water to evaporate it. But if a dehumidifier worked this way, the appliance would be working at cross-purposes. Once a dehumidifier removes moisture from the air, you have to remove the moisture from the room.

Paying for reliable convenience features can wind up saving money and maintenance in the long run—even if you buy a machine with more capacity than you really need. Unlike air conditioners, where the costs of running high- and low-efficiency units varies widely, electric operating costs vary only marginally between the least and most efficient dehumidifiers. A Department of Energy analysis of the operating costs of dehumidifiers with a capacity of 20 pints per day running 1300 hours per year (from morning to night during the summer) found only a $9 per year difference between the least- and most-efficient models.

Dehumidifier Capacity

The industry standard for dehumidifier capacity is specified in pints per day. Unlike air conditioners and many other appliances, they are not rated by energy efficiency (an EER rating). Among the 300-plus models from 30 manufacturers that were rated by the Association of Home Appliance Manufacturers (AHAM), an industry trade group, the pints-per-day ratings range from about 10 to 50. Smaller units in the 15-pint-per-day range should handle most rooms and even full basements (up to about 500 square feet) that are only moderately damp.

Increase capacity by about 25 percent if the area is very damp and moisture regularly condenses on walls during the summer. Units rated in the 20- to 25-pint-per-day range should be able to handle larger areas—for example, a 60 x 25-foot basement. Units with capacities near the top of the range should only be necessary in very large, very wet areas. Large machines collect as much as 50 pints of water per day and generally include some warning system that automatically signals when the water collection pan is full and has to be emptied.

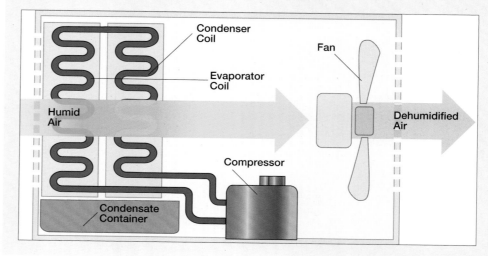

Servicing a Dehumidifier

USE: ▶ screwdriver • vacuum ▶ soap or disinfectant • lubricating oil

1 Remove the access panel, and you're likely to find a dirty air filter that can be replaced (or washed) at least once a year.

2 Behind the panel is a removable water tank where condensation collects. Follow manufacturer's directions for cleaning.

3 Even if you regularly empty the tank, you should periodically wash it with soap and water or a disinfectant.

DEHUMIDIFIERS & THERMOSTATS

Thermostats

Among many variations on the standard thermostat, one of the most valuable is the automatic setback. It can save about 3 percent of fuel costs for every degree of heating or cooling you do without over 24 hours, which makes these programmable units among the most cost-effective energy-saving improvements available. You can program these units to save while you're asleep and at work, and return to 68°F shortly before you rise and before you get home. A 10° setback twice a day can cut fuel bills by about 20 percent.

Setback thermostats are available in many configurations, including the traditional round shape. Some have enough built-in memory to allow complex programming that includes different weekday and weekend schedules. Several companies also make special thermostat controls with large numerals for visually-impaired people, and with a surrounding ring that makes a distinct click as you move up or down the temperature scale.

The standard round design is still made; it is available as a setback unit that changes temperature levels in preset time cycles.

This thermostat has a large-number sleeve, and a knurled outer ring that makes it easier to turn the dial.

Modern setback thermostats have enough memory to update dozens of displays, aside from time and temperature.

Some units are available with large type and either tactile or audible feedback as you change the setting.

4 Behind the exhaust grille, you will find condenser coils. Clean dust on and between the coils using a vacuum.

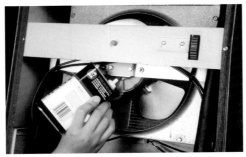
5 To foster smooth and quiet operation, oil the fan motor (typically with only three drops) following manufacturer's directions.

6 Clean the intake and exhaust grilles; install a new air filter or washable filter if called for; and reattach the access grilles.

12 Ventilation

316 FOUNDATIONS
- Ventilation Basics
- Ventilation Paths
- Crawl-Space Venting
- Ventilation Requirements
- Installing Foundation Vents

318 BATHS
- Venting Baths
- Design Options
- INSTALLING BATH VENTS
- INSTALLING TIMER SWITCHES

320 KITCHEN & LAUNDRY
- Venting Kitchens
- Range Venting
- Laundry Venting
- INSTALLING DRYER VENTS
- Dryer Vent Options

322 WHOLE-HOUSE VENTILATION
- Ventilating with Fans
- INSTALLING WHOLE-HOUSE FANS
- House Fan Design Options
- INSTALLING CEILING FANS
- Reducing Fan Vibration

324 ROOFING
- Venting the Roof
- Roof Vent Combinations
- Maintaining Attic Air Flow
- Installing Strip-Grille Vents
- Plug Vents & Perforated Soffits
- Through the Roof
- INSTALLING ROOF VENTS
- INSTALLING GABLE VENTS
- INSTALLING RIDGE VENTS

12 Ventilation

VENTILATION BASICS

Water, in any form, can be one of the most destructive forces on a house. While most buildings are designed to cope with water from the outside—even if you need to replace a few new shingles or some caulking after a storm—many are not well-equipped to deal with moisture that is trapped inside in the form of vapor-laden air. It is good for houses to be able to "breathe". The East Coast is dotted with 200-year-old houses that have been breathing like crazy from the day they were built. Houses need to breathe out some cool, conditioned air in summer and some warm, heated air in winter. It costs you money on your heating and A/C bills, but it will take some of the 7 to 10 gallons of potentially harmful moisture produced daily inside a home with it.

This moisture comes from many sources: cooking, washing clothes, watering plants, taking showers, and humidifying the air during winter. Unless indoor air and the moisture it contains has some way out of the house, you're in for lingering odors, stale air in general, mold growing on wall paint, and even enough moisture condensing on cold windows to make puddles that peel paint and rot window sills.

You may picture insulation company ads of dollar bills flying out leaky windows. But those ads don't show all the dollar bills that must fly back in to pay for a vent fan in the stagnant kitchen, new tile to replace the buckled floor in the bath, new paint or wallboard to replace the mold-encrusted walls in the cellar—only some of the problems caused by trapping moisture inside.

Ventilation Systems

There are several ways to alleviate ventilation problems, which often are most troublesome in the more energy-efficient houses and apartments built since 1980 or so. Many of these homes are so airtight that they trap too much moisture and stale air. The solution is to vent excessive heat during the summer (especially in the attic) and excessively warm, moist air during the winter. One easy, low-tech way to do that in winter is to open a few windows a crack. It costs a bit more in heat, but the fresh air can correct moisture and stale odor problems.

Some parts of the house may need more attention. Moisture can condense during winter months in a poorly ventilated attic. In the summer, heat can build up in that same attic, resulting in higher A/C bills and a shorter life span for the roofing shingles. Think of your roof as a passive heating and cooling system that needs a constant flow of fresh air to function properly.

Roofing Vents

For roofing vent systems, the most important thing to know is how much air is exchanged. About 1.5 cubic feet of air exchanged per minute will adequately vent a typical attic in both winter and summer. In some situations, particularly in modern, tightly constructed houses, you can improve energy efficiency by exchanging temperature between a flow of stale indoor air being exhausted and fresh air pulled in with a fan.

It's difficult to figure out air-exchange rates without sophisticated equipment, but knowing the differences between venting systems can help you make informed choices. Roof vents, for example, come in five types: continuous ridge vents, soffit vents, gable vents, turbine vents, and motorized fans. Some vents are active (electrical), and some are passive. Both can work in other areas in the house, particularly where moisture is produced, such as kitchens, baths, and laundry rooms.

Ventilation Paths

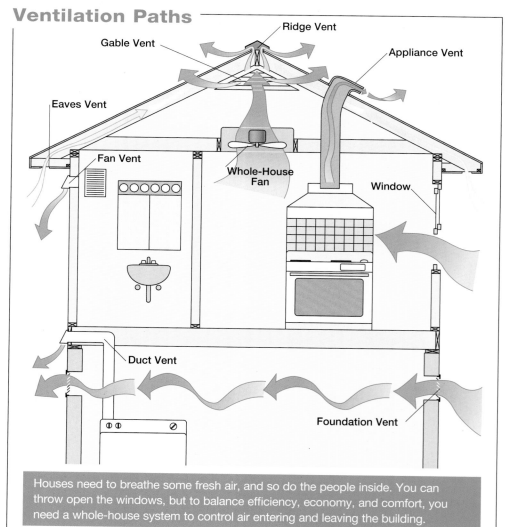

Houses need to breathe some fresh air, and so do the people inside. You can throw open the windows, but to balance efficiency, economy, and comfort, you need a whole-house system to control air entering and leaving the building.

FOUNDATIONS

Crawl-Space Venting

Dirt floors require more ventilation than concrete floors. If the floor of a crawl space is concrete and the walls are insulated, you can ventilate with a series of small foundation vents. The number of vents depends on the total square feet in a given space. A general rule is to have 1 square foot of vent area for every 150 square feet of floor space. Sliding metal vents are designed to replace the space of one 8 x 8 x 16-inch concrete block. A ventilated crawl space needs to be screened, with either pressure-treated-wood or PVC lattice and a welded wire netting.

Plastic lattice provides venting in many shapes and colors—and when it gets dirty you just wash it down with a hose.

PT (pressure-treated) lattice has built-in resistance to water damage and rot, even near ground level.

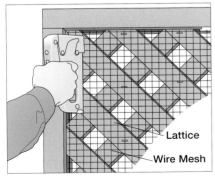

Use galvanized wire mesh or plastic screening to keep insects and animals from entering.

Ventilation Requirements

BASEMENTS & CRAWL SPACES
There are many formulas contractors use to gauge ventilation requirements—and a few DIY rules of thumb—such as providing 1 square foot of vent per 150 square feet of floor space. But every case is different. You'll need more venting in a damp climate with a shaded site and a dirt floor, and less in a dry climate with more sun and wind and a concrete slab on the crawl space floor.

ROOF & ATTIC VENTILATION
For flat, sloped, and gable roofs, install 1 square foot of vent area per 300 square feet of attic space (1/300) if there is a vapor barrier under the insulation. Install 1/150 if there is no vapor barrier. Distribute the vents evenly between eaves and ridge vents. For gable roofs, include two louvers at opposite ends near the ridge. Same for hip roofs, plus 1/600 at ridge, with all vents interconnected.

Installing Foundation Vents

USE: ▶ drill • hammer • trowel • cold chisel • masonry bit • work gloves ▶ vent grille • mortar

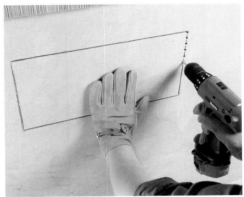

1 To break through the foundation, drill closely spaced holes along the vent outline with a masonry bit.

2 Use a cold chisel along the outline to weaken and clear away the masonry. Always wear eye protection.

3 Smooth out rough edges around the opening as needed for fit, and set the vent in a bed of cement.

4 Secure and seal the vent edges with a surface coat of cement. Don't clog the fins on adjustable vents.

VENTILATION BASICS 317

12 Ventilation

VENTING BATHS

No room will produce more water vapor than a bathroom. For this reason, bathrooms are usually finished with a moisture-resistant gypsum board or cement panel. All this extra protection from water vapor will probably escape your eye—until the mold and mildew start to develop and the tile grouting turns black. If you have a bathroom with persistent mold and mildew problems, chances are that your real problem is inadequate ventilation.

Bathroom vents are typically electric fans recessed in walls that vent directly outside (on a first-floor bathroom) or through the roof soffit (on a second-floor bathroom). The fans are either connected to a timer to ensure full clearing of water vapor or connected to the light switch. If your bathroom vent doesn't have an independent switch, it's a good idea to have an electrician install one.

Choose an exhaust fan based on the size of your room. For bathrooms under 50 square feet in area, choose a fan that can move 50 cubic feet of air per minute, or CFM. For rooms that are 50 to 100 square feet, choose a fan whose CFM rating is a close match to the square footage of the room—so a 75 square foot room would require a fan rated at 75 CFM. For larger rooms, assign 50 CFMs to each fixture and then add the fixtures together—a room with a toilet, bathtub, and separate shower would require a fan rated for 150 CFM.

Vent Placement

A bathroom vent should be able to exchange the entire volume of air from the room in 30 minutes. The best place to put the vent is near the ceiling or within the ceiling itself, to ensure that warm moist air is adequately vented. The ductwork from the vent can be routed either directly to an outside wall or through the attic to a soffit vent.

You'll have to remove insulation from the access area along the joist where you'll place the bathroom vent. You can pack the outside of the ductwork itself with insulation—a good idea for keeping down condensation inside the ductwork.

Direct, through-the-wall installations are usually the easiest, particularly when you save the electrician some work by picking a spot in

Design Options

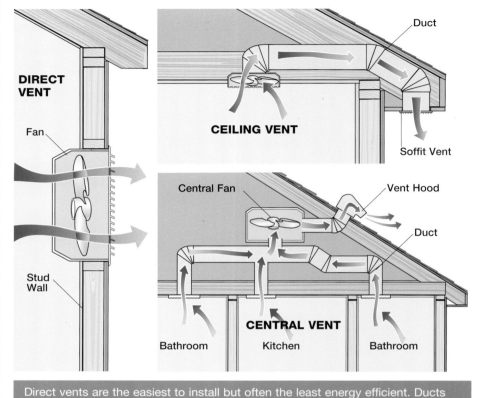

Direct vents are the easiest to install but often the least energy efficient. Ducts from ceiling vents should be insulated to prevent condensation problems. Central fan systems reduce noise while serving several rooms at once.

Installing Bath Vents

USE: ▶ reciprocating saw • drill • screwdriver • caulking gun • measuring tape • pencil ▶ cover grille

1 Mark an unobstructed release point outside the house, and screw the fan housing to the nearest stud inside.

2 Connect the duct to the inside housing and outside vent. Flexible ducts allow an offset between inlet and outlet locations.

BATHS

the same framing bay as an electrical outlet. Mount the fan in an open area where it can easily collect moisture.

All this may change if you have a plaster interior wall or if you have to punch through a brick exterior wall. Although it can be done, it's wiser to make a ceiling installation and route the fan duct to an outlet either in the roof or in the roof soffit.

Vent to the Outside

If you take the ceiling route, don't simply dump warm, moist air into an attic where it will condense. Use the space between ceiling joists to route ducts to a soffit or roof vent.

The soffit outlet is preferable because it doesn't put a hole in the roof and because warm, moist air in the duct will condense where it is exposed in an unconditioned space such as an attic. When the duct walls are cool, moisture will form before the air is exhausted and then drip back down the duct into the fan and back into the room. Avoid those problems by taking the duct straight up from the ceiling grille into the bay between ceiling joists and then turning it toward the outside wall and running it with a slight slope downhill to the soffit outlet. The duct will also be out of the way if you decide to use the attic or crawl space floor for storage.

Installing Timer Switches — MONEY SAVER

USE: ▶ circuit tester • screwdriver • switch box • timer switch with coverplate • NM wiring • wire nuts

When moisture lingers as you're ready to leave the bathroom—after a shower when you're dry but the mirror is still foggy—replace the fan's standard on-off switch with a timer switch. (You can find them in an electrical supply house or a store that sells saunas.) It has a knob that you can rotate to set an extended run time. Then, sometime after you leave the room, the fan shuts off automatically. You don't have to remember to turn it off, and you won't waste energy running it all day.

1 Once you're sure power to the old switch is cut, remove the holding screws and disconnect the wires.

2 Connect the new switch to the wires. Make sure the wire loops around the terminal and that the screw is tight.

3 Most timer switches are spring-loaded. Just turn the switch to select the amount of run time, and leave the room.

GREEN SOLUTION

• switch box • vent hood with backdraft damper • vent kit • fan switch with speed control • flexible duct • duct clamps • insulation • NM wiring • screws • wire nuts

3 Make code-approved electrical connections to the housing, and install the fan motor—often a plug-in component.

4 To finish the inside, tuck loose-fill insulation (or use spray-foam) to seal around the housing, and clip on the cover grille.

5 Cover the exhaust end of the duct with a vent hood. Some backdraft dampers prevent air leaks when the fan is off.

12 Ventilation

VENTING KITCHENS

Kitchen ventilation requires extra consideration because cooking creates grease as well as water vapor. While oil may have a higher boiling temperature than water, it does not need to boil to produce vapor, which will condense into a greasy film over all kitchen surfaces if not properly vented.

Vent Hoods

Ovens are often sold with a vent hood as an attachment; hoods are either vented or ventless. A vented range hood is the most effective method for removing cooking odors, unwanted moisture, and grease directly at their source. Ventless hoods function primarily to collect airborne grease. Filters on ventless hoods need to be changed every so often, and they do not contribute to the exchange of air in a closed kitchen.

If you have a ventless hood in your kitchen, or no vent at all, consider installing a ceiling fan. A ceiling fan can work in conjunction with an open window or doorway. The cross-ventilation this creates helps reduce the build-up of cooking odors.

Settlers in Texas' hot and dry Hill Country solved their ventilation problems with a dog-trot—a breezeway through the middle of the house, seen here in Lyndon Johnson's grandfather's cabin, located at LBJ National Historical Park in Johnson City.

Wall & Ceiling Vents

Another approach besides installing a range hood is installing ceiling- and wall-mounted vents. Wall vents should have a swinging louver on the outside vent to prevent air and insects from getting inside. When selecting the vent fan (or even an unvented range-hood unit), be sure to match the fan's capacity with the size of your kitchen. Fans are rated by the cubic feet of air that they can vent per minute—abbreviated CFM. To size a fan correctly, one rule of thumb is to use 150 CFM per linear foot of range surface along walls, and 180 for ranges on islands. Using this rule, a 30-inch-wide range (2.5 linear feet) times 150 CFM would require a range hood fan rated at about 375 CFM if it were installed along a wall, and a 450 CFM fan if it were installed on an island.

Wall vents and ceiling vents can become cold spots during very cold weather, and there is no real way to avoid this weakness. The trade-off is between ventilation and insulation, and you should take that into consideration when planning kitchen ventilation. A perfectly insulated kitchen with no air flow is neither desirable nor healthy. Because kitchen vents should operate only when needed, they should have a backdraft louver.

Other options in kitchens and baths (or both) is to use a centralized system with one motor that is large enough to pull air from several locations. One advantage of this type is noise reduction because the fan motor is not in the wall or ceiling, but mounted in the attic. For the whole house, large fans located in central areas, such as stairwells, can pull enough air through the house to create a quick air change.

Vents can be a focal point of the kitchen with multiple built-in fans and filters.

Range Venting

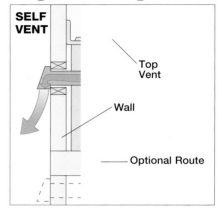
SELF VENT

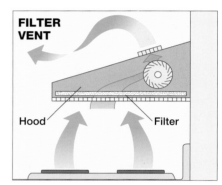
FILTER VENT

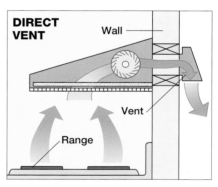
DIRECT VENT

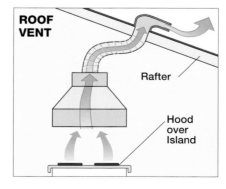
ROOF VENT

KITCHEN & LAUNDRY

LAUNDRY VENTING

Clothes dryer vents always vent directly to the outside of the house because the air that comes out of the clothes dryer is a combination of heat, moisture, and fine lint. Several types of vent materials will do the job, including flexible plastic that is generally the easiest to install. But noncombustible metal duct is best for fire protection and required by most building codes.

You should take that into consideration when planning the installation because a clogged dryer vent can be a fire hazard when lint builds up and prevents the escape of air. For this reason, dryer vents should be as short as possible, and run directly from the back of the dryer to the outside wall.

Most dryers heat with electricity, but some produce heat with natural gas. If you have a natural-gas dryer, it is important to make sure that the vent system is up to code.

To cut down blockages inside a vent, position the outlet at the same level as the dryer's vent, and avoid sharp bends in the line and circuitous routes from inlet to outlet. In basements, route dryer hoses to vent above the foundation sill, so you drill through wood instead of masonry.

FIRE WARNING

The U.S. Consumer Product Safety Commission estimates that 14,000 fires related to clothes dryers occur every year. A main cause: lint buildup in the dryer filter or vent line.

SOLUTIONS
- Clean the lint filter before or after drying every load of clothes.
- Make sure the exterior vent outlet and cover flap are not clogged.
- If the dryer runs hot, disconnect the vent to check for hidden clogs.
- To trap less lint, use smooth-wall duct instead of ridged material.
- Check manufacturer's instructions. Many UL-rated machines require metal vent duct, not plastic.
- Install a smoke detector and fire extinguisher in the laundry room.

Installing Dryer Vents — GREEN SOLUTION

USE: ▶ saber saw • screwdriver ▶ dryer vent • flexible duct • vent hood with damper • duct clamps • caulk

1 Pick a spot between studs, and trace and saw a hole the diameter of the vent line.

2 Transfer your cuts through the outside wall, and mount the vent cover.

3 Inside, one end of the flexible duct fits over the vent cover pipe.

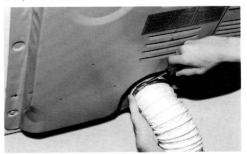

4 The other end fits over the dryer outlet. Attach both ends using clamps.

Dryer Vent Options

Dryer vents should be routed the shortest possible distance from the dryer to the outside of the house. However, you may want to extend the vent hose in a basement to reach framing that is easier to drill through than masonry. Where vents may be unsightly on the outside of the house, you can hide them behind shrubbery. But do not use long, twisting lengths of vent tubing only for cosmetic reasons. Also, do not vent the dryer directly into the laundry room, even on winter's coldest days. You might save a little heat, but the open vent will deposit gallons of water vapor and a cloud of lint into the air.

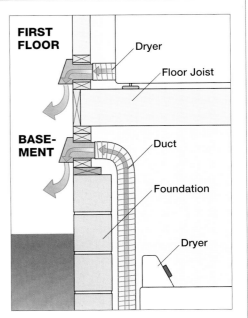

12 Ventilation

VENTILATING WITH FANS

You can easily vent your entire house with a big electric fan installed into the gable end of the attic. Depending on the size of the unit and the speed of the fan, air can be completely exchanged throughout a house in a matter of minutes. During hot weather, the fan will keep your house cool until the peak of the afternoon, when the air conditioner can take over. In the winter, it can vent moisture more quickly than passive vents and help prevent the formation of ice dams. If the fan system is connected to a series of vents that can be opened or closed to regulate the flow of air, then the bathrooms, kitchen, and attic can be vented individually or together.

Installing a whole-house fan requires more work, but the project is within the range of many do-it-yourselfers. The fan will probably need its own electrical line, and you may want an electrical contractor to install it, especially if it is a 220-volt line. (See "Electrical," pp. 94–157.) For fans mounted in gable ends, be sure the fan and electrical connections are designed for exposure to rain and snow. Water is usually not a problem with fans installed in attic floors over a stairwell, but be careful about locating one near a gable-end vent.

Ceiling Fans

Ceiling fans are a good choice for circulating air, especially in rooms with high ceilings. Ceiling fans can also work in conjunction with whole-house attic fans for a highly efficient air exchange system. Warm air rises naturally and remains at the top of a room until it cools and then descends. A ceiling fan operating at the same time as an attic fan will help to remove moist or stagnant air from a room much more efficiently than an attic fan acting alone.

Ceiling fans with multiple speeds are the best option for doing the twin tasks of circulating and ventilating. Some fans come with three speeds—the slowest speed will keep warm air circulating downward, the middle speed will create noticeable cooling drafts in the room, and the highest can actually clear a room's air. A kitchen with an open window or door and a ceiling fan set on high can eliminate the aftermath of a burnt roast or a fish fry in a matter of minutes.

Installing Whole-House Fans

USE: ▶ crosscut saw • utility saw • wire cutters • wrench • circuit tester • hammer • screwdrivers • dust

1 Mark an opening in the ceiling below the framing of the attic floor, and cut through the drywall with a utility saw.

2 Span several ceiling timbers to make a sturdy work platform, and cut through the joist that crosses the fan opening.

House Fan Design Options

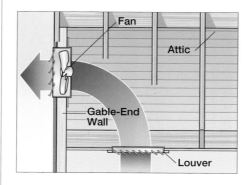

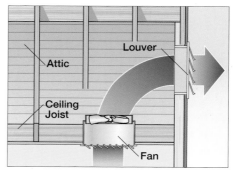

Installing Ceiling Fans

USE: ▶ drill • hammer • long-nose pliers • saber saw • screwdriver • wire cutter • wrench • circuit tester

1 Cut an opening for the special ceiling-fan junction box, and screw the box to a nailer cut to fit between joists.

2 Run required electrical cables into the box, position the box over the hole, and then fasten the nailer between the joists.

WHOLE-HOUSE VENTILATION

MONEY SAVER

▶ mask ▶ whole-house fan • fan switch with speed control • switch box • saber saw • 2x joists • joist hangers • nails • NM wiring • rubber washers • screws • wire nuts

3 Set double headers across the cut joist one piece at a time to facilitate nailing, and double the full-length side joists if required.

4 Mount the fan housing on the opening. Make the electrical and motor connections according to manufacturer's instructions.

5 On most models, louvers in the cover grille mounted below the fan open automatically when the fan is turned on.

ELECTRICAL SAFETY

▶ **Don't overload the circuit.** Attic fans and ceiling fans come with information on electrical requirements, including specific wattage ratings. Some may require power for a fan motor and a light, and some large fans may require 220 volts. To plan a safe electrical installation, be sure the total circuit demand does not exceed the rating of your fuse or circuit breaker.

Reducing Fan Vibration

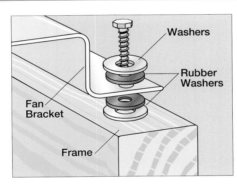

MONEY SAVER

▶ ceiling fan • fan switch w/speed control • mounting bracket • switch box • NM wiring • octagonal electrical box • nails • screws • wire nuts • wood blocking

3 Bring the wiring through holes provided in the mounting bracket, and secure the bracket to the nailer and joists.

4 Attach the fan motor to the bracket following manufacturer's directions. Make secure connections that can't vibrate loose.

5 Install the fan housing and blades to finish the job. The fan should have its own wall switch if there is no pull chain.

VENTILATING WITH FANS 323

12 Ventilation

VENTING THE ROOF

A well-vented roof keeps a house cool and cuts down on the air-conditioning bills in the summer, prevents the formation of ice dams and condensation in the winter, and helps prolong the life span of asphalt shingles. A good ventilating system is simple to install and inexpensive—it may even eliminate the need for mechanical air conditioning entirely.

Ridge Vents

A recent innovation in roof ventilation systems is the continuous ridge vent. This system allows the natural flow of air along the roof rafters and the wind outside to create negative air pressure that will also draw air from the attic. Continuous ridge vents are the most energy-efficient method of exchanging attic air.

The ridge vent covers a slit that runs along the roof. It "caps" the opening with screening and a small roof of plastic that keeps rain from entering the vents.

Install a ridge vent by removing the capping shingles along the roof ridge to expose the roofing felt; then, cut away the roofing felt 3 inches from the top to expose the roof sheathing. Snap a chalk line 2 inches from the top of the ridge on each side of the roof. Cut out the sections with a circular saw set to the exact depth of the roof sheathing (usually ½ to ⅝ inch). Attach the ridge vent with caulking and roofing nails.

Ridge vents work by allowing air to flow from the soffit vents out through the peak of the roof. Turbine vents also draw air up through the roof, but here the heat rising up into the vent turns the fixture's turbine blades and so helps pull more air out of the attic space. The hotter the air becomes, the faster it turns the turbine.

Other Roof Vents

Other common types of vents include fixed grilles (usually in soffits and high on gable ends), which allow air to pass through louvers and a variety of power fans. The venting ability of these systems depends on the size of the opening being vented. Building codes usually require the area of attic vent openings to equal at least 1/300 of the total (1/150 if no vapor barrier in ceiling) square feet of attic space being ventilated.

Roof Vent Combinations

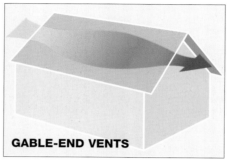
GABLE-END VENTS

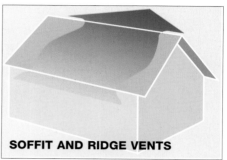
SOFFIT AND RIDGE VENTS

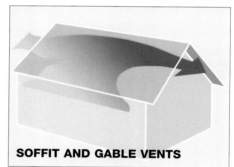
SOFFIT AND GABLE VENTS

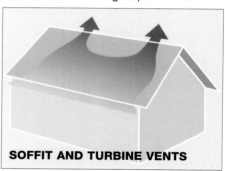
SOFFIT AND TURBINE VENTS

Attic vent combinations. Top left: gable vents provide end-to-end ventilation. Top right: soffit and ridge combination. Bottom left: soffit and gable. Bottom right: soffit and turbine combination. Check local codes for minimum venting requirements.

Maintaining Attic Air Flow

Air baffles keep loose fill added to attic floors from blocking air flow through soffit vents. To add baffles, first scrape away excess insulation, then set the forms between rafters and staple the flanges into the roof sheathing.

ROOFING

Soffit Vents

Soffit vents come in three basic configurations: round, rectangular, and perforated. The round variety, called plug vents, are easier to install than rectangular, or strip-grille, vents. You need only an electric drill and an auger bit, or a hole saw, to cut the hole for plug vents, whereas continuous vents require a circular saw to cut the opening. It is easier and safer to drill an overhanging section than it is to cut it with an upside-down circular saw—especially if you're working on a ladder. If you are installing rectangular vents, have someone hold the ladder steady.

Another option for installing soffit vents is to remove the plywood, cut the holes (circular or rectangular) while the plywood is secured to a worktable, insert the vents, and renail everything back in place. This is a viable option if the soffit has dry rot and is in need of replacing anyway. If you're planning to install rectangular vents, this approach tends to be best.

Continuous perforated soffits are manufactured in preformed sheets of vinyl or aluminum, and can be installed once the old soffits have been ripped out. Perforated soffits eliminate the need for cutting plywood, but they may not come in the size or color that you require. Home centers usually have manufacturers' catalogs listing the sizes and colors of their products.

Clearing an Air Path

After you've decided how to vent your soffits, inspect the areas between the ceiling joists to make sure insulation is not blocking air flow. Insulation in the soffit space will completely defeat the purpose of soffit vents, and a section-by-section inspection is the best way to ensure air flow through the vents.

Homes with blown insulation are more likely to have this problem because the insulation is loose and tends to drift. One way to keep blown cellulose away from soffit vents is to construct barriers to act as dams, separating the ceiling area (where insulation is needed) from the soffit area. Make the dams from rigid material, and cut them to fit between the joists.

With fiberglass insulation, simply fold back the batt or tuck it under itself to keep the end section away from the soffit.

Installing Strip-Grille Vents

USE: ▶ circular saw • wood chisel • drill • hammer • screwdriver ▶ strip-grille vents • screws

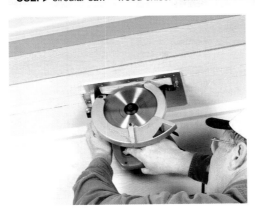

1 Use a circular saw (very carefully, in this position), to make parallel cutouts for the long edges of the continuous vent strip.

2 Slice through the short ends of the cutout with a sharp chisel, or drill a small starter hole and use a reciprocating saw.

3 Some vent strips have flanges that must tuck under the exposed edges of plywood along the cut; others screw down.

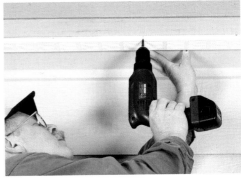

4 Attach surface-mounted strip vents to the soffit using ½-in. screws spaced about a foot apart.

Plug Vents & Perforated Soffits

Plug vents are the easiest vents to install. Drill a hole in the soffit between pairs of rafters, and press the vents in place.

Perforated soffit vents typically are made of vinyl panels that are perforated to supply ventilation through the entire soffit area.

VENTING THE ROOF

12 Ventilation

THROUGH THE ROOF

Even in areas with moderate climates year-round, air temperatures in attics can soar on sunny days. That may not seem like a problem if the space is used only for storage. But the floor of an unfinished attic is the ceiling of a finished living space below. And even if the common floor/ceiling is insulated, oven-like temperatures of 125° F and higher can radiate through to the living space and increase the load on air conditioning.

But reducing attic temperatures to increase cooling efficiency is only one reason for ventilating an attic. You also need attic airflow to carry away moisture that rises from the living space. Even if the floor-ceiling is insulated and protected with a vapor retarder (commonly 6-mil polyethylene sheeting), some of the moisture produced in cooking, washing, and other household operations inevitably seeps into cracks and seams. Again, that may not seem like a problem if the space isn't lived in. But moisture collecting in the attic can condense on wood framing to foster rot, and drip puddles onto insulation and ceiling drywall.

To prevent these problems, the best approach is to treat unfinished attics as outdoor space. Wrap the living areas below with layers of insulation and vapor barriers, but flood the attic with fresh air. There are several ways to provide thorough ventilation. Most of the systems are easiest to install on new construction, where they can be conveniently woven into weatherproof layers of shingles and siding. But the three main options of soffit vents (shown on the previous page), as well as roof vents, gable vents, and ridge vents, also can be installed on existing homes.

To handle the installations, you will have to do some of the work from the attic and some from a ladder to reach the triangle of siding just under the roof ridge, or from the roof itself.

If you do the job yourself, make a safe working platform that bridges open framing in attic floors; use ladders safely; and work on roofs only if they have a low slope. Bear in mind the most important safety rule of all: if you feel uncomfortable or unsafe working high off the ground, don't. Hire a contractor, and devote your DIY energy to a ground-floor project.

Roof vents come in many forms, but generally include some type of weather-shedding hood and a flange around the base that forms a seal between the vent and the surrounding roofing material. These vents are a good choice if you want to solve "dead-zone" problems—for example, where an attic above a one-story wing of the house joins the solid wall of the adjacent second story. Those attics can't be vented at each end and tend to trap air. A roof vent (or turbine vent) installed at the ridge near the closed end of the attic can solve the problem. To make these and any other roof vent installation watertight, the most basic rule is to cover the vent flanges with roof material on the high side and along the edges of the vent, and cover the roofing material with the vent flange on the low side.

Gable vents require a similar installation, except on an existing building you'll have to cut through siding instead of roofing to reach the plywood sheathing.

Ridge vents, which ventilate the full length of the attic, require a little more work. In addition to stripping shingles along the ridge, you will have to trim back the wooden roof sheathing on each side of the ridge. The narrow gaps will create a continuous flow from soffit vents up and out of the roof.

Some roof windows have a vent flap in the frame so that you can circulate air even in bad weather with the unit closed.

Installing Roof Vents

USE: ▶ reciprocating saw • 4-ft. level • pry bar

1 *Determine the vent location between rafters inside the attic, and drive a nail to mark the location outside.*

Installing Gable Vents

USE: ▶ circular saw • hammer • level • staple gun

1 *Frame out a rough opening, and install headers between studs in the end wall to form the required rough opening.*

Installing Ridge Vents

USE: ▶ chalk-line box • circular saw • hammer

1 *Strip ridge-cap shingles to expose about a 2-in. section of roof deck on both sides of the ridge.*

ROOFING

MONEY SAVER

• putty knife • hammer • measuring tape • pencil • utility knife ▶ roof vent • replacement shingles • galvanized roofing nails • roofing cement

2 Mark the opening required by the vent manufacturer, centered around the nail. Then cut back the shingles.

3 Once you've trimmed the shingles, cut through the roof deck with a reciprocating saw to make the vent opening.

4 To shed water, slip flanges on the upper section of the vent underneath shingles. Seal nails with roofing cement.

MONEY SAVER

• caulking gun • measuring tape • pencil ▶ gable vent • 1x4 casing • screening • 2x framing lumber • caulk • galvanized nails or screws • heavy-duty staples

2 Use nails driven from inside to locate and mark the opening at the top of the gable-end wall outside.

3 Strip back tar paper, or air barrier paper, as needed. Then cut through the plywood sheathing with a circular saw.

4 Caulk under the edges of the vent, and nail it per manufacturer's instructions. The back should be screened.

MONEY SAVER

• pry bar • putty knife • straightedge • utility knife ▶ galvanized roofing nails • replacement ridge-cap shingles • ridge vent • roofing cement

2 Peel back shingles and tar paper; snap a chalk line as a guide; and cut through the plywood roof deck along the ridge.

3 Nail down the ridge vent on top of the ridge. The flexible vent material conforms to the roof, while baffles allow air to escape.

4 Conceal the vent without decreasing its effectiveness by installing a new line of overlapping ridge-cap shingles.

THROUGH THE ROOF

13 Floors & Stairs

330 PREPARING OLD FLOORS
- Subfloors ◆ Removing Old Trim ◆ Floor-Trim Options
- Plywood on Sleepers, Subfloors & Wood

332 SOLID WOOD
- Materials ◆ Installing Solid-Wood Flooring

334 ENGINEERED-WOOD FLOORING
- Basic Types ◆ Installing Floating Floors
- Preparing Floors ◆ Laying Parquet Floors

336 WOOD-FLOOR REPAIRS
- Removing Stains ■ REPLACING BOARDS
- Six Ways to Stop Squeaks ◆ Plugging

338 REFINISHING
- Restoring Wood Floors ■ PREPARING THE ROOM
- Refinishing Wood Floors

340 RESILIENT FLOORING
- Materials ◆ Installing Sheet Flooring
- ■ REPAIRING TILE FLOORS

342 TILE FLOORS
- Installing Tile ◆ Thinset vs. Thickset ◆ Materials
- Preventing Tile Trouble ■ REPLACING TILES
- Mortar Mixes ◆ Grout Mixes ◆ Cutting & Nipping
- Sill Transitions

346 FINISHING & TRIM
- Installing Baseboards ◆ Base Molding
- Staining ◆ Finishes

348 CARPET
- Materials & Weaves ◆ Pad vs. Cushion-Backed
- Pads ◆ Installing Carpet ◆ Spot Repairs
- Stretchers & Kickers ■ PATCHING CARPET
- ■ SPOT PATCHING ◆ Tuft Patching

352 STAIRS
- Stair Building ◆ Stair Anatomy ◆ Stair Formulas
- Adding Pull-Down Stairs ◆ Carpeting Stairs
- Common Problems ◆ Replacing Treads
- Tightening Posts ◆ Stopping Stair Squeaks
- Tightening Balusters ◆ Replacing Balusters

13 Floors & Stairs

SUBFLOORS

We expect a lot of floors: we walk all over them; grind in spilled food, dirt, and grease; and park tons of furniture on them. And that's not even counting the special abuse small children and pets can inflict. Little wonder floors sag, squeak, and become a bit dingy after a while.

Sure, floors are functional—so much so that it's easy to overlook their importance as a design element in any room. Painting walls and ceilings or adding new furniture by themselves can do a lot to perk up a room. But if the floors need work because they are stained, scraped, cracked, or just plain dull-looking, then the redecorating job is only half done, and it will show. Laying the new floor covering does require more skill than a simple paint job, but fortunately the work is well within the range of most do-it-yourselfers.

A Stable Base

That new floor covering won't last very long if the supporting structure underneath it isn't in good shape. For example, floor joists—the floor's underlying framework—have to be straight, solid, and level. You can feel low spots where joists have sagged or settled as you walk across a floor. Don't expect new flooring to hide the peaks and valleys, though; in the long run, the unevenness will damage it.

Subflooring is another important part of the floor structure. Nailed to the floor joists, the subfloor is made of plywood sheets or wood planks. Sometimes another layer of plywood, called the underlayment, is also nailed down over the subflooring to provide a more stable base for the floor covering. The problem with both subflooring and underlayment is that nails can work loose, causing the floor to lift slightly and creak as you walk over it. The noise is a nuisance, but for vinyl, ceramic tile, and even wood, that slight flexing can crack the flooring.

When laying new flooring, you don't always have to rip out the old floor covering (unless it is carpeting). For example, you can lay carpet directly over old solid-wood flooring, but you'll have to install an underlayment over it for new ceramic or vinyl flooring (after refastening loose boards). If old vinyl or ceramic tile flooring is in bad shape, it probably will have to be removed, or at least covered with a new underlayment. But usually you can lay new flooring over either one after repairing and smoothing over damaged spots. Remove floating floors (page 82) before laying new flooring.

If you've torn up the old flooring, check the underlayment. If it's level, smooth enough, and thick enough, you can probably reuse it after covering chips and dips with leveling compound. If necessary, replace it with ¼ to ½-inch-thick, exterior-grade AC plywood for vinyl and wood floors. Ceramic and stone tiles are heavier, so they need a heavier underlayment. Cement backer board is best for ceramic tile, especially in kitchens and bathrooms.

Floor-Trim Options

To mark cuts around complicated molding, use a contour gauge. It duplicates even intricate shapes.

To recess new flooring, cut the trim short with a saw resting on a sample of the flooring and subflooring.

Removing Old Trim

USE: ▶ flat pry bar • chisel • hammer • permanent marker ▶ thin scrap wood • masking tape

1 Use a small flat bar to pry quarter-round trim away from the baseboard. Work slowly from one end to avoid splitting.

2 Use the same procedure to separate top molding from the base. You may need a chisel or second bar for leverage.

3 Set a thin piece of wood, such as a shingle, behind the baseboard to avoid marring the wall as you pry.

PREPARING OLD FLOORS

Plywood on Sleepers

If a floor is dry and level, you can install solid-wood strip flooring on pressure-treated 1x4 or 2x4 sleepers glued to the floor. Seal the concrete, and then apply adhesive made for bonding concrete to wood in ribbons about ⅛ inch thick and 4 inches wide. Do the perimeter first; then fill in the floor with sleepers cut 18 to 48 inches long. Overlap any short sleepers 4 to 6 inches, and secure each with two concrete nails. Allow about 10 to 12 inches between rows.

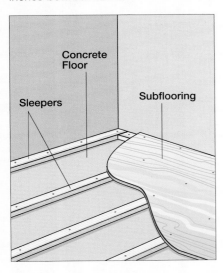

Plywood on Subfloors

Resilient flooring, parquet, and laminated wood all can be installed on dry, sealed concrete. But in damp basements (and for better insulation) lay flooring over a built-up subfloor. A built-up subfloor also is needed when using solid wood flooring wider than 4 inches. Over an old subfloor—for example, plywood or old boards uncovered when you removed wall-to-wall carpeting—run new sheets of plywood subflooring perpendicular to the old sheets.

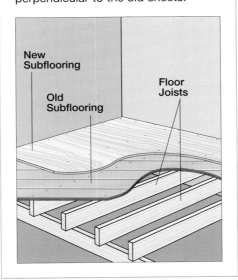

Plywood on Wood

A sound wood floor makes a good base for plywood underlayment. Lay the first panel across the floorboards. If the end of the panel falls over a seam, cut the panel to fall in the middle of a board. Use ring-shank nails every 6 inches along panel edges (keep them ⅜ inch in from the edges) and every 8 inches in the field. Nails should penetrate the flooring but not the framing. Leave 1/16 inch between the ends of sheets of plywood.

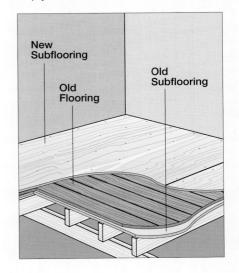

4 Some baseboards may pop off the wall. To remove stubborn nails, set a metal bar between the nailhead and the wall.

5 With the metal bar in place, tap the baseboard (on scrap wood) to force out the nailheads so you can pull them.

6 Label the location of each piece as you remove it. On a room with many corners, this will make reinstallation easier.

13 Floors & Stairs

SOLID-WOOD FLOORING

Solid-wood floors offer a richness that's hard to beat, but properly installing the flooring requires some expertise with tools and wood. You can choose from strip or plank flooring, either prefinished or unfinished, in a variety of woods. Strip flooring generally measures between 1½ and 3¼ inches wide, with 2¼-inch strips being a popular size. Planks are at least 3 inches wide, and are often secured with countersunk screws topped with wooden plugs in a matching color. Manufactured wood-veneer products designed for do-it-yourself installation (as shown on pages 82-83) look like solid wood, but there are limits to the repairs that can be made.

Ordering & Storage

Most solid-wood flooring is graded according to color, grain, and imperfections, as shown at right. You will have to weigh the cost against the wood quality. When ordering 2¼-inch-wide flooring, multiply the number of square feet in the room by 1.383 to find the amount of board feet needed, including the waste. For flooring of other sizes, ask your dealer how to compute the quantity.

When the wood is delivered, stack it in the room where the floor will be installed for several days—this allows it to acclimate to the moisture content of the room. Never store wood flooring in a poorly ventilated, damp place, because the wood may warp or twist as it picks up moisture. In addition, damp wood will shrink as it dries after installation, leaving behind unsightly gaps between boards and perhaps even cracks.

Installation Tips

Solid-wood flooring is generally installed parallel to the long dimension of the room. To allow room for the wood to expand and contract with seasonal changes in humidity, leave a gap around the perimeter of the room equal to the thickness of the boards. This gap will be covered by the baseboard trim. Finally, don't nail down the flooring without a dry run. Take the time to lay out the pieces of the floor, arranging them so that long boards alternate with shorter pieces, and end-to-end joints are evenly distributed over the floor.

Materials

GRADES OF OAK FLOORING

Type	Description
Clear	Mostly heartwood, uniform appearance, minimum number of character marks such as burls and small tight checks.
Select	May contain sound sapwood and slight imperfections.
No. 1 Common	Contains prominent variation in color and characteristics; checks and knotholes permitted.
No. 2 Common	Contains natural variations as well as minor defects and a limited number of pieces without tongues.

Hardwoods are the most durable but the most expensive; oak is most common.

Softwoods are less durable and less expensive; pine is most common.

Stain an oak floor with a pale tone, or apply a clear sealer for the lightest finish.

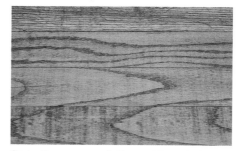
Stain the same oak floor with a dark tone for a completely different look.

Flooring can come prestained and prefinished, and in thin parquet patterns.

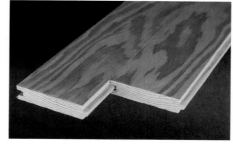
Flooring can come in slabs of cedar or redwood for porches and decks.

SOLID WOOD

Installing Solid-Wood Flooring

USE: ▶ floor nailer • flat pry bar • backsaw • chalk-line box • framing square • hammer • permanent marker ▶ flooring • flooring nails • masking tape • wedge

1 Use a flat pry bar to pry molding and baseboard off the wall. Label it to make installation easier when the floor is laid.

2 Use a piece of flooring as a guide, and shorten vertical trim where necessary to make room for the new floor.

3 Check the room for square, and snap a chalk line to help you set the first piece. Bury uneven margins under baseboard trim.

4 Set the first board on your line, slightly away from the wall to allow for expansion. Predrill and nail this board, setting the nails.

5 Take the time to lay out several rows before nailing. It helps you plan staggered seams and match wood grains and tones.

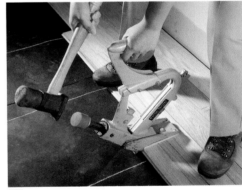
6 Rent a floor nailer to edge-nail boards. You load it with nails and pound with a heavy hammer on the driving arm.

7 Some boards have a crook and won't fit easily. Apply pressure by driving a wedge between the board and a nailed block.

8 To finish rows, measure the required length on a full board next to the end gap. Remember to leave wall clearance.

9 To close up joints on the last row, use a pry bar against a wood block on the wall to tighten the joint, and face-nail the board.

13 Floors & Stairs

BASIC TYPES

You have two basic choices when buying engineered flooring: a hardwood veneer glued to a plywood base or a synthetic, wood-like laminate over a fiberboard base. Both offer the look and feel of solid wood but are quicker and easier to install. Engineered wood flooring is surfaced with various types of hardwood and comes prefinished with a factory-applied top coat. That cuts out a lot of work, letting you wrap up the job that much sooner.

Another important advantage of engineered flooring is that it's more stable than solid wood. Because it isn't as prone to swelling and buckling, use it over concrete floors in basements and places that would be too damp for solid wood. But engineered flooring does have one big drawback—the wood veneer is thin. You can't just sand and refinish the flooring when it has been damaged or worn down.

Installation

Plywood-backed veneered flooring can be glued directly to an underlayment or installed as a floating floor. But laminate-over-fiberboard flooring should only be installed as a floating floor. Be sure to leave a ½-inch gap around the edges of the floor for expansion.

Floating floors are not nailed or glued directly to the underlayment or floor below. Instead, glue the tongue-and-groove edges of the boards together—in some systems snap together—and install over a ⅛-inch-thick pad of high-density foam.

Preparing Floors

You can install a manufactured wood floor over either an existing solid-wood floor or new plywood underlayment. To get the best results when installing a new veneer floor over plywood, lay the new plywood panels at a 90-degree angle to the seams of the existing floor. (With a floating floor, orientation of the new floor panels doesn't matter.) Plan the subfloor layout to maximize the use of full sheets and minimize seams. Before you nail down the sheets, check the old floor for high spots, which should be planed or chiseled flat.

USE: ▶ saber saw • power sander • putty knife

1 Instead of piecing plywood at door openings, make a cutout with a saber saw; a full sheet provides more stability.

3 To provide a completely smooth surface for the new finished floor, cover the screwheads with putty.

4 Drive screws about 6 in. apart along joints between sheets. Use a drywall knife to fill the joints with putty.

Installing Floating Floors

USE: ▶ hammer • flat pry bar ▶ flooring • foam-backed pad • flooring adhesive • scrap wood • rags

1 Manufacturers' installation directions will vary, but they generally include laying a foam-backed pad on which the floor floats.

2 Pieces of finished flooring are locked to each other with glue in the mating seams but not fastened to the subfloor.

3 Use scrap wood or an extra piece of flooring to seat one panel of flooring against another. Hammer blows damage the joints.

ENGINEERED-WOOD FLOORING

- power drill/driver ▶ plywood • screws • putty

Laying Parquet Floors

USE: ▶ notched trowel • saber saw • hammer • framing square ▶ parquet flooring • adhesive

2 Use screws instead of nails, set about 6 in. apart, along the edges of the plywood. Drive them flush or just below the surface.

1 Typical parquet floors (solid or veneer) rest in a bed of adhesive. Apply them with a notched trowel that leaves ridges.

2 Make layout lines to square the flooring in the space. Try to leave partial pieces of equal size around the edges.

5 Finish the subfloor preparation by sanding the dried putty. Even small ridges of putty can disrupt a veneer floor.

3 To fit a stubborn piece, use a scrap section of flooring with its tongue or groove aligned to avoid damage.

4 Make small cutouts so full tiles turn the corners of obstructions. Use a saber saw with a fine-toothed blade.

4 When you apply too much glue and it oozes onto the finished surface, wipe the excess away with a damp rag.

5 You can remove base trim, install the new floor, and reinstall the old trim to cover the expansion gaps.

6 On multipart base molding, you may want to pull only the base quarter-round and reset it over the finished floor.

BASIC TYPES

13 Floors & Stairs

FIXING FLOORS

Because people will naturally take the quickest route from one room to another, wood floors remain pristine in corners while taking abuse in the heavily traveled areas.

Sometimes you will have to replace floorboards, but as most hardwood flooring is interlocked with tongue-and-groove joints, you can't pry up just one or two boards without damaging others. When you do install a piece of new flooring, lightly sand it along the grain, and then match the surrounding finish by applying stain in light coats. Obviously, it also helps if you use a replacement board with a grain pattern that blends in—not a clear board in a grainy section or a boldly streaked board in a nearly clear section of the floor.

Dents & Cracks

If the damage is in one spot, such as a deep dent that can't be erased with surface sanding, a plug offers the easiest solution. Select a drill bit with a diameter slightly larger than the damaged spot; then, drill until you hit undamaged wood. Insert a plug as described at right.

You can fill deep cracks and splits in hardwood floors with wood fillers. Some fillers may be stronger than the wood itself—dense mixtures that resist scratches but also don't take wood stain. A few may leave a yellowish streak even more noticeable than the original crack, however. To prevent this, test the filler to see how it accepts stain, or opt for a softer, more porous, powdered filler that can be sanded, stained, and sealed like natural wood.

Stains

If you have an area that has been stained by a pet, you can remove both the stain and odor with wood bleach or household bleach after sanding the area down to bare wood. Soak the sanded area with full-strength bleach, let the wood dry, sand again, and soak the wood a second time. Because bleach kills odors but also lightens the wood, stain the patch before resealing the surface. Apply several coats of polyurethane to seal any odors in the wood, and blend the surface with coats of wax.

Removing Stains

There are dozens of recipes for removing stains, but one of the oldest and most reliable is household bleach.

Replacing Boards

USE: ▶ power drill • wood chisel • hammer • putty knife ▶ replacement board • finishing nails • putty

1 When a board is beyond repair, remove the bad section. Start by drilling a row of holes just past the damaged area.

2 Use a sharp chisel to square up the edges of the holes. Create a slight undercut so the new piece fits tightly.

Six Ways to Stop Squeaks

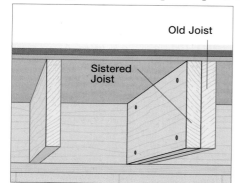

Double up a weak joist using lumber of the same size. Secure it with construction adhesive and screws.

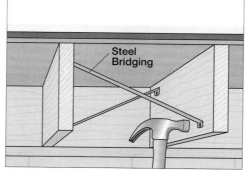

Attach steel X-bracing or a solid piece of framing between joists. This stabilizes the flooring and provides more support.

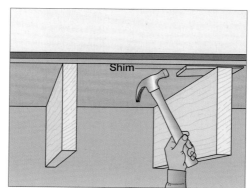

If floor joists aren't tight against the subfloor in the squeaking area, wedge shims in the gap, and tap them into place.

WOOD-FLOOR REPAIRS

Plugging

Once a dark stain is lightened, sand lightly (or use steel wool), and then blend the repair with stain and paste wax.

To repair deep stains—such as a burn—or to conceal small gouges, drill out the damage, and insert a plug.

You can use dowels or cut your own plugs. Use the same wood if you have extra, and finish the new plug to match.

MONEY SAVER

- scrap wood

3 To fit a new piece into the interlocking system of tongues and grooves, you have to remove the bottom lip of the replacement.

4 Angle the replacement into position by hand, and use a block of scrap wood or flooring to seat the trimmed edge.

5 Predrill the replacement board to prevent splitting, drive finishing nails, putty the holes, and finish to match.

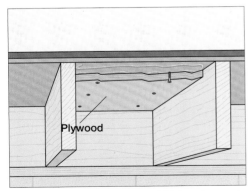
Glue and screw a section of ¾-in. plywood between joists just under the squeak, and screw through it into the floor.

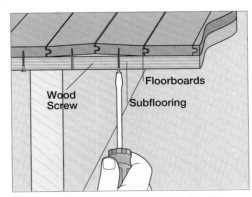
Pull down loose or bulging boards with screws driven from below. The screws should stop ¼ in. below the finish floor.

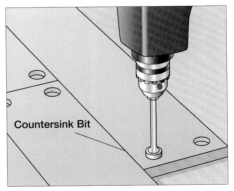
If all else fails, and you can't anchor squeaking boards from below, drill surface countersinks, drive screws, and add plugs.

FIXING FLOORS

13 Floors & Stairs

RESTORING WOOD FLOORS

When a wood floor is so worn that wax does little to improve it, it's time for refinishing. Instead of hiring a pro for the job, you can save quite a bit of money by refinishing the floor yourself. You'll have to sand off the old finish (making more dust than you ever thought possible), apply a sealer, and then finish with two coats of polyurethane or floor varnish.

With a rented floor sander you can sand about 200 to 250 square feet of flooring in a day. If you use a water-based acrylic sealer and water-based urethane floor finish, you could seal and top coat the floor in one day. (The floor still must cure for about a week, however, before it can stand up to heavy traffic.)

If your floors haven't been refinished for some time, don't automatically assume that they must be sanded down. Varnished or polyurethaned floors that are in reasonably good condition sometimes can be restored by cleaning with a good paint cleaner. Rub out any heel marks with steel wool or fine sandpaper, smooth out rough spots with fine sandpaper, vacuum away all dust, and apply two coats of varnish or polyurethane.

Sanding Wood Floors

If your floors do need sanding, you'll have to clear all furniture and objects from the room; everything must go because you need to get at the entire floor, and you have to remove the dust between sandings. You must also clean all room surfaces before refinishing, and any extraneous objects will just collect dust that could contaminate the final finish.

Mask doors, heat registers, and any outlets to other rooms to prevent the spread of dust, even if your sander comes equipped with a dust bag. Check the floor for exposed nailheads or raised boards, which can easily rip a sanding belt, and clean the floor of waxy materials, which will clog the belt. Fill holes, nicks, or dents with putty to match the finish. Because sanding creates so much dust, it's best to wear a dust mask. You also should wear goggles or safety glasses.

Using Power Sanders

For most do-it-yourselfers, stopping and starting a power sander are the hardest operations. Once the big drum of a floor sander gets going, it's fairly easy to keep the machine moving ahead at a steady pace. You can slow down over stained or dirty areas, but not too much because the belt keeps chewing through the wood at the same quick pace.

As soon as the belt starts turning, the sander should be moving across the floor. With most drum sanders, when you get to the end of the room and the belt is still turning, you have to tip the machine up quickly. Otherwise, the belt will cut a noticeable furrow in the wood that's difficult to blend out with the edger. If you can rent a lever-action type of floor sander (the kind most professionals use), you will not have to tip up the machine at the end of a pass. Instead, a lever on the handle raises the sanding drum inside the housing.

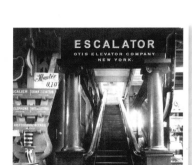

The first step-type escalator made for public use was invented by Charles D. Seeberger, and introduced by Otis Elevator at the Paris Exhibition of 1900, where this ornate beauty won first prize.

With edgers, which spin sanding discs instead of rotating belts in a loop, you have to supply the off-and-on touch by hand. You use a medium-grit sandpaper to clean the outermost edges of the floor, and then blend the straight-line drum pattern with the rotary edger pattern using fine-grit paper. If you like, after you return the rented equipment, you can use a random-orbit electric sander to get a better blend and to touch up hard-to-reach areas missed by the larger machines.

Preparing the Room

Before sanding a floor, strip the room of anything removable. Take down drapes, or pin them up from the floor and wrap them in plastic bags. Seal up switches, outlets, and heat registers with plastic and masking tape. Because sanding creates inflammable dust, turn off any pilot lights in the area.

Cover up switches, outlets, heat registers, and any other openings with heavy plastic sheeting and masking tape.

GREEN SOLUTION

To protect yourself from the dust generated by sanding, use a dust-mist respirator or a disposable dust mask.

To seal a doorway, you can buy commercial barriers with zippers, or just fasten plastic sheets or tarps around the trim.

REFINISHING

Refinishing Wood Floors

MONEY SAVER

USE: ▶ drum sander • edger • rotary buffer • orbital sander • scraper or wood chisel • vacuum • sponge mop ▶ sandpaper for all equipment • sealer

1 Use a scraper or chisel to reduce raised edges. Also check for splinters or nailheads that can tear sanding belts.

2 You can use medium-grit paper to sand most floors. On very rough surfaces, make initial passes on the diagonal.

3 Make your first pass with the edger, blending its circular sanding traces with the linear traces of the drum.

4 Use a hand scraper to clean into corners and against baseboard trim. (Another option is to remove the trim.)

5 After vacuuming, make a second pass using fine sandpaper. Most residential floors need only two passes.

6 Finish the second pass with your edger, and then vacuum again to remove dust, which can mar the finish.

7 Apply the first coat of sealer or stain to product instructions. Work slowly to avoid trapping air bubbles.

8 Many sealers require a light sanding or steel-wooling between coats. Handle this step with a rotary buffer.

9 Make sure the surface is clean (you may want to use a tack rag) before applying at least one more coat.

RESTORING WOOD FLOORS

13 Floors & Stairs

SHEETS & TILES

Not all resilient flooring is the same—the most obvious difference is between the type laid in large sheets up to 12 feet wide, and the familiar 9-inch or 12-inch square tiles. But there are other important considerations. The more vinyl the flooring contains, for instance, the more durable and costly it tends to be. Solid vinyl is best, followed by vinyl composition. Printed vinyl is inexpensive, but only a thin layer protects the floor from wear.

Both tile and sheet vinyl flooring come in a mind-boggling array of solid colors and patterns. The big plus with tiles is that you can get creative by laying out patterns using contrasting colors, including light or dark borders, diagonals, a checkerboard layout, and so on. And individual tiles can be removed and replaced, making repairs on the flooring easy.

You lay out vinyl tile floors in much the same way as ceramic tiles. (See pp. 334–335.) A sturdy flat underlayment is important for both tiles and sheet vinyl; it's usually plywood sheets laid over the subfloor. The tiles can be either self-stick or dry back (with adhesive applied separately).

The main advantage of sheet vinyl, on the other hand, is that there are few seams. That's better for kitchens and other rooms where spills and splashes are common. For rooms wider than the 12-foot sheets, you have to decide where the seam will fall.

Materials

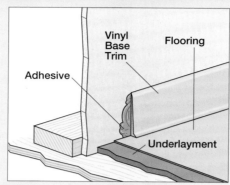

You can run flooring near the wall, and add either wood trim or molded vinyl base secured with adhesive.

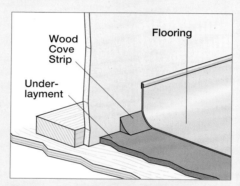

Another trim option is to install an angled block of wood to reinforce the curve and run the flooring up the wall.

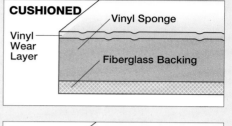

CUSHIONED — Vinyl Wear Layer, Vinyl Sponge, Fiberglass Backing

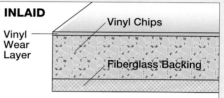

INLAID — Vinyl Wear Layer, Vinyl Chips, Fiberglass Backing

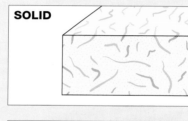

SOLID

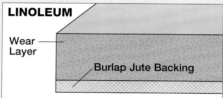

LINOLEUM — Wear Layer, Burlap Jute Backing

Installing Sheet Flooring

USE: ▶ scissors (for paper template) • utility knife • hook knife (optional) • trowel • rolling pin or commercial floor roller ▶ sheet flooring • paper for template

1 DIYers are wise to take the extra time and effort to make a paper template, particularly on a project with many corners.

2 To hold the paper template in place while you work, cut out several 1-in. triangles, and tape across them.

3 Lay the paper template over the sheet vinyl, and tape it down. Use a utility knife or hook knife to trim the vinyl to size.

RESILIENT FLOORING

Repairing Tile Floors

USE: ▶ utility knife • straightedge • putty knife ▶ new vinyl patch • tile adhesive • seam sealer • masking tape

MONEY SAVER

1 To repair torn vinyl, cut a square of new vinyl, and tape it over the damaged area. Align the patch so that the patterns match.

2 Use a straightedge and sharp utility knife to cut through both layers of vinyl. This ensures a perfect fit for the patch.

3 With both layers sliced through cleanly, remove the tape, and lift out the underlying damaged section.

4 Lift the existing flooring where possible, and apply a thin layer of flooring adhesive under the edges.

5 Spread adhesive across the underlayment, and lay in the vinyl patch piece. Press it firmly in place.

6 Use a two-part seam sealer to fuse the edges of the new patch to the old vinyl. Work it in with the applicator tip.

• masking tape • vinyl flooring adhesive • embossing leveler

4 Before laying vinyl over existing embossed vinyl, spread embossing leveler over the floor to fill the depressions.

5 When the leveler sets, spread flooring adhesive over the floor using a blade or trowel. You may need a notched trowel.

6 Lay the sheet vinyl out in the room, and use a rolling pin or commercial roller to smooth and bond it to the floor.

SHEETS & TILES

13 Floors & Stairs

SETTING CERAMIC TILE

Ceramic tiles make durable floors, well suited to high-traffic, rough-use, and potentially wet areas such as entry halls, baths, and kitchens. In deciding color and design, remember that floors with busy patterns or several colors tend to look smaller. Similarly, dark colors shrink the floor visually, while light colors expand it. Small tiles give the illusion that the floor is larger. Large tiles make the floor look smaller.

Layout

The key to a good tile job is layout. The standard procedure is to center the pattern in the room, which means drawing diagonals on the floor to locate the center and then snapping a guideline through that point parallel to the longest straight wall. The premise is that stock-sized tiles won't fit without some cuts and that it makes sense to have full tiles in the middle of the room and equal cuts around the edges.

But there are at least two exceptions. The first is when walls aren't square. In this case, you need to adjust the layout so that angled cuts will lie along the least visible wall. The second exception occurs in rooms with many wall jogs. Here, use full tiles starting from a highly visible corner just inside the door, and let the odd sizes fall in hard-to-see areas.

Setting Tile

Adhesive manufacturers' recommendations typically specify a consistently even bed of adhesive, spread with a notched trowel. For tiles smaller than 8 inches square, use a trowel with ¼-inch deep notches. Use a trowel with ⅜-inch deep notches for larger tiles. Spread the adhesive over the area in sections measuring 3 square feet with the flat edge of the trowel. The adhesive should be at least the depth of the trowel's notches. Then, comb the adhesive into parallel ridges with the notched edge, holding the trowel at a 45-degree angle. When you press the tile into the adhesive, the ridges will flatten and completely cover the back of the tile. Slide the tiles perpendicular to the ridges as you press them into place for the best contact. Clean oozed adhesive out of the joints between tiles so that you'll have room for the grout.

For a tiled surface to look good, grout seams have to be the same width and perfectly straight. With large tiles, you can create a nearly perfect grid using tile spacers (although it's time-consuming to set them between each tile); just remove them before applying grout. Small tiles are usually sold on sheets already spaced for grout—you can't get them too close together, but you could install individual sheets of tiles too far apart. You might want to try a dry layout to see whether you will have trouble estimating joint width where full sheets meet. Lay out several, and step back to see whether you can spot the sheet seams. If they stand out, ask the tile supplier for a few spacers that you can use to make sure the sheet-to-sheet joints match the other seams.

Thinset vs. Thickset

Thickset is the time-consuming, old fashioned way to set tile. With this installation, also called a mud set, you trowel out a thick bed of mortar to level the floor and support the tiles. It requires some experience with masonry work but makes a very durable floor. The thinset method is faster and easier for DIYers. You comb ridges of adhesive with a notched trowel onto a smooth underlayment surface (such as cement board) and simply press the tiles in place.

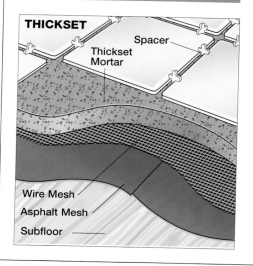

Installing Tile

USE: ▶ snap cutter • tile nippers • wet saw (optional) • notched trowel • framing square • chalk-line box • measuring tape • rubber float or squeegee

1 You can lay thinset tile in adhesive directly on plywood, but cement board makes a more durable underlayment.

2 Measure the area, and snap lines to create a square layout, with full tiles in the main field and equal pieces at the edges.

3 Spread a layer of adhesive with a notched trowel that leaves ridges, according to manufacturer's instructions.

TILE FLOORS

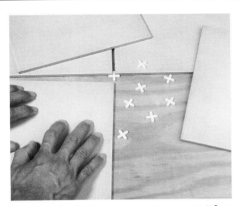

Use small plastic spacers to account for grout joints in your dry-run layout. You can also use them during the job.

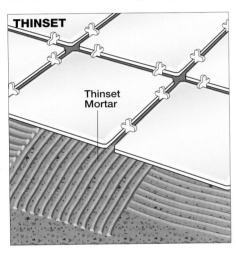

THINSET — Thinset Mortar

Materials

◆ **CERAMIC**
Ceramic tiles are fired from a bisque of natural clay. Glazed tile is coated with fired-on color; the glaze can be glossy, matte, or textured. Don't use high-gloss glazed tiles on bathroom floors because water will make them slippery.

◆ **QUARRY TILE**
This term applies to any hard, red-bodied clay floor tile of consistent dimensions no less than ⅜ inch thick. The color is usually a deep, brick red, although other colors are available, depending on the clays that are used. Glazed quarry tile comes in a wide range of colors.

◆ **MOSAIC**
Mosaic tile (2 inches square or smaller) comes in many shapes and sizes. The tiles are usually mounted on sheets for easy installation. In porcelain tiles, the color runs through the body of the tile, so scratches are practically invisible.

◆ **STONE**
Stone tiles are cut from slate, marble, granite, limestone, and other materials, and sawed and ground to a consistent thickness. Granite and marble usually are sold polished and sealed, while slate has a more natural, textured look.

• sponge paintbrush • soft cloth • work gloves ▶ floor tile • joint spacers (optional) • tile adhesive • grout • sealer (optional)

4 Set the tiles into the adhesive, leaving room for grout between tiles. You may want to use plastic joint spacers.

5 Apply grout over the floor with a rubber float. Work the mix on an angle to force it into seams; then wipe away the excess.

6 To protect grout joints, the weak links in a tile floor, you can take the extra step of applying a sealer to the grout only.

13 Floors & Stairs

PREVENTING TILE TROUBLE

If ceramic tile is such an indestructible building material, why would a tile floor need maintenance and repair? Often, the fault lies not in the tile but in the floor below. Follow a manufacturer's requirements for the supporting floor, and the tile is likely to require only cleaning, not mending.

Wood-frame floors give a little as you walk on them, which makes them more comfortable than concrete. But tile is rigid, which sets up a conflict when it's laid over anything but a masonry base—unless you make the frame almost as rigid as solid concrete. For example, you can double up the existing joists, gluing and nailing a second 2x8 to each one.

In most houses with wood floors, sheets of plywood laid over the joists are ½-inch thick. For tile, it's wise to add an additional layer of ¾-inch-thick plywood to make the floor strong enough to bridge the spaces between joists without flexing. When adding a second layer of plywood, make sure to stagger the new layer so the seams don't line up.

Most tile is designed to have a specific amount of space for grout. Some have protrusions that keep the tiles uniformly separated; some are sold with spacers. In any case, you should follow the instructions for grout seams, even though they may appear to be a little wide. If the grout pattern seems to overshadow the tile, you can increase or decrease its prominence with color.

Mortar Mixes

	Latex Portland Cement Mortar	Dry-Set Mortar	Epoxy Mortar	Organic Mastic Adhesive
Form	Mix at site	Mix at site	Mix at site	Ready-mix
Bed thickness	3/32–1/8" 1¼" (thickset floor)	3/32–1/8"	3/32–1/8"	3/32–1/16"
Application	Somewhat difficult	Somewhat difficult	Very difficult	Easy
Pros & cons	Strong, flexible, resistant to freeze/thaw, but more expensive than dry-set	Strong, resists freezing, 3-day wait before grouting	Strong, resists water and chemicals, tricky application	Inexpensive, weaker, not waterproof

Grout Mixes

Uses	Commercial Portland Cement	Sand Portland Cement	Dry-Set	Latex Portland Cement	Epoxy	Silicone or Urethane
Glazed wall tile		x	x	x		x
Ceramic mosaic tile	x	x	x	x	x	x
Quarry, paver & packing-house tile	x	x		x	x	
Dry or limited water exposure	x	x	x	x	x	x
Wet areas	x	x	x	x	x	x
Exteriors	x	x	x	x	x	x

Replacing Tiles

USE: ▶ power drill • hammer • cold chisel • pry bar • small trowel • notched trowel • vacuum • work gloves • rubber gloves ▶ replacement tile • tile adhesive

1 To remove a damaged tile without damaging others, drill a row of holes and score a line with a hammer and cold chisel.

2 Use a pry bar to work along the chisel line and pry up sections of tile. Once you remove a piece, the rest will come easily.

3 Clean out loose grout and dust, and spread a new layer of tile adhesive. Use a notched trowel to make ridges.

TILE FLOORS

Cutting & Nipping

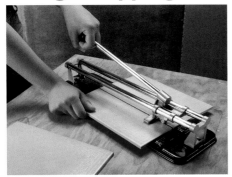

Align your cut mark on the cutter bed, apply pressure to the handle, and drag the cutter wheel across the tile.

Cut tiles with a snap cutter (for straight cuts) or tile nippers (for curved cuts). You may want to rent a tub saw (also called a wet saw) if you're tiling a large area and have many cuts to make. Snap cutters work by scoring the tile, which weakens it and allows it to snap apart in a straight line. They come in sizes for small and large tiles. Nippers work by nibbling away small pieces of tile one bite at a time. Use them to form small irregular shapes and curves.

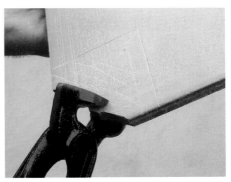

Nippers look like pliers with a sharp edge on the jaws. Make very small cuts to avoid breaking the tile.

Sill Transitions

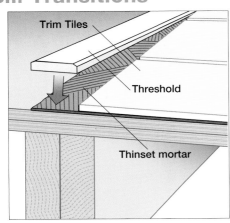

Trim Tiles
Threshold
Thinset mortar

Ease the transition from a tiled room with a threshold, or saddle. You can use wood, marble, or synthetic material. Most are formed with a beveled edge to reduce the chance of tripping where floor levels change, and are thicker than the tiles that butt against them. You may need to trim the bottom of an existing door to fit a new threshold. Attach the threshold with adhesive, leaving a gap for a grout joint between the threshold and the tiles.

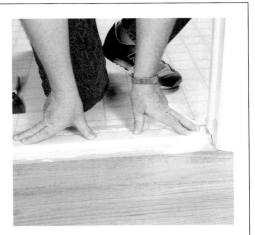

MONEY SAVER

• grout • scrap wood • sealer (optional)

4 Set a replacement tile into the bed of adhesive. Center it in the patch area to create even grout joints.

5 With the tile in position, use a block of wood to protect the tile surface, and seat the tile evenly with the surrounding floor.

6 When the tile adhesive sets, mix a small batch of grout, force it into the seams with a trowel, and wipe off the excess.

13 Floors & Stairs

FINISHING A WOOD FLOOR

Wood finishes come in many varieties, each with its own features. Choose carefully—ease of application is one consideration, but don't forget maintenance. Hard finishes that stay on top of the wood will typically require less maintenance than those that penetrate the wood's pores.

No matter what material you select, follow the manufacturer's instructions to the letter. Many finishes contain toxic or inflammable chemicals. Water-based finishes are less hazardous to work with than solvent types, but even these require precautions.

Some open-pore woods, such as oak, may require filling before the final finish is applied if you want a smooth surface. Floors made of maple, a closed-grain wood, don't need filling. Both neutral and colored fillers are available.

One of the most effective ways to apply filler is to use burlap. Rub against the grain to force the filler into the pores of the wood. Then, rub with the grain to remove any excess before the filler dries on the surface. Before applying the filler, thin it as needed with the recommended solvent.

Baseboard Trim

Baseboard trim provides a decorative look and finishes off a room. The most popular designs are shown at right, but many other styles are available. All base trim runs along the bottom of the wall, covering any horizontal gap between the wall and floor. A shoe molding normally is used with base trim to close any horizontal gap between the base and the floor.

Lumberyards stock both softwood and hardwood trim. Softwoods such as cedar, pine, fir, larch, and hemlock are usually available, but this depends on where you live. Redwood is more plentiful on the West Coast, for example, than in Florida.

You can buy standard, unfinished trim in any length up to 16 feet, with lengths available in 2-foot increments. The miter is the most common joint that is required for joining pieces of base trim, so a good backsaw and miter box or power miter saw are essential to the job. (See also "Back-Cutting Joints" and "Coping Joints" on pages 390–391.)

Once the base is cut, you can fasten it with finishing nails, brads, and/or glue. If you use glue, use just enough so that the glue will adhere to the material but not so much that it squeezes out when the piece is pressed into place. As a rule of thumb, try to use as few nails as possible to hold the trim in position—the job will look much neater and much more professional.

Sometimes trim must be sanded to remove minor imperfections. Do this very carefully so that you will not change the shape of the trim with the abrasive. Use a fine-grit, closed-coat sandpaper.

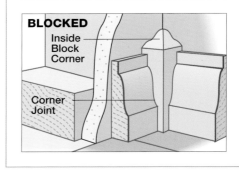

Base Molding

BASIC — Base Trim

BUILT-UP — Base Cap, Base Shoe

BLOCKED — Inside Block Corner, Corner Joint

Installing Baseboards

USE: ▶ power miter saw or backsaw with miter box • power drill/driver • hammer • nail set • measuring tape • pencil ▶ baseboard trim • finishing nails

1 You can rely on a ruler, but DIYers are wise to rough-cut boards and set them in position to mark the miter cuts.

2 For accurate miter cuts (45° cuts on mating boards), use a power miter saw that swivels or a miter box with fixed stops.

3 Fasten baseboards with finishing nails. You may want to predrill pilot holes near the ends to avoid splitting.

FINISHING & TRIM

Staining

While a clear finish is fine for most wood floors, you can stain the wood first to alter its natural color. Just make sure the stain you choose is compatible with your finish. Before you apply stain, give the wood a final sanding and vacuuming. Test the color on a scrap piece of wood before working on the whole floor. Brush it on (or use a paint pad or clean cloth), let it set, and then wipe it off. You need to let most applications dry for 24 hours, and use steel wool to knock down raised grain before sealing.

Use stain to darken wood, or unify the appearance of a floor made of boards with a lot of grain variation.

Finishes

◆ **POLYURETHANE**
Poly is durable and easy to apply. Water-based poly dries more quickly than oil-based but is not as durable. Use two or three thin coats instead of one thick one.

◆ **PENETRATING SEALER**
This finish colors and seals in one step. Apply it with a brush or paint pad, let it soak into the wood for several minutes, and then wipe off the excess.

◆ **VARNISH**
Varnish creates a hard, durable finish but with a slight yellowing. It comes in high or medium gloss or a flat finish—choose a type formulated for floors.

◆ **PICKLING**
Pickling accents the wood grain. Brush on a coat of paint cut about 30% with thinner, and, after about 15 minutes, wipe off the excess along the grain.

◆ **WAX**
Applied over a sealer, multiple coats can produce a lustrous patina that polyurethane cannot. Wax is not as durable, though—but it's good for spot repairs.

Clear sealers are normally brushed on, but you can also spray them on to apply several thin coats to intricate shapes.

Aside from staining or clear sealing, you can use other treatments such as pickling to create colored finishes.

4 Drive the nails almost flush with the wood surface to avoid marring the wood with hammer blows.

5 Use a nail set to drive nails home. On floors that will be carpeted, it's wise to finish the baseboard in advance.

6 Work around the room one board to the next. Leave the leading edge rough cut until the joint fits; then, measure for length.

FINISHING A WOOD FLOOR

13 Floors & Stairs

WALL-TO-WALL CARPETING

Most brand-name wall-to-wall carpeting (such as DuPont's Stainmaster and Solutia's Wear-Dated Gold Label) combines nylon fibers—which have excellent durability and resistance to abrasion, crushing, and mildew—with improved fibers that are less shiny, antistatic, and have been treated with a stain-resistant coating. Besides the fiber type, you have to take into account pile depth, pile density, and texture when selecting carpeting.

Bear in mind, for example, that increased density (the number of yarn tufts in a given area) usually means a better appearance, feel, and durability. Also, a dense, low-pile, level-loop design, which presents a pebbly, uniform surface, is generally the easiest to vacuum. More irregular surfaces, like twist and shag carpeting, tend to trap the dirt in the pile.

Before ordering carpet, make a scale drawing of the room on graph paper, setting each square equal to one square foot. Include all doors and obstacles—the more accurate the drawing, the easier it will be for the dealer to recommend the amount of carpet you need.

If your carpet must be seamed, place the seam in a low-traffic area. Seams are less noticeable when they run parallel to light rays and should run toward the room's primary source of light (such as a south-facing window). If your carpet has a pattern, you will have to buy additional yardage so that you can match the pattern.

Materials & Weaves

◆ **NYLON**
The longest-wearing synthetic material, nylon resists stains and is resilient, although not as springy as wool. It's easily cleaned, but the colors tend to fade when exposed to sunlight.

◆ **POLYPROPYLENE**
Polypropylene (or olefin) resists fading because the pigment is built into the fiber, but it is probably the least durable and soil-resistant of the synthetics—don't use it where it will receive heavy wear. It's generally found only in loop-pile designs.

◆ **POLYESTER**
This material approximates the softness of wool and has bright, clear, fade-resistant colors and excellent stain-resistance. It is less expensive and less resilient than nylon. It's generally found only in cut-pile constructions.

◆ **WOOL**
Wool is soft, durable, fade-resistant, and rich-looking, but it's expensive and less resistant to stains than synthetics. Of all the carpeting materials, wool is the most resilient—meaning it springs back after you walk on it.

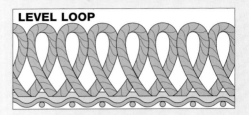

LEVEL LOOP

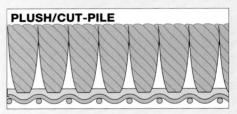

PLUSH/CUT-PILE

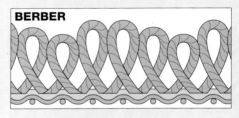

BERBER

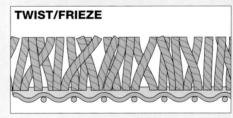

TWIST/FRIEZE

Installing Carpet

USE: ▶ hammer • utility knife • power stretcher & kicker (optional) ▶ carpet • carpet pad • tack strips • double-faced seam tape

1 Install tack strips around the room, leaving about a ⅜-in. gap next to the baseboards. The sharp nailheads face up.

2 Roll out the pad, and trim the edges just inside the tack strips. Fasten the pad with staples (wood) or adhesive (concrete).

3 To fit carpet into corners, make a relief cut with a utility knife, press one side in place, and cut away the overage.

CARPET

Pad vs. Cushion-Backed

A carpet plus pad makes the most comfortable and durable floor.

There are two types of wall-to-wall carpet. Standard carpet is installed on a pad and secured to tack strips around the room after stretching (such as with a power stretcher). Cushion-backed carpet has the foam backing already attached to its underside. It is laid in latex or fastened to the floor with a special double-faced tape; it is easier to lay because it requires no stretching. It is also less expensive than conventional carpeting but will generally not last as long.

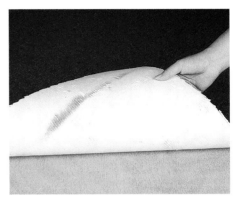

Carpet with a foam backing is easier to install but less durable.

Pads

Carpet padding adds luxury and warmth, helps cut noise, and prevents wear. The least expensive is bonded polyurethane; only dense polyurethane should be used in high-traffic areas. Prime urethane and grafted prime foam shouldn't be used with a stiff-backed carpet such as a berber. (Use a cellular sponge-rubber padding instead.) Other foam pads are graded by thickness and density—choose the best you can afford.

4 Where you have to join carpeting (or padding) one of the easiest methods is to use double-faced seam tape.

5 You can rent special tools like this kicker, which grips the carpet to work out any wrinkles and get a tight fit.

6 The last step is to tuck the edge of the carpet into the gap between the tack strips and the baseboard.

WALL-TO-WALL CARPETING

13 Floors & Stairs

SPOT REPAIRS

Carpet may seem to be a very durable material, but it's actually one of the less sturdy floor coverings. Stains, burns, snags, and beaten-down pile—especially in heavy-traffic areas—can all ruin the carpet's appearance. With proper care, you can avoid these problems, but when necessary, it's possible to make simple repairs yourself.

Scraps from the original installation are the best material to use for repairs, but you can also make patches for holes by taking bits of carpet from out-of-the-way areas like the backs of closets or underneath furniture you never move. Small areas often can be repaired by pulling out old tufts and setting in new tufts with a tuft setter, a machine that is available from most carpet dealers. It's important to be aware that the repair will probably be quite visible, especially if the carpet is worn or the color has faded. Allow some time for the repaired area to blend in with its new surroundings.

Carpet Repair Basics

The key to a successful carpet repair is to work slowly and patiently. To patch a small rip, first fold back the torn section, and then apply a good-quality latex seam adhesive. Carefully tuck the torn part back into the carpet, and press it down with a smooth rolling action, using a large bottle or rolling pin. If any of the adhesive happens to ooze up, be sure to clean it off immediately with water and detergent. When the adhesive has dried, replace any loose or missing pile.

If the rip is larger, you'll have to release the tension on the carpet using a knee-kicker in the corner nearest the rip. Lift the corner off the tack strip, and roll it back. Using heavy thread that matches the color of the pile, mend the rip by sewing with 1-inch long stitches spaced ¼ inch apart. Depending on the direction of the rip, run the stitches either parallel to rows of pile or perpendicular to them. Check frequently to make sure that you are not stitching any strands of the pile down on the face of the carpet. Then, carefully work a thin, wavy strip of latex adhesive into the stitched backing, and cover the damp surface with a paper towel. Lastly, roll the carpet back to the wall and rehook it back on the tack strip.

Caring for Carpets

Periodic vacuuming is the surest route to long carpet life. Without a once- or twice-weekly vacuuming (whether the carpet looks as if it needs it or not), the grit and dirt that settle on the top layer of the carpet will eventually be ground into the fabric, where it will do damage. You'll probably notice after vacuuming that the carpet looks shaded. That's because the piles of most carpets are directional. Carpets with a pile that is woven in different directions, such as a textured saxony, won't have this problem.

Beyond regular vacuuming, it's a good idea to clean your carpets annually, to rid the fibers of deep grime and to restore the carpet to its original color. You can do this yourself or hire a professional carpet cleaner. Hot water is the most common method of rug cleaning. With any wet cleaning (such as shampooing), there's a danger of overwetting the carpet, though, which can distort it. Try to shampoo the carpet on clear dry days, so that the floor will dry more quickly.

Because most stains sit on top of carpets treated with soil-resistant finishes, you usually have enough time to mop them up, and deep penetration doesn't become the problem it used to be. Commercial spot removers can handle a wide variety of stains, but as a last resort for stubborn stains, you should call in a professional cleaner.

Stretchers & Kickers

To make your carpet fit tightly and lie flat, you'll want to rent one or both types of carpet-stretching tools. A power stretcher grips the carpet and stretches it without tearing. The tool comes with long extensions for bracing against the opposite wall. You crank the handle to increase pressure. A kicker is shorter, and lifts the carpet and hooks it on the tack strip when you bump it with your knee. It's especially useful in tight areas where you can't fit the stretcher.

Patching Carpet

USE: ▶ hammer • utility knife ▶ carpet remnant • carpet tacks • double-faced seam tape

1 To make a large patch, box the area with scrap carpet, and drive tacks to maintain the tension and avoid wrinkling.

2 Cut a patch piece to fit within the box. If the carpet has a pattern, you may want to cut an irregular patch following the design.

CARPET

Spot Patching

USE: ▶ carpet patch tool ▶ carpet remnant • adhesive • double-faced seam tape

1 Use a circular carpet patch tool to remove a cookie-cutter section around a deep stain, burn, or tear.

2 Peel the cover tape from a piece of double-faced adhesive patch tape. Cut it larger than the hole, and fold to insert.

3 Use the circular cutter to cut a patch piece from a remnant, and press it firmly in place over the adhesive.

Tuft Patching

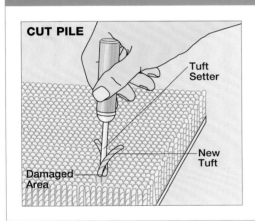

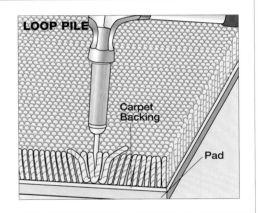

To patch a small area of cut-pile carpet with replacement tufts, cut the damaged pile down to the backing with scissors, and pick out the pile stubs with tweezers. Apply latex cement to the carpet backing. Set replacement tufts into position using a tuft setter; drive the setter through the carpet backing with a few taps from a hammer. For loop-pile carpeting, poke one end of a long piece of yarn into the backing, and make successive loops, adjusting the length.

MONEY SAVER

3 Set the patch over the damaged area, and cut through the bottom layer of carpet using the patch as a guide.

4 Remove the damaged section, lift the edges of the carpet, and install double-faced seam tape on all sides.

5 Set the patch in position, and press it firmly onto the tape before pulling the temporary tacks and releasing the tension.

SPOT REPAIRS

13 Floors & Stairs

STAIR BUILDING

The easiest staircase to build is a set of straight-run stairs; in most circumstances, a straight-run design will be your first choice. While spiral stairs are space savers and prefabricated models are fairly easy to install, they aren't recommended for a main stairway because they compromise safety and make moving large objects up and down all but impossible. In terms of materials, straight-run stairs require nothing more than a few stringers and the steps.

All the steps must be the same size—the most critical components of stair design are the ratio between unit rise and unit run and their consistency from one step to the next. Local building codes regulate the acceptable dimensions. In "Stair Formulas" at right, the minimum and maximum requirements are listed, but these are not necessarily optimums. For most people, a 7-inch rise and 11-inch run are the most comfortable (the maximum rise generally allowable is 7¾ inches; the minimum run is 9 inches).

Building codes also address landings and the room people will need to get onto and off of the stairs safely. Headroom is another important issue—it's defined as the vertical distance measured from an imaginary line connecting the front edge on all of the treads on up to the overhead. Again, most codes establish a minimum (80 inches) from that line to any object above. This is required to prevent you from knocking your head against the ceiling or other obstruction.

Stair Formulas

1 Maximum riser height is 7¾ inches. Divide total rise of stairs by 7¾ inches; then divide total rise by the number of resulting risers to determine unit rise.

2 Minimum tread depth (run) is 9 inches—10 or 11 inches is safer. Simple formulas to match tread depth to riser height: rise + run = 17 to 18, or alternatively rise x run = 70 to 75.

3 Typical code requirements specify a minimum of 80 inches of headroom in all parts of stairway.

Stair Anatomy

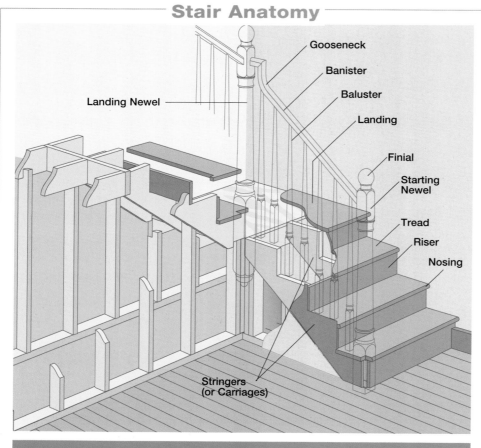

Most stairs are supported by saw-toothed boards called stringers, although there are exceptions, such as circular stairs. Because it is difficult to plan stairs, DIYers (and most contractors) leave this job to stair subcontractors.

Adding Pull-Down Stairs

USE: ▶ reciprocating saw • power drill/driver • saw

1 Mark the stair opening, and cut out the drywall ceiling. You need to brace a center joist at both ends, and cut out a section.

5 To support the stair assembly in the new opening while you attach it, screw a supporting wood lip to both sides.

STAIRS

Carpeting Stairs

There are several options when carpeting stairs. One is to use the narrow tack strips used for wall-to-wall carpeting. Install the strips with the points facing up, so they bite into the carpet—but not through it—and hold the carpet in place. You could also use tack strips made especially for stairs—like those for wall-to-wall carpet, except they have opposing barbs to hold the carpet in a corner from two directions at a time. If you don't mind seeing the fasteners, use decorative stair rods to retain carpet runners at each turn on the stairs.

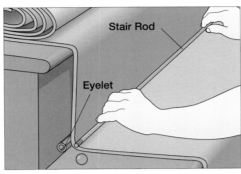

You can pin carpet runners with decorative stair rods, staple them, or wedge them into the step/riser joint with a stair tool.

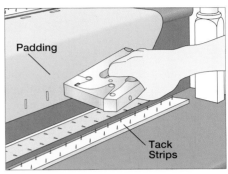

You can pin padding with staples, and secure carpeting with standard tackless strips or special tackless stair clips.

• hammer • socket wrench • framing square • measuring tape ▶ pull-down stairs • 2x lumber (for framing & nailer) • framing connectors • nails • lag screws • shims

2 Install half of a double header at each end of the opening. Nail up the end of your cut joist, and then add the second board.

3 Although some stair manufacturers don't call for the detail, it's wise to double up full joists on each side of the opening.

4 To finish framing the opening, add a short joist between headers to box in the stair opening. Use metal framing hangers.

6 Following manufacturer's directions, bolt the stair frame onto the opening. You may need to use wooden shims.

7 Measure from the fold-down hinge to the floor in a straight line—first on top of the leg, and then on the bottom.

8 Transfer the two measurements to the extended leg, draw a line between them, and cut off the excess portion.

STAIR BUILDING

13 Floors & Stairs

COMMON PROBLEMS

Squeaking stairs may be one of the better low-tech burglar alarms around, but all that creaking gets to be a nuisance after a while. And when the stair tread finally loosens up enough, it could cause someone to trip and fall. The same is true for loose handrails and balusters, too. You don't want the railing to break just when it's needed most—when someone falls on the stairs.

In most cases, repair is quicker and cheaper than replacement, plus a new part will probably look out of place on a well-worn staircase. For balusters, you may be able to find a suitable replacement at a building-supply store. Or you can commission a woodturner to make a replacement baluster. If several balusters need replacing, the best choice may be to tear them all out and install a new set.

Loose handrails are a common problem. Handrails along the wall side of stairs usually attach to brackets that are screwed to the wall. When tightening a wall-hung handrail, make sure the brackets are screwed into studs or blocking between studs; then, install longer screws or reposition the loose bracket so that you can drive the screws into new holes. Handrails on the open side of stairs usually are attached to a newel post at the bottom and to the balusters along the handrail's length. Repairing one of these handrails usually tends to be just a matter of tightening the connections at the newel post.

Replacing Treads

Worn, cracked, or badly warped stair treads should be replaced. Begin by removing molding or nosing and balusters from the tread. Pry the tread loose enough to remove nails. Put a piece of wood under the pry bar to protect risers and skirtboards from damage. If you can't pry out the tread, drill a starter hole at the rear of the tread, and cut it out with a keyhole or reciprocating saw. On landings made from floorboards, you can add a new nosing piece secured with screws.

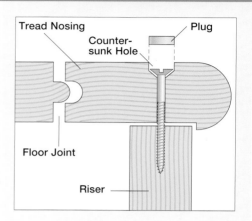

Tightening Posts

Because of their location at the end of the stairway, and because they can't be braced side to side, newel posts are especially prone to loosening. To firm up a loose post you may be able to tighten the bolt holding the post base from beneath the stairs. If the connection to the banister is loose, you can drill through the post and secure the joint with a screw or lag screw. Countersink the hole, and conceal the screwhead with a wood plug.

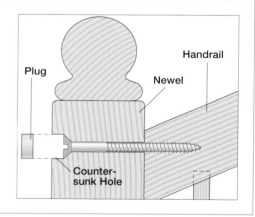

Stopping Stair Squeaks

Squeaking is generally caused by a loose joint where your weight makes one piece of wood rub against another, or shift on the nails holding pieces together. It's easy to pinpoint problems by walking on different parts of a squeaking tread. It's also easy to fix the problem on stairs where the construction is exposed underneath. There are several ways to stop squeaks on exposed stairs. Without that access, however, you're forced to make repairs from above, driving a screw and plugging the head.

To support the joint between a horizontal tread and a vertical riser, add small blocks fastened with glue and screws.

Screwing on a shelf bracket also reinforces the tread-riser joint. Use screws that won't protrude through the boards.

STAIRS

Tightening Balusters

You can tighten a connection between the baluster and the railing by wedging the joint and pinning it with a screw.

Add glue to the wedge, tap it into the loose joints with a hammer, and cut off the excess with a trim saw.

To secure the connection with a screw, first predrill a pilot hole to avoid splitting the baluster. Countersink the screwhead.

Replacing Balusters

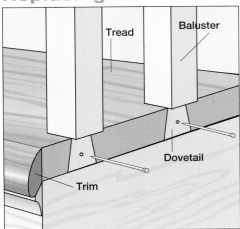

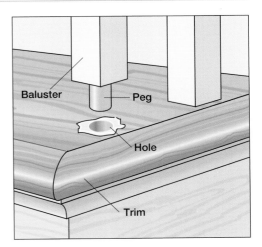

Traditional railings have balusters set into the treads. There are two basic types: dovetails and pins. Because dovetails are cut into the side of the tread, you have to remove a piece of trim to expose and replace them. Pinned balusters are fastened into a hole in the top of the tread. To remove either type, it helps to cut the damaged baluster in half and remove each section. Some modern railings have a molded track top and bottom. Replace these by matching the angle cuts and nailing in place.

Tighten loose joints by driving a wooden wedge into the seam between treads and risers. You can add glue to the wedge.

You can also use a wedge on the edges of treads and risers where they are recessed into a diagonal stringer.

Drill a pilot hole, and screw the riser to the tread. Measure carefully to center the screw in the tread board.

COMMON PROBLEMS

14 Walls & Ceilings

358 MATERIALS
- Underneath It All ♦ Paneling ♦ Wallcoverings
- Drywall ♦ Scaffold Setups

360 PANELING
- Paneling Options ♦ Installing Panels ♦ Panel Trim
- Solid Paneling ■ REPLACING AN INTERLOCKED PLANK
- ■ PATCHING PANELING ♦ Scribing Joints

364 DRYWALL
- Tools & Materials ♦ Installing Drywall
- Finishing Panel Seams ♦ Finishing Corners
- Drywall Repairs ■ FIXING SMALL HOLES
- ■ FIXING LARGE HOLES ■ FIXING CORNERS

370 PAINTING
- Paint Basics ■ "GREEN" PAINTS ♦ Paint Choices
- Paint Application Options ♦ Preparing Walls
- Sealing over Stains ♦ Painting Walls
- Paint Problems ♦ Special Finishes

376 WALLPAPER
- ■ MAKING SPOT REPAIRS ♦ Stripping Old Wallpaper
- Hanging Wallpaper

378 WALL TILE
- Patterns ♦ Grouts ♦ Installing Tile ■ GROUT REPAIRS
- ■ REPLACING TILE ♦ Restoring Bath Tile

382 GLASS & MIRRORS
- Installing Glass Blocks ♦ Mounting Mirrors

384 SOUNDPROOFING
- Controlling Noise ♦ Sound Transmission
- Making Sound-Absorbing Walls
- Sound-Resistant Construction Options

386 CEILINGS
- Wood Planking ♦ Installing a False Beam
- Installing a Suspended Ceiling

14 Walls & Ceilings

UNDERNEATH IT ALL

The most common kind of wall in residential construction is the wood-framed wall, or stud wall. The framework of studs provides space for wiring and outlets, plumbing, ducts, and insulation. Studs also support a surface that covers these utilities. Before the 1940s, this surface was generally plaster, but sometimes solid-wood paneling, brick, or stone.

The walls and ceilings of most houses today are covered with drywall. Also known as plasterboard, wallboard, or by the trade name Sheetrock, it provides a solid foundation for interior finishes such as wallpaper, paneling, or tile. But because its paper-covered face can easily be painted, drywall often serves as the finished surface.

Skilled laborers once spent many hours working with wet plaster to lay up new walls, smoothing several layers over a base of wood or metal lath to achieve smooth interior surfaces. Drywall crews today finish off a room in a fraction of the time, using preformed 4x8 (or larger) panels. These sheets are quickly cut and nailed or screwed in place on the wall studs—or even attached directly over an old plaster surface that is beyond repair.

Decorative Options

Besides resurfacing with new drywall, you can change the appearance of existing walls with paneling, which gives a room the warm glow of wood. Sheet paneling is inexpensive and easy to put up; board paneling takes a little more labor and costs more, but many people feel that the depth of real wood tones make the effort worthwhile. Solid wood also has the advantage of installation right over the studs. Wall tile is another fairly expensive and labor-intensive option, but ceramic tiles should last a lifetime. Also, tiling a half-wall or sink surround requires more patience than expertise.

If you're looking for something more ornate than painted walls, wallpaper (more properly, wallcovering, which includes solid or fabric-backed vinyl) is much simpler than it used to be, with adhesive-backed products eliminating messy paste-mixing and brushing. If you'd like to exercise some artistic skill, add decorative touches with a stenciled border or one of the decorative paint techniques on pages 374–375.

Paneling

Simulated wood boards are available as 4x8-ft. sheet paneling, which is easy to install, and a good way to resurface damaged walls. (Local codes may require a layer of fire-resistant drywall underneath.) Solid-wood boards provide a more traditional look. When solid wood is overwhelming in a small room, try paneling just one wall or applying half-height wainscoting.

Wallcoverings

VINYL-COATED • VINYL

Wallpaper isn't as popular today as surface treatments and glaze finishes, but there are many situations where wallpaper is a good choice. Most pre-pasted papers that DIYers use have a thin vinyl coating (left). Many are available with matched border strips. In high-moisture areas, you can use heavy-duty vinyls (right). These wallcoverings have a tougher (advertised as scrubbable) surface.

MATERIALS

Drywall

Gypsum drywall (often called by the USG brand name Sheetrock), is made from a slurry of powdered gypsum rock and water that is poured between sheets of paper and dried. Drywall is easier to install than plaster, the material it replaced. The panels can provide stability and fire protection in large spaces. Laminations of thin sheets can even be applied over curving surfaces.

ANAGLYPTA / FIBERGLASS

Some old-style papers such as Anaglypta (left) are available today in reproductions. The paper is much thicker and stiffer than conventional wallcoverings, and typically embossed with a pattern. The material is costly but substantial enough to bridge minor wall imperfections. The modern equivalent is fiberglass wallcovering (right), which also is fire- and moisture-resistant.

Scaffold Setups

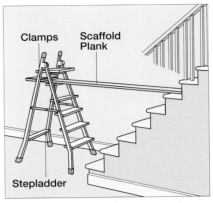

In a stairwell, reach lower areas from a scaffold-grade plank clamped to a stepladder and resting on a stair.

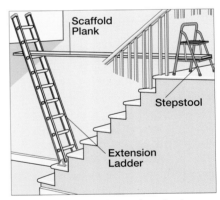

To reach higher, clamp the plank between an extension ladder and a stepstool on the landing.

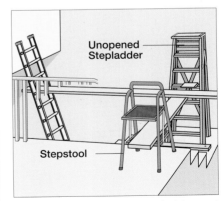

In a narrow stairwell, use a ladder and stepstool to support one plank that in turn helps to support a second plank.

UNDERNEATH IT ALL

14 Walls & Ceilings

PANELING OPTIONS

Paneling may have a slightly bad rap because some low-end products are simply artificial-looking—a poor likeness of wood grain glued to a flimsy backer board. Most paneling is anything but a cheap coverup, though. Solid plank paneling, sheet paneling covered with prefinished wood veneers, and combinations of plank and veneer paneling can be quite expensive. With plank paneling, you can use hardwoods such as birch, maple, and oak, or softwoods such as pine, cedar, or cypress. Sheet paneling, made from real wood veneers bonded to plywood, can be bought with or without a finish applied. Choose light-colored woods (such as birch or maple) in either plank or sheet to keep a room bright and make the room seem larger, or go with more traditional, darker woods. All have a subtle patina that can't be duplicated by simulated finishes.

Built-Up Walls

Custom-designed, built-up panel walls give a room a touch of elegance. Start with a base layer of ¼-inch-thick hardwood veneer panels attached to studs or furring strips. Over that base you can apply any number of raised panel designs—say, outlining a large box grid with 1x4s, subdividing them with 1x2s, and trimming with quarter-rounds and other molding profiles. But this kind of paneling project requires a lot more carpentry skill and experience than installing sheet paneling.

Panel Trim

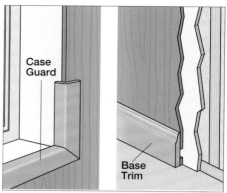

Cap molding is installed at the top of wainscoting or at horizontal seams. Seam cover trim hides vertical joints.

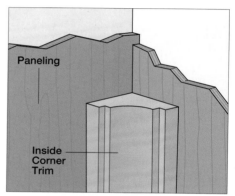

In addition to base trim, special panel trim pieces protect panel ends. Use round-edged case guard at windows.

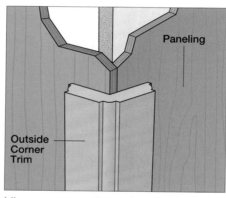

Like corner guard on drywall, this trim covers and protects the seam where two panels meet at a corner.

Almost any stock trim, including simple corner round or a beaded molding, will cover seams at inside corners.

Installing Panels

USE: ▶ 4-ft. level • T-square • hammer • table saw, circular saw, saber saw, or handsaw (to cut boards) • saber saw or keyhole saw (to cut holes for electrical

1 Use a 4-ft. level or other straightedge to check the walls for level and plumb. Mark low spots—these will need to be shimmed.

2 Nail up 1x3 or ¾ furring strips and check them for level; they provide an even nailing surface for the paneling.

3 The surfaces of the furring strips should be plumb from top to bottom. Use pairs of shingle shims to fill low spots.

PANELING

Solid Paneling

You can cover a room quickly and economically with sheet paneling, but it won't have the richness or detail of solid wood. Solid paneling is usually built around a frame of ¾-inch-thick boards that can run up to 12 inches wide. Many combinations of thinner panels and moldings can be used within the frame. Solid paneling also can be applied board to board. There are many styles of interlocking planking.

Knotty-pine planking is one of the most common applications, often installed over horizontal furring nailed to the stud wall.

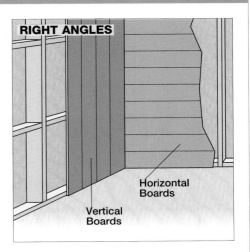

RIGHT ANGLES — Horizontal Boards, Vertical Boards

PLANK PANEL PROFILES

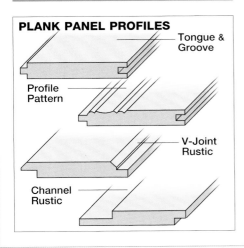

Tongue & Groove · Profile Pattern · V-Joint Rustic · Channel Rustic

Solid hardwood paneling with inset frames is probably the most expensive but also the most elegant wallcovering.

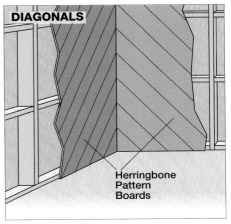

DIAGONALS — Herringbone Pattern Boards

boxes) • clamps • scriber • eye protection • gloves ▶ sheet paneling • furring strips • panel adhesive • common nails • finishing nails • shims

4 Install filler strips to provide support for the panels and seams. Leave small gaps at the ends of the strips to prevent buckling.

5 Set the glued panel with the bottom shimmed off the floor. Pull the panel away from the wall until the glue gets tacky.

6 Remove the blocks, reset the panel, and nail it in place. Color-matched panel nails in the grooves won't be noticeable.

PANELING OPTIONS

14 Walls & Ceilings

BRIGHTEN A DARK LAIR

For a cavelike room with small windows and dark paneling, there are three basic options for bringing light to the gloom: replacing, painting, or restaining. In many cases, replacing is the easiest option, because you can usually nail new paneling right over the old.

Printed panels with a synthetic finish can't be stained and painted very easily, so replacement with a lighter-tone paneling is the best option. If the old paneling was nailed up without adhesive, you may also be able to restore the plaster or drywall surface below if you prefer. Pulling off glued paneling will likely rip the drywall or plaster, and may expose problems covered by the paneling. Before deciding, carefully pry away the paneling in a couple of out-of-the-way places and inspect the wall—then you'll know what you're getting into.

If the paneling is either solid wood or a wood veneer, you can scuff-sand and paint it; or with more effort, you can lighten the wood with bleach and/or power sanding, and then apply a new coat of light-toned stain. If the paneling has a sheen, sand off the finish first so that the bleach will soak into the wood grain. Try commercial wood bleach or a 50:50 solution of household bleach and water. Numerous washings or applications may produce the results you want; if not, the paneling will have to be replaced. After bleaching, sand the surface to remove more color and smooth it—it will be roughened by the bleaching.

Replacing an Interlocked Plank — MONEY SAVER

USE: ▶ circular saw • pry bar • block ▶ replacement plank

1 Cut the damaged plank up the center using a circular saw with the blade set to the thickness of the panel.

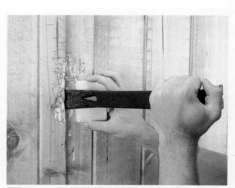

2 Remove the plank in sections with the flat end of a pry bar, using a block of wood to give you leverage.

3 Use a circular saw to cut the bottom groove edge off the new plank so it can be fitted in place next to the old planks.

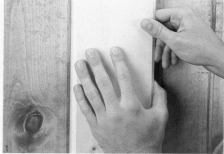

4 Fit the tongue of the new plank into the old groove, and seat the opposite side on top of the tongue next to it.

Patching Paneling

USE: ▶ saw • utility knife • caulking gun • hammer ▶ panel patch piece • furring strips • panel adhesive • masking tape • finishing nails

1 Choose a patch piece of paneling that closely matches the color, grain, and groove pattern of the area that needs replacing.

2 Tape the patch over the damaged area, and use a utility knife to score the patch outline onto the damaged wall panel.

3 Remove the patch, and keep scoring carefully with the utility knife until you cut all the way through the old paneling.

PANELING

Touch-ups

To conceal skin-deep scars in either solid plank or sheet paneling, use a color-matched repair stick.

To patch deeper scars, cut away loose wood fibers, apply a wood filler, smooth, and touch up with wood stain.

Scribing Joints

If a panel will butt against a corner that's not straight or against an uneven surface such as a fireplace mantle, you'll have to use a scriber (or a contour gauge) to duplicate the shape and trace it onto the panel. Hold the panel in place plumb on the wall, with the edge an inch away from the corner. Use a compass (or scriber) to transfer a layout line to the panel; then cut the line with a saber saw. To minimize chipping, fit the saw with a fine-toothed blade or with a special panel-cutting blade that has teeth oriented to cut only on the plunge.

A contour gauge has small, sliding pins that make an accurate impression of a complex contour, such as a countertop.

Use a scriber to transfer the shape of an irregular surface (such as an out-of-plumb wall) onto a piece of paneling.

To cut along an irregular line or complex contour, use a saber saw fitted with a panel-cutting blade.

MONEY SAVER

4 Glue furring strips behind the hole to support the patch piece. Clamp the pieces in place, and let the adhesive dry completely.

5 Apply construction adhesive to mating surfaces of the furring and the patch piece, and set the patch in position.

6 Seat the patch piece firmly; then countersink color-matched paneling nails around the perimeter, and putty the holes.

BRIGHTEN A DARK LAIR

14 Walls & Ceilings

DRYWALL BASICS

The standard drywall panel for new construction and remodeling is ½ inch thick. Special ½- or ⅝-inch-thick ceiling panels are recommended for ceilings, especially those with 24-inch-on-center framing. Lighter ⅜-inch panels are good for resurfacing work. When a plaster wall is beyond repair, you can nail the thin drywall right over it. Special ¼-inch panels are used for curved surfaces.

Drywall is generally available in 4-foot widths and 8-, 10-, 12-, and 14-foot lengths. It may seem easiest to hang 4x8 sheets vertically, running the long dimension from floor to ceiling. And it's true the 4x8 sheet, which weighs about 60–70 pounds, is easier to maneuver and install than longer sheets—especially on ceilings. But whenever possible, you should install sheets horizontally on walls, and always choose the longest practical length to minimize the number of end joints. You'll need a helper, but horizontal installation saves work during the time-consuming process of taping and finishing joints. Also, vertical joints are harder to tape—you have to stoop to do the lower section and stretch to get the top. The long horizontal seam is easy to reach.

When you handle drywall during installation, remember to treat it gently because it breaks easily if dropped or hit, and you will crush a corner if you put the full weight of a sheet on it. Stack drywall flat to prevent warping if you don't plan on using it right away.

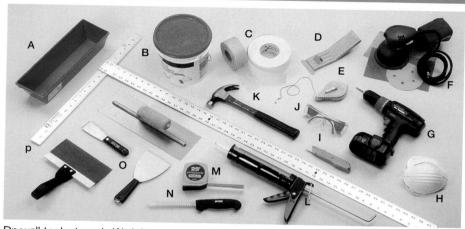

Tools & Materials

Drywall tools: trough (A); joint compound (B); tape (C); panel lifter (D); chalk-line box (E); power sander (F); drill/driver (G); dust mask (H); utility knife (I); safety glasses (J); hammer (K); caulking gun (L); measuring tape (M); drywall saw (N); taping knives (O); T-square (P).

You may need a lot more than nails, screws, tape, corner bead, and joint compound to drywall. In addition to the specialized tools above, you have a choice of gypsum materials. The residential range includes ¼-inch panels for resurfacing existing walls to ⅝-inch panels for heavy-duty applications. There are also special board treatments, for example, a fire-code board that makes sense in utility rooms, and a moisture-resistant board for kitchens and baths.

Material	Quantity
Drywall panels	For 4x8 panels: divide room perimeter by 4, deduct about ⅓ sheet for a door, and ¼ sheet for a window
Joint compound	About 1 gal. per 100 sq. ft. of drywall
Joint tape	About 400 ft. per 500 sq. ft. of drywall
Nails or screws	About 1 fastener per sq. ft.

Installing Drywall

USE: ▶ measuring tape • T-square • utility knife • pencil • drywall hammer or power drill/driver • caulking gun • panel lifter • eye protection ▶ drywall panels

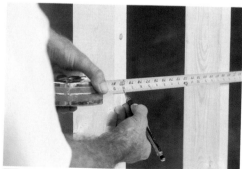

1 Before cutting panels, measure along the framed wall to be sure that the drywall edges (every 48 in.) will fall on a stud.

2 To cut drywall, score the surface with a utility knife, snap-break the panel along the cut, and slice through the paper backing.

3 Ceiling panels are difficult to install without a helper; if you're working solo, use a deadman (p. 359) to hold up the panel.

364 WALLS & CEILINGS / DRYWALL

DRYWALL

Hammers vs. Screwdrivers

It's convenient to nail on drywall, but you'll get more holding power and fewer repairs by using pointed, sharply threaded drywall screws, especially on ceilings. The traditional nailing tool (used by some but not all pros) has a wide head on one end and a cutting hatchet on the other. DIYers generally do better with a basic clawhammer. Power drill/drivers need a special screw-setting head.

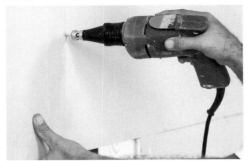

Practical power drill/drivers have a belt clip, a screw-holding head, and a torque clutch to disengage when the screw seats.

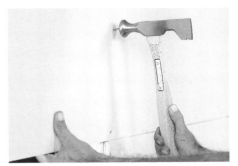

The hatchet end of a traditional drywall hammer is rarely used for cutting. DIYers are better off (and safer) with a utility knife.

Cutting for Outlets

USE: ▶ measuring tape • drywall saw • utility knife • pencil ▶ drywall panel • outlet box

1 Always mark box locations on the floor to avoid burying them accidentally. Measure from side wall and floor to mark the panels.

2 Transfer the measurements onto the front of the panel using a 4-ft. T-square. You can also trace a spare outlet box.

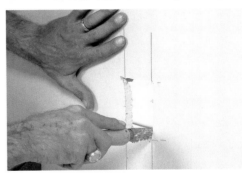

3 Use a pointed drywall saw to cut out the hole. Don't push on the panel to fit a miscut; trim the hole with a utility knife.

• construction adhesive • drywall nails or screws

4 You can apply construction adhesive to studs to make a stronger, continuous bond that is less likely to pop nails or screws.

5 Install the top wall panel first, butting it against the ceiling. If you're working alone, a pair of nails can support the panel for nailing.

6 Butt the bottom panel against the top using a panel lifter; the gap near the floor will be covered by baseboard trim.

DRYWALL BASICS

14 Walls & Ceilings

FINISHING DRYWALL

Nailing up drywall doesn't require much skill, but covering the seams to create a smooth, continuous surface does. Pros make finishing drywall look easier than it really is, but with a little practice, and by avoiding common mistakes, do-it-yourselfers can get good results, too.

Don't take shortcuts or rush the job. Taping should be done in three stages. Trying to get by with two thick coats can cause the compound to take too long to dry, shrink excessively, crack, or droop. Failure to allow enough drying can have similar effects.

Gaps at the seams between sheets should never be wider than ⅛ inch. If you try to fill a larger gap, the joint will likely fail. Drywall is cheap—take down the poorly fit piece and install a new one correctly. And don't use dehydrated joint compound or a mix that has been stored in freezing temperatures.

When smoothing out compound, the knife blade must ride smoothly on the drywall surface. When a knife nicks a nailhead or other imperfection, a ridge appears in the compound. So before applying the compound, go over the surface with a wide knife to locate any culprits. Give each nail a final, solid thwump—don't break the paper but do leave a dimple—and trim any small tears away with a utility knife.

Taping Drywall Joints

Knife blades must be clean when you start and throughout the process. One hardened speck on the blade will leave a distinctive groove, requiring extra spackling and sanding. Don't allow debris or hardened specks to ruin the mix, and don't dip a dirty knife in the bucket of compound—use a hawk or mud pan to hold a working supply. Scrape the compound off the blade before each smoothing stroke.

Any dry spots under the tape are likely to bubble under successive coats. The key is to watch for a continuous and uniform color change as you smooth the tape. Also avoid running the blade over a seam again and again. You will remove too much compound.

Imperfections not reduced by scraping or sanding will cause problems in successive coats. But don't oversand, or you'll scuff up the paper tape and drywall surface paper. That furry result persists through finishing and painting.

Finishing Panel Seams

USE: ▶ 3-, 6-, and 12-inch drywall taping knives • sanding pole • mud tray or hawk • tape dispenser

1 Use a drywall taping knife to apply a first coat of joint compound about 4 in. wide over the seam between panels.

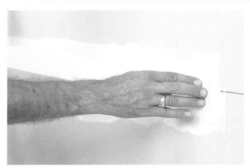

2 Embed paper drywall tape in the first coat of compound by smoothing it against the wall with the drywall knife.

Special Tools

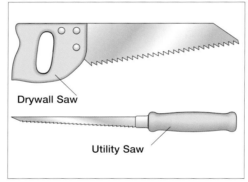

A drywall saw is good for cutting out door and window openings; use a utility saw for cutting out holes for utility boxes and pipes.

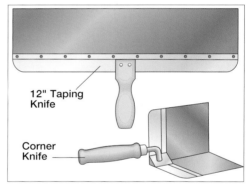

A 12-in. taping knife is used to apply wide finish coats; a two-faced corner knife works on both walls of an inside corner.

Finishing Corners

USE: ▶ drywall hammer • 6-inch drywall knife (or outside corner knife) • sanding pole • mud tray or hawk •

1 Nail on corner guard to strengthen outside corners. Use a drywall nail every 6 in., or set it in place with a corner crimper.

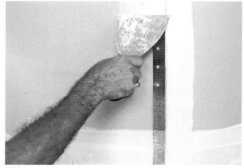

2 Apply a first coat of joint compound, and remove the excess; after drying and sanding, apply two more coats.

DRYWALL

(optional) • stepladder or scaffolding (as required) ▶ joint compound • drywall tape • sandpaper

3 *Once the coat dries, apply a second coat of compound with a wider taping knife. Sand lightly between coats where needed.*

4 *Also finish nail- and screw-heads with three coats; each one should be dry and sanded before you apply the next coat.*

5 *Sanding long seams is much easier with a sanding pole loaded with 120-grit sandpaper or a sanding screen.*

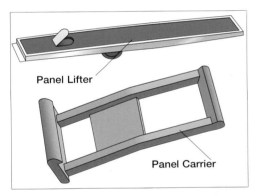

A panel lifter keeps a drywall panel a few inches off the floor; a panel carry allows one person to move a 4x8-ft. sheet.

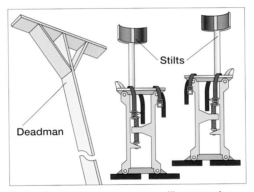

A deadman can prop up a ceiling panel when you're working solo; drywall stilts let you finish seams without moving a ladder.

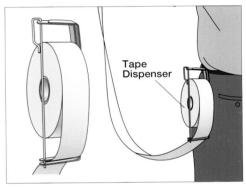

A tape dispenser hooks to your belt, giving you a ready supply of drywall tape while you work your way around a room.

tape dispenser • corner crimping tool (optional) • stepladder or scaffolding (as required) ▶ corner bead • joint compound • drywall tape • sandpaper

3 *Finish inside corners with paper tape. Make a crease down the center, and set it into the corner over embedding compound.*

4 *Finish inside corners by applying a second coat of compound over the tape, working on one side at a time.*

5 *After a light sanding, apply a finish coat. Try to smooth the fresh compound instead of oversanding after it dries.*

FINISHING DRYWALL 367

14 Walls & Ceilings

DRYWALL REPAIRS

Drywall is tough, but it can be torn, chipped, cracked, or even punctured accidentally, especially when moving furniture around. Large holes require more time and effort to fix, but you can make minor repairs quickly with only a few simple tools. For example, a deep scrape sometimes tears the surface paper on the wallboard. The first step is to remove any loose or frayed paper facing by neatly trimming the paper to a straight edge with a utility knife. Then fill in the shallow paperless section with compound.

Chronic Cracks

Many house frames move enough seasonally to disrupt drywall joints on a regular basis. If standard taping hasn't held in the past, try another approach. Instead of using fiberglass mesh tape (popular with do-it-yourselfers because it is easier to apply) and conventional all-purpose joint compound, use paper reinforcing tape and setting-type joint compound (powder). Although it's tricky to mix correctly and harder to work with because it dries so fast, you will get a harder and more durable bond, which reduces the chance of the crack reoccurring.

Start by removing any reinforcing tape and scraping out the crack. Fill the crack, and embed the tape with setting compound. Then, if you want, use ready-mix for the next two coats. If cracks like these persist at the ceiling-wall joint, consider installing crown molding.

Fixing Small Holes

USE: ▶ taping knife • sandpaper ▶ patching material • joint compound

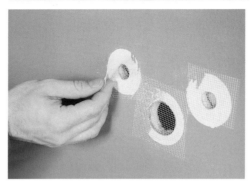

1 You can plug small openings with screening, or apply surface mesh to support and reinforce repair compound.

1 Over larger holes, you can use a self-sticking patch kit that has a reinforcing panel. Just peel off the backing and apply.

MONEY SAVER

2 Instead of falling into the hole, joint compound embeds in the mesh. Multiple coats are needed to create a smooth surface.

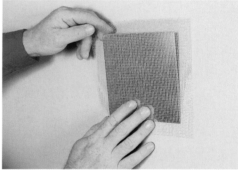

2 Once the mesh is fixed to the wall, you can spread joint compound over the entire patch; then sand, prime, and paint.

Fixing Large Holes

USE: ▶ drywall saw or utility knife • caulking gun • power drill/driver or hammer • 6-inch taping knife • sanding sponge ▶ drywall patch • 1x3 scraps or furring

1 Cut out the damaged area, leaving a clean-edged rectangular shape. Cut 1x3 braces to fasten inside the new cutout.

2 Set the braces with construction adhesive and drywall screws. Hold or clamp the brace as you drive the screws.

3 Apply construction adhesive to the side braces before setting the patch. Add top and bottom braces on larger holes.

DRYWALL

Fixing Corners

USE: ▶ utility knife • hacksaw • floor scraper • metal snips • power drill/driver • drywall knives ▶ corner bead • drywall screws • joint compound • sandpaper

MONEY SAVER

1 To fix a broken outside corner, first make a rectangular cutout with a utility knife around the damage to prevent tearing.

2 Use a hacksaw to cut through metal corner guard above and below the damage. You also can use metal shears.

3 Use a pry bar and the claw end of a hammer to pull the nails that are driven through the corner guard into the drywall.

4 Use metal shears to cut a replacement piece of corner guard. Clear away torn paper and gypsum in the replacement area.

5 Screw the new piece of guard in place on the corner. Make sure that the new piece aligns with the old guard.

6 Use a taping knife to apply drywall joint compound. Install three coats, allowing each to dry, and sanding lightly in between.

MONEY SAVER

• drywall nails or screws • joint compound • construction adhesive

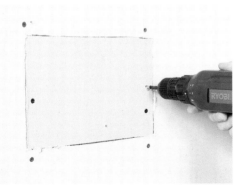

4 Place the patch piece on the braces, move it back and forth to set in the adhesive, and secure it with drywall screws.

5 Finish the seams of the patch with drywall tape (either paper or fiberglass) and three coats of joint compound.

6 Finish-sand the final coat with a small-celled sanding sponge. Prime the fresh compound before repainting the wall.

DRYWALL REPAIRS

14 Walls & Ceilings

PAINT BASICS

Most paints are either water-based or alkyd-based. (Synthetic alkyd resins have replaced oils.) Water-based paints, which include latex, vinyl, and acrylic paints, are the easiest to work with—they cover well, dry quickly, and tools can be cleaned with soap and water. Alkyd paint is more expensive, takes longer to dry, and must be cleaned with paint thinner; it is generally tougher than water-based paint and stands up better to scrubbing. Both paints come in various sheens from flat to high-gloss.

A paint mixer attaches to a power drill.

Safety Gear

Always wear a respirator when you paint without adequate ventilation. Use a tight-fitting spray-paint mask or a cartridge-style respirator with replaceable filters. Disposable paper dust masks aren't sufficient to keep you from inhaling paint vapors, which can make you sick. You should also wear a respirator whenever you spray paint from a can or paint sprayer. It's a good idea to wear goggles, as well, to protect your eyes. Also protect yourself by checking product label cautions. Some sealers and cleaners can produce vapors that can be ignited—for example, by a stove pilot light.

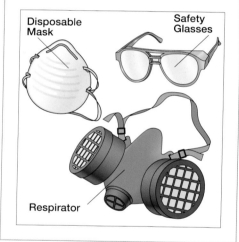

GREEN SOLUTION

"GREEN" PAINTS

Most paints contain volatile organic compounds (VOCs), which are organic chemicals that vaporize, or become a gas, at room temperature. While everyone's sensitivity differs, breathing excessive amounts of VOCs can lead to a number of medical problems, including eye, ear, and nose irritation; nausea; and liver, kidney and central-nervous system damage.

Avoid problems by using zero-VOC or low-VOC paints. Zero-VOC products are usually made from botanicals, milk, and essential oils. All of the major paint manufacturers have reduced the levels of VOCs in their products. The Environmental Protection Agency standards for low-VOC paints specify that paints must contain less than 250 grams per liter (g/L) for flat paints and less than 380 g/L for nonflat paints.

For truly "green" paints, though, look for products that comply with more stringent requirements, such as those endorsed by Green Seal and the Master Painters Institute, two organizations that have created standards for paints and coatings. The low-VOC paints that emit the lowest levels of VOCs contain less than or equal to 50 g/L.

Paint Choices

PAINT	APPLICATIONS	PROS/CONS
Latex primer	New plaster or drywall, uncoated wallpaper, finished wood, new brick	Easy clean-up, quick-drying, almost odor-free; doesn't perform well on unfinished wood
Alkyd primer	New plaster or drywall, finished or unfinished wood, any new masonry	Best primer for wood, good for all paints; doesn't perform well on drywall, needs solvents for cleanup
Primer-sealer	Unfinished wood, mildew stains	Quick-drying, good for bleeding knots; needs alcohol for cleanup
Latex paint	Plaster/drywall, primed wood, vinyl trim, steel, aluminum, cast iron	Easy clean-up, quick-drying, inexpensive; not as strong as alkyd, needs primer over wood, adheres poorly to gloss finishes
Alkyd paint	Plaster/drywall, unprimed wood, vinyl trim, steel, aluminum, cast iron	More durable than latex, adheres to all types of paints; slow-drying, cleans up with solvents, needs primer for drywall and plaster

PAINTING

Paint Application Options

Once you pick out your paint, you need to select applicators best suited to the job. Use a synthetic brush (polyester or nylon) for all latex-based paints. Natural fibers (bristle) will absorb the water in latex paint, making the brush heavy and bushy. Use a polyester or natural bristle brush with alkyd paint. A 2½-inch-wide trim brush with a long handle is ideal for most painting projects. The best brushes are flagged and tipped, two extra manufacturing steps that let the brush hold more paint and spread it more evenly.

A paintbrush that will last through many projects has a hardwood handle, a tight metal ferrule at the neck, and thick bristles.

Some DIYers like paint pads. They are not flexible the way a brush is, but can work along trim and in corners.

Rollers & Sleeves

Rollers make short work of large, flat surfaces—and you can attach a long handle to reach ceilings and high walls. A 9-inch-wide roller is standard, but smaller and larger rollers are made for special applications. Roller sleeves (the replaceable part) come in a variety of naps, from very short for smooth surfaces to long for rough surfaces. Generally, it's best to use a ⅜-inch-long nap for walls and ceilings. Longer naps produce a heavier paint stipple; shorter naps produce less surface pattern.

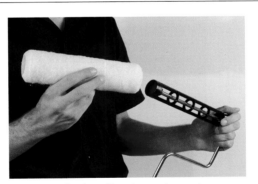

Use a synthetic roller sleeve for water-based paints and a wool, wool/nylon, or mohair cover for alkyd-based paints.

Roller pans should be sturdy to avoid spills and deep enough so you can almost submerge the sleeve.

Painters & Sprayers

Power painters and sprayers get a job done quickly. Power painters allow you to keep rolling without reloading your roller. Sprayers are good for hard-to-brush surfaces and create a very smooth finish. The trick to spraying is to keep the tool moving whenever paint is flowing. You also need to keep the tool parallel to the surface you're painting, even as you swing back and forth across the surface.

A power sprayer uses compressed air to spray a fine mist of paint. You can rent the gun and an electric-powered air compressor.

Power painters pump paint straight through to the applicator under pressure. A trigger allows you to limit the flow.

14 Walls & Ceilings

BEFORE YOU PAINT

Painting would be much easier if you could find a contractor to do the prep work, and you handled the painting yourself. Yet you probably wouldn't save much money, even if you could find someone willing to do it. Preparation—the tedious and time-consuming scraping, patching, and sanding—is the worst part of the job, but the most important for a good-looking and long-lasting finish. Poor surface preparation is the number-one cause of paint failures.

Cleaning

Take time to dust and vacuum all surfaces first. Then wash the walls with soap and water, especially in kitchens and bathrooms. If there are little gray-green dots of mold, add up to a quart of bleach to 3 quarts of an ammonia-free detergent solution. Protect nearby floors with newspapers and drop cloths. Allow the bleach solution to remain on the wall for 15 minutes, and then rinse thoroughly.

Smoothing Surfaces

Paint won't hide imperfections—in fact, it accentuates them, making any previously unnoticed bumps or dents stand out. To avoid surprises, shine a bright light at a low angle across the surface to spot problem areas before you begin to paint. Apply joint compound to fill in any depressions, and sand off bumps, ridges, or other raised imperfections.

Priming

Follow the paint manufacturer's recommendations for priming. Some surfaces, such as unpainted wood, require primer with certain paints but not others. Failure to use a primer will reduce paint adhesion and lead to flaking and peeling paint. Previously painted walls and ceilings generally do not require a primer unless you plan a radical color change (such as white over red) or you have stains to cover. But if you've patched with joint compound, the unpainted compound will absorb paint differently than surrounding painted areas, leaving a blotchy finish. Although you can usually apply two topcoats, it's better to use a less-expensive primer, perhaps tinted to a similar color as the top coat, covered by one top coat.

Preparing Walls

USE: ▶ paint scraper • utility knife • taping knives • palm sander • paintbrushes • eye protection

1 Scrape any cracked or flaking areas with a paint scraper until you reach paint that is solidly fixed on the wall.

2 Where the paper surface of drywall is torn, trim the tear free using a utility knife, and apply joint compound over the damage.

Sealing over Stains

Remove surface stains as best as you can with a detergent/water solution or an appropriate solvent/spot remover. To prevent remaining discoloration from bleeding through, seal the area with a stain-killing primer, such as pigmented white shellac. These sealers dry fast and won't slow you down. Similarly, coat knots in paneling or trim so that resins won't bleed through.

Pigmented white shellac has tremendous hiding power. It also makes a good primer on metal.

Painting Walls

USE: ▶ angled sash brush • wide paintbrush or paint roller and tray • extension pole for paint roller

1 Use a trim brush to cut in a 2- to 3-in. strip of paint around all trim and in areas you can't easily reach with a roller.

2 Dunk the paint roller at the deep end of the pan; then roll it over the ribs at the shallow end to distribute the paint evenly.

PAINTING

▶ joint compound • 120-grit paper for palm sander • primer

3 Use a wide blade to fill large bare patches with joint compound. Apply several thin coats instead of one thick layer.

4 Use a power palm sander or a sanding block to smooth the patches. Oversanding can scuff the drywall surface.

5 Prime all repaired areas. This prevents the dry joint compound from sucking water out of the paint, which creates dull spots.

Repainting Trim

To help new paint adhere to old trim, lightly sand glossy surfaces. Also sand away small imperfections, feathering to areas of sound paint so that you don't leave ridges. If need be, take old trim down to raw wood by stripping, sanding, or heating and scraping. If paint was applied before the late 1970s, test for lead before removing the finish. Contact the National Lead Information Center www.epa.gov/lead.

Use a heat gun to gradually soften layers of paint. Work on one small area at a time, and keep the gun nozzle moving.

Use a stiff scraper to remove layers of paint heated by the gun. Use a razor-edge scraper to clear grooves in molding.

▶ paint • drop cloths • cleanup rags

3 One basic and economical approach to painting is the load-on method. Start by applying thick layers in vertical stripes.

4 Roll left and right on the diagonals across the thick stripes. This spreads the paint evenly on the wall.

5 When the original stripes have been spread over a section of wall, roll straight up and down to create a uniform stipple finish.

BEFORE YOU PAINT

14 Walls & Ceilings

PAINT PROBLEMS

Sometimes you think you have done everything right, down to following all the directions on the paint can, and yet the job still goes wrong. The fresh paint looks nice for a while, but then the trouble starts—flaking, cracking, and other problems appear from a cause or causes unknown. Some problems become evident almost immediately. Wrinkling or sagging, for example, can occur shortly after the paint is applied but before it dries, indicating that the paint was applied too thickly. Apply two thin coats instead of just the thick one to avoid the problem.

Peeling & Blistering

Peeling is a problem more often associated with exterior painting. If peeling occurs indoors, especially in an older home, and you have not painted over a glossy surface, the culprit is often calcimine paint, an old-fashioned mix with a low binder content. If you live in an older home with the original paint in place, wash surfaces with hot water and scrub off the calcimine residue.

Blistering, a problem that is similar to peeling, can occur if a latex paint is exposed to very high humidity or moisture before the paint has dried sufficiently. Showering after you paint is fine, but not if you've just painted the bathroom.

First developed in the 1930s as an underlayment for plaster, gypsum drywall soon made plaster obsolete—it went up faster and easier, required no lath, and presented a flat surface without painstaking finish work.

Dull/Brown Spots

Dull spots are often caused by a failure to prime patched areas. Both joint compound and unfinished drywall absorb paint differently than the surrounding painted areas. Applying too much compound can cause dull spots in these areas even if you have primed them.

Brown spots on a painted surface—sometimes appearing either glossy, soapy, or sticky—occur when ingredients in latex paint leach out to the surface. The culprit is excessive moisture, and brown spots are most likely to form on bathroom ceilings, especially right over the shower. The spots will wash off with soap and water but may occur once or twice again before disappearing.

Mold & Mildew

Mold and mildew spots often occur in bathrooms, but you'll also find them in other rooms by windows where ceilings join exterior walls. Inadequate insulation makes these interior surfaces cold, causing condensation. Mold feeds on the moisture and eventually forms gray-green spots that bleed through paint. Soap or detergent solution won't kill mold. Use a solution of household bleach and water with a non-ammonia detergent. (Never mix ammonia and bleach.) Allow it to soak into the mold for 15 minutes before scrubbing and rinsing.

To clean up severe mold damage such as this, you may need several applications of full-strength bleach.

Special Finishes

You can transform the look of a room with one of many decorative painting techniques. While some require considerable skill and artistry, others (such as those shown at the right) are easily mastered. Before you tackle a real wall, it's wise to experi-ment with these finishes on a sample piece of drywall. These finishes have become increasingly popular, in part because there is no single correct way to do them. Feel free to try different applicators and mixing ratios.

Sponging paint on a wall with a natural sea sponge creates a dappled finish, either subtle or bold, depending on the colors.

Ragging involves applying a layer (or layers) of paint over a surface with a bunched-up rag, creating a textured look.

PAINTING

Spot-Painting Patches

USE: ▶ paint scraper • paintbrush • paint roller and tray ▶ primer • paint • sandpaper & sanding block

To make spot repairs blend in without repainting the entire wall, start with a careful sanding to eliminate any ridges of dried joint compound. Then prime the patched area so the top coat will be evenly absorbed over the patch and surrounding surface. Use a fairly dry roller so the painted area won't dry to a well-defined edge. Also select a sleeve that will produce a finish closely matched to the existing wall stipple. Then coat the patch area with paint, and begin to feather the painted edge. Do this by applying normal pressure as you cross the patch, and lighter pressure, lifting the roller, as you coat the nearby wall.

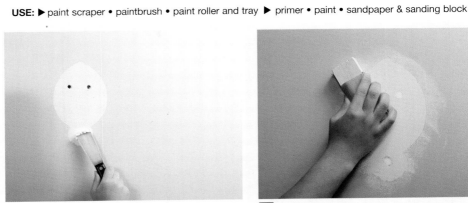

1 Fill nailholes and small dents with lightweight joint compound. Leave a slight mound—the compound shrinks as it dries.

2 Use 150-grit sandpaper to sand down the dry joint compound until it is flush with the wall surface.

3 Because joint compound absorbs paint differently than drywall, prime the patched area before topcoating.

4 Roll paint onto the patch, and work the topcoat across the area. Keep a wet edge on the paint to avoid leaving ridges.

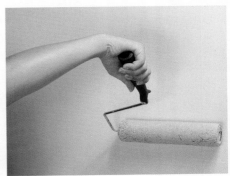

5 Feather out the edges of the new paint with light, lifting strokes. To cover, you may need to repeat this process.

Combing paint (also called dragging) involves making narrow lines in a painted surface with a paint comb or other tool.

Stippling involves bouncing a special brush straight up and down, creating dots that blend when viewed from a distance.

Spattering is a technique of adding a partial topcoat of a contrasting color or glazing agent. Just tap the brush on the stick.

PAINT PROBLEMS

14 Walls & Ceilings

WALLPAPERING TIPS

Wallpaper comes in two basic types, those that are prepasted and those that need pasting. DIYers should use prepasted wallcoverings because the application is straightforward: you dunk it and hang it. Planning, cutting, and measuring is the same for both types, but the alternative to prepasted paper—mixing and brushing on the adhesive—increases the chance for errors, including premature drying, lumps under the surface, and air pockets that are difficult to fix.

Pick an unobtrusive starting point. You can make minor corrections where paper turns a corner, but it's nice to have a finish line that's separated from the starting line—such as a door in the corner, preferably one that's normally left open. On a highly visible wall, misaligned patterns would be glaring, but it would take an eagle eye to spot it behind a door.

After soaking, prepasted papers should be set aside for the time specified by the manufacturer. Keep it consistent from sheet to sheet because the paper may expand or shrink a little while it absorbs the adhesive. If one strip is hung right away next to a sheet that had a long soak, you could get enough of a mismatch to notice. Overworking the paper and stretching it into position can cause the same problem. Instead of tugging on it, which can stretch some materials by ¼ to ½ inch, shift the sheet by pushing on it with your hands.

Making Spot Repairs — MONEY SAVER

USE: ▶ utility knife • square • wallpaper syringe • seam roller ▶ patch • adhesive • masking tape

1 To patch a damaged area, align a patch piece over the pattern, tape it down, and cut through both layers.

2 Remove the patch and the damaged piece from the wall; the cutout part of the patch should fit exactly in the hole.

1 To fix an air bubble, first make a small puncture hole in the bubble with a razor knife.

2 Use a special syringe to inject adhesive so that the bubble can be rolled flat with a seam roller.

Stripping Old Wallpaper

USE: ▶ wallpaper scarifier • paint roller and tray • paint scraper or drywall taping knife ▶ wallpaper remover

1 Use a wallpaper scarifying tool to score the surface of the paper. This will allow the remover to soak in thoroughly.

2 You can steam off paper, or apply a chemical stripper. It works through the surface and loosens the glue underneath.

3 Use a drywall knife to scrape away pieces of the wallpaper. You may need multiple applications on stubborn spots.

WALLPAPER

Hanging Wallpaper

USE: ▶ 4-ft. level • chalk-line box • tray • utility knife • scissors • seam roller • paste brush (if needed) ▶ wallpaper • adhesive (if needed)

1 Use a level or a chalk-line box to mark lines where the seams will fall. It pays to plan the layout of strips before you paste.

2 Dunk prepasted wallpaper in a pan of lukewarm water; unpasted papers need to have adhesive spread on them with a roller.

3 Fold the soaked roll onto itself—a process called booking—to make it easy to carry and place the strip on the wall.

4 Unbook the paper, and position it on the wall. Align it at the ceiling line and close to your plumb guideline.

5 Unroll the strip, and press it in place by hand. Allow enough material to turn the corner, which may not be plumb.

6 As you hang more strips, smooth them with a brush, starting at the top corner near the guideline and moving down and across.

7 Trim away excess paper, called the allowance, at the top and bottom with a guide (a drywall knife) and a utility knife.

8 You can raise the seam to make small adjustments and match the pattern. You can butt seams or overlap them.

9 You can overlap seams by hand-cutting through both strips for a perfect match. Finish edges by using a wallpaper roller.

14 Walls & Ceilings

WALL-TILE BASICS

Ceramic tile is not only attractive, it's a practical solution for bathrooms and kitchens because it's waterproof, durable, and easy to clean. It also works well on floors. (For applying tile to floors, see "Floors & Stairs," p. 90.)

Tile is made from clay that has been fired. Glazed tile, available in matte or shiny finish, has a hard surface that is impervious to stains, but it can be scratched; it is the standard tile around sinks and tubs. Unglazed tile, made only in matte finish, picks up stains from grease and oil but resists scratching; it is often the choice for floors.

Flat tiles are called field tiles; those with finished edges or shaped to fit around corners are trim tiles. Tiles larger than 4x4 inches are sold loose; smaller tiles can be purchased in sheet form, with a few square feet of tile bonded to a thin webbing on the back.

Hanging Backer Board for Tile

The type of material required as backer board for tile depends on the tile's location and the type of adhesive you will use. If you're using organic mastic adhesive and the area will not get consistently wet, standard (or water-resistant) drywall panels are adequate. However, if the area will be subject to regular soakings (such as a shower stall or kitchen-sink area), you should use a cement-based backer board. This material also works well in areas that call for a heavy underlayment, such as a tiled half-wall around a woodstove, where the surface must be fire-resistant.

Cement backer board, which is made from portland cement and fiberglass mesh, comes in ½- and ⅝-inch thicknesses. You can purchase 4x8-foot stock, but the panels typically measure 32 or 36 inches wide by 5 or 8 feet long. Fasteners must be spaced more closely than with ordinary drywall—8 inches apart along the studs. You'll also need special fasteners for cement panels: use 1½-inch galvanized roofing nails or 1¼-inch galvanized-steel screws. Regular drywall nails or screws will rust through. Cement-panel joints must be taped as well: use tile-setting mortar or tile adhesive instead of joint compound. Tape the joints with a fiberglass mesh tape that is specially designed for backer board.

Patterns

PAINTED BORDERS

CUT SQUARES

RECTANGLES

MOSAICS

Thinset

In baths and kitchens, as well as other high-moisture areas, you can spread a thinset bed over cement board.

You can also spread a thinset bed over drywall. Use a towel with a notch depth specified by the tile manufacturer.

Grouts

You can grout seams when the tile has set up. (Set-up time depends on the set method and adhesive used.) You can choose a grout based on the type of tile, expected wear and tear, and color. Many manufacturers offer dozens of color choices. Use a shade close to the tile to minimize the grid effect of seams and hide irregular spacing between custom tiles. Generally, lighter grouts highlight the grid, and darker grouts highlight the tile—a good choice on layouts where accent tiles form a pattern.

If you add colorant to a grout mix, set aside a sample, and let it dry to get a true reading of the final color.

WALL TILE

Installing Tile

USE: ▶ T-square • drill/driver • taping knife • notched trowel • sponge-faced float • tile cutter • tile nippers • sponge ▶ tile • adhesive • grout • fiberglass tape

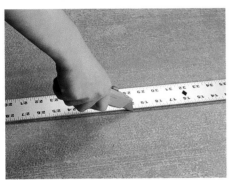

1 Cut cement backer board the same way you cut drywall: score it with a utility knife, and snap it along the score line.

2 Backer boards can be nailed up with galvanized roofing nails or screwed in place for more stability with galvanized screws.

3 Finish the backer-board seams with fiberglass mesh tape and a coat of tile mortar or adhesive.

4 Mark plumb and level lines to begin the layout. Your starting point should leave equal-sized partial tiles in corners.

5 Apply a coat of thinset adhesive to a section of the wall with a notched trowel. Don't cover your layout lines.

6 Press each tile firmly in place, using a slight twisting motion to bed it in the adhesive. Leave equal spaces for grout.

7 Cut tiles as needed for corners and the top of the wall. Use tile nippers to make irregular cuts (such as for pipes).

8 After the adhesive cures, spread grout using a sponge-faced float or squeegee. Sweep at angles across the tile seams.

9 After firming the joints, wipe off the excess using a damp sponge. Allow a dry haze to form; then polish with a cloth.

WALL-TILE BASICS

14 Walls & Ceilings

CLEANING TILE

For day-to-day cleaning, simply wipe the tile down with warm water and a sponge. A baking soda and water solution or a mild solution of white vinegar and water remove light build-ups of dirt, grease, and soap scum. For more stubborn cleanups, use a strong solution of all-purpose cleaner or a commercial tile cleaner. Rinse thoroughly with clean water.

Ceramic tiles with metallic glazes should be cleaned and polished with a metal polish. If you need to scour the surface, use a woven-plastic pot scrubber rather than steel wool, which can leave black marks and metal hairs that cause rust stains in grout joints.

If tile cleaner does not seem to be able to remove a stain, try one of the following cleaning suggestions. For tar, asphalt, and oil stains, use lighter fluid followed by household cleaner. For ink, coffee, tea, blood, or dyes, use a 3-percent hydrogen peroxide solution or full-strength household bleach. For liquid medicine and shellac, use denatured alcohol. For rust, use a commercial rust remover followed by a household cleaner. For nail polish, use nail polish remover. For chewing gum, chill the gum with an ice cube, and then peel it off the tile surface.

Cutting Tile

A tile cutter or diamond-blade wet saw will make straight cuts in tile; special drills are required to make a hole in the middle of a tile. For the curved or irregular cuts needed to fit tiles around supply pipes, sink cutouts, or other contours, use tile nippers. As the name implies, these tools nip away tiny bits of tile. Working with nippers takes a strong wrist and plenty of patience. To use them, hold the tile glazed side up and take small, ⅛-inch bites to break off tiny pieces. If you take too large a bite, the tile is likely to break.

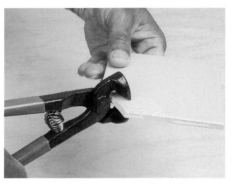

Tile nippers have sharp jaws to crack through a glazed finish. By taking small bites, you can form irregular shapes.

MONEY SAVER

GROUT REPAIRS

Water can puddle on tile all day without leaking, but even a hairline crack in grouted seams lets water seep through. If you catch cracks in time, you can scrape out existing joints (a can opener works well) and regrout, which will save you the cost of retiling and prevent damage to the tile substrate. You may have to experiment with grout samples to get a good color match—aging affects the color—even if you use a leftover supply of the original material.

If adjacent tiles are loose, new grout will not hold them in place. To repair, remove the tile and tile adhesive, and reset the tile.

Replacing Tile

USE: ▶ cold chisel or glass cutter • hammer • putty knife • rubber gloves ▶ tile • adhesive • grout

1 Remove a damaged tile by scraping out the grout, scoring the tile with a chisel or glass cutter, and breaking it out.

MONEY SAVER

2 After scraping off the old adhesive, spread new adhesive on the replacement tile and in the empty space.

3 Press the new tile tightly in place. You can use masking tape to keep it centered while the adhesive sets.

4 When the adhesive is dry, regrout the tile, tool the joints, and wipe away the excess. You can seal grout with silicone.

WALL TILE

Restoring Bath Tile

USE: ▶ razor knife • bucket • sponge-faced float • sponge • screwdriver • rubber gloves ▶ grout • sealer • caulk

1 Razor out old grout with a utility knife or a grout saw; for narrow joints, use an awl or a nail driven into a dowel as a tool.

2 Old caulk in the corners and at the tub seams must also be cut out and removed with a razor knife.

3 Mix a batch of new grout according to the manufacturer's instructions. You may want to seal the tiles first to prevent staining.

4 Apply the grout with a sponge-faced float or a squeegee. Do not spread grout into the corners at the tub-tile seam.

5 After tooling the grout joints, wipe off the grout with a damp sponge. Allow a dry haze to form; then polish with a damp cloth.

6 Use silicone caulk to seal the seam at the tub, the joints at the corners, and wherever the tile meets a different material.

7 Tool the caulked joints with your fingertip. Allow grout and caulk to cure before using the shower or bathtub.

8 To recaulk the escutcheon, first remove the faucet handle by unscrewing the setscrew in the middle of the handle.

9 Spread a bead of caulk around the rim of the escutcheon plate, set it back in place, and reattach the faucet handle.

CLEANING TILE

14 Walls & Ceilings

WALLS OF GLASS

Building walls with blocks of glass may sound like a dumb idea—something like making a child's building set out of bone china. But modern glass blocks are rugged enough to make large exterior walls that let in light and keep out the weather. Interior partitions or half-walls of glass block can create rooms within rooms, defining spaces in a house without isolating them from one another. Glass block is also a popular material for custom shower enclosures and partitions in bathrooms, because it distorts images enough to ensure privacy.

Installation

Glass block is traditionally installed with solid mortar joints like bricks or concrete block. However, good glass block work is a specialized skill, more so than laying concrete block or brick. It is normally a job for a mason; make sure before hiring a contractor that he or she has experience laying glass block.

Some blocks are designed to make installation easier for DIYers. Glass-block kits with clear plastic spacers are stacked and sealed with silicone caulk instead of mortar. Other types include glass blocks that snap into a metal framework and glass-like blocks made of plastic that clip together with interlocking flanges. And if you don't want an entire wall of glass, you can buy prefabricated glass block windows that install in a rough opening the way a regular window does.

Glass Blocks

Glass blocks offer a lot of design options, including finishes that control the amount of light transmitted and different surface textures that blur or completely distort images for privacy. The blocks are made by fusing two sections together to create an airspace that gives glass block an R-value of about 2.0, which is about the same as an insulated glass window and better than the thermal value of a 12-inch-thick concrete wall. Locating glass block on a south-facing wall adds greatly to overall energy efficiency, thanks to the solar energy it transmits during the day.

OPAQUE FINISH

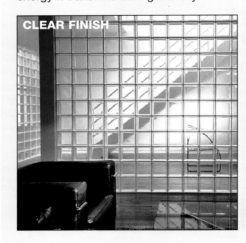
CLEAR FINISH

ICE FINISH

Installing Glass Blocks

USE: ▶ power drill/driver • framing square • measuring tape • utility knife • caulking gun ▶ glass blocks • metal channel • gaskets • caulk

1 Kits are available that make installations easier. A structural channel screwed to the framing conceals the glass blocks' edges.

2 Your framing dimensions need to account for both the width of the glass blocks and the caulked joints.

3 Set the blocks into the channels, and use a sample piece of the flexible gasket material to check your spacing.

GLASS & MIRRORS

Mounting Mirrors

Most mirrors are ¼ inch thick and therefore quite heavy. Handling large mirrors safely requires special equipment such as glass suction cups. Whenever you mount a mirror larger than 3 to 4 square feet, use adhesive in addition to any mechanical supports. The best supports are J-clips or J-channels, which also hold the mirror in place until the adhesive cures. The channels also protect edges from damage and prevent water from getting behind the mirror, where it can ruin the silvering on the back. Always use mirror mastic. Other adhesives, such as those containing silicone, may react with the silvering. If you are setting a mirror on a backsplash or similar hard edge and not using a J-channel, place rubber setting blocks under the mirror edge about one-quarter of the distance in from each end. Mirrors that are mounted without frames should have finished edges.

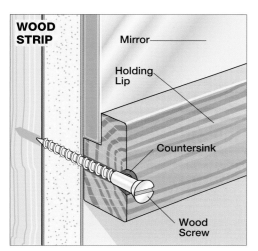

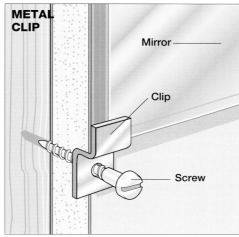

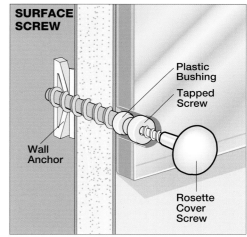

4 When you complete one full course, set a full length of the gasket. It is molded to fit into ridges along the blocks.

5 The gaskets are set in all horizontal and vertical seams. Exposed seams that are normally grouted are caulked instead.

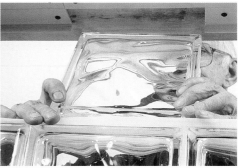

6 Leave out one block-sized section of channel on the top course. Insert the last blocks here, and slide them into position.

WALLS OF GLASS

14 Walls & Ceilings

CONTROLLING NOISE

Drywall, insulation, and even air spaces inside walls and ceilings all help dampen sounds that pass from one room to another. Sometimes just adding a rug or stuffed furniture will make a room quieter, but in houses built with thin walls (and no insulation in interior walls), drastic measures may be needed.

An easy first step is to use acoustic sealant or caulk to seal all joints and openings between the noisy area (say, the basement) and the rooms you want to keep quiet. If the framing is exposed on one side of the wall or ceiling (or if you are willing to remove the drywall), install sound attenuation blankets in the wall cavities.

When you install drywall, use ⅝-inch-thick sheets instead of the usual ½-inch. And don't secure it directly to the framing; any vibrations that strike one side of the wall, floor, or ceiling will be transmitted through the framing to the other side. Instead, attach resilient metal channels to the framing, and then attach the drywall to the channels. Depending on the degree of noise and the results you are trying to achieve, you may want to install a sound-deadening board on top of the drywall, and cover it with another layer of drywall.

Instead of the second layer of drywall on the ceiling of a noisy room, try installing an acoustical ceiling below the drywall. Small tiles can be glued on directly, stapled to wood furring strips, or snapped into metal channels; larger panels can be suspended in a metal grid.

Sound Transmission

Sound is carried from room to room by impact, through openings (such as heating ducts), and by vibrations transmitted through the structure. The intensity of a sound is measured in decibels (dB). A jump of 10 dB represents a doubling of perceived sound. The ability of materials to absorb sound is measured by the noise-reduction coefficient (NRC), which is an average of sound reduction for the material across a range of frequencies; the higher the number, the more sound it absorbs. Construction techniques and materials both effect sound levels.

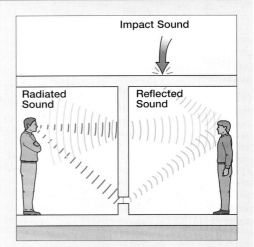

Sound level	Intensity
Threshold of pain	120–130 dB
Train, circular saw	100–110 dB
Factory, traffic noise	80–90 dB
Face-to-face conversation	60–70 dB
Average office	50 dB
Quiet radio	40 dB
Quiet conversation	20–30 dB
Soundproof room	10 dB
Threshold of audibility	0 dB

Material	NRC
Carpet on foam pad	0.55
Carpet on slab	0.29
½" drywall	0.16
Plywood-paneled wall	0.15
Plaster wall	0.09
Bare wood floor	0.09
Painted concrete block wall	0.07
Bare brick wall	0.04
Bare concrete floor	0.02

Making Sound-Absorbing Walls

USE: ▶ hammer • power drill/driver • caulking gun ▶ 2x6 and 2x4 lumber • blanket insulation • drywall screws • construction adhesive • metal hat channel (opt.)

1 To hold extra insulation and two rows of studs, build the partition with a 2x6 plate nailed to the subfloor with 16d nails.

2 Set one row of 2x4 studs in a conventional 16-in.-on-center layout, flush with one side of the partition wall.

3 Nail another row of standard studs between the others, but nail them flush with the other side of the partition wall.

SOUNDPROOFING

Sound-Resistant Construction Options

You can take several steps during construction to reduce noise transmission. It's important to insulate partition walls, as well as ducts and pipes—this reduces the noises of moving air and water, and saves energy. Wall design is also important. Different configurations rate different sound transmission classes. (See the different STCs at right.) Thicker walls with separated layers break the transmission of sounds and reduce noise room to room.

Reduce plumbing sounds from drain lines in framing cavities by wrapping the pipes with batts of insulation.

SOUNDPROOFING WALLS

STC 30
⅝" DRYWALL ON BOTH SIDES OF STUDS SET AT 16" ON CENTER
STC 35
TWO LAYERS ⅝" DRYWALL ON STUDS SET AT 24" ON CENTER
STC 43
⅝" DRYWALL, ONE SIDE ON METAL HAT CHANNEL
STC 46
⅝" DRYWALL ON STAGGERED STUDS 12" ON CENTER OF 2x6 BOTTOM PLATE
STC 50
⅝" DRYWALL, ONE SIDE ON HAT CHANNEL, WITH 3½" INSULATION BETWEEN STUDS

Batts of insulation also reduce sounds from rushing air in ducts and reduce temperature loss through duct walls.

Reduce sound transmission through partition walls by caulking small gaps, even when they will be covered by trim.

4 Weave blankets of insulation through the two rows of studs to reduce sound transmission through the wall surfaces.

5 Install a first layer of drywall vertically over the studs, using nails or screws. You can leave the seams unfinished.

6 Install a second layer of drywall horizontally, and finish the seams. This layer can also be installed on resilient metal channel.

14 Walls & Ceilings

FINISHING CEILINGS

Nearly anything you can put on a wall—paint, wallpaper, paneling, trim—can also be used effectively on ceilings. The work will be more difficult, though, because everything has to be done over your head. Painting, of course, is the easiest and most common way of finishing off a ceiling, but textured drywall and suspended ceilings offer some special advantages, especially in remodeling situations.

Ceiling Textures

A ceiling texture adds a low-tech, inexpensive design feature to any interior. These textures may appear difficult to achieve, but with the right equipment, they're quite simple—they can even help disguise mistakes made while installing drywall and taping seams.

Blown ceilings require equipment most people will have to rent. A popcorn-like material (polystyrene or vermiculite) is fed from a hopper into an airless spray gun attached to an air compressor. Once blown onto the ceiling, it is usually left to dry without retouching. This texturing takes some practice to apply properly. If you don't have a place to practice, hire a painting or drywall contractor to do the job for you. You can also create a textured ceiling by spraying diluted joint compound through an airless paint sprayer. However, you must add just the right proportion of water, and mastering the spraying technique takes practice.

Wood Planking

There are many places to use wood paneling aside from walls. For example, beaded, tongue-and-groove boards are commonly used on the ceilings of porches. It may seem more difficult to work on a ceiling than on a wall, but that job may go faster once you set up because there are no windows and doors in the way.

Whether you use a softwood (such as pine) or a hardwood (such as oak), you can surface-nail planks in place. It's wise to predrill at the ends of boards, even when the lumber is ¾ in. thick and you are using finishing nails. You can set the nailheads and fill the holes with putty.

One time-saving option is to drive your nails in the seams between planks where they will be concealed. Many planks have a beaded or grooved edge that creates a V-shaped seam. The idea is to set a finishing nail at an angle, placing it deeply into the edge of the leading panel, and driving it home with a nailset. As you continue, the edge of the next board in line will cover the nailhead. It also helps to use a bead of construction adhesive along the supporting wall studs or furring strips.

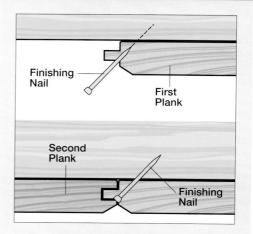

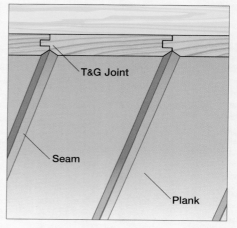

Installing a False Beam

USE: ▶ table saw • measuring tape • hammer • router • paintbrush • power drill/driver • nailset • eye protection ▶ 1x pine boards • scrap wood • nails • wood

1 You can make an elegant false beam from three pine boards. Start by cutting 45-deg. miters on the board edges.

2 Use scrap wood as a spacer, and clamp the sides of the beam in place. Then add glue and nail on the bottom edge.

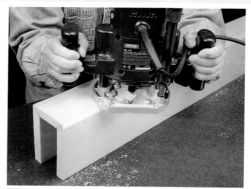

3 You can leave the beam edges as plain miter cuts, or use a router to create a decorative bead along the exposed edges.

CEILINGS

Installing a Suspended Ceiling

USE: ▶ level • hammer • utility knife • straightedge ▶ edge track (molding) • metal channel (main tees and cross tees) • hanger wires • nails • ceiling tiles

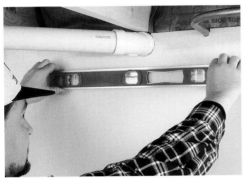

1 To conceal the pipes, electrical conduit, and joists above the basement, start by establishing a level line on the walls.

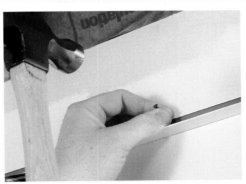

2 A typical hung-ceiling system includes edge track that captures the edges of the tiles. Nail it on your level lines.

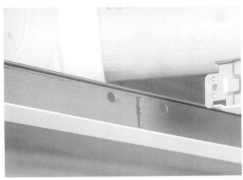

3 Lightweight, molded aluminum channels carry the tiles and conceal the edges in the main tile field over the room.

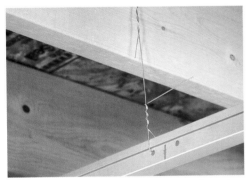

4 The channels run from wall to wall, and tie into the edge tracks. Use wire wrapped around nails to support them.

5 If you have to trim ceiling tiles, use a sharp utility knife held against a straightedge, such as a framing square.

6 With the perimeter tiles in place, finish the ceiling by inserting full tiles into the grid pattern of suspended channels.

glue • wood putty • stain • 1x2 or 1x3 nailer

4 Finish your false beam to suit the room decor. After filling nailholes and sanding, you can stain it or paint it.

5 One easy way to install a false beam (it will conceal the copper plumbing pipe) is to install a nailer on the ceiling.

6 Set the prefinished false beam around the ceiling nailer; attach it with finishing nails; and set the nails to finish.

FINISHING CEILINGS

15 Trimwork

390 BASICS
- Finish Work ◆ Biscuit Joinery
- Common Molding Profiles
- Back-Cutting Joints ◆ Coping Joints
- Decorating with Trim
- Installing Molded Trim ◆ Joinery Cuts

394 CROWN MOLDING
- Installing Crown Molding
- Attaching Crown Molding

396 BASE TRIM
- Basics ◆ Installing Three-Piece Base Trim
- Making a Scarf Joint

398 DOOR & WINDOW CASINGS
- Assembling a Jamb
- Installing Simple Colonial Casing
- Three-Piece Victorian-Style Casing
- Victorian Bellyband Casing
- Symmetrical Arts & Crafts Casing
- Neoclassical Fluted Casing

402 WALL FRAMES
- Drama for Your Walls
- Installing Wall Frames
- Wall-Frame Assembly Jig

404 WAINSCOTING
- Panel Products ◆ Installing Sheet Paneling
- Paneling Options ◆ Installing Wainscoting

406 CHAIR RAILS & PLATE RAILS
- Milling and Installing a Plate Rail
- Installing a Chair Rail

15 Trimwork

FINISH WORK

Trim is both practical and decorative. It covers rough edges and seams between different building materials inside and out, and it adds a distinctive touch that gives a house architectural detail and character. Installing trim can be a rewarding job if you master the art of making various kinds of simple miter cuts and getting a tight fit. To cut miters, you need a good miter box and a backsaw with a sufficient number of teeth per inch to make fine cuts without splintering the molding. Power miter saws make quick work of cutting even difficult angles.

Trim is sold by the foot, in lengths ranging from 6 to 14 feet. Try to get lengths that will span each wall, corner to corner, to avoid unsightly splices. Keep in mind that many softwood varieties of trim can be either finger-jointed or clear. Finger-jointed is less expensive because it has splices and must be covered with paint. Clear trim can be stained.

Types of Molding

Home centers carry pine, oak, and poplar moldings in a great variety of shapes that are designed for specific locations and uses. The first trimwork to be installed are the casings that go around doorways and windows. Next are base and shoe moldings, which trim the wall at the floor. Cove or crown molding is used along the wall at the ceiling, and corner molding for both inside and outside corners is used to hide seams and protect corners.

Because techniques for milling are not perfectly standardized, it's best to buy all the pieces of trim from the same milling lot to avoid fractional differences in size.

Biscuit Joinery

USE: ▶ biscuit joiner • bar clamps • measuring tape • pencil ▶ biscuit wafers • yellow wood glue

1 Before cutting, position the pieces of wood as you will join them, and mark reference lines for the biscuits.

2 Biscuit slots are cut with a biscuit joiner, a tool that cuts to a specific depth and width sized for the biscuit you will use.

Common Molding Profiles

BASE MOLDINGS

Base moldings can be as simple as plain square stock (S4S) or more elaborate with built-up components like shoe and base cap. Colonial base (top) is a more traditional style, while Ranch base (center) is common in today's homes. Contemporary base (bottom) is a stock item or easily made with a router or saw.

CHAIR RAILS

These moldings do what their name implies—protect walls from damage caused by chair backs bumping up against them. They have evolved into purely decorative trim that visually divides open wall space or creates a demarcation between paint, wall-paper, or paneling. There are many styles available that range from simple to sublime.

Back-Cutting Joints

To make tight miter joints, use a sharp saw blade and a miter box or miter cutter to produce a clean angle cut. Test-fit the joint, and if it doesn't close tightly, use a block plane to shave a thin amount of wood from the bottom part of the cut. Be careful not to plane along the cut line on the face of the joint. This back-cutting ensures that the two cut faces will touch.

Use a plane to back-cut miter edges, shaving with the grain of the wood.

For exact 45-deg. joints, sharp saw and plane blades are a necessity.

BASICS

3 Either white or yellow wood glue can be used in biscuit joinery. Test-fit biscuits to be sure the slot is free of sawdust.

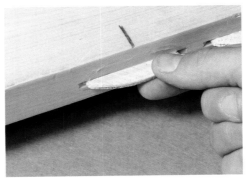

4 Biscuit wafers, made from compressed beech wood, expand when glued to fit the slot. The result is a tight, concealed joint.

5 After filling the slots with glue, the biscuits are inserted into both slots and the wood sections are clamped.

CROWNS

Aptly named, crown moldings are often the most elaborate and costly moldings in a room; they can add a crowning touch to almost any space. Styles range from simple coves ⅞-inch wide to custom-made crowns up to a foot or more edge to edge. They may be used alone or layered with other moldings for dramatic effect.

CASINGS

Casings serve mainly to cover openings or joints between walls and window or door jambs. Casings are usually installed with a "reveal," a small setback from the door or window side of the jamb that leaves part of the jamb exposed, adding texture to the installation. A ⅛- or ⁵⁄₃₂-inch reveal is typical for most designs.

BUILT-UPS

Although there are many ready-made molding profiles, "trimming out" can be a carpentry art form. By combining different types of molding and square stock, you can create unique built-up trim profiles to suit your taste or your home's decor. The illustration at right is just one example, using four basic molding styles.

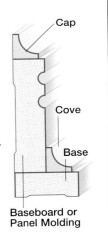

Cap
Cove
Base
Baseboard or Panel Molding

Coping Joints

For inside corners, coped joints allow you to match two intricate molding faces. Cut the first piece square, and butt it into the corner. Cut the other piece on a 45-degree angle, leaving the face exposed. Use a pencil to outline the cut edge, and follow this when you make the second cut with a coping saw, back-cutting it slightly so the cut face mates tightly with the corner piece.

Cut a 45-degree face onto the inter-secting trim piece; then scribe the cutout.

Use a coping saw to back-cut along the scribed line; file the edge if necessary.

15 Trimwork

DECORATING WITH TRIM

Trim does have a functional side—it covers gaps, rough edges, and transitions between building materials. But when skillfully laid out, it can do much more. Wide baseboards, wainscoting along walls, picture molding, crown molding, and false beams on ceilings all add architectural detail and a decorative touch that can't be achieved with paint or wallpaper.

The overall design, type of trim used, and variety of wood you choose will have a lot to do with the results you achieve. (For more details on trim profiles, see pp. 382–383.) Clean, tight mitered joints are also important to the overall effect. For that you'll need a good miter-box saw or a power miter saw like the pros use (even if you have to rent it).

Installation Tips

Mark the locations of studs on the floor before the walls are drywalled so you know where to nail the trim. Note also the locations of any in-wall braces you may need to secure the trim.

Despite good equipment and careful layout, some trim joints still need a little adjusting. To close up a joint, try a trick called back-cutting. Use a sharp block plane to undercut the back edges on both pieces of trim so the boards won't touch until the visible surfaces meet.

You can usually nail through trim without splitting the wood, but predrill if you are nailing near the edge, into oak (or other hardwoods), or nailing very thin or narrow trim.

Molding Details

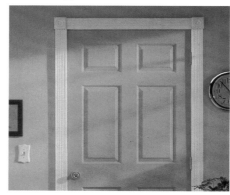

You can use combinations of beaded molding and decorative medallion blocks to create distinctive door trim.

More elaborate door surrounds are available in precut hardwood kits. Period moldings are now also made from foam.

Cornice moldings for ceilings can be built up in stages from stock lumber or ordered in paintable foam sections.

Elaborate cornice moldings can be painted, stained, or treated with a variety of faux-finish surface glazes.

Installing Molded Trim

USE: ▶ power miter saw or backsaw & miter box • pencil • caulking gun • hammer • drywall taping knife • paintbrush ▶ molded trim • construction adhesive •

1 Hold a section of the molded trim in place, and mark guidelines along the top and bottom edges with a pencil.

2 Following manufacturer's directions, install a bead of adhesive just inside your lines on the wall and ceiling.

3 Press the molding into the beads of adhesive, and fasten the lightweight sections with finishing nails.

BASICS

Originally made from plaster, ceiling medallions are now made of light foam glued and joint-compounded in place.

Complex cornices, medallions, mantels, and other special trim pieces are also available in exotic hardwoods.

Joinery Cuts

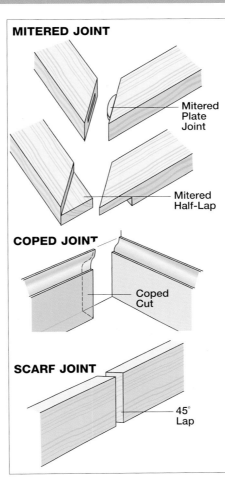

MITERED JOINT
- Mitered Plate Joint
- Mitered Half-Lap

COPED JOINT
- Coped Cut

SCARF JOINT
- 45° Lap

There are many ways to join wood trim—for example, with miters, half-laps, and biscuits. On some baseboard corners, you may have to make a coped joint. If you have to butt baseboards, cut mating edges at 45 degrees so that if the joint opens slightly, you won't see a gap. With reproduction trim made of rigid polyurethane foam, you can use a utility knife or trim saw to make cuts. Install this trim with adhesive and nails.

Foam moldings are cast from original, carved-wood trim. The cladding accepts paint.

---

finishing nails • sandpaper • primer & paint

4 Where molding sections meet and along the top and bottom edges, use drywall compound to fill seams and gaps.

5 Spread the compound smoothly, and when it dries, lightly sand the wall and ceiling seams with fine sandpaper.

6 Wipe away sanding dust; cover the fresh compound with a prime coat of paint; and finish with a finishing coat.

15 Trimwork

INSTALLING CROWN MOLDING

To calculate the rough lengths of molding you'll need, start by measuring all the walls about 2 inches below the ceiling. If the adjoining surfaces are true, simply follow the wall and ceiling lines. (You can test for bumps and depressions using a long, straight 2x4.)

Install the crown with glue and nails driven into wall studs. You can drill pilot holes through the top of the molding and drive 16d finishing nails into the top plate of the wall frame or into the ceiling joists.

If the wall and ceiling aren't true, you can do one of two things. The first is to force the crown into position using a wood block and toenails. This may work when the ceiling is fairly flat with only a few shallow depressions. The molding itself won't be straight, however, which may magnify the problem.

The second thing you can do is to test a scrap piece of molding at several locations to find the lowest point of the ceiling. Then level that low point along the walls where you'll be working, and strike chalk guidelines. Install the molding along the guidelines, and fill any gaps between the top of the molding and the ceiling with joint compound or caulk. This tends to flatten out the crown molding, though.

Installing Corners

In most cases, the best approach is to install all outside corners before inside corners. But you need to plan the installation carefully, and allow at least an inch or so of extra wood in case you make an error and need to recut coped joints at inside corners.

When two outside corners are separated by an inside corner, save the piece coped at both ends for last. In rooms with walls longer than your molding stock, of course, you'll need to piece the lengths together with a scarf joint. (See p. 389) Once you've cut the cope on inside corners, create a tight fit by adjusting the position of both corner pieces and by sanding or filing the cope as needed. To adjust the tacked-in-place square-cut piece up or down for fitting, tap it with a hammer and block.

Installing Crown Molding

USE: ▶ chalk-line box and measuring tape • crown

1 Once you've established your guideline location, snap a chalk line for the molding.

5 Cut the coping miter on another board, and transfer the dimension from Step 4, measuring from the miter tip.

9 Use a flat rasp as needed to clean up the upper section of the coped cut or to increase the back-cut angle.

Attaching Crown Molding

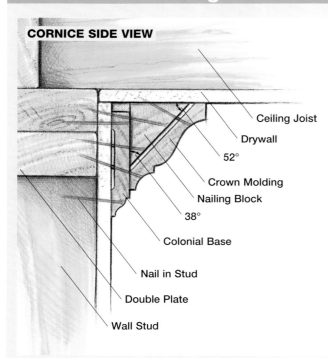

CORNICE SIDE VIEW
- Ceiling Joist
- Drywall
- 52°
- Crown Molding
- Nailing Block
- 38°
- Colonial Base
- Nail in Stud
- Double Plate
- Wall Stud

With most molding installations, you can nail into wall studs to make solid connec-tions. On some cornice installations, however, upper nails will reach into ceiling joists every 16 inches only if the joists run perpendicular to the wall. Where joists are parallel with the wall, you need either nailing blocks or a continuous nailer for support. You may want to install the blocks for extra support in any case.

CROWN MOLDING

molding and nailing blocks • power miter saw and coping saw • files • power drill and bits • hammer • finishing nails • nail set ▶ wood glue • caulk and caulking gun

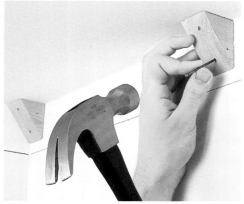

2 Nail or screw the support blocks to wall studs and the wall top plate every 16 in.

3 Install the full-length square-cut cornice, fitting it into the corner. Don't nail within 3 or 4 ft. of the corner yet.

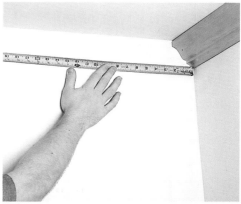

4 Measure out from the corner (plus an extra couple of inches) to find the rough length of the coped molding.

6 Use a coping saw to start the profile cut. Angle the saw to back-cut the coped piece.

7 Rotate the saw as needed to maneuver the thin blade along the profile of the miter.

8 Use an oval-shaped file (or a round file in tight spots) to clean up curved sections of the profile.

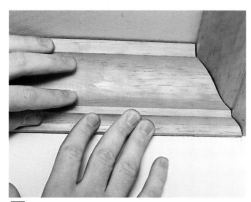

10 Test-fit the coped piece in place, supporting the other end to be sure the board is level.

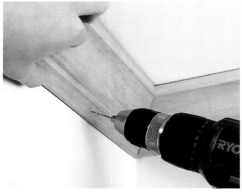

11 Adjust and fit the pieces; support the coped piece in place; and drill near the coped end to prevent splitting.

12 To finish, drive and set finishing nails to secure the corner pieces, and sand or caulk the joint as needed.

INSTALLING CROWN MOLDING

15 Trimwork

BASICS

Base trim includes a variety of board and molding applications to the bottom of a wall. The trim serves a practical purpose in that it covers the inevitable gap between the wall surface and floor, but it also serves a design function in that it provides a strong visual line around the base of a room and acts as the foundation for the rest of the trim. Of course, the decision as to what type of base trim you will use is tied into the trim motif of the room as a whole. Certain base treatments are more appropriate with some trim styles than others, but a few different options provide an adequate selection for most situations. If your trim package is based on stock molding profiles, there is simple, one-piece baseboard stock available. But if you are committed to a style that features wider, more complex moldings, the base trim should be of taller, heavier stock—usually a three-piece assembly. Keep in mind that specialty millwork suppliers can offer a wide variety of base profiles that you otherwise will not find. So, to expand your options or just to be inspired, it is worth exploring these resources.

One-Piece Baseboard

Most lumberyards and home centers offer baseboard moldings in two different styles to match their stock casings—colonial and ranch (also called "clam-shell"). The height of these moldings can run from 3 to 5½ inches, and most are about ½ inch thick. Select a profile and size that is compatible with the rest of the trim details in the room. If the floor is to be carpeted, the simple baseboard is all you will require. But for a tile or hardwood floor, you should also plan to install a flexible shoe molding to cover inevitable gaps between the different materials.

Built-up Base Trim

Most traditional trim styles feature a three-piece assembly consisting of a flat or molded base, a decorative cap molding, and flexible shoe molding. Some elaborate styles add additional layers or embellishments to the mix, but once you understand the basic principles and techniques for the installation, you can add or subtract elements to suit your taste.

Base Height. The height of the baseboard trim should be in proportion to the trim in the rest of the room, but it should relate to the size and height of the room as well. A room with 8-foot-high ceilings can accept a base that is 5 inches high, but a room with 9- or 10-foot ceilings needs a more substantial base—perhaps one that is 8 or 9 inches high. If you are in doubt as to the appropriate height of the molding, cut some scrap stock to various dimensions, and place it on the floor in the room to better judge the proportions.

Covering Mistakes. Even though the central portion of built-up base trim is relatively rigid, the layered construction provides a means for accommodating irregularities in both the wall and floor surfaces. Both the cap molding and shoe molding are flexible enough to conform to slight dips and humps so that most gaps can be eliminated.

Installing Three-Piece Base Trim

Three-piece base calls for installing a flat baseboard with an added base cap and shoe molding. This installation gives you the opportunity to customize the base treatment. Begin by attaching the baseboard to the wall studs using finishing nails. You can use butt joints at the inside corners. Install the cap molding next. For tight joints, cut coped joints at the inside corners. Install shoe molding once the finished flooring is in place. Use coped joints at the inside corners and miter joints at outside corners. Attach the shoe to the baseboard, not the finished flooring.

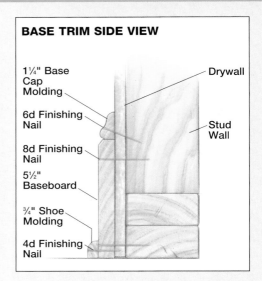

BASE TRIM SIDE VIEW
- 1¼" Base Cap Molding
- 6d Finishing Nail
- 8d Finishing Nail
- 5½" Baseboard
- ¾" Shoe Molding
- 4d Finishing Nail
- Drywall
- Stud Wall

Detail of one-piece base trim with stock Colonial door casing.

Detail of three-piece base trim with traditional casing.

BASE TRIM

Making a Scarf Joint

USE: ▶ power miter • clamps • baseboard • shoe molding • wood glue and filler • power drill and bits ▶ hammer • nail set • sanding block • sandpaper

On straight runs in large rooms, you may not be able to buy stock molding long enough to reach wall to wall in one piece. In this case you have to join the lengths, creating additional joints. The easiest solution is to make two square cuts and butt one piece against the other. But shrinkage may cause a noticeable gap, as it can at inside corner joints. The best approach is to cut an overlap between the pieces, called a scarf joint. Simply slice 45-degree cuts through the thickness of both pieces so that the surface of one covers the cut section of the other.

1 Start by making a square cut (or miter if you plan to cut a cope) on the far end of the board you need to piece.

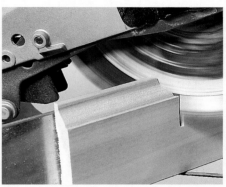

2 Cut both the 45-deg. miters that will form the scarf joint. (Locate the scarf over a wall stud.)

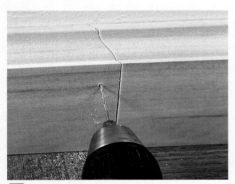

3 Test-fit the scarf joint (without glue), and predrill through both pieces at the joint.

4 Apply wood glue to both sides of the joint. (You will need to wipe off some excess after nailing.)

5 Drive finishing nails through the predrilled holes, and set the heads just below the wood surface.

6 Fill the nailholes with wood filler. You may need to wait and fill the holes again: some fillers shrink as they set.

7 Clean up any traces of glue and filler around the joint with sandpaper before finishing the wood.

8 To further disguise the joint, consider nailing a shoe molding to the bottom of the baseboard.

BASICS

15 Trimwork

Assembling a Jamb

USE: ▶ drill, router or circular saw • hammer or nailer • clamps • combination square • chisel • hammer • glue and nails ▶ jamb stock

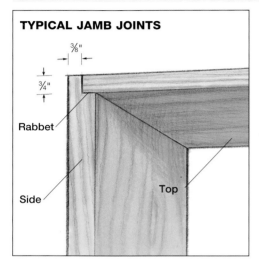

TYPICAL JAMB JOINTS
- Rabbet
- Side
- Top
- 3/8"
- 3/4"

1 Cut the head jamb to length, and mark the joint outline with a square for a ¾-in. jamb leg.

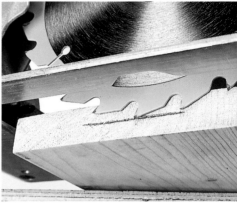

2 Set the blade depth on a circular saw to reach halfway (⅜ in.) through the thickness of the board.

3 To control the rabbet cut, firmly clamp a guide board and the jamb to a stable bench.

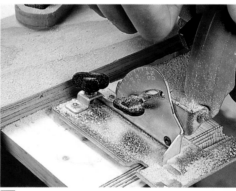

4 Make the innermost cut with the saw along the guide. Then make multiple passes to kerf the remaining wood.

5 You can use the saw to remove all the wood, or clean up the thin strips between kerfs using a sharp chisel.

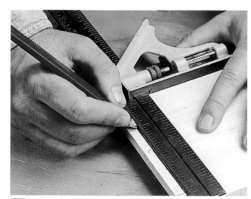

6 When the rabbet is cleaned up and ready for assembly, mark a nailing line on the outside of the joint.

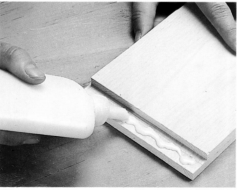

7 Add carpenter's glue to the mating surfaces of the jamb parts just prior to assembling them using nails.

8 Square up the jamb frame before fastening. You may want to set the pieces around a square block for support.

398 TRIMWORK / DOOR & WINDOW CASINGS

DOOR & WINDOW CASINGS

Installing Simple Colonial Casing

USE: ▶ hammer • combination square • power miter saw • nail set ▶ clamshell or ranch casing • finishing nails • wood glue

1 Mark the reveal by sliding the blade on the combination square. Cut a miter on one end of the casing.

2 Align the short side of the miter with one reveal mark; transfer the opposite mark to the casing.

3 Use 4d finishing nails to tack the head casing to the head jamb of the door. Leave the nailheads exposed.

4 Cut a miter on a piece of side casing. Rest the miter on the floor or spacer (for carpet or finished floor).

5 Mark the length of the side casing pieces by running a pencil along the top edge of the head casing.

6 Apply a small bead of glue to both miter surfaces. Nail the side casing to the jamb and wall framing.

7 Drive a 4d finishing nail through the edge of the casing to lock the miter joint together.

8 Use a nail set to recess the nailheads about ⅛ in. below the wood surface.

Closing Gaps

Use shims to help the miter joints close tightly. Trim the shims.

ASSEMBLING A JAMB 399

15 Trimwork

Three-Piece Victorian-Style Casing

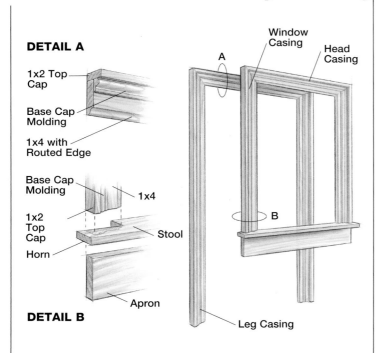

Victorian Bellyband Casing

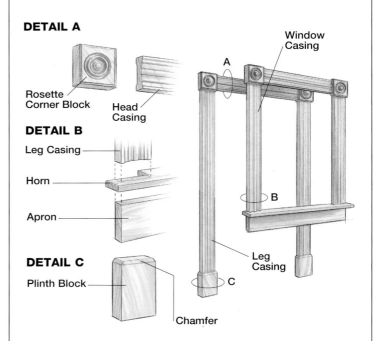

Although three-piece built-up casing looks large compared with most stock Colonial casing, its scale actually is about halfway between the scale of modern molding and the overwhelming trimwork of the Victorian era. It may be a little heavy in rooms with standard 8-foot ceilings. The easiest approach is to prepare built-up lengths of this casing on a workbench.

Installation Details. Mark a ⅛-inch reveal line along the edges of all three sides of the window or door frame. Then square off the bottom of the leg casings, and stand one leg casing in place. Make a mark on its inside edge at the point where the inside edge of the leg casing intersects the reveal line on the head casing. Repeat the same process for the other leg casing before mitering at the mark. Then tack the components into place.

Cut and fit the head casing next. If the profiles line up, you can pull the casings off the wall and install them permanently using glue and nails. Use carpenter's glue between the casing and the jamb and in the miters, and dots of panel adhesive between the casing and the wall. When mitering large casings, clamp the casing firmly so that it won't move during the cut. It's difficult to get perfect miter joints with large casings, so you may have to fill some gaps with caulk.

Victorian bellyband casing is the hallmark interior detail in many older homes. Due to the distinctive rosette blocks used at the corners of the casings, you can install this style of molding simply by butting the casing legs against the blocks. You still need to make a careful layout and allow for reveals. They can be a little tricky on corner-block jobs because the casing legs are always slightly narrower than the blocks.

Installation Details. One layout approach is to use a combination square to draw reveal lines on the jambs. The inside edges of the casings will follow these lines. You should set each corner block with its lower inside corner flush with the corner formed by the side jamb and head jamb. This arrangement is easy to lay out and install, but you should check that the casing follows the reveal on the jamb and is centered on the corner block. To be sure about this part of the layout, you might want to tack the blocks into position first and center the narrower casings on the wider blocks to establish the reveal along the jamb. Nail the pieces with glue in the joints and dabs of panel adhesive on the back of the casings.

The height of the plinth blocks must reflect the total height of the base treatment. (The base treatment used with this style of casing should be at least 6⅞ inches tall.)

DOOR & WINDOW CASINGS

Symmetrical Arts & Crafts Casing

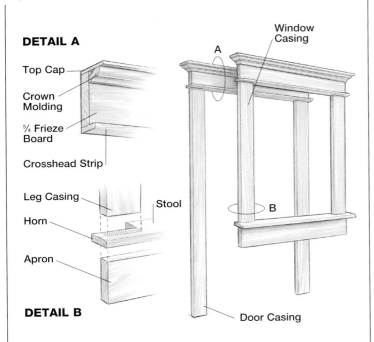

Neoclassical Fluted Casing

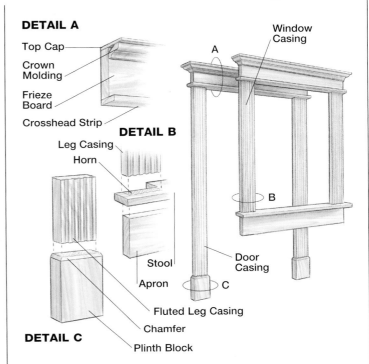

A hybrid approach to trimming openings resembles the Arts and Crafts style in some respects but blends with a simplified version of the Neoclassical approach.

Installation Details. Start by measuring the distance between the outside edges of the leg casings, and cut the ¾ ×6 frieze board to this measurement. Cut a length of 1½-inch-wide bullnose stop molding ¾ inch longer (for a ⅜-inch reveal) as the crosshead strip. Next, cut a strip of crown molding 6 inches longer. Center the length of crown molding on the top edge of the board, with 3 inches hanging off each end. Mark the ends of the crown on its bottom edge. These marks represent the short points of the compound miter cuts that form the outside corners at each end of the crosshead cap. This detail allows the crown molding to return to the wall on both ends of the frieze board.

When you have cut the returns, predrill them to avoid splitting; then glue and nail the molding to the face of the board first, and attach the returns. Now you can make the top cap. If you're using lattice, a typical reveal is ⅜ inch. Measure the distance between the outside edges of the returns, and add ¾ inch. Cut the lattice; then center and fasten it over the crown moldings.

Many people can picture flutes as a feature of great stone columns on public buildings. But this design motif, which looks like a series of shallow troughs in the material's surface, is widely used in wood.

You're likely to find that stock fluted casings are too small and special-order casings are too expensive. So you may want to make your own or have a local woodworking shop make them for you.

Installation Details. Using a combination square, establish reveal lines along all three sides of the jamb, and install the plinth blocks flush with the inside edges of the leg jambs.

Next, square off the bottoms of the leg casings. Position one of the leg casings on top of one of the plinth blocks, and mark the spot where the reveal line on the head jamb hits the casing. Then square-cut each leg casing to length, and install it. Measure from the top outside edge of one leg casing to the top outside edge of the other leg casing, and use this measurement (plus an allowance for a small overhang on each side) to build a decorative crosshead over the opening.

On windows, you have a couple of options. You can stop the leg casings at a window stool or run them down to plinth blocks on the floor.

15 Trimwork

WALL FRAMES

Wall-frame trimwork divides walls into large, aesthetically pleasing units. The size of the frames is determined by the design and dimensions of the room. It is best to draw a design on graph paper before attempting installation.

The size of the frames can vary within reason to accommodate windows and doors, but it is best to keep the margins—the spaces above, below, and between frames as constant as possible. A margin of 2¾ to 3 inches looks best.

The most frequently used molding for making wall frames is base cap molding, also called panel molding. Its dimensions are usually 11/16 x ⅜ inches.

Do not install wall frames piece by piece on the wall. It's quicker and more accurate to assemble wall frames on a table using a jig prior to installation.

Wall-Frame Assembly Jig

To make an assembly jig, screw a square corner of a piece of scrap plywood onto a larger piece of plywood (at least 2 x 2 feet), leaving the equivalent width of the molding on two adjacent sides of the larger piece. (See "Installing Wall Frames," below.)

This will allow you to press glued edges of mitered molding tightly together on the jig and sink brads into each side of the joint using a hammer or a pneumatic brad nailer. Begin by joining long sides and short sides. Then join pairs of the two-sided units.

Try to make each wall frame using the same molding stock from the same source (and the same bundle if possible). Otherwise, there is high probability that the profiles of the molding won't line up exactly.

Cut a small block the size of the top margin, and use it to scribe a line below the chair rail (or subrail, if you're using one) to represent the top edge of the wall frames. Hold the block against the bottom of the rail; hold a pencil at the bottom edge of the block; and slide the block across the rail along all the wall runs. Measure and mark the top corners of each frame on this line.

If you intend to nail the wall frames into place by hand rather than using an air nailer, predrill the wall frame for 6d finishing nails, and insert the nails before positioning the frame on the wall. Because you'll be nailing between studs over most of the wall (so you

Drama for Your Walls

Wall frames create an illusion of depth and density because 1) they are three-dimensional and 2) they divide the wall area into smaller, denser segments, which always creates more visual interest. The three-dimensional quality of wall frames is fundamentally different from that of the alternative treatment: raised panels. Despite the name, raised panels actually produce a concavelike, or receding, effect whereas wall frames are more convex, protruding outward. In terms of sculpture, concave units create negative space while convex units create positive space. Raised panels, therefore, deliver a uniform sense of volume, mass, and density, while wall frames create a higher level of tension and dramatic interest.

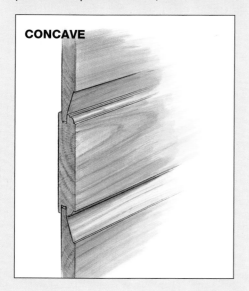
CONCAVE

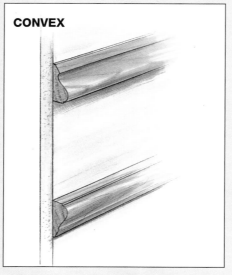
CONVEX

Installing Wall Frames

USE: ▶ power drill with screwdriver bit and ¾" dry-

The most accurate way to attach each wall frame to the wall is to use a spacer block below the rail rather than trying to align the frame with a guideline.

In many cases you will not be able to nail directly into wall studs to attach the frames. This is one carpentry job where a lack of a suitable nailing base should not be a problem. The frames are very lightweight and decorative in nature, you can use a combination of adhesive and nails to safely hold them to the wall.

4 *Measure and mark the top corners of each wall frame along the guideline. Then apply adhesive to the first frame.*

WALL FRAMES

won't have solid nailing surfaces), you'll need to use panel adhesive to stick the frames to the wall. Using a caulking gun, apply dots (not lines) of adhesive to the back of each frame.

Hold the block you used to draw the reference line up against the chair rail or subrail; butt the frame against it; and nail the top edge. Use a level to plumb the sides; then nail the sides and the bottom edge. Once you've installed all the units in a room, caulk the perimeter of each wall frame, both inside and out. Once installed, you have a few options for including the frames in the room's decorating scheme. Painting the frames a color that contrasts with the rest of the wall will make them stand out. Painting everything all one color creates a more subtle effect. In general, darker areas recede and lighter areas appear to move toward you.

Wall-Frame Assembly Jig

This type of jig helps you hold the glued, mitered ends of two pieces of molding firmly together while you fasten them. You'll need two such jigs: one for acute angles, as shown, and one for obtuse angles.

Position the two strips so that they form a point near the corner of a work surface, such as a sheet of plywood that measures about 2 x 2 feet.

Back the point of the strips away from the corner, as shown. This gives you room to drive fasteners into the joints of the wall frames. Join the jig strips by screwing or nailing them to the work surface.

A jig like this, secured to a work surface, allows you to hold the frame pieces for a tight fit and secure nailing.

---

wall screws • pneumatic nailer or hammer and nails • framing square • measuring tape ▶ pencil and level • scrap plywood and one-by lumber • basecap molding

1 Attach a small piece of plywood (with two adjacent factory edges) to a 2 x 2-ft. sheet. Leave a margin of 1⅜ in.

2 Apply glue to the mitered ends, and fasten them using an air nailer or hammer and brads. Keep your fingers clear.

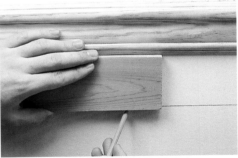

3 Cut a block the same width as the top margin; place it against the chair rail; and scribe a guideline on the wall.

5 Using the spacer block from Step 3, align the wall frame with the corner marks, and fasten it in place using 6d nails.

6 With the top edge fastened, plumb the vertical sides of the frame using a level, and attach them using 6d finishing nails.

7 Double-check that the bottom of the frame is level, and then fasten it. Fill all holes, and caulk all around the frame.

15 Trimwork

PANELING PRODUCTS

Sheet paneling—of real or simulated wood—is a popular wall surface for do-it-yourselfers. It adds warmth to any room in the house and is particularly desirable in recreation rooms because it holds up under hard use. Paneling goes up quickly, as well, without the time-consuming finishing steps required by drywall.

Building codes regulate whether paneling may be applied directly to studs. Sometimes, a layer of drywall must be installed beneath thin paneling to support it, as well as to add a measure of fire resistance.

As nice as sheet paneling may look, it can't duplicate the richness of a room lined with real wood, however. Boards allow you to create your own pattern, and they may be nailed up directly over studs with no drywall needed, unless fire codes prohibit this. Board paneling is more expensive than sheet paneling and can take more time to install and finish, but the results are worth the effort.

Prefinished Paneling

Among the least expensive paneling is prefinished hardboard. Sometimes referred to by the trade name Masonite, hardboard paneling often has a top layer that is factory-finished with a wood-grain pattern. Hardboard panels usually measure 4x8 feet and range from 1/8 to 1/4 inch thick.

Prefinished plywood paneling is available in a wide selection of colors, patterns, and thicknesses. Sheets are most often 4x8, ranging from 5/32 to 1/2 inch thick (with 1/4 inch being the most common). The face of each panel is printed, embossed, or color-toned with wood grain or other decorative effects. Some are laminated with an overlay that offers even more variety. In addition, some types are grooved to simulate individual wood boards.

Board Paneling

Board paneling can be bought in a variety of hardwoods and softwoods milled especially for this use. It can be from 3/8 to 1 inch thick and comes in various widths up to 12 inches. The boards have either tongue-and-groove or shiplapped edges. In addition, the surface edges of the boards may be beveled for decorative effect.

Installing Sheet Paneling

USE: ▶ saber saw or keyhole saw • 4-ft. level • hammer • measuring tape • straightedge • caulking gun

1 Butt one end of the first sheet into a corner; then adjust the other end against a level so that the sheet is plumb.

2 Measure the height and width of utility boxes for cutouts. Precut them with a saber saw or keyhole saw.

Sheet-Paneling Options

Type	Application
Standard 303	The name for a variety of APA siding patterns
T1-11 siding	A widely used APA-303 panel with vertical grooves
Pine or fir	Usually exterior, must be painted or stained
Bead-board plywood	Used for porch ceilings or interior wainscoting; paint-grade only

Installing Wainscoting

USE: ▶ saber saw or keyhole saw • chalk-line box & plumb bob • scriber • straightedge • 4-ft. level • drill

1 Walls need to be furred out with furring strips to ensure an even nailing surface. Shim any low spots.

2 After the furring is completed, nail the first board into place. Scribe and cut its edge if the adjacent wall is not plumb.

WAINSCOTING

▶ sheet paneling • finishing or ring-shanked nails • prefinished trim as required • caulk

3 Using a caulking gun, apply a bead of construction adhesive in a zig-zag pattern along the length of each stud face.

4 Nail the panel into place with paneling nails. Special ring-shanked nails are available in colors to match the paneling.

5 For a coordinated look, nail prefinished matching molding along the base and around windows and doors.

Plank-Paneling Options

Type	Application
Shiplap	Exterior or interior, vertical or horizontal, edges overlap
Tongue and groove	Exterior or interior, vertical or horizontal, edges lock in place
Clapboard	Usually exterior horizontal siding, edges overlap
Hardboard	Exterior or interior, needs painting or staining
Board and batten	Usually exterior, rough-sawn vertical, battens overlap edges

Concealing Nails

An easy and effective way to fill nail-holes and repair surface dings is with a crayon-like colored wax pencil.

• hammer • measuring tape • pencil ▶ plank paneling • finishing nails • furring strips • wood trim as required

3 Boards will have lapped edges or tongue-and-groove slot fittings. Attach boards to the furring with finishing nails.

4 Measure for cutouts as they occur. Mark the cutout on the board; drill a starter hole; and finish cutting with a saber saw.

5 Fit sections together over the cutout. If necessary, use an electrical junction box extender for the outlets.

PANELING PRODUCTS

15 Trimwork

CHAIR AND PLATE RAILS

Picture rails, chair rails, and plate rails are all horizontal applications to a wall surface, each with a particular functional aspect. While chair and plate rails are sometimes installed as part of a wainscoting, they are not limited to that type of application. Each of these elements can be used on its own to provide a strong trim component to a room.

Chair Rail

Chair-rail molding can consist of a single piece of stock or a combination of profiles mounted at a height of 30 to 36 inches above the floor. Originally, chair rail was designed to protect wall surfaces. But as a decorative element in a room, the chair rail has exceeded that particular function and is used to create a strong horizontal line around a room, dividing the wall height into distinct areas.

Chair-Rail Profiles. There are moldings that are sold specifically as chair-rail stock, but you can use a wide variety of other moldings, including a flat piece of 1x3.

Some chair-rail applications feature a projecting cap that provides a flat surface on the top of the rail that returns to the wall. Whatever the configuration, it is important to consider the intersection of a chair rail with vertical trim elements, such as door and window casings. If the casing protrudes farther into the room than the rail, you can simply let the rail butt into the side of the casing. But if the rail protrudes beyond the casing, fashion it into a finished return to give a finished appearance.

Plate Rail

A plate rail is a variation on the chair-rail theme, but mounted higher on the wall—normally 60 to 72 inches from the floor. Consisting of a narrow shelf with a shallow groove parallel to its front edge, the plate rail is frequently used as a cap for an Arts and Crafts-style wainscoting. Decorative plates or other types of artwork can be propped against the wall with their lower edge engaged in the groove. If you wish, you could omit the groove and use the shelf to display small collectibles or other items. A plate rail can also be used in conjunction with other styles of wainscoting, or simply as a trim element used on its own.

Milling and Installing a Plate Rail

USE: ▶ router and edge guide • corebox bit • hammer • nail set • nail gun and nails (optional) • power miter

1 Install an accessory edge guide to the router, and use a corebox bit to mill a groove in the top surface of the plate-rail shelf.

2 Mark a level line on the wall at the height of the top of the plate rail. Cut a rail to length with appropriate end joints.

Installing a Chair Rail

USE: ▶ spirit level • hammer • nail set • nail gun and nails (optional) • miter saw • files and rasps ▶ chair

Corners in chair-rail applications should be handled the way you would crown or baseboard molding—miter cuts for outside corners as shown in step 2 and coped joints for inside corners as shown in steps 3 and 4. When cutting outside corners, hold the first piece in position against the wall and mark the inside of the miter cut with a pencil. Do the same for the second piece and test the fit. Make the neces-sary adjustments and test again. When satisfied, attach the molding to the wall.

1 Mark a level line, around the room to indicate the top edge of the chair rail. Use a 4-ft. level for this.

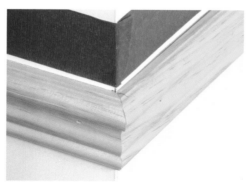

5 Cut the miter angles on both halves of an outside corner joint to test the fit before nailing the first piece to the wall.

6 Make a square cut on the end of a one-piece chair rail to allow it to butt into the edge of a door or window casing.

CHAIR RAILS & PLATE RAILS

saw ▶ 1-by stock • apron stock • wood glue • finishing nails

3 Cut the appropriate miters on shelf stock, and apply glue to the joint surfaces. Use small nails to pin the parts together.

4 Run a bead of glue along the top edge of the rail; then place the shelf in position. Use 8d finishing nails to fasten the shelf.

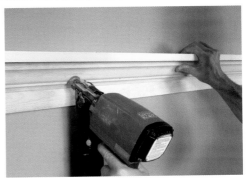

5 Cut apron molding to size; then nail it to both the rail and shelf to finish the bottom of the shelf.

rail stock • finishing nails • wood glue • masking or painter's tape

2 Apply masking tape to the wall just above the layout line, and mark the locations of wall studs.

3 For an inside corner joint, run the first piece square into the corner; then cut a coped joint on the second piece.

4 Detail of completed inside corner joint on one-piece chair rail. Repeat the process for a built-up assembly.

7 When a chair rail includes a cap, create an elegant termination point by notching the cap around the casing.

8 Cut the cap stock to notch tightly around a door or window casing; then shape a rounded transition on the end.

9 Detail of notched and rounded end of chair-rail cap at door casing. You can also chamfer the end piece.

CHAIR AND PLATE RAILS   407

16 Cabinets & Counters

410 TOOLS & MATERIALS
- Special Tools ◆ Sheet Materials

412 BASIC TECHNIQUES
- Cutting & Edging Plywood ◆ Doweling
- Cutting Dadoes ◆ Making Biscuit Joints

414 CABINET CONSTRUCTION
- Cabinet Basics ◆ Cabinet Anatomy
- Framed ◆ Frameless ◆ Modular

416 INSTALLING CABINETS
- Wall Preparation
- Installing Wall Cabinets
- Installing Base Cabinets

418 DOORS
- Latches ◆ Raised-Panel Doors
- Flush, Overlap, and Lip Doors

420 DRAWERS
- Drawer Basics ■ INSTALLING METAL GUIDES
- Cutting Dovetails ■ INSTALLING WOOD GUIDES

422 LAMINATE COUNTERS
- Countertop Materials ◆ Installing Sinks

424 SOLID COUNTERS
- Solid Surfacing ◆ Solid Wood
- Installing Tile ◆ Masonry

426 FINISHING
- Protecting the Wood ◆ Finishes
- Before Painting ◆ Sanding ◆ Finishing

428 RESURFACING CABINETS
- ■ REFACING CABINETS ■ FINISHING TOUCHES

430 CABINET REPAIRS
- ■ UNSTICKING DOORS ■ REHANGING DOORS
- ■ UNSTICKING DRAWERS ■ REBUILDING DRAWERS

432 CABINET SPECIALTIES
- Fold-Down Doors ◆ Pull-Out Platforms

16 Cabinets & Counters

CABINETRY BASICS

Cabinetry projects amount to high-end carpentry—not art exactly, but they tend to be both more complicated and more exacting than other home improvement jobs. After all, it takes a fair amount of skill, and the right tools, to make cabinet drawers that slide effortlessly, doors that swing closed smoothly, and joints between the cabinet's many wood parts that are strong, straight, and tight. Well-crafted cabinet-work can make a room, though, and in kitchens especially, new cabinets and countertops can increase the value of a home.

You can avoid a lot of work by purchasing factory-made cabinets and counters—you just attach the units to the walls and hook up the plumbing and electrical outlets. But whether you go with prefab cabinets or build them yourself, the first step is selecting the materials. On cabinet faces, there is a basic choice between solid wood such as oak and composites such as particleboard covered in plastic laminate.

Counter materials must be resistant to damage from impacts, burning, cutting, and moisture. Granite may be the ideal material, but it's also the most expensive. Modern plastics—both thin laminates and thicker solid forms—are more affordable choices, and by far the most commonly used. Less common are concrete (with many of the positive aspects of granite or marble but without the looks) and solid wood, which makes attractive but high-maintenance countertops.

Special Tools

What tools you'll need for installing cabinets depends upon how you'll be doing it. Cabinets from a kit will require a power drill/driver, clamps, a 4-foot level, and a few other simple tools. If you're making the cabinets yourself, a table saw, router, and electric sander will be indispensible items.

For a custom laminate countertop, you'll also need a saber saw. Laminates should be cut only with very sharp tools. To avoid exposing the dark substrate material under the color surface, countertops and edgings normally are installed with a small overlap and then trimmed using a router equipped with a ball-bearing guide to keep the high-speed bit's friction from scorching the laminate.

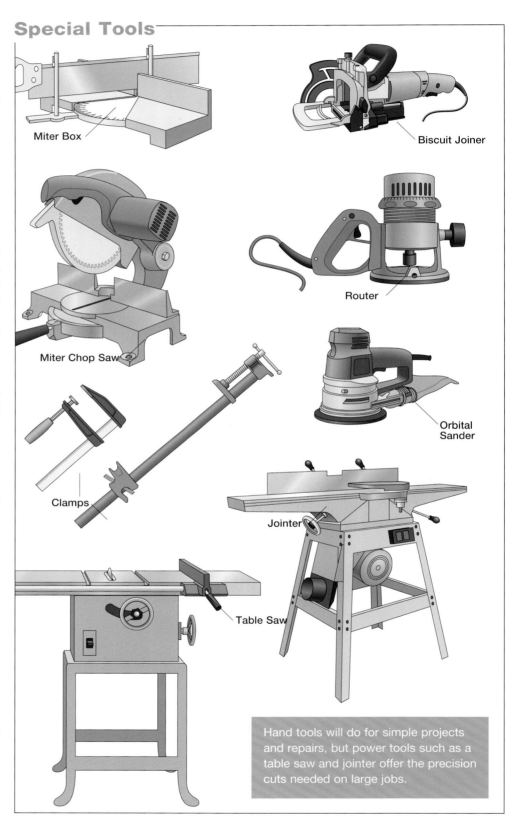

Hand tools will do for simple projects and repairs, but power tools such as a table saw and jointer offer the precision cuts needed on large jobs.

TOOLS & MATERIALS

Table Saw Safety

Turn the saw off when you finish a cut; unplug it before adjusting the blade. Never reach across a moving blade; stand to one side when cutting. Never feed in stock or clear wood scraps freehand; use a push stick and feather boards. Cut knotty or warped wood by hand.

The blade guard is a movable plastic shield that covers the saw blade. Use it whenever you make a through cut—it will keep flying dust and debris out of your face and prevent your hands from accidentally touching the blade.

A feather board is a thin board, the ends of which are cut like a comb. It clamps to a table saw to keep stock under control and moving in one direction. Use one to keep the stock from lifting up and another to keep it against the rip fence.

A push stick has a handle at one end and is notched at the other to fit against the edge of your stock. It should be used to push stock through when cutting, enabling you to keep your hands well away from the blade. If you make one, use solid wood or thin plywood.

Sheet Materials

The easiest way to check whether sheet materials are warped or damaged is to lay them flat on a concrete floor (preferably before you buy them). Remember that these sheets are almost always 4x8 feet—if you don't have a truck or van that can transport them, try to find a lumberyard or home center that will deliver them. (Most will deliver large orders.) Tying them to the roof of your car can be not only nerve-racking but dangerous. Another option may be to have the lumberyard or home center cut the panels for you. Just supply them with your cut list or a set of plans. It's inexpensive, it's great if you don't have a table saw, and you'll be able to fit the stock into a much smaller vehicle.

PLYWOOD, made from thin layers of wood veneer, is sold in 4x8-foot sheets. It's much cheaper than solid wood and (because of its lightness and flexibility) better suited to cabinetry.

APPLICATIONS: Construction (softwood) plywood is used to form the carcasses of cabinets and for the undersides of countertops. Thin (¼-inch) hardwood plywood is used for decorative purposes, much like a veneer—not for frames.

PARTICLEBOARD, waferboard, and oriented-strand board (OSB) are usually cheaper than plywood and are better for making counters than cabinets.

APPLICATIONS: These boards are often found beneath countertop laminates and furniture veneer. They don't take screws or nails well (tending to split and crumble); it's better to join them with rabbeted joints and glue.

HARDBOARD, also known as fiberboard, is made from wood chips and fibers, held together with phenol-formaldehyde glue.

APPLICATIONS: Hardboard comes in several types. Standard has one smooth side; prefinished has one painted side. Plastic-laminated has a laminate cemented to one side, making it useful for DIY cabinetmaking.

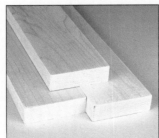

LUMBER is used less than manufactured sheet materials as the sole material for cabinets because of the expense. It does, however, make beautiful, extremely sturdy cabinetry.

APPLICATIONS: Cabinets are often made from a cheap grade of wood covered in veneer: wafer-thin sheets of high-quality hardwood, such as mahogany, walnut, and bird's-eye maple. The solid knotty pine of traditional country cabinets is easy to work with.

16 Cabinets & Counters

Cutting & Edging Plywood

USE: ▶ circular saw (with carbide-tipped blade) • straightedge • utility knife ▶ plywood panels • adhesive-backed veneer strip

1 Before you cut plywood with a saw, first score along your cut line with a utility knife to prevent the surface from splintering.

2 When using a power saw, cut plywood with either a carbide-tipped or a fine-toothed plywood-cutting blade.

3 Use adhesive-backed veneer tape (applied with pressure or heat) to cover end grain and create the look of solid wood.

Doweling

USE: ▶ doweling jig • power drill • mallet • bar or pipe clamps ▶ wood dowels • wood glue • glue stick

1 To connect edges of panels or boards with dowels, use a doweling jig to bore matching holes in mating edges.

2 Doweling jigs have guide holes in several diameters to guide the right-sized bit into the edge of the board.

3 Add glue to the dowels, tap them into one set of holes, and then glue the exposed ends and clamp the mating board in place.

Cutting Dadoes

USE: ▶ table saw (optional) • circular saw with rip guide • mallet • wood chisel • C-clamps ▶ wood

1 To cut channels in wood, the best option is a table saw, but you can use a circular saw with a rip guide.

2 Set your blade depth, and adjust the rip guide to cut first along the inner edge, then along the outer edge of the channel.

3 To clear wood from the channel, make several passes with the saw or chisel the scrap. Fasten the board to the worktable.

BASIC TECHNIQUES

Making Biscuit Joints

USE: ▶ biscuit joiner • bar or pipe clamps • pencil ▶ biscuits (plate splines) • water-based glue

1 *Biscuit joinery is the easiest way for DIYers to make professional wood joints. Start by marking centerlines over the joint.*

2 *Biscuit cutters line up with your marks. Adjust the blade depth to cut a groove in the middle of mating edges.*

3 *Biscuit wafers glued in place bridge the clamped joint. The compressed wood wafer expands when glued, so work quickly.*

Alternatives for Cutting Rabbets

With a table saw, you can cut dado grooves with a special cutter in one pass or make several passes with a standard blade.

With a circular saw, raise the blade; cut the edge of the dado using a guide; and make repeated passes to clear the groove.

With a router, you can install a bit that's as wide as the groove you need. Test the width on a crosscut before dadoing.

Alternatives for Shaving & Sanding

Electric-powered planers take the effort out of planing boards. You can set the blade for heavy trimming or light smoothing.

The old standard, a hand-powered plane, also will smooth and true up the edge of a board. A sharp blade is essential.

A belt sander also can trim and smooth. Use a coarse-grit belt to remove wood and a fine-grit belt to finish the edge.

CUTTING & EDGING PLYWOOD

16 Cabinets & Counters

CABINET BASICS

The basic cabinet component is a carcass—the frame that supports the counter and outlines doors and drawers. These are generally constructed of fir or birch plywood or particleboard, joined together with rabbeted joints and glue. Base cabinets are braced on bottom risers made from 2x4s (which become the toekick).

Doors and drawers fill the carcass openings, and the drawer faces and doors can be built three ways—as overlays (slightly larger than their slots), full overlays (the door and drawer faces nestled against each other), or insets (set flush into the carcass openings).

Designing

If you don't feel comfortable creating your own kitchen layout, there are several options. For an involved project with a big budget, hire an architect or interior designer to draw up your plans. A good contractor can also draft something to your specifications, and he or she will be familiar with local codes. Some home centers provide design services on computers with a CAD program—computer-assisted design software that enables you to play around with different design elements and see realistic pictures of the results.

Stock or Custom?

Stock cabinets are the equivalent of an off-the-rack suit—you won't be getting the highest quality or custom alterations, but you'll save big. The range of frame types is limited, but there are plenty of options available: different styles of doors, sliding shelves, breadboards, wine racks, and the like. Custom-made cabinets, on the other hand, are made to order by a cabinetmaker to fit whatever style and layout you desire. This enables you to have a more unusual design (one to fit an odd layout) or to match your old cabinets or something you see in a magazine. In between these two choices are semicustom cabinets (also known as custom modular), which are mass-produced like stock cabinets but offer more flexibility in terms of materials, layout, and options—you can even alter heights and depths if you want. These cabinets often take longer than both stock and custom cabinets to arrive, though, and must be ordered far in advance.

Cabinet Anatomy

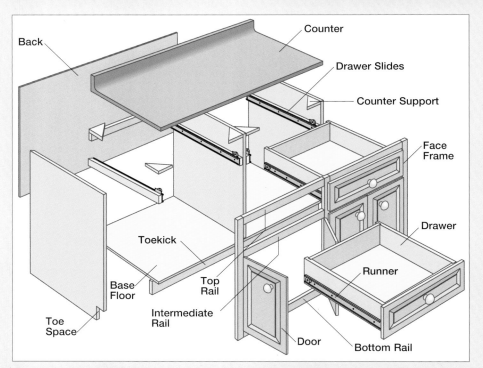

Cabinets are basically empty boxes with interchangeable components. Traditional framed cabinets (shown) have a face frame attached to the front, with doors hinged to this frame and drawer fronts to match. Hardwood is often used for the decorative fronts, while plywood or particleboard is used to construct the boxes.

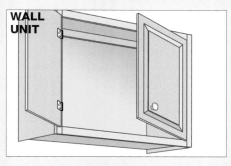

WALL UNIT

WALL CORNER UNIT

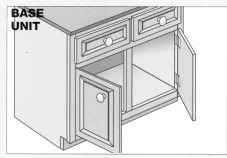

BASE UNIT

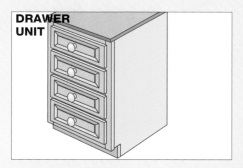

DRAWER UNIT

414   CABINETS & COUNTERS / CABINET CONSTRUCTION

CABINET CONSTRUCTION

Framed

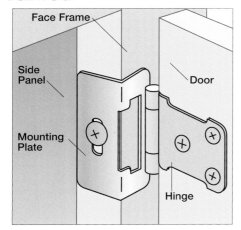

Face-framed base cabinets have a carcass that supports the counter and frames all the openings. The typical way of treating the openings is with overlays: the doors and drawers are larger than the opening. When all the doors and drawers are closed, thin strips of the carcass are visible. This lends itself to contrasting colors or wood tones and decorative hinges. Door and drawer panels also can be set into the openings of the face frame, creating one flush surface.

Frameless

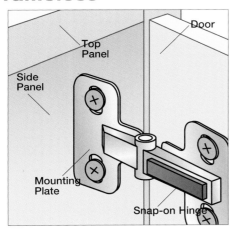

Frameless cabinets (sometimes called "European") are basic boxes that have no face frames—the frames are covered by the doors and drawers. These more expensive, less durable units call for inside hinges (with holes measured in millimeters instead of inches) that can be finicky to adjust and keep aligned. On the plus side, doors can open a full 180 degrees, and overall access is better because there are no face frames in the way, only the walls between sections.

Modular

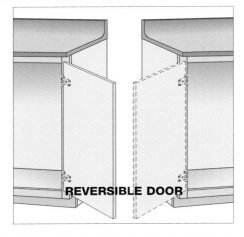

Modular cabinets come in independent units, finished on both sides, that you can fit into your kitchen layout in a number of ways. Of all the types, they're the simplest to install. They can stand alone for use as a vanity or be placed at a counter end without requiring you to order finished sides. They also may be ganged in a continuous lineup by screwing them together through their vertical face stiles. Installers often hide the screw-heads beneath the hinge leafs.

CABINET BASICS

16 Cabinets & Counters

PREPARING SURFACES

Before installing cabinets, make sure the walls are plumb and level. Ideally, you should use a long (more than 4-foot) level. You can also check walls with string and blocks. Cut three blocks of wood to the same width. Nail two of them at the same level on opposite ends of the wall, drive nails partially into them, and run a taut string between the nails flush with the top of the blocks. When you run the third block along the wall under the string, the string will bow out at high spots and gap at low spots. High spots can be shaved down with a rasp or 80-grit sandpaper; low spots should be filled with joint compound applied flush with the rest of the wall. Sand these patches down when dry.

Wall Preparation

USE: ▶ spirit level or straightedge • stud finder • power drill/driver (optional) • measuring tape • pencil

1 Use a level or straightedge to find low spots on the wall; mark them with a pencil; and fill them flush with joint compound.

2 An electronic stud finder can locate the framing members behind the wall. It's wise to drive a nail or two to verify stud location.

Installing Wall Cabinets

USE: ▶ spirit level • power drill/driver (optional) • handscrew clamps • stepladder • screwdriver • rubber mallet • utility knife • pencil ▶ wall cabinets • filler strips

1 A glass-door corner cabinet anchors this installation of uppers. Even with the doors off, you may need a helper.

2 Plumb the first cabinet carefully (shimming where necessary) so adjoining sections will fit closely and run level.

3 Secure the cabinet with screws driven through backing strips (generally at the top and bottom of the cabinet) into wall studs.

Installing Base Cabinets

USE: ▶ power drill/driver with hole saw or saber saw • handscrew clamps • rubber mallet • spirit level • utility knife • pencil ▶ base cabinets • filler strips • shims

1 You may need to install a ledger to support the counter, for example, where rounded corner units hold storage carousels.

2 Before setting the lower cabinets, use a hole saw (or saber saw) to make access holes for plumbing supply and waste lines.

3 Check the base units for square against the wall and each other. The corner cabinet (right) holds revolving storage trays.

INSTALLING CABINETS

• screwdriver ▶ 1x2 support ledgers • wood screws • joint compound

3 Locate the highest point on the floor where you will install the cabinets; this is the base where all measurements originate.

4 Measure up from the high spot for the countertop height and down from the ceiling low point to locate the uppers.

5 Use a 4-ft. level to install a 1x2 ledger that will help to support the upper cabinets during the installation.

• 1x2 support ledgers • trim • wood screws • shims • stain (to match trim to cabinets)

4 Set the next cabinet in line on the support ledger, and clamp its top and bottom to the corner cabinet.

5 Check each adjoining cabinet for plumb; clamp it; and drive screws through the reinforced back panel.

6 Use a power drill to drill holes for the hardware; if they are predrilled, you just tighten the knobs using a screwdriver.

• 1x2 support ledgers • side trim • toespace trim • wood screws • stain (to match trim to cabinets)

4 On some installations, you need to shim the spaces between units. A clamp keeps the face trim joint tight.

5 Check for level to maintain uniform support for the counter. Fasten the cabinet with screws through the reinforced back frame.

6 Glue and nail long pieces of trim that match the cabinet finish along the toe-space under the lower cabinets.

PREPARING SURFACES 417

16 | Cabinets & Counters

CABINET DOORS

At first glance, making a cabinet door seems like a pretty simple project, but there is more to it than just cutting a slab of wood to fit over the opening. You have to choose hinges, latches, and handles, and determine how the door will fit over—or into—the opening. Then there is the design of the door itself, the materials you will use to make it, and the job of aligning and attaching it all.

Some other larger concerns include planning the layout. Just how will doors swing out from the cabinets, and can any drawers located over them be pulled out when the doors are open? Strength is a factor too because a door is supported at only two points (the hinges). You have to be sure that both the hinge and the material used to make the door will stand up to the weight of the open door.

Door Construction

Safety latches on lower cabinet doors will keep contents out of the hands of curious toddlers.

Most cabinet doors are not made of a single, solid piece of wood because even slight warping would make closing them almost impossible. (Overlap doors are a little more forgiving than inset types, which will still close but will leave gaps.) Plywood covered with veneer offers more stability than solid wood because it does not warp as easily.

Most high-quality cabinets have doors that are built up from frames and panels. These doors have wooden frames (made from ¾ or one-by stock) with an inside panel of wood, sheet material, or glass.

When they are made correctly, the frames' joints are extremely strong and stable, and the door will stay square. Because the panels are not glued firmly in place but are fitted into grooves, they are free to expand and contract within the frame without affecting how the door sits in its opening. If you can make sturdy joints (using mortises and tenons, dowels, or biscuit joints), you can make a simple frame-and-panel door.

Latches

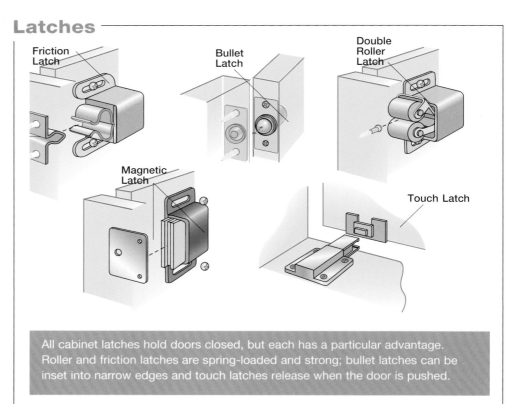

All cabinet latches hold doors closed, but each has a particular advantage. Roller and friction latches are spring-loaded and strong; bullet latches can be inset into narrow edges and touch latches release when the door is pushed.

Raised-Panel Doors

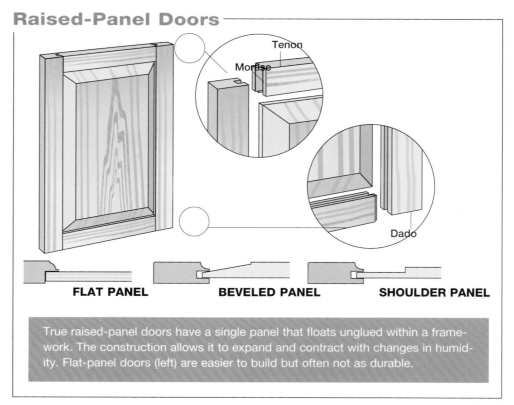

True raised-panel doors have a single panel that floats unglued within a framework. The construction allows it to expand and contract with changes in humidity. Flat-panel doors (left) are easier to build but often not as durable.

DOORS

Flush Doors

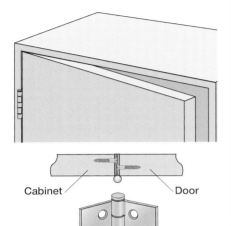

Butt or leaf hinges are used for flush-fit cabinet doors. Each leaf requires a mortise cut into the frame and door edge.

Flush (or inset) doors are easy to make—they are simple rectangles cut to fit into openings in the cabinet frame. The door faces are flush with the frame surface.
- STYLE: In a small kitchen, flush doors help the cabinets appear to recede, while heavy or elaborate surface trim carves up the facade and makes them stand out. But flush doors have drawbacks: misalignment is most noticeable, and light-colored cabinet faces are interrupted by dark joint lines.
- DURABILITY: Even gradual shifting or slight warping can make a flush door catch on the frame. There is no overlap or margin for error as there is with overlap doors.

Overlap Doors

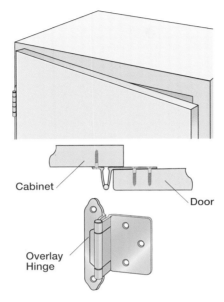

Overlay or surface mount hinges are used for overlap doors. Because the door is not inset, these are the easiest hinges to adjust.

Overlap doors are cut slightly larger than their openings, and they sit on top of the cabinet frame. The overlap takes care of misaligned seams—the seams are all concealed by the doors. But the face frames still show and will need to be finished to match the doors.
- STYLE: The door panels—particularly their edges—as well as the hinges, become prominent in this style. If you use plywood or another composite material, you'll have to conceal a lot of rough edge grain with paint or veneer tape.
- DURABILITY: Problems with the doors are simple to fix: just adjust the hinges or, if necessary, replace them.

Lip Doors

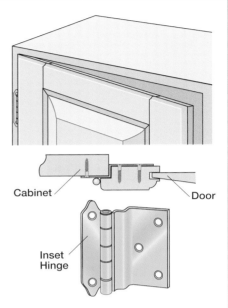

Inset hinges are shaped to fit around a rabbeted edge and are used to hang lip doors. You see only a small slice of the hardware.

Lip doors might be the best compromise among these three styles. With these, half the door thickness sits inside the cabinet opening, and half overlays the cabinet frame. When the exposed edges of solid wood doors are rounded over, the panels blend into the frame. One side of the hinge, a decorative leaf, is exposed on the frame. The other leaf, a wider support, is concealed on the back of the door.
- STYLE: These doors look more elegant because they integrate with the face frames.
- DURABILITY: Lip hinges are easy to install. Both sides of the hinge are surface-mounted, so there is no drilling or chiseling mortises.

16 Cabinets & Counters

DRAWER BASICS

It's easy to tell a well-made drawer from a cheap one. Sturdy drawers are made from solid wood using dovetails, and have full-extension, under-mounted metal guides with ball bearings. Cheaper drawers are made from laminated particleboard, are nailed or epoxied together, and have plastic guides. Drawers for custom-made cabinets are not constructed until after the carcass is built, to ensure a perfect fit.

The best drawer joints are dovetails and other similar joints, which would hold together firmly even without nails (or, worse yet, staples). They make what are called mechanical connections. Next best are connections with a structural bridge from one board to another: for example, doweled joints, or the more modern equivalent, biscuit joints, in which a small wafer is glued into slots aligned in both boards. Simple butt or lap joints won't hold up under repeated use. Joints like these are often reinforced with nails or staples, which can eventually work loose. Mechanical connections might not even come apart if you try and force them. (See "Cutting Dovetails," middle right, for a technique to create this joint.)

This 1940s-vintage kitchen-of-the-future may not look that different from many kitchens today—except there is no microwave, no dish-washer, and no trash compactor. And that futuristic linoleum that covers the floor is also used on the curvaceous island countertop.

Types of Drawers

As with cabinet doors, there are three basic types of drawers: flush, overlap, and lip. Flush drawers have faces that fit into their openings flush with the cabinet's carcass. These drawers require the most precise cuts to fit properly, and even a little swelling can make them stick. Both overlap and lip drawers allow more room for error because part or all of the drawer face covers the surrounding cabinet frame.

Drawer options include pull-out platforms for stacking large items like pots and pans, false drawer fronts that hold slide-out cutting boards, and vertical drawers custom-made to fit into an unusual space—these are great for storing narrow items such as cutting boards and trays.

Installing Metal Guides

USE: ▶ measuring tape • power drill/driver

1 Modernize older cabinets by installing metal drawer rollers. Most require a sturdy side rail that fits flush with the face frame.

Cutting Dovetails

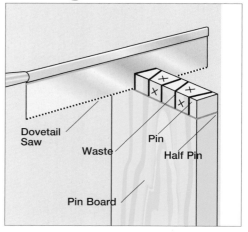

Installing Wood Guides

USE: ▶ router • backsaw • miter box

1 When older wooden drawer glides wear out, make new ones with a runner strip fastened to a side rail.

Drawer Anatomy

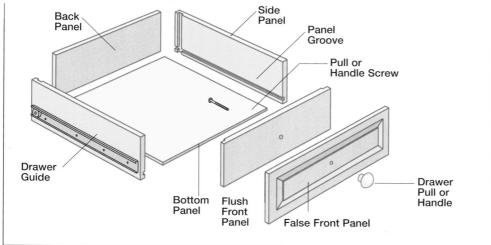

DRAWERS

MONEY SAVER

(optional) • torpedo level • pencil • screwdriver ▶ drawer • slide hardware • 1x3 or 1x4 side rails • wood screws

2 Mark a level line on the side rail; then use the track as a template to mark screw holes. Drill pilot holes, and attach the guide.

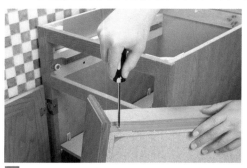

3 Measure depth on drawer sides of the guide placement on side rails; then install mating roller guides on the drawer sides.

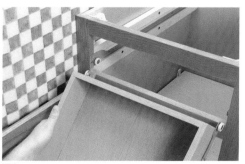

4 Before installing all mounting screws, test-fit the drawer. Tip the drawer into the opening to connect the guide rails.

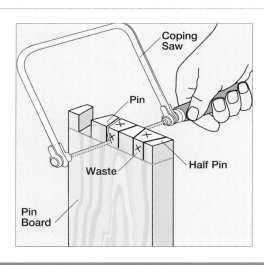

Dovetails are the best joint to use for drawers because they can take a lot of stress in many directions. Each joint has two pieces: the leg piece is cut into keystone-shaped pins, which fit into sockets that are cut into the tail piece. Careful layout and cutting are necessary for a perfect fit that doesn't jam or gap. Mark the layout of the pin piece with a T-bevel set at an angle of 75–80 degrees (use more of an angle for softwoods than hardwoods). The wider part of the pins should be toward the back side. First make the marks on the end of the piece: three dovetails is typical, and their width is an aesthetic choice. Then lay out their height (the same as the thickness of the wood or slightly larger). Make the vertical cuts for the pins with a dovetail saw (pictured at far left), and then make the horizontal cuts with a coping saw (left). Use this piece as a template to mark the tail. Clean the cuts with a chisel, and test-fit them before you glue and assemble. You can buy special jigs for cutting pins and tails at the same time with a router.

MONEY SAVER

• screwdriver • power drill/driver (optional) • torpedo level ▶ 1x3 or 1x4 side rails • 1x1 wooden runners

2 Install side rails with runner ends flush with drawer opening. Once past the runner, of course, the drawer has no support.

3 Measure and rout new drawer sides to the same dimension as side-rail runners. Rout to final depth in successive passes.

4 Test-fit drawer in the opening. Runners should fit easily in routed grooves. Rub runners with paraffin to smooth operation.

DRAWER BASICS

16 Cabinets & Counters

COUNTERTOP MATERIALS

If there is one place to splurge in kitchens, it's on the counters. The ultimate splurge is granite, the most expensive material available—maybe four or five times more costly than plastic laminates. It's very strong and dense, and resists almost all staining, burns, and scratches from cutting. The entire counter becomes an excellent chopping block and hot plate. Granite is much more durable than marble, another pricey choice. Most grades of marble stain easily, yellow over time, and are a pain to keep clean day by day. Stone countertops also need a strong base to support their weight.

Quartz composite counters combine stone chips with binders and resins. The result is an extremely durable product that resembles stone. Solid surfacing material is made of polyester or acrylic. The material is impervious to water and easy to repair—simply sand away dents and burns. The material is available in solid colors, patterns, and looks that resemble real stone. The price tags for quartz composites and solid-surfacing counters approach those of real stone. Installations usually require a professional.

Plastic laminates, available in a huge number of colors and finishes. They are a popular choice because they are durable, reasonably stain resistant, and relatively inexpensive. However, burns are impossible to repair. You can attach laminate to a substrate yourself. (See opposite.) Or have a counter fabricator or home center attach the laminate for you so that you can install the counter later.

Installing Post-Form Counters

USE: ▶ saber saw • straightedge • power drill ▶ wood glue • caulk • drywall screws • veneer strip

1 Use a saber saw to cut the counter-top to length; a framing square clamped to the counter can act as a guide.

2 Exposed ends are covered with end-cap pieces. If they don't have an adhesive, attach them with wood glue.

3 Position the counter tightly against the wall, and attach it to the cabinets from underneath with drywall screws.

4 Apply a bead of silicone caulk in the joint between wall and backsplash and between any joints in the countertop.

Installing Sinks

USE: ▶ power drill/driver • keyhole saw or saber saw • screwdriver • caulking gun ▶ sink clips and screws • caulk • wood bracing and blocking

1 Before setting a self-rimming sink, turn it upside-down, and apply a thick bead of waterproof caulk under the lip.

Once the plumbing is roughed in and the counter is ready, you can install the sink. Your choices range from cast iron to stainless or enameled steel. (Be sure to get at least 20-gauge.) Counter-mounted sinks are either frameless/self-rimming (which just need to be fastened down) or framed/rimmed (which require a strip to seal them to the counter). Install the faucets and drain before you install the sink.

2 Center the sink in the countertop opening, set L-shaped sink clips into the perimeter groove, and tighten the screws.

LAMINATE COUNTERS

LAMINATE

Laminate is the material to use if you're doing the job yourself, and even so it's tricky. The sticking point—literally—is contact cement. This glue is applied to mating surfaces, and when they make a firm connection, they are permanently attached. There is no maneuvering room the way there is with most glues.

Another difficulty with laminate is its hardness. You need this durability on counters, but the brittle sheets chip unless cut with very sharp tools. To avoid exposing the dark substrate material under the color surface, countertops and edgings normally are installed with a small overlap, and then trimmed to a fine joint with a router. But if your bit does not have a ball-bearing guide, friction from the high-speed rotation is likely to scorch the edge. And if you push the router fast enough to avoid scorching, the joint is likely to chip: it takes practice.

The two main advantages to laminates are price (starting at about $3 a square foot for 1/16-inch-thick counter material; $15 to $25 per linear foot installed) and color variation. There are hundreds of options, including different patterns and textures. Color-through laminates cost more but eliminate dark seams. They still chip if you drop a heavy pot, but the damage isn't as noticeable.

Plastic laminate has good resistance to staining, moisture, and abrasion, but a pan hot off the burner will leave a scorch mark that is usually impossible to patch. Also, repeated knife cuts, though small and shallow, eventually discolor and create a cloudy area.

> ### CAUTION
> Contact cement should be used in a well-ventilated area. Exposure to its fumes can cause irritation to your nose, throat, and lungs, and fumes will explode if ignited. It will also irritate your skin if not washed off with soap and water. Wear eye protection and rubber gloves when handling contact cement. Local codes may require latex-based, VOC-compliant cement.

Installing Laminate

USE: ▶ router • roller • clamps • utility knife • straightedge • paintbrush ▶ lattice • contact cement

1 Trim laminate to size by scoring the rigid material with a utility knife or a special laminate scoring tool.

2 Clamp a straightedge just beyond your score line. Work on a sound, flat surface, and snap off the excess.

3 Read adhesive label cautions (and check the safety tips below) before rolling on sheets or brushing on edging.

4 Position the edge strip over the double-thick plywood counter edge, and roll the strip firmly to get good adhesion.

5 Set thin lattice strips over the counter while you position the main sheet. Remove the strips in sequence.

6 Use a router to trim laminate edge joints. Use a beveled bit with a ball-bearing guide wheel to avoid scorching.

16 Cabinets & Counters

SOLID FACTS

Solid surfacing for countertops is the same through its full thickness—a blend of resins with mineral fillers. There are more colors and patterns than when it was first introduced, and the selection continues to grow. It can be worked like wood; the edges can be molded into elaborate details, and you can make minor repairs by sanding and buffing. Larger gouges can be fixed by bonding in new material—at least on truly solid sheets. So-called solid-surface veneer costs less and has a similar seamless appearance, but does not offer the same, full-thickness advantages. The downside to all solid surfacing is price (installed quotes of $50–$120 and higher per linear foot) and difficulty: creating a multisection counter with invisible seams the first time out is unlikely.

Ceramic tile counters are not just extremely durable—it's no simple thing to smash one of them—but are just as resistant to staining and burning as granite. Unlike granite, tiles are also available in numerous colors and styles. But while the tiles provide durability, all that grout in between them can be trouble. Water discolors grout and can eventually drain through to the base, causing delamination. (The only alternative is to painstakingly paint the grout lines with silicone.) It's also a challenge to get an evenly spaced layout on splashboard, counter, and rounded facing, particularly on wrap-arounds. The cabinet carpentry must be extremely rigid for best results and strong enough to hold up tiles embedded in adhesive over exterior-grade or marine plywood.

Solid Surfacing

Many consumers will recognize this material by one of the best-known brand names, Corian. Unlike plastic laminate, a thin layer that rests on a plywood or particleboard countertop, solid surfacing is one material throughout. Generally, it is installed by contractors who custom-fit counters to your cabinets. Sections can be fastened together with nearly invisible seams, even over wide corner units where sheets of laminate would create a noticeable joint. Also, where solid surfacing is cut or edged—for example, at a sink—the exposed edge needs no extra finishing.

Solid-surface countertops need no edging and show no seams—but you can install decorative inlays and trim.

Installing Tile

USE: ▶ power drill/driver or screwdriver • tile

1 Typical ¾-in. plywood counters are reinforced with a second layer to create a 1½-in.-thick base.

5 Mark layout lines for the field tiles, and work one small area at a time, embedding the tiles in adhesive.

Solid Wood

A maple butcher-block countertop is beautiful, but will it last? In its original application—as an actual butcher's block—knife scars were periodically removed by cutting a slice off the thick block. That's impractical on kitchen counters, where the wood will scratch and eventually stain even when sealed with a penetrating oil. Hard-surfaced clear sealers will chip, and pieces can be picked up in food preparation.

MATERIALS
- **WHAT YOU'LL NEED.** You'll need at least a full inch-thick base for the butcher block to withstand typical kitchen chores. Edges can be glued up and routed for decorative overhangs. Special FDA-approved, non-toxic finishing oils are also available.
- **ESTIMATING.** Butcher block is sold in 18-, 24-, and 36-inch widths, and in custom sizes.

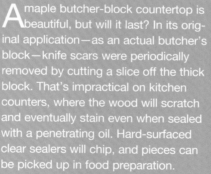

SOLID COUNTERS

cutter • tile nippers • notched trowel • rubber float • sponge & bucket • spacing lugs ▶ ceramic tile • thinset adhesive • grout mix

2 Most cabinets are sold with triangular brackets at the corners. Mount the counter by screwing up through the brackets.

3 Make a dry layout to square up tiles on the counter and plan the installation. Use spacers to allow for grout joints.

4 Use a notched trowel or a plastic spreader that leaves ribs of adhesive according to manufacturer's directions.

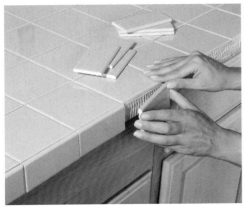

6 To eliminate a hard corner, tiles used for counter edging can have a rounded-over top called a bullnose.

7 To grout tile joints, spread the soupy grout mix over the entire surface, working mainly at a diagonal to the seams.

8 When grout hardens, use a clean, damp sponge to remove the grout haze. You will need to make several passes.

Masonry

Solid masonry counters are among the easiest to clean, the most durable, the most elegant—and the most expensive. Many types of stone can be used, including marble, which requires more maintenance than most, and granite, which resists scratches, heat, and most stains. A contractor can install a jointless, wraparound counter made of concrete. Lower cabinet frames often need to be reinforced to carry the load.

MATERIALS

- **WHAT YOU'LL NEED.** Most stone countertops are ordered and delivered completely finished. Sink cutouts are made with a special blade and a wet saw. For concrete counters, check with local tile or kitchen-cabinet suppliers.
- **ESTIMATING.** Precise measurements are required to fabricate masonry counters. Price varies widely by the size, shape, and type of material.

SOLID FACTS 425

16 Cabinets & Counters

PROTECTING THE WOOD

Dirt, grease, oils, and even food would quickly stain wood cabinets and counters not protected by some sort of finish. The most basic finishes have only one ingredient. Linseed oil, for instance, adds tone and protects wood at the same time. Other finishes call for several ingredients—for example, a sealer to close the wood grain, stain to add color, and a protective top coating such as varnish.

Types of Finishes

Fast-drying finishes are the most convenient, combining a small amount of solids (the actual coating) and a large amount of thinner. When the thinner evaporates, solids are left in place to protect the wood. But brush strokes may set in the surface as it dries, and lap marks will show if the wet edge dries during application. This is most troublesome on a large surface. You have to move quickly to pick up the wet edge. Fast-drying films are also relatively thin because they have a small proportion of solids.

Slow-drying finishes, such as boiled linseed oil, have more solids and less thinner. These dry to tackiness when they evaporate, and the remaining film hardens gradually—6 to 12 hours is generally needed before the surface can withstand contact without marring. The hardening continues for a much longer period, from a day to a week, depending on the finish.

Finishes on surfaces used for food preparation should be lead-free and free of other metals, such as cobalt or manganese salts, which are used as drying agents.

Finishes

- **SANDING SEALERS** and wood conditioners are applied to wood with large pores (such as pine) before finishing to make stains look even; sometimes sanding sealers are applied to help smooth out rough surfaces. A sealer may also be applied after staining to prevent it from bleeding into the finish. Shellac and oil varnish are common sealers; there's also lacquer sanding sealer, which is used before lacquering.

- **WOOD STAINS** add transparent color to wood and bring out subtle grain patterns. Stains contain pigment, a solvent (oil- or water-based), and a binder. Pigmented stains are more opaque and useful for disguising inferior stock. Aniline dyes are sometimes used to impart rich, if artificial, colors—these dyes are more translucent and generally reflect the natural color of the wood.

- **GEL STAINS,** thicker and easier to use than regular wood stains, come in oil-based pigmented and water-based dye formulations. Because of their consistency, they're quite easy to handle and apply without the drips and lap marks of regular stains. The consistency makes them particularly useful if you're refinishing cabinets that are already on the wall.

- **POLYURETHANE,** a synthetic varnish made from plastic resins, provides a clear, hard finish more resistant to yellowing than oil varnishes. Polyurethane dries quickly and is tough enough for floors. Both oil- and water-based varieties are fast-drying, resistant to moisture and heat, and easy to apply. They come in varying glosses, or sheens.

Before Painting

Fill all nicks, depressions, or holes with putty. Cabinet manufacturers sell color-matching putty and paint for touchups.

Before painting, take off doors, drawers, and hardware; wash everything with a mild detergent. Use a liquid deglosser on varnished cabinets to dull the shine. Scrape loose paint off painted cabinets, and sand with 180-grit sandpaper—painted hardware can be soaked in paint remover. Holes should be filled with putty and sanded flush. Prime bare wood before painting.

If you're covering the cabinets with opaque paint, sand filler with 220-grit sandpaper so marks won't show through.

FINISHING

Sanding

To remove paint traces or surface scars, use a belt sander with progressively finer grit papers. Don't bear down on veneers.

A powered pad or orbital sander will do a good job of finish-sanding large areas without showing directional traces.

Finish up by hand-sanding in corners and around intricate detail molding. Start with 80-grit and work up to 220-grit paper.

Finishing

For better control of color absorption, especially on softwoods, apply stain with a soft, lint-free cloth, and rub it into the wood.

Use a quality natural-bristle brush or a foam pad to apply polyurethane finishes. Smooth out brush marks as you work.

If you use a natural finish like boiled linseed or tung oil, you can build up shine and keep wood clean with lemon-oil polish.

Tools

You can finish and refinish most wood surfaces with a simple hand brush, but there are some exceptions, such as a lacquer finish, that must be applied with a sprayer—a brush leaves lap marks. Among many sanding tools that speed up the job, the array of new detail sanders have triangular heads that can dig into corners. One of the most important tools is a tack rag—to make sure the surface is completely clean.

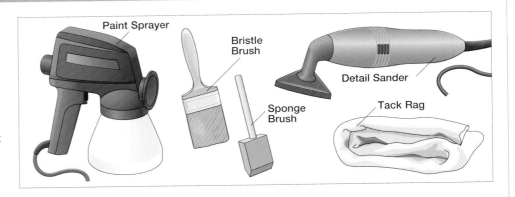

PROTECTING THE WOOD 427

16 Cabinets & Counters

REJUVENATING CABINETS

In many kitchens, you'll tire of the appearance of your cabinets long before they wear out. Rather than replace them completely, you can resurface them and make them appear new. The idea is to apply a combination of cabinet doors, drawer fronts, and thin veneers over the visible parts of the old cabinets. You'll need to be familiar with basic woodworking techniques, particularly if you buy stock sizes of solid woods and veneers and custom-cut and fit each piece in place. An alternative is to hire one of the growing number of kitchen cabinet resurfacing firms.

Using a Resurfacing Kit

A resurfacing project can be handled as a considerably less challenging DIY job if you use a kitchen cabinet resurfacing kit. The step-by-step sequence using a kit may vary somewhat from the usual custom resurfacing job. But the basic approach and process is usually the same. If your kitchen is similar to the typical American kitchen (which, according to one resurfacing company, has 15 cabinet doors and 11 drawer fronts), resurfacing with a kit should take from 16 to 24 hours and save up to 70 percent of the cost of installing all new cabinets. The kits, which you assemble from parts, come in a variety of door sizes and raised panel designs and cost about $600 for an average kitchen.

The overall job can be divided into three basic parts: replacing cabinet doors, adding veneer, and replacing drawer fronts. In brief, here's the sequence for changing laminated particleboard cabinets into oak cabinets: remove all the old cabinet doors and hardware; match the new door fronts to the cabinet openings (or add framing strips to decrease the size of the openings); apply self-sticking wood veneer over the visible frame edges and exposed side panels; then stain, seal, and finish the ¾-inch-thick, solid oak door fronts and oak veneers to suit.

Follow a similar sequence on the drawer fronts: first remove the drawers and hardware; cut the existing drawer fronts flush with the top, bottom, and sides of the drawers; glue, screw, and clamp the new, solid oak fronts over the cut-down drawers; then finish to suit and attach new hardware.

Before...  **...and after**

Resurfacing systems (available from contractors and in kit form) save money on a kitchen remodel by saving the underlying cabinet structure and replacing door and drawer fronts.

Retrimming Bases

USE: ▶ hammer • pry bar • backsaw • wood rasp • coping saw • nail set ▶ nails • wood putty

1 Use a small pry bar (or chisel) to pry off old rubber molding. Once an edge is free, pull the strip as you pry.

2 Prepare the wall for new molding by scraping away old adhesive. Don't use a chemical softener or remover.

3 Nail new wood molding into wall studs with finishing nails. Set the nail-heads, and fill holes with putty.

4 You can use one-piece molding with a decorative top bead, or build up the base with quarter-round moldings.

RESURFACING CABINETS

Refacing Cabinets

A kitchen resurfacing job can be a major overhaul where every visible piece of wood, including narrow door and drawer edges, is recovered. Sometimes new door and drawer fronts will do. But even a full facelift is easier to manage and less costly than a new set of cabinets. The idea is to save sturdy cabinet frames and clad them in a new skin, which can be a colorful plastic laminate or a light-toned natural wood. The job can include new counters, drawer fronts and drawer liners, trim, hardware, and other details. Here are some of the highlights.

Many older kitchens have solid, well-built cabinet frames but a dated or dingy appearance that is ripe for resurfacing.

MONEY SAVER

Once the doors, drawers, and hardware are removed, every visible edge of the frame can be covered with laminate.

Drawer liner kits like this one cover the old bottom and sides with a thin, non-slip, easy-to-clean laminate.

With a new liner installed, you can forget about liner paper or contact paper that inevitably starts to curl and peel.

New door and drawer fronts generally are available in different woods and laminates with a variety of panel designs.

Finishing Touches

Cover rough plywood edges with self-stick veneer. Apply with moderate heat to activate the glue and pressure.

Resurfacing contractors can replace or conceal every old section of your cabinets. In a typical sequence, cabinets are stripped down to their frames, and exposed surfaces, such as sections of the framework revealed between doors, are sanded and refinished. You get to select new materials, from solid wood to plastic laminate, for the new doors and drawer fronts. Take the opportunity to spray-paint old cabinet interiors.

MONEY SAVER

Install new drawer fronts with glue and screws. Dry-clamp the new wood in position to transfer holes for hardware.

REJUVENATING CABINETS

16 | Cabinets & Counters

FIXING DOORS

If a once-perfect door begins to stick, chances are that either the wood has begun to warp or the hinge screws are coming out of the wood. Your first step should be to tighten the screws and see whether the door realigns; you might need to remove the door, fill the holes with wood putty, and rescrew them. If it's not the hinges, you'll have to either plane the door, straighten out its warp, or shim the hinges.

Straightening a Warped Door

There are several ways to straighten a warped cabinet door—or at least to try. Some woods have grain with a built-in bias that is almost impossible to straighten out once it has curved. You can try using a combination of water and heat by soaking down the concave side of the warp with wet cloths and drying out the other side with a heat lamp. The door will need to be forced straight and held there—for example, with pieces of 2x4 and clamps.

Another approach is to suspend the door between two supports with the concave-shaped warp turned upside down. The idea is to overload the warp at its highest point with something heavy, such as a concrete block, and leave it in place until the warp flattens out. In fact, you should let the warp overcorrect because some of the original bend is bound to return once you unload the weight and reinstall the door. It's also a good idea to add a diagonal back brace to stiffen a panel. Use glue and screws to secure the brace.

Unsticking Doors

Half the battle is knowing exactly where the door sticks. Find out by coating the edge with powdered chalk.

MONEY SAVER

Start planing where chalk has rubbed off on the cabinet, and gradually work away from the high spot.

Rehanging Doors

If the door has a gap near the top, remove the hinge screws, and make a cardboard filler to build out the mortise.

MONEY SAVER

If the door is too tight near the top, remove the door, and use a chisel to slightly deepen the hinge mortise.

Diagnosing Doors

Hinges and catches typically are strong enough to resist even constant opening and closing. The weak links are the hinge screws. When they begin to work loose, the door starts to bind, which applies more stress and causes more binding. Tightening loose screws will help. But you'll get more permanent results by installing longer screws that bite into fresh wood for more holding power.

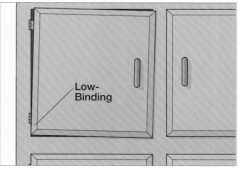

When a cabinet door binds and sticks at the bottom, plane the bottom corners and tighten screws in the upper hinge.

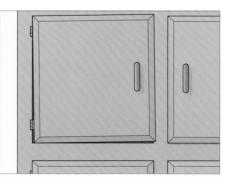

When a cabinet door binds and sticks at the top, plane the top corners and tighten screws in the lower hinge.

CABINET REPAIRS

REPAIRING DRAWERS

A cabinet with drawers that are sticking isn't just a nuisance, it's an eyesore—often the faces of the drawers that are stuck shut will all be at slight angles to one another. Sticking drawers can be blamed on loose or worn runners or cleats or on structural problems with the drawer itself. For minor or occasional sticking, seasonal humidity may be the culprit.

To check an uncooperative drawer, you'll need to empty it and pull it out to examine the hardware. Wooden runners might have shiny spots indicating uneven wear—they can be easily lubricated or planed. Metal or plastic runners and cleats need to be checked for level, for loose or missing fasteners, and for broken parts, which should be replaced. Loose screws should be replaced with larger screws, or the holes should be filled. (See below.) Bearings can be cleaned with ammonia and relubricated if they're not rolling properly.

If a drawer is jamming because its bottom has bowed out (due to being overfilled), you can take out the bottom and reinstall it upside-down. To do this, you may have to carefully knock apart the corner joints to free the bottom. Check for small brads or nails that are often inserted as insurance against glue failure, even in well-made dovetail joints, and pull these first. If the wood splits during this process, remember that it too can be reglued when you reassemble the drawer. The bottom should not be glued into its slot, so it should be easy to remove and reinstall.

Unsticking Drawers

Sometimes lubricating the cleats and runners does the trick—try running a bar of soap or a candle over them.

MONEY SAVER

If lubrication doesn't work, you can plane or sand the runners slightly, test-fitting the drawer as you go.

Rebuilding Drawers

Remove nails on bottom or corners. If necessary, separate corner joints, slide out bottom, and reglue sides square.

MONEY SAVER

Slide drawer bottom back into its slot, but don't glue in place. Drawers can be cut low in back for bottom repairs.

Filling Screw Holes

Cabinet hardware is designed to support only limited loads; it can easily be overstressed if doors are ill-fitting. You can improve holding capacity by filling in the existing holes, giving the screw threads solid material to turn through. Use a paste-type wood filler, dowel, or several wooden toothpicks to fill the hole. Trim off the protruding ends, and drive in new screws with wide threads.

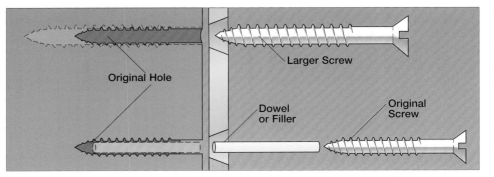

Use a larger screw, if possible, in a worn screw hole. Otherwise, wood filler or a wooden dowel (or a toothpick or two) can be trimmed and dipped in glue to fit snugly into a worn screw hole. The idea is to provide fresh material into which screws can bite.

16 Cabinets & Counters

Fold-Down Doors

On wide cabinet doors that are hinged to drop down, making access easier or providing a working tabletop, provide maximum support with a continuous hinge, often called a piano hinge. The hinges need a lot of screws but make a strong connection. To keep the folding door in a level position, also install a folding support bracket on each side. To prevent accidents with doors, consider folding supports that are self-balancing or have adjustable tension springs to slow the door's rate of descent.

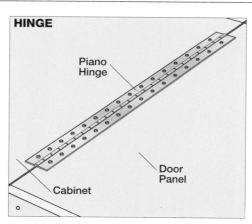

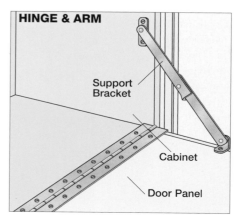

Installing a Pullout Platform

USE: ▶ power drill/driver or screwdriver • clamps • saw • measuring tape or rule • pencil ▶ pullout platform kit • screws • shelf lumber

There are some items that you may need to move out of a cabinet without lifting them up and hauling them—a stereo, for instance, or (more typically) a television set. Special hardware (generally sold with a platform) allows a shelf to slide like a drawer on guides that fit into cleats in the cabinet. Many kits include a swivel feature so you can rotate your shelf when it is extended. Or you can add a second, swiveling shelf on top of the pullout platform. Check to make sure that weight on an extended shelf does not tend to tip the cabinet forward.

1 Mark the center of the pullout kit, and align it with the center of the cabinet opening. Allow room for doors to close.

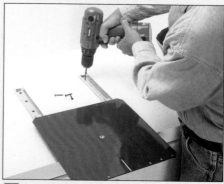

2 Extend the pullout base to expose the back of the glides, and screw them down to the fixed cabinet shelf.

3 Extend the pullout platform, and check the width of your shelf to be sure that it will rotate in the cabinet opening.

4 Cut and center your main shelf, and screw up through the platform into the bottom of the shelf to fasten them together.

5 From normal viewing height, even with the shelf extended and swiveled, the hardware can't be seen.

CABINET SPECIALTIES

Slide-Away Doors

Special double-acting door hardware can make doors slide back into a cabinet. The hardware is a combination of hinges and glides that are joined together. The hinges allow you to open the doors with a normal swing. If you open them all the way, in line with the sides of the cabinets, a guide panel on which the hinges are mounted can slide on tracks back into the cabinet. These hinges are often used for entertainment centers or anywhere that open doors would be in the way.

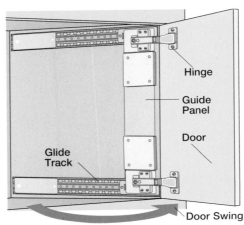
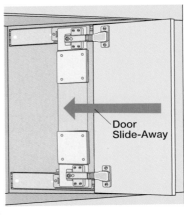

Display Details

Cabinets can have a variety of glazing, from plain window glass that will shatter dangerously when broken to elaborate and colorful leaded panels. If you make your own, the safest approach is to use plastic, or tempered glass, the kind required by building codes on patio doors. It breaks into tiny pebbles instead of large shards. It also is safest to have your glass drilled for hinges by the glazier who cuts it. The process requires a special bit turning at slow speed and constant lubrication.

To display the contents of a cabinet, you can use clear tempered safety glass or opt for using frosted glass.

In this kitchen, fold-out hoppers edged in wood that matches the cabinets have glass inserts so you can see what's inside.

Knock-Down Hardware

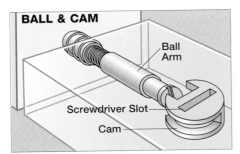

With ball-and-cam hardware, the rounded cam section is turned with a screwdriver to tighten the ball arm.

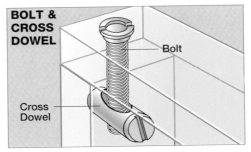

A bolt and cross-dowel fastener copies an old woodworking trick—inserting a nail through a dowel to add holding power.

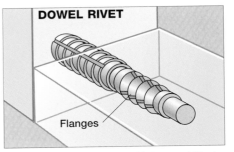

Dowel rivets are inserted into predrilled holes under pressure. The flanges on the dowel dig into the wood.

17 Shelving & Storage

436 BASIC ASSEMBLY
- Shelving Types
- Making Wide Plywood Shelves
- Materials
- Shelf Joinery
- Reinforcing Shelves

438 SHELF SUPPORTS
- Bracket Options
- Brackets & Z-Brackets
- Standards
- Fastening
- Hanging Wood-Framed Standards

440 BASIC BUILT-INS
- Customized Storage
- Plugging
- Filling Holes
- Constructing a Built-in Shelving Unit

442 CLOSETS
- Adding Closet Space
- Closet Layout
- Framing a Closet
- Closet Systems

444 UTILITY STORAGE
- Ideas for Workshops
- Shop Storage
- ATTACHING CASTERS
- Rack & Prefab Systems
- Assembling a Wall System

446 STORAGE OPTIONS
- Adding Storage Space
- BASIC TOY BOX
- Cedar Closets
- MORE STORAGE IDEAS

17 Shelving & Storage

SHELVING TYPES

Shelving is an easy and economical way to add extra storage space in almost any part of your home—along walls, inside closets, and even in the basement or garage. Building shelves doesn't usually require a lot of skill or specialized tools, so this is one project just about any do-it-yourselfer can handle. And unless you decide to use hardwood—which looks great but costs a bundle—it won't cost a lot to install them either.

Solid wood shelving is the way to go when you want to show off the wood or your work. But the cost per board-foot often rules out using classic hardwoods like oak, cherry, or walnut. Softwoods, such as fir or pine, are a better bet; they can be painted or coated with polyurethane to bring out their natural beauty.

Plywood and particleboard offer a couple of advantages when it comes to shelving, though. They cost less than solid wood, and can be bought faced with decorative surfaces. They also come in sheets, which makes them ideal for a really wide shelf. Inexpensive, manufactured storage units ready for assembly often are made from melamine-coated particleboard.

Wood trim will help match your new shelves to the rest of the room or add some interesting detail. Trim is also a handy way to hide seams, gaps, exposed edges of plywood, and other blemishes. You can get trim in either hardwood or softwood. If you plan on finishing a project with stain or sealer, make sure the trim matches the wood you used for the rest of the project.

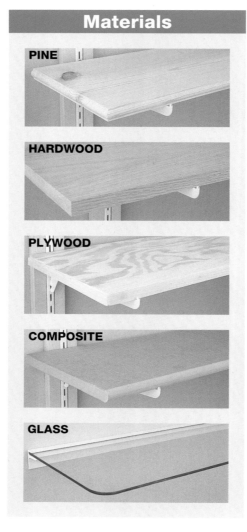

Materials: PINE, HARDWOOD, PLYWOOD, COMPOSITE, GLASS

Shelf Joinery

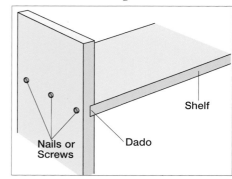

A dado joint is formed when a shelf fits into a channel of the mating piece. The strong joint also keeps the shelf stable.

Reinforcing Shelves

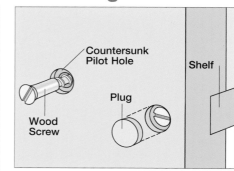

To reinforce joints with screws, drill pilot holes from the outside, countersink screws, and hide them with wood plugs.

Making Wide Plywood Shelves

USE: ▶ circular saw with rip guide • measuring tape • sawhorses • power drill/driver • hammer • orbital sander ▶ ¾-in.-thick plywood • 1x2 clear pine

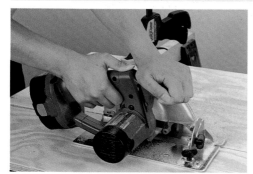

1 Use a circular saw to rip sheets of plywood into wide shelves. Stock lumber such as pine normally is no wider than 12 in.

2 You could conceal the rough front edge with veneer tape, but a narrow strip of pine is more rugged and adds extra width.

3 Drill pilot holes in the leading edge of the pine, and spread carpenter's glue where the other edge will meet the plywood.

BASIC ASSEMBLY

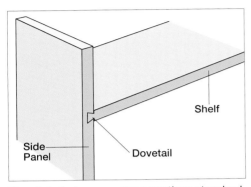

Dovetail dadoes are stronger than standard dadoes, with flared ends on the shelf to prevent withdrawal.

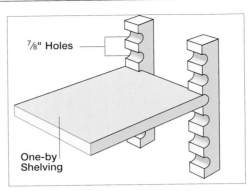

Adjustable shelves can be made in many styles. Here, a shelf with rounded ends fits into supports with rounded grooves.

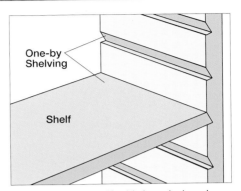

An adjustable shelf with beveled ends slides into 45° kerfs cut at even intervals into the side panels.

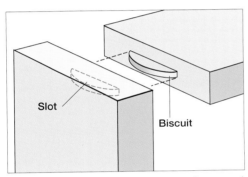

Biscuits are flat, elliptical wood wafers used to reinforce joints. They're glued into slots cut into mating pieces with a biscuit joiner.

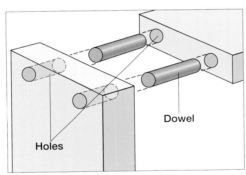

Joints reinforced with dowel pegs and glue are strong but more difficult to make and center than biscuit joints.

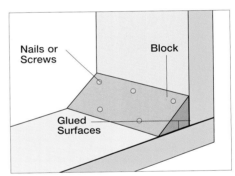

Glue blocks are glued and fastened to the inside surfaces of shelf butt joints to reinforce them.

• wood glue • 10d nails • wood putty • sandpaper

4 Set the pine against the plywood; attach it with clamps; and fasten it through the pilot holes with 10d finishing nails.

5 Wipe off excess glue; set the nails; and mix up a small amount of putty to fill holes on the exposed edge of the pine.

6 Sand the trim seam and the edges of the pine. The storage shelves can be 16 in. deep or more, with a durable, finished front.

SHELVING TYPES

17 Shelving & Storage

BRACKET OPTIONS

There are two basic types of ready-to-hang shelving supports: stationary shelf brackets and shelving standards. Stationary brackets come in many sizes and styles, and range from utilitarian to decorative. Shelving standards are slotted metal strips that support various types of shelf brackets, including horizontal cantilevered brackets, adjustable arm brackets, adjustable end-clip brackets, and continuous Z-brackets.

Mounting Brackets

For maximum strength, anchor shelf supports to wall studs. If your shelf will carry a light load, you can anchor its supports between studs with mollies or toggle bolts. Attaching supports directly to the studs is always better, though, because sooner or later something heavy will wind up on the shelf. Use masonry anchors to attach shelf supports to brick or concrete walls. For extra holding power, attach shelf supports to a ledger secured to wall studs with 3-inch-long wood screws.

Metal shelf standards can be mounted directly to walls or, for a more decorative look, you can insert the standards in grooves routed into the wood itself or into hardwood strips. (See below.) Cut the standards to fit using a hacksaw, and attach them to wall studs with 3-inch screws. Use a carpenter's level to make sure that both standards are plumb and that the corresponding mounting slots are level. Mount standards 6 inches from the ends of shelving to prevent sagging. For long wall shelves, install standards every 48 inches.

Many kitchen and closet storage systems use wire grids that attach to walls with molded plastic brackets. If you anticipate light loads, you can mount these brackets to drywall using the screws and expansion anchors usually included with such systems. For heavier loads, use drywall screws to fasten the brackets directly to studs.

In the 1920s, Sears sold thousands of house kits complete with studs and joists, nails, paint – the works. Closets were part of the package, including No. 9266 triple-unit clothes closet, "the best closet arrangement known in architecture."

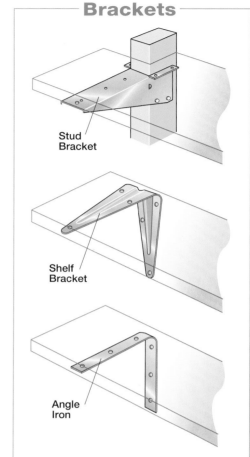

Brackets

- Stud Bracket
- Shelf Bracket
- Angle Iron

Screw shelf brackets to wall studs with the longer arm attached to the wall and the shorter arm to the shelf.

Hanging Wood-Framed Standards

USE: ▶ table saw, circular saw, or router • power drill/driver • 4-ft. level • measuring tape • eye protection ▶ 1x4 lumber for frame • metal shelf standards

1 Use a table saw (or circular saw or router) to cut a groove down the center of the ¾-in. standard frame.

2 Make the groove large enough to hold the metal standard. Predrill in the grooves so that you can screw the frames to wall studs.

3 Center the standard frames on wall studs; level the bases; plumb them with a level; and screw them to the wall.

SHELF SUPPORTS

Z-Brackets

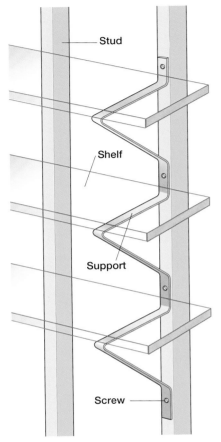

Continuous Z-brackets, often used for workshop or utility shelving systems, support multiple shelves.

Standards

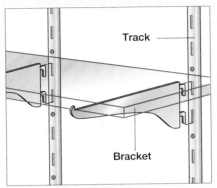

Slotted metal standards are leveled and screwed to wall studs. Shelf support brackets clip into the slots.

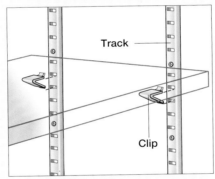

Adjustable end-clip standards can be surface-mounted on side panels or screwed into side-panel grooves.

Fastening

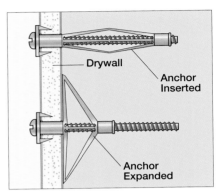

Screw shelf supports to wall studs whenever possible; otherwise, use hollow-wall anchors.

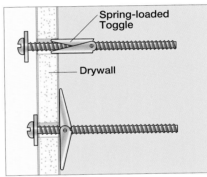

Some anchors expand when you tighten the mounting screw. Toggles have spring-loaded extensions.

- shelf brackets • boards for shelving • screws

4 Seat each standard in its groove, and tack it with a screw. Temporarily mount two standards, and double-check for level.

5 Once the slots are dead level, drive the rest of the screws. The standard should sit flush with the surrounding wood.

6 Fit shelf brackets into corresponding slots in the standards, adjust spacing to suit, and mount the shelves.

BRACKET OPTIONS 439

17 Shelving & Storage

CUSTOMIZED STORAGE

Built-in storage units are an excellent way to make the most of existing storage space in your home. Ready-made or custom-made built-in shelving units, entertainment centers, kitchen cabinets, medicine cabinets, window seats, and under-bed drawers are not only inexpensive and easy to assemble, they allow you to add a unique, personalized touch to your living spaces. (If you rent your home, however, make sure to check your lease and consult your landlord before embarking on one of these projects. Alterations may not be allowed.)

Built-in Shelving

A built-in shelving unit can create valuable storage capacity from an overlooked wall space, such as the area between windows or between a door and its adjacent corner. To construct the shelving, you'll need 1x10 or 1x12 lumber for side panels, top and base panels, and shelves; four 2x2 strips for spreaders; trim molding to conceal gaps along the top and bottom of the unit; 12d common nails and 6d finishing nails. If the unit will be bearing heavy loads, use hardwood boards, and make sure that the shelves span no more than 36 inches. To make installation easier, cut the side pieces an inch shorter than the ceiling height. (This way, you'll be able to tilt the unit into position without scraping the ceiling.) Paint or stain the wood pieces before assembling the unit. Hang the shelves from pegs or end clips inserted into holes drilled in the side pieces.

Plugging

USE: ▶ power drill/driver with plug-cutter bit • table saw or other saw ▶ lumber for plugs

1 You can cut wood plugs from dowels, or make your own from any wood with a plug-cutter bit.

2 If you don't have a drill press, try to keep your portable drill vertical, and drill to about the same depth on each plug.

3 Remove plugs by running the board through a table saw (or use a circular saw or handsaw) set to the plug depth.

4 Add glue to the plug, and tap it into the hole. Trim excess wood with a saw or plane, and sand when the glue dries.

Filling Holes

USE: ▶ putty knife or flat-bladed screwdriver • container for mixing (if needed) ▶ wood putty • sandpaper • stain (if needed)

1 Wood putty comes premixed and as a powder. You can mix powder types to suit: thicker for holes; thinner for surface damage.

2 Apply wood putty by pressing it in the hole with a putty knife, flat-bladed screw-driver, or your fingertip. Leave a little excess.

3 Small patches dry quickly and sand easily. Because putty shrinks, you may need to fill deep holes with two applications.

BASIC BUILT-INS

Constructing a Built-in Shelving Unit

USE: ▶ pry bar • saw • measuring tape • hammer • drill/driver • square • miter box and backsaw ▶ shelving lumber • 2x4s • trim • common and finishing nails

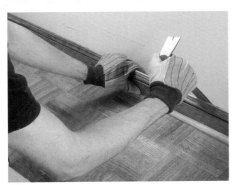

1 Use a pry bar to carefully remove the base trim where you will install the built-in unit. Save it for reinstallation later.

2 Build a box out of 2x4s for the base of the unit. Use it to elevate the bottom shelf, and anchor the built-in unit to the wall.

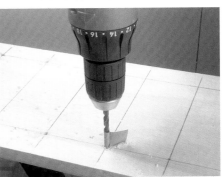

3 Cut 1x10s or 1x12s for the sides. For adjustable shelves, mark your bit (or use a depth guide), and drill rows of peg holes.

4 Make a support frame from 1x4s for the ceiling, and tack the side panels. A finished shelf conceals the top frame.

5 Add glue and nails to secure the side panels to their support frames. You should add several support cleats along the sides.

6 Nail the bottom shelf in place. Make the shelf flush with the support box so that you can nail on baseboard trim.

7 Cut trim pieces to fit around the top of the unit (matching any existing wall trim), and fasten with 6d finishing nails.

8 Replace the cut baseboards around the unit. Miter-cut new matching trim pieces to fit around the base of the unit.

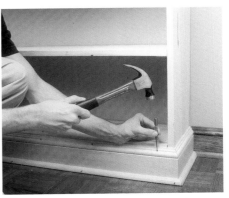

9 Use 6d finishing nails to nail in the base trim; set the nails; putty the holes; and finish the unit with paint or stain.

17 Shelving & Storage

ADDING CLOSET SPACE

What homeowner or apartment dweller hasn't complained about having too little closet space? Fortunately, there are almost always ways to find a bit more closet space. You may have to put in some time and effort, but in the long run, it's always easier than buying a new house with bigger closets.

The easiest and most obvious solution is one of the many commercial closet organizing systems now on the market. But constructing your own version of a commercial closet organizer is far more economical. With a combination of shelves and plywood partitions, you can divide a closet into storage zones, with a single clothes pole on one side for full-length garments; double clothes poles on the other side for half-length garments like jackets, skirts, or slacks; a column of narrow shelves between the two for folded items, handbags, or shoes; and one or more closet-wide shelves on top.

Building a New Closet

Sometimes, taking away from your living space and adding a new closet is the only way to increase closet space. This will involve framing new walls, finishing them with drywall, adding matching trim, and hanging an interior door. The closet's interior should be at least 24 inches deep and, for a bedroom closet, 48 inches wide. Try to locate the closet at a corner so that you only have to build two walls.

Closet Layout

It's always nice to have a lot of closet space. But given the high square-foot cost of a house, it is expensive to turn over a lot of floor space to storage space. The best plan is to minimize the closet footprint and maximize the closet efficiency by organizing the interior. For example, a typical closet has one shelf and one hanging pole. It also may be a full 24 inches deep, even though most clothes and many other kinds of storage will fit in a narrower area. Pick up 2 or 3 inches in depth at each closet in the house, and your kitchen is a foot wider or your family room a foot longer. One simple way to increase capacity is to install two storage shelves above the hanging pole. Allow a foot or so of clearance under the door opening for the first, full-depth shelf, and install a second, half-depth shelf above it. You also can divide up closet space with partitions so that one section has full-height hanging storage and another has two, half-height hanging poles for shorter items. The best plan is to build (or buy) a closet system with adjustable components to suit changing needs.

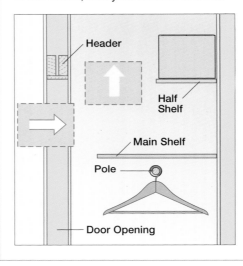

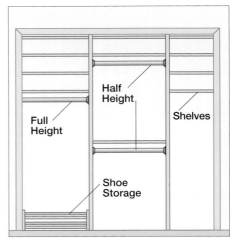

Framing a Closet

USE: ▶ circular saw or crosscut saw • hammer • measuring tape • framing square • chalk-line box • 4-ft. level • drywall tools • stepladder ▶ 2x4s for studs, wall

1 Mark the closet outline on the floor. Nail down the 2x4 soleplate every 2 ft. Mark the door, allowing space for jambs.

2 Transfer the closet outline to the ceiling using a chalk-line box and level. Nail on the outside studs and top plates.

3 Set studs 16 in. on center, and toenail them to the top plate and soleplate on all four sides. Nail crossbraces between studs.

CLOSETS

Closet Systems

There are many types of closet systems. One of the most versatile is wire racks. Mix-and-match components include several stock lengths of shelving with integral hanging bars, plus support brackets and clips that allow you to install these systems in almost any configuration. They also are easy to alter. You can create the same kind of compartmentalized storage with custom and stock wood systems, or hire one of many specialized closet companies to build your system.

If you have a spare room to convert, you can design a wood storage closet system to create custom storage.

Use wire-rack components to build compartmentalized, multitiered storage into a standard closet space.

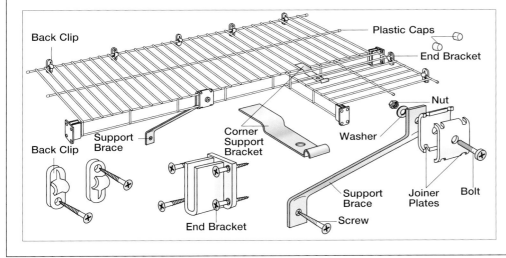

Use interlocking wire bins and box storage racks on tracks to increase the capacity of a closet and free floor space.

4 *Nail jack studs on each side of the doorway to carry the header. Install a double header across the opening.*

5 *Remove the section of soleplate in the doorway so that you can run finished flooring into the closet space.*

6 *Install drywall inside and out, and finish and sand the joints. Install the door, closet pole, and fittings.*

ADDING CLOSET SPACE 443

17 Shelving & Storage

IDEAS FOR WORKSHOPS

Workshops and other utility areas such as garages, attics, and basements can benefit from storage upgrades as much as any other room in the home—perhaps even more so, as utility areas are prone to clutter. Convenience, flexibility, and safety are the things to keep in mind when reorganizing your work space. Try to provide storage space for tools and hardware as near as possible to where they'll be used. In addition to a sturdy workbench, utility shelving is a mainstay in any workshop. You can buy ready-to-assemble units or make your own using ¾-inch particleboard or plywood shelves and ¾x1½-inch (1x2) hardwood stock for cleats (nailed to the wall), ribs (nailed to the front underside of the shelves), and vertical shelf supports (nailed to the ribs).

Don't forget about Peg-Board. To make a Peg-Board tool rack, attach washers to the back of the pegboard with hot glue, spacing the washers to coincide with wall studs. Position the Peg-Board so that the rear washers are located over studs. Drive drywall screws through finish washers and the Peg-Board into studs. (Use masonry anchors for concrete walls.)

Finally, try to take advantage of any otherwise wasted space. The area in your garage above your parked car is the ideal spot for a U-shaped lumber storage rack, made of 1x4 stock and connecting plates. The space in front of the car could be used for a storage cabinet or even a workbench.

Shop Storage

You can buy a ready-made workbench for your shop or build one yourself that is almost as sturdy. For a basic bench, use 4x4s (or doubled 2x4s bolted together) for the frame, and a double layer of ¾-inch plywood glued and screwed together for the work surface. There are hundreds of different storage products—from huge, roll-away tool carts to tiny bin dividers—but you can also build your own to use your work space efficiently. To protect children, provide locked storage for hazardous materials.

Move a roll-away cart around the shop where you're working, and store it under a bench. Most are rated by weight capacity.

An old-time storage technique is still a good idea: screw lids to supports, and use jars to keep small items in plain sight.

Lumber storage can be portable, too, on a homemade rolling cart like this one—one side is for sheet goods and one is for boards.

Attaching Casters — MONEY SAVER

USE: ▶ power drill/driver • measuring tape or ruler • socket wrench • open-end wrench • pencil ▶ casters • bolts • nuts

1 Inset a caster at each corner of your storage container, and mark the bolt-hole locations for drilling.

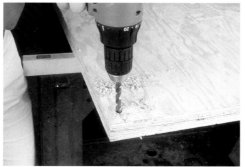

2 Drill pilot holes through the base of the container—in this case, two layers of plywood glued and screwed together.

3 Bolt the casters to the wood base. Use one wrench to hold the nut while you turn the bolt head with a socket wrench.

UTILITY STORAGE

Rack & Prefab Systems

When you need a few storage shelves but don't want to make a permanent installation on the wall, use metal rack or prefabricated wood units. Metal rack systems are held together with nuts and bolts. You can install shelves at any level using holes in the corner posts. The systems, often sold in kits, are not stable without the X-pattern bracing on the back. Prefab wood storage units assemble easily and hold a variety of tools and supplies.

Metal rack shelf units reinforced with X-braces are sturdy enough to hold heavy supplies, adjustable, and easily altered.

Prefab wood storage units, typically sold as knockdown kits, can be purchased with many shelf and drawer combinations.

Assembling a Wall System

USE: ▶ screwdriver • hammer or rubber mallet ▶ wall-system kit

1 Before starting a kit project, read through the instructions, and make sure you have all the panels and hardware.

2 Panel sections may fit together with wooden dowel pegs. You tap them into one panel, and press them into mating panels.

3 Some panels join with threaded fasteners. The stud end mounts in one panel, and fits into a predrilled hole.

4 With threaded fasteners, once the stud end is in position, you use a screwdriver to engage and tighten the joint.

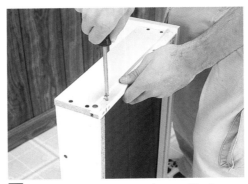

5 Assemble drawer sections with dowels and fasteners. Kits typically include full-extension drawer guides.

6 A modest-sized unit such as this can be assembled in under an hour. The plastic laminate panels require no finishing.

17 Shelving & Storage

ADDING STORAGE SPACE

Few of us have the luxury of too much storage space. Yet if you look closely, odds are you will find places where the space you do have is being wasted or underutilized. Here are a few spots where dead space can be converted into handy places to put things.

Narrow hallways that end at a closet or door offer a surprising amount of storage space, especially if your ceilings are high. Install shelving around the doorway—or better yet, build a deep platform over the door supported by ledgers nailed to each side wall.

In the space under the eaves below a sloped attic roof or under a staircase, you can build a made-to-fit bookcase or built-in cabinets. This is especially good in a child's bedroom, where it won't seem so low to the ground.

Garments hung on hangers usually do not take up a closet's full depth. The space you gain by moving the clothes pole toward the back of the closet may accommodate wire-rack shelves attached to the back of the closet door.

To utilize the overhead space in your garage, build deep storage platforms supported by ledgers screwed to wall studs and threaded rods hooked to ceiling joists or rafters. You can also hang tools from the walls by mounting pegboard. (See p. 436 for how to mount Peg-Board.) You can buy sets with a variety of hooks and brackets for tools. For small items, such as jars of nails, make shallow shelves by nailing 1x4 boards between the exposed studs.

Basic Toy Box

To make this toy box, first cut all the parts to the dimensions in the materials list. Rabbet the front and rear panels, and attach them to the side panels with glue and finish-ing nails. Square up the box with clamps; let the glue dry. Attach the trim and corner guards with glue and brads. Glue and nail cleats to the inside bottom of the front, back, and side panels, and glue the bottom panel to the cleats. Paint or stain the wood. Attach the piano hinge to the box, and then to the lid. Attach the lid supports and handles.

MONEY SAVER

MATERIALS

- 2 front/rear panels ¾ x 17¼ x 38½
- 2 side panels ¾ x 17¼ x 18
- 2 top trim strips ¾ x 1½ x 20
- 2 top trim strips ¾ x 1½ x 40
- 2 base trim strips ¾ x 2½ x 20
- 2 base trim strips ¾ x 2½ x 40
- 4 corner guards ¼ x 1 x 1 x 13¼
- 2 cleats ¾ x ¾ x 37
- 2 cleats ¾ x ¾ x 15½
- 1 bottom panel ¾ x 16⅞ x 37
- 1 lid ¾ x 21 x 42
- Hardware: 36" piano hinge, 2 toy-box lid supports, 2 handles

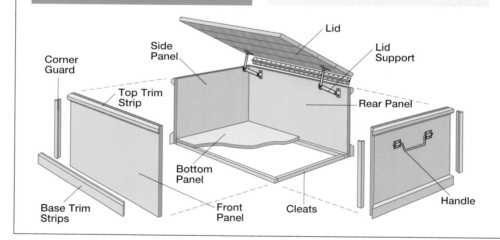

Cedar Closets

Both solid cedar boards and composite cedar panels have only moderate resistance to insects, and are used more for their pleasant aroma and appearance. The sheets of pressed red and tan particles are no less aromatic than solid wood, but the panels are 40 to 50 percent less expensive, and are easier to install. Solid boards require more carpentry work, and are likely to produce a fair amount of waste unless you piece the courses and create more joints. To gain the maximum effect, every inside surface should be covered, including the ceiling and the back of the door. The simplest option is to use ¼-inch-thick panels, which are easy to cut into big sections that cover walls in one or two pieces. Try to keep cedar seams in boards or panels from falling over drywall seams. No stain, sealer, or clear finish is needed; just leave the wood raw. The cedar aroma will fade over the years as natural oils crystallize on the surface. But you can easily regenerate the scent from the panels by scuffing the surface with fine sandpaper.

STORAGE OPTIONS

More Storage Ideas

Need more storage space? On this page are eight ways to build more storage capacity into attics, basements, and other utility spaces: 1) on racks suspended from floor joists; 2) on furring strips screwed to joists; 3) in roll-out cabinets under stairs; 4) on shelves between studs in wall cavities; 5) on shelves built under steeply sloped roofs; 6) on shelves cut to fit into the spaces between rafters; 7) on high shelves suspended from rafters near the roof ridge; 8) on plywood panels fitted between truss webs.

MONEY SAVER

1. JOIST HANGING RACK
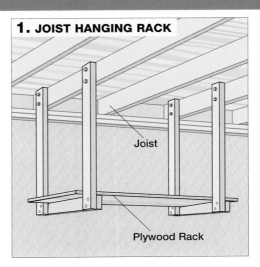

2. JOIST FURRING RACK
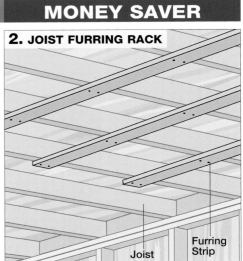

3. STAIR PULLOUTS
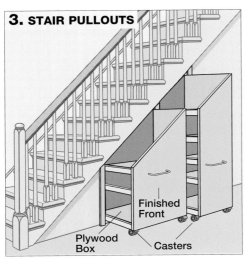

4. IN-FRAME SHELVES
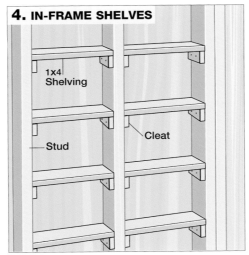

5. GABLE-END SHELVES
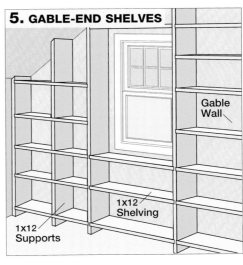

6. EAVES SHELVES
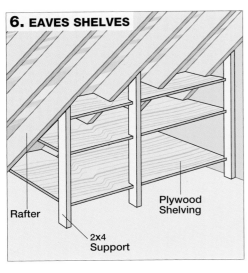

7. RAFTER-HUNG SHELVES
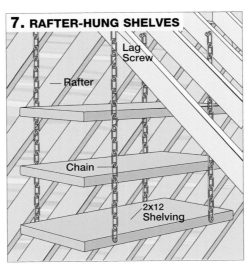

8. TRUSS-MOUNTED SHELVES
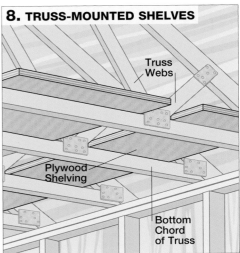

ADDING STORAGE SPACE 447

18 Roofing

450 TOOLS & EQUIPMENT
- Roofing Language ◆ Anatomy of a Roof
- Special Tools ◆ Ladders & Scaffolds

452 DESIGN OPTIONS
- Roof Design ◆ Roof Types ◆ Ventilation
- Rise & Run ■ OVERHANG OPTIONS

454 FLAT ROOFS
- Typical Flat-Roof Systems ◆ Flat-Roof Details
- ■ REPAIRING A FLAT ROOF ◆ Roof Decks

456 ASPHALT SHINGLES
- Choosing Shingles ◆ Installing Shingles
- ■ SHINGLE REPAIR ◆ Preparing for Reroofing
- Deterioration ◆ Installing Double-Layer Roll Roofing

460 SLATE
- Working with Slate ■ REPAIRING SLATE

462 MASONRY
- Clay Tile ■ REPAIRING CLAY TILE
- Concrete Tile ■ REPAIRING CONCRETE TILE

464 WOOD SHINGLES & SHAKES
- Installing Wood Roofing
- ■ REPAIRING SHINGLES & SHAKES

466 METAL
- Standing Seam ◆ Corrugated Panels
- ■ TWO WAYS TO REPAIR METAL ROOFS

468 SKYLIGHTS & ROOF WINDOWS
- Installing a Skylight ◆ Design Options
- ■ SOLAR TUBE

470 FLASHING
- Installing Step Flashing ◆ Valleys
- Flashing & Counterflashing ■ FLASHING REPAIRS

472 CLIMATE CONTROL
- Winter Problems ◆ Ice Dams
- Ice Control ◆ Temperature Extremes

474 GUTTERS & LEADERS
- Maintaining Gutters ■ REPAIRING GUTTERS

18 Roofing

ROOFING LANGUAGE

Standing out in the yard with a pair of binoculars at the ready, you might be mistaken for a bird watcher by your neighbors. How would they know you are only following the advice of the Asphalt Roofing Manufacturers Association on the best way to inspect your shingles. If your knees get a little wobbly when you climb a ladder, using binoculars is not a bad idea. Of course, you could ask two or three roofing contractors to take a look instead. But it is good policy to know something about the condition of your roof—and the language roofers use—before asking for estimates. The following chapter covers each type of roofing material in turn, from asphalt shingles to slate. But before getting into the particulars, it's worth taking time to nail down a few basic terms.

Roofing Speak

One square of shingles is the amount needed to cover 100 square feet of roof surface. This is the standard measure you'll find in contractor's estimates, and it is the way to order shingles and most other roofing materials. To cover that area you may need more of one type of shingle than another, depending on their size and configuration. But a square of standard asphalt shingles (the material most used on residences) is composed of three bundles of 27 shingles each.

Coverage refers to the number of layers of roofing protection provided. For example, standard modified bitumen for flat roofs or asphalt shingles for sloped roofs provide one layer. Dimensional asphalt shingles that show a more textured, shakelike roof, may provide two layers of coverage.

The slope of a roof is expressed as a ratio: inches of rise (vertically) per inches of run (horizontally). For example, a low-slope, 3-in-12 roof gains 3 inches of height every foot. On a 16-foot-long run from the eaves up to the ridge, the roof would rise 4 feet. Just measure a set distance along the side wall in from the eaves (run) and then a straight line up to the roof (rise) to find the slope on your roof. You could use slope to help calculate an order or to determine what type of roofing to use. For example, on standard asphalt shingle bundles you might read that the manufacturer doesn't recommend installation (or has special requirements) on roofs with a slope of less than 4-in-12.

As a safety guideline, you may use slope to decide if a roof is walkable. That means you can work on it without scaffolding. For most people, the cutoff point is a 6-in-12 slope, which means that a 16-foot-long run would rise 8 feet from the eaves to the ridge. But use some common sense too: for example, wear sneakers, and go up only when the roof is dry. And if you feel uneasy about being up there, even on a low-slope roof, stay on the ground.

Estimating the Order

There are several ways to estimate roof surface area. The most obvious and reliable is simply to go up on the roof with a tape measure.

From the ground, you can measure the floor

Anatomy of a Roof

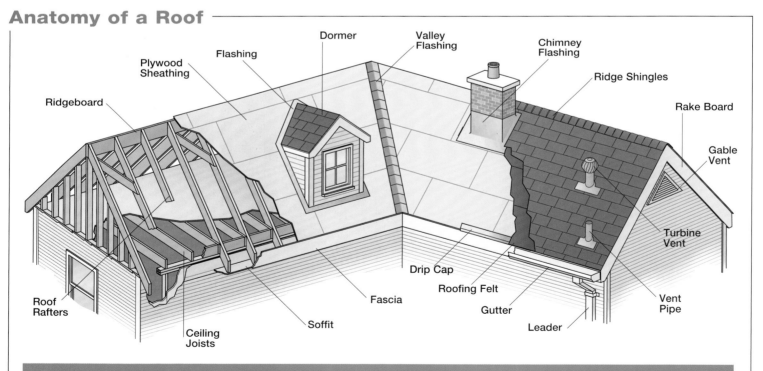

The elements of most roofs are similar to those of the gable roof, shown above. Rafters carry the weight down to the house frame. Plywood sheathing supports water-shedding shingles. Flat roofs, the exception, have joists like floors instead of rafters.

TOOLS & EQUIPMENT

plan of the house, add on overhanging areas, and then multiply by one of many conversion factors based on the roof slope—for example, by 1.03 for a nearly flat roof with a slope from 1-in-12 to 3-in-12, by 1.12 for a 6-in-12 slope, and by 1.45 for a steeply pitched roof with a slope of 12-in-12. Add on 10 percent to cover ridge tabs, hips, valleys, and starter courses, and round up to the next full square.

Multiplying by conversion factors is not necessary if you measure on the roof. But you should figure in extras—for example, four full shingles for every 5 linear feet of hip or ridge, or about 30 linear feet per square of shingles, which builds in some waste. To save money, buy nails in bulk; figure to use about 2 pounds of standard 1¼-inch galvanized roofing nails per square.

When ordering roofing felt, remember most manufacturers assume a 2-inch overlap, and one 432-square-foot roll will cover about 400 square feet. If you decide to overlap more, you'll need more rolls of felt.

Special Tools

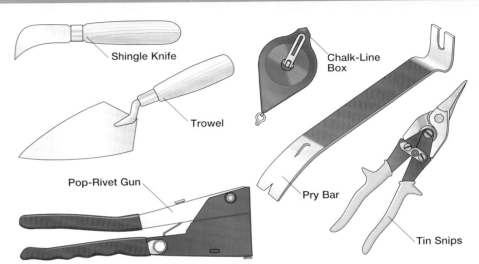

Roofing tools include the basics like a hammer, of course, and a few you might not have in the toolbox: a shingle knife, trowel, pop-rivet gun, chalk-line box, pry bar, and metal shears. Professional roofers generally use air-powered nailers to speed up the job.

Ladders & Scaffolds

Take a warning from the construction industry, where the most serious accidents are the result of falling. In plain language, you need to be particularly careful working on a roof, as you will be anywhere from 10 to 30 feet off the ground. You can make the job safer, though, by using the variety of ladders, scaffolds, and fall-arrest devices on the market today. Use extension ladders in good condition rather than stepladders to gain access to your roof. Scaffolding and working platforms make a roofing job even easier, and can be rented for the duration of a job. The simplest are metal cleats or roof brackets anchored to the roof itself. Ladder jacks, pump jacks, and scaffolding provide movable platforms for working. Those rated for construction will support you and piles of shingles at the roof edge.

ROOFING LANGUAGE

18 Roofing

ROOF DESIGN

Roof designs have changed over the years, reflecting a general trend toward cost savings, efficiency, and ease of installation. Victorian-era mansard slate roofs, with geometric patterns of multicolored tiles, are perhaps the most artistic and difficult roofs to build (and fix). If you're lucky enough to own a house with a roof like that, you may question your luck after getting a bill for fixing a leak.

With modern roofs, the major cost in repair and replacement is labor. Older homes with slate roofs have the dual expense of labor and material. Sometimes homeowners who need to replace a slate roof will opt for asphalt shingle replacement, because the price of new slate roofing can be as much as 10 times that of asphalt shingles.

Roofs can be pitched at almost any angle, from nearly flat to almost vertical. While a roof's pitch is mostly a matter of design, the roofing materials that can be used are beholden to the pitch, if they are to shed water, snow, and debris. Flat roofs cannot be covered with asphalt shingles; mansard roofs cannot be made with built-up roofing.

Climate

Climate and weather patterns are the primary concerns for the general design of the roof. A flat roof, for instance, is impractical in a part of the country with a great deal of snow or rain—that's why houses in New England generally have steeply pitched roofs.

The type of roofing material needed to meet weather conditions and the slope of the roof are secondary concerns. Clay tile roofs, for example, look beautiful on Spanish-style

Roof Types

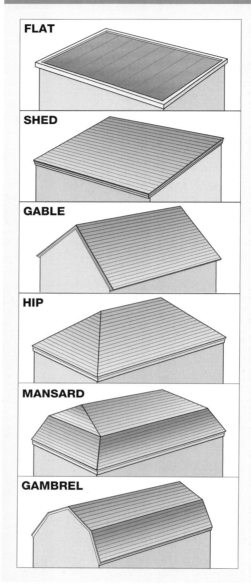

- ◆ **FLAT ROOFS** are easiest to install and fix for do-it-yourselfers. Made with built-up roofing material or roll roofing, they last 10–15 years.

- ◆ **SHED ROOFS** are similar to a flat roof, only they are pitched at an angle. They can be roofed in metal, roll roofing, shingle, or tile, depending on the pitch.

- ◆ **GABLE ROOF** is a simple A-frame roof with a single ridge, no hips or valleys. It can be sheathed in any roofing material, except roll roofing.

- ◆ **HIP ROOFS** have all roof sections sloping toward the roof ridge. Hips are capped with shingles to keep out water.

- ◆ **MANSARD ROOFS** slope steeply at first, then flatten out. Typical in Victorian designs. Done in slate or shingle, sometimes copper metal.

- ◆ **GAMBREL,** the basic barn roof design, has a slope that flattens near the top. This allows for more storage space in an attic. Needed extra support is provided by interior cross beams. Roofed in most materials.

Ventilation

Ridge vents run along the top of the roof. They allow optimal attic ventilation by creating a strong flow from vents in the soffit overhang. In many cases, they are woven into the shingles so you don't even notice them.

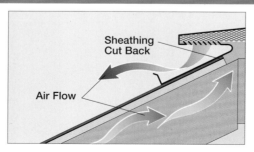

GREEN SOLUTION

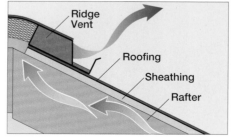

DESIGN OPTIONS

homes in the Southwest. However, clay tile does not perform very well in colder climates, where ice and snow can back up into the tiles and cause roof damage. A slate roof with a steep pitch, on the other hand, performs very well in cold climates, but may not shed heat very well in hot climates.

Seasonal weather should also be taken into account when choosing a color for your roofing material. Asphalt shingles, for example, come in a variety of colors, from white to pitch black, with shades of gray, green, and red also available. As you would guess, a black roof pitched at a low angle on a house in Florida is not the brightest idea for keeping a house cool. In hot climates, white shingles, Spanish tiles, and wood shakes or shingles will help deflect the sun's heat. In cold climates, slate and dark asphalt shingles will absorb the sun's heat.

Trusses

If you're nervous about cutting rafters from scratch and you don't mind not having open attic space and dormers, roof trusses are a good framing option. Roof trusses are difficult to install, but you can get them made to order, which eliminates cutting. Trusses are essen-tially framed triangles, with 2x4 or 2x6 chords (the outside lines of the triangle) and 2x4 webs (supporting members, most often in W- or M- shapes inside the triangle) held together by gussets, which are flat metal or plywood plates. The two top chords and the long bottom chord form the shape of a gable roof.

You erect trusses right on the top plates, with one truss per stud bay. If you've framed the walls at 16 inches on center, for example, then the trusses will occur every 16 inches on-center as well. This is why you lose the attic space—every 16 or 24 inches there's a bunch of intruding supports from roof to floor. Also, you can't cut into a truss to install dormers, because cutting any one of the framing members compromises the structural integrity of the entire truss. For this reason, trusses are common in storage buildings and garages.

You order trusses from a local truss manufacturer or building-supply store, by specifying the desired length of the bottom chord. There may not be a wide variety of ridge heights for you to choose from, however.

In a warm climate where you rely more on air conditioning than heating, use light colored shingles; they reflect more heat.

Aluminum paint (applied mainly over flat roofs) is another way to reduce heat build-up. It also reduces surface cracking.

Rise & Run

The rise and run of your roof should suit house style and regional climate. Generally, cold, wet locales require houses with steeply pitched roofs to shed snow and rain. In hot, arid locales, low-slope and flat roofs are more suitable.

GREEN SOLUTION
Overhang Options

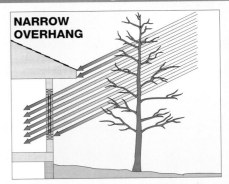

A narrow overhang can be the best design where you want the greatest heat gain and maximum light through the windows.

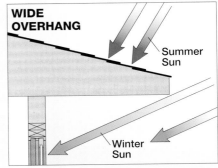

A wide overhang is good in wet climates to shed water, of course, and to block more of the harsh rays of the summer midday sun.

18 Roofing

FLAT ROOFS

Flat roofs look good on modern houses in architecture magazines, but they have problems. They would be great in a climate where it rarely rained. And some houses just don't work with a steep gable or Prairie-style hip roof, so they're stuck with a water-collecting top, which is bound to need repairs.

Old-style flat roofs were built up from as many as five layers of asphalt felt paper, each one sandwiched between beds of molten tar, applied with the kind of labor-intensive steps builders don't have the time for today. But if you have an old built-up roof, don't be too eager to tear it off. First, removal of so much heavy material is a major project. Second, if the roof was applied over a solid frame and covered with gravel (called ballast) to keep it flat and shielded from sunlight, the surface could last 40 years or more. Third, most of these flat roofs just don't spring a leak somewhere in the middle of the interlaced layers, where it is difficult to make a long-lasting repair. They normally open up along the edges, at seams protected by metal flashing, which are easier to fix.

Fixing Edge Leaks

Along the edge of a flat roof, water enters where metal flashing is raised. Hairline breaks can be sealed with a liberal coat of roof cement and reinforcing fabric. Larger and longer openings need to be reset. After cleaning out any debris, trowel as much tar as you can fit under the raised edge and re-nail the flashing to the roof. Add a coat of tar on top of the metal and over the new nails, extending it past the flashing onto the roof a few inches. Then embed in the tar a layer of fiberglass roofing tape, and add a final, top coat of tar.

Repairing Newer Flat Roofs

With modern coatings such as modified bitumen, bubbles, punctures, and other openings are easier to fix—with one caution. The rubbery sheets of these roofs are not joined to each other with tar. They are fused by heating the material until it begins to melt, making it easy to add a waterproof cover. The old section can be cleaned, scarified, and then heated to fuse with a patch piece. But because of fire danger, this is best done by a professional roofer.

Typical Flat-Roof Systems

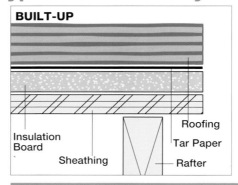
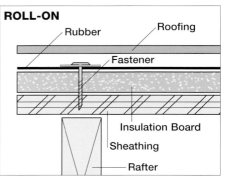

Built-up and roll-on roofs. Built-up roofs (left) have several layers of roofing felt and asphalt under a final layer of asphalt and gravel. Modern roll-on roofs (right) have a single layer of synthetic rubber, attached with nails and roof cement.

Flat-Roof Details

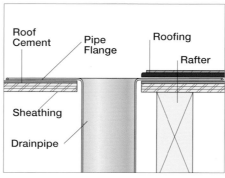

Interior drains are placed at low spots, and carry water through part of the structure en route to an exterior outlet.

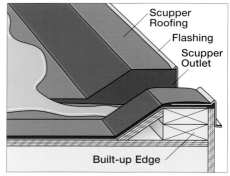

Scupper troughs are cut into the built-up roof edge so that collected water can drain to an exterior downspout.

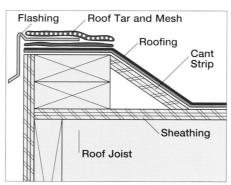

Edge flashing is essential to protect the roof from seepage. Typically, the flashing is coated with roof fabric and tar.

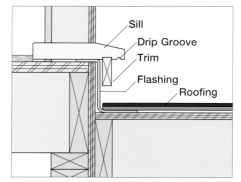

Where a flat roof meets a second-story wall, the entire seam needs flashing. Its upper edge must be protected.

FLAT ROOFS

Repairing a Flat Roof — MONEY SAVER

USE: ▶ spade • utility knife • putty knife • trowel • broom ▶ roof cement • roof patch • gravel

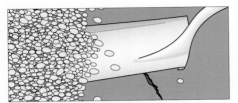

1 With a flat spade, scrape away surface gravel from the damaged area.

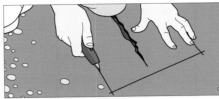

2 Cut away a rectangular piece around the damaged section with a utility knife.

3 Fill up the removed patch with a generous amount of roof cement.

4 Tamp down the patch material, generally more roofing or fiberglass mesh.

5 Using a mason's trowel, cover over the patch with roofing cement.

6 Cover the patched area with gravel, called ballast, that protects the surface.

3 Common Flat-Roof Problems

1. Too many roofers leap to the ultimate solution for a leaky flat roof: a whole new roof. Sometimes, of course, that's the best advice. But in general, don't opt for removal and replacement until you take a shot at fixing spot leaks, particularly when they occur anywhere near a protrusion through the roof surface. That includes the edges, interior drains (drainage holes in the roof overhang that connect to down spouts), the perimeter of skylights, chimneys, and plumbing vent pipes.

2. You're likely to need a new roof—if not immediately, then in a year or two—if there are bubbly areas that compress when you walk on them. They might be the size of your foot or larger. Bubbles indicate that some of the layers of roofing have delaminated. If you step down and hear the sucking sound of water, you'll need the roof much sooner than later—there's water trapped inside your roof, rotting it away.

3. Most flat roofs retain some water in puddles. Standing water can result because the roof was not built with enough slope or because the building—or even a few supporting joists—have gradually settled over the years. These depressions can cause problems where water that doesn't immediately drain stands against seams. In the summer, large puddles also can become stagnant. Depressions in modern flat roofs made of rubberlike sheets can be filled before they begin to leak by bonding on pieces of new roofing contoured to raise the low spot.

Roof Decks

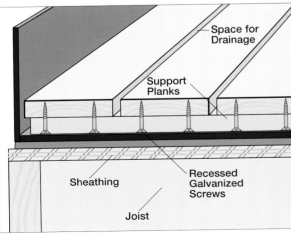

Install a doorway to a flat roof (and railings according to local codes), and the roof can become a deck. To protect the roofing and still have access in case of leaks, cover the surface with duck boards, which are removable sections of spaced planks laid on sleepers.

18 Roofing

CHOOSING SHINGLES

Four out of five residential roofs are made of asphalt shingles or continuous sheets of asphalt called roll roofing, which is usually reserved for low-slope roofs not visible from the ground.

You may be asked to chose between regular asphalt shingles or fiberglass, with a bottom mat of fiberglass mesh that is lighter, stronger, and longer-lasting than asphalt. Fiberglass-mat shingles are a good choice for reroofing jobs, as they reduce the load carried by the rafters without giving up durability. About 80 percent of all shingles sold for new homes and reroofing are the fiberglass-mat variety, including almost all the heavyweight, overlay-type shingles.

Shingle Weight

Shingle weight is an important factor on both new roofs and on reroofing jobs because heavier shingles last longer, carry a longer warranty, and generally offer a better fire rating. Of course, they are more expensive than lighter shingles. The weight rating (240 pounds for a standard shingle) denotes the total weight of a square of shingles (enough to cover 100 square feet of roof).

The heavyweight shingles are those over the 240-pound rating—generally 300 pounds per square or more. Individual shingles in this category often are configured in layers—like a shingle on a shingle—that simulate the dense pattern of slate or wood shakes. The heavyweights are a good choice for new homes and additions, but a questionable choice for reroofing jobs, where the weight can overload the roof structure.

Color

You don't often see a bright green, blue, or red roof—even though asphalt shingles are available with granules in those colors—because they can become a bit oppressive after a few seasons. Off-white or light gray shingles make a house look larger, but will mar more easily than dark shingles and show wear sooner (even though they will not wear out any faster than dark shingles). Light colors on the roof can reflect more sunlight than dark, which will keep the house cooler in summer and reduce air-conditioning costs. If gaining heat is more important, a dark shingle would be the most energy-efficient choice.

Installing Asphalt Shingles

USE: ▶ ladder or scaffolding • hammer • chalk-line box • shears • utility knife • work gloves ▶ metal drip

1 Roofers have different techniques, but all start with a sound roof deck where nail heads are driven flush.

2 Nail on a metal drip edge at the edge of the eaves with roofing nails. This protects the fascia boards from rot.

6 Start each new course at a 6-in. offset, to stagger the seams in adjacent courses. Vertical chalk lines aid in alignment.

7 To trim shingles in valleys and other areas, use shears or a utility knife. This is a closed valley with interwoven shingles.

TOOLS AND MATERIALS

◆ **WHAT YOU'LL NEED.** You need only basic tools to install asphalt shingles. You can cut them with shears or a utility knife, and nail them in place with a standard hammer. For reroofing jobs, a pry bar and flat shovel are handy for removing old shingles.

◆ **TYPES OF SHINGLES.** Dimensional shingles (bottom) are thicker than regular flat-tab shingles (top) and are not uniform in color, which produces a three-dimensional look similar to slate. Heavier shingles also last longer and have better fire ratings.

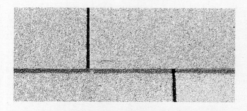

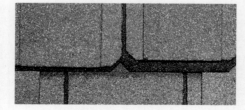

ASPHALT SHINGLES

edge • roofing nails • roofing felt • shingles • roof cement

3 Roll roofing felt on top of the decking, nailing it every 10–12 in., 3 in. from the edge. Overlap rows by several inches.

4 For the starter course, snap a chalk line, and lay the shingles with the tabs pointing up. Put one nail through each tab.

5 The first course covers the starter course with the tabs pointing down. Each shingle should have at least four nails.

8 To shingle around a vent stack, trim to overlap only the upper half of the vent collar, and seal underneath with roof cement.

9 To shingle the ridge, cut slightly tapered tabs from whole shingles, and wrap them across the ridge, nailing on both sides.

10 To save time, let full shingles extend past the roof overhang, and trim all of them at once using shears.

Shingle Repair — MONEY SAVER

USE: ▶ pry bar • hammer • caulking gun ▶ new shingles • roofing nails • roofing cement

1 Remove all damaged shingles, and pull any protruding nails with a pry bar.

2 Nail new shingles except the topmost course, which must be cemented.

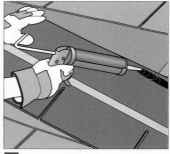

3 Apply roofing cement to the underside of the topmost course of shingles.

4 Slide the new shingle into place, and tamp down on the surface so it sets firmly.

CHOOSING SHINGLES

18 Roofing

REROOFING

Most asphalt shingles should last without leaking for 15 to 20 years. Some last longer—even 25 or 30 years. After about 15 years, however, you might start checking for signs of wear. But don't jump the gun. There is no advantage in reroofing a building ahead of time—before the shingles have started to deteriorate and years before they are ready to spring a leak.

Reroofing consists of applying new shingles over the existing roofing material. This is less expensive and easier than a tear-off job, which requires that the old roofing be stripped off and hauled away.

Tear-Off or Reroof?

The first step in determining whether or not to reroof is to check the rake of the roof to find out how many roofing layers there are. (The drip edge sometimes is applied before reroofing and may hide evidence of previous layers.) Once you determine the number of layers, check local roofing codes for the maximum number of roofing layers allowed—the figure will be different depending on the type of roofing and pitch of the roof. For asphalt shingles, codes usually allow the original plus two layers of reroofing.

All rotten boards under the old roofing must be replaced. Go to the attic and examine any suspicious spots, including voids and separating plywood. Check for rot by poking with a screwdriver or awl. If rot is limited to a few places, you need only remove the old roofing and replace the boards in those spots. Where necessary, build up the roof above the replacement sheathing with extra layers of shingles to make a flush surface for the new roof.

Checking Wear

Here are the four progressive stages of shingle wear to look for. First, you may notice the tiny chips embedded in the surface of asphalt shingles, called granules, accumulating in gutters and at downspout outlets. Second, you will see bare black patches of tar appearing as more granules are lost. This is hard to see on a dark roof, even with binoculars, but obvious on shingles with white or gray surface granules. Roofs at this stage probably will not leak—not yet. But in a few years they probably will.

In the third stage, exposed sections of the shingles, called tabs, have lost most of their surface granules and start to become brittle. You have to touch the tabs to detect this condition. Within a few seasons, however, the tabs will start to curl noticeably. Even then, the roof may not leak. But now is a good time to reroof—before the curling becomes excessive, and gets in the way of new shingles. In the fourth stage, brittle shingle tabs crack and break. Bare, black patches that appear as the tabs break off are unmistakable. You also may see nailheads holding down the shingles beneath the broken tabs. At this point, you are likely to have small leaks that may start to rot the wooden roof deck and rafters even if you do not see large water stains on the ceiling. After all, nails put holes in the shingles, and are placed so that they would be covered by those broken tabs.

Preparing for Reroofing

USE: ▶ pry bar • hammer • trowel • utility knife • paint scraper ▶ shingles • roofing nails • flashing

1 Before roofing over old shingles, strip a few sections down to the deck to check the decking for water damage.

2 Use a pry bar to pry off the old ridge cap. A new ridge cap will need to be installed over the new layer of shingles.

3 The new layer needs a level surface; broken or bent tabs must be replaced. Cut them away with a sharp utility knife.

4 Cut a tab from a new shingle, and nail it in place with two or three roofing nails to fill the space in the existing roof.

5 Scrape away old roof tar around plumbing vent stacks to clear the way for a new piece of molded flashing.

6 Install new flashing made from molded plastic or metal over the new roofing. The top edge will be shingled.

ASPHALT SHINGLES

Installing Double-Layer Roll Roofing

USE: ▶ chalk-line box • broom • hammer • trowel ▶ roofing felt • roofing nails • roofing cement

1 Snap a chalk line 35½ in. from the eaves; roll out the first layer on top of the roofing felt; and nail at 12-in. intervals.

2 After the first layer is installed (with courses overlapped about 6 in.), spread roofing cement on the first course.

3 Roll out the first course of the second layer, nailing it in place every 12 in. with roofing nails.

4 At course overlaps, trowel on roof cement. Some roll roofing is designed to overlap up to half the previous layer.

5 Successive courses cover the strip of roof cement. Some roll roofing is available with light-colored granules.

6 Long roofs may need vertical seams. These should be nailed and cemented like the horizontal laps.

Signs of Deterioration

The signs of shingle wear are easy to spot and indicate both the degree of decay and the roof's expected life.

1. Loose granules will appear in the gutters while the shingles still look like new.

2. Spottiness of shingles, with bare areas where granules have worn away, indicates the next stage of decay.

3. Shingle tabs will eventually curl at the edges and become brittle. Roofs at this stage are about to spring leaks.

4. The final stage of a shingle's life span are broken tabs. At this point you need to reroof or install a new roof from scratch.

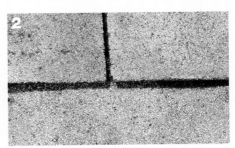

REROOFING

18 Roofing

ROOFING WITH SLATE

Perhaps no other roofing material lasts as long as slate. Incredibly, some old churches and homes from America's colonial period still have their original slate roofs. Slate is still in demand for upscale custom homes, churches, and country-club clubhouses, but most new homes today are roofed with asphalt shingles instead because they are so much cheaper.

Like asphalt shingles, slate comes in many colors, sizes, grades, and weights. Due to its weight—three times that of asphalt—roof rafters and roof sheathing need to be up to code to support slate's heavy load. Slate can be placed over a layer of composition shingles only if the slope of the roof is 4-in-12 or more, and only if a structural engineer has confirmed that the roof framing can bear the weight of 7 pounds per square foot.

Today, only a handful of roofing companies specializing exclusively in slate are in business. Most roofing contractors will do occasional slate roofs. And because it is difficult to cut and apply, slate roofing is not an easy job for the do-it-yourselfer. If you do decide to do the work, expect to make a considerable investment in time and materials.

For a longer-lasting roof than slate, how about slabs of rock? Used as early as the eighth century, the type of roof found on this Scandinavian barn will last for centuries rather than decades.

Installing Slate

Slate should be installed tilted slightly upward at the eaves, extending ½ inch beyond the rake and 1 inch beyond the eaves. Use a piece of lath to shim the starter course, which is made up of slates set lengthwise.

Slate is often laid on top of one layer of 30-pound roofing felt. Some contractors prefer to use individual felt strips under each course to provide additional cushioning. Two copper or brass nails, installed in predrilled holes, hold each slate. The slates are set so that their beveled edges show, with about a ¼-inch gap between slates. The gaps from one course to the next should be offset by at least 2 inches.

To cap each side of the ridge, use slates that are the same width. Alternate the overlap at the peak from one side of the roof to the other. Slates are fastened to the ridge with two nails.

Working with Slate

USE: ▶ nail set • hammer • work gloves ▶ slate roofing • scrap wood

1 To cut a slate roof tile, first use a nail set to punch a series of holes along your cut line on the back of the tile.

2 To complete the cut, place the slate between two pieces of wood at the score line, and tap lightly with a hammer.

3 Smooth out any rough spots along the edge with gentle taps from a hammer, supporting the slate as you work.

MATERIALS

◆ **WHAT YOU'LL NEED.** To work on a slate roof you need three specialized tools: a nail ripper (which is used to cut an old nail flush to the decking surface), a slate hammer (which has a sharp edge for cutting shingles and a point for poking new nailholes), and a T-bar (to use as an edge for trimming shingles).

◆ **SELECTING SLATE.** Slate comes in a variety of colors, from green to gray to red. The colors are generally muted. The material is sold in uniform lengths but varying widths. The thickness may also vary to some degree. A batch of slate sold in one color will also have natural variations. This is considered desirable and will create a slightly mottled appearance as opposed to one color.

Slate is applied in an overlapping pattern with staggered seams. You can use one color, or a mix.

SLATE

Repairing Slate

USE: ▶ nail ripper • hacksaw blade • hammer • screwdriver ▶ new tile • S-hook • scrap wood

1 To remove a cracked slate, hook the nail ripper onto the nail shaft and hammer the ripper to cut the nail.

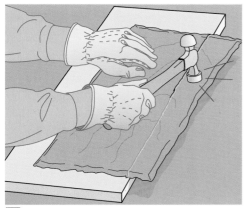

3 To cut the new tile, punch a series of holes in the back of the slate with a nail set and tap with a hammer.

5 Insert the new slate where the old slate was removed, gently prying up on the course above to make room.

MONEY SAVER

2 Using a hacksaw blade instead of a ripper, you can reach under a damaged slate to cut the nails.

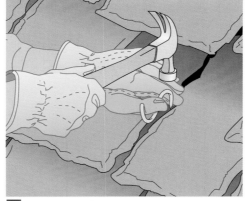

4 Hammer an S-hook between slates to hold the bottom edge of the replacement slate in position.

6 To avoid breaking existing slate as you slide in the new piece, you may need to use a temporary wedge.

Slate Tips

◆ Use solid-copper or brass nails driven through factory-punched holes in each slate.

◆ Practice cutting slates on the ground—it's safer than cutting on the roof. Always wear eye protection.

◆ Rent a wet saw for complicated edge cuts. This will save time and allow for accurate cuts.

◆ Give new slate roofs extra pitch at the eaves. Nail a ¼-inch-thick strip of wood along the eaves so the bottom course turns upwards slightly. This gives the roof a classic appearance and provides drip protection for the fascia.

◆ Slate cannot be nailed over asphalt or other types of roofs and cannot be fastened with pneumatic nail guns. Slate must be nailed firmly but gently, by hand, so that nailheads seat without over-stressing and cracking the slate.

◆ Slates with hairline cracks should be discarded; cracks only worsen with exposure to the weather.

ROOFING WITH SLATE

18 Roofing

CLAY TILE

A clay tile roof is what most people picture when they imagine a house in the American West. An adobe stucco house, designed in the Spanish mission style, would look incomplete with a roof made from anything but naturally colored, terra-cotta, red clay tiles. Like slate, clay tiles are heavy, weighing as much as half a ton per square. If you're considering installing clay tiles, check with a local roofing supply source and your municipal building department to make sure your roof's structure can handle the tiles you choose. Do not tile a roof with a slope flatter than 3-in-12.

Installing Clay Tile

Clay tiles will come with instructions, which will tell you whether or not they require underlayment. Some tiles are nailed directly to sheathing, while others require battens to be laid first. These battens are 1x2 strips of redwood or pressure-treated pine that are spaced at intervals to match the tile exposure. (14 inches is typical.) Further preparations may be called for, such as one 2x2 along all ridges and hips, 1x2 starter strips along eaves and rakes, or 1x3s nailed to rake rafters to allow the tiles to extend further sideways.

Use nails and flashing that will last as long as the tiles—copper is preferred. You should apply a metal drip edge along the eaves before the underlayment is installed. Take special care in the valleys: put down 90-pound mineral-surfaced roll roofing, then W-metal (ridged) flashing at least 2 feet wide. Cover hips and ridges with a double layer of felt.

MATERIALS

- **WHAT YOU'LL NEED.** Copper flashing and nails are best with clay tiles. To cut tiles, you'll need a circular saw equipped with a masonry blade and a good pair of safety glasses for eye protection.

- **SPECIAL INSTALLATIONS.** If your roof has a steep slope of 7-in-12 or more or if you live in an area subject to high winds, fasten every third or fourth course of tiles with metal clips, observing local codes.

Repairing Clay Tile — MONEY SAVER

USE: ▶ pry bar • hammer • wet saw (optional) ▶ new tile • tile clips • scrap wood

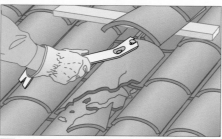

1 Wedge up tiles above the damaged course with wood strips; remove old tile nails with a pry bar; and remove broken tile.

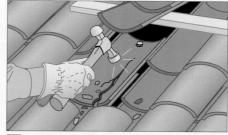

2 Wedge up overlaying tiles; break the old concave tile underneath with hammer; and remove debris. Do not hit good tiles.

3 Attach the tile clip to the roof support on one end and to the underside of replacement concave tile at the other end.

4 New convex top tile should fit in without stressing adjacent tiles. If need be, shave edges with a wet saw.

MASONRY

CONCRETE TILE

If you want the look of slate or clay but don't have the money to invest in it, you can substitute a cheaper alternative: concrete tile. These roofing tiles—sometimes referred to as synthetic slate or fiber-cement roofing—imitate these looks at a much lower cost while keeping much of their durability. Made from cement on a fiberglass backing, the tiles are light, attractive, and warranted for 40 years. Besides the lower cost, another advantage of concrete tile is its uniformity—you are less likely to get defective tiles or mismatching colors when purchasing synthetic slate. Concrete tile also has an excellent fire rating.

Concrete tiles are manufactured in a number of geometric shapes, rather than all being flat like real slate. This allows for a much greater variety in your design and will give the roof a stylish look that can greatly enhance the value of your home.

Installing Concrete Tile

It is not advisable to put concrete tiles over old asphalt shingles, because the slates will outlast the underlying asphalt. Any potential problems with old roof sheathing, roof rafters, fascia boards, or soffits should be addressed before a house is roofed (or reroofed) with concrete tile.

As with natural slate and clay tile, concrete tiles are installed on clean decking, with freshly laid roof felt if an underlayment is required. (The manufacturer will specify.) Copper flashing is preferable to aluminum, so the flashing does not wear out before the roofing material. Use W-metal flashing in the valleys.

Some tiles are nailed directly to the sheathing, while other ones must be attached to 1x2 wood battens. The strips are then nailed in place at intervals matching the tile exposure, which is usually about 14 inches. Wood strips along the eaves, rakes, ridges, and hips may also be needed.

Repairing Concrete Tile — MONEY SAVER

USE: ▶ hammer • pry bar • pliers ▶ new tile • copper nails • hanger strip

1 Break cracked tile with a hammer, taking care not to break adjacent tiles. Clear out debris to expose the support.

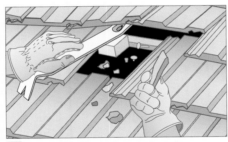

2 Raise the upper course, and set in wedges to provide access to the supporting batten. Clean batten for nailing.

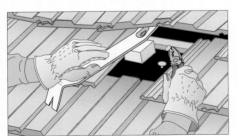

3 Pull any old nails that protrude—they can break new tiles. Repair any openings in the felt with roof cement.

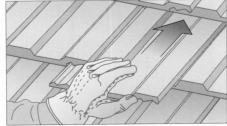

4 Slide a new concrete tile into place under the top course, and clip it into a hanger strip. (See clip detail on p. 454.)

MATERIALS

◆ **WHAT YOU'LL NEED.** Concrete tile is brittle and can be difficult to cut without a wet saw fitted with a masonry blade, a tool you can rent. Wear safety glasses when cutting through masonry material.

◆ **TILE STYLES.** Concrete tiles are made in the classic barrel shape of Spanish clay tile, in curved S-shapes, and interlocking flat tiles that resemble slate shakes (above).

18 Roofing

WOOD SHINGLES

Wood shingles and shakes usually are made from western red cedar, a long-lasting, straight-grained wood. The grain is what gives this wood surprising strength, whether it's cut thick or thin. Straight-grain wood generally does an excellent job of shedding water, even after years of weathering. Shakes and shingles also resist heat transmission twice as well as composition shingles.

However, wood shingles require more maintenance than other roofing options, especially if you live in a region with a harsh climate. In such areas it is advisable to treat the roof with a preservative every 5 years or so. Regular cleaning is also recommended to clear away the debris that traps moisture and, in turn, breeds fungus, mildew, rot, and insect borers.

Wood shingles are not fire-resistant; some local codes may require that wood roofing be pressure-treated or installed over fire-retardant plywood. Some localities have banned wood roofing altogether; check with your local codes before proceeding. (You might also want to check with your home insurance company to see if your premiums will be affected.)

Shingles or Shakes?

Shingles are thinner than shakes and are sawn smooth on both sides. Shakes are often split by hand and have an irregular surface. They are thicker and therefore more durable than shingles, which last no more than 20 to 25 years. Shakes are either taper split (which are split on both sides) or hand split and resawn (which have one split and one sawn face). There are also straight-split shakes, which do not taper and are not intended for residential use.

Both shakes and shingles are available in number 1, 2, and 3 grades. Grade 1 is cut from heartwood, and is both knot-free and more resistant to rot than the other two grades. Grade 2 has a limited amount of sapwood; grade 3 shakes, knotty and mostly sapwood, should be used only for outbuildings.

Shingle length is determined by the desired exposure (the length of the shingle exposed to the weather). Exposure is determined by pitch: shingle widths vary from 3 to 9 inches. One advantage of shingles is that you can add a new layer of shingles over an old one.

Installing Wood Roofing

USE: ▶ hammer • carpenter's pencil • spacing jig • staple gun ▶ roofing nails • heavy-duty staples • skip

1 Wood roofing can be applied over rafters and horizontal skip sheathing.

2 Today, most wood roofing is applied over decking and a layer of roofing felt.

5 The starter course should be two shingles thick and overlap eaves by 1 in.

6 When the roof butts against a second story, install step flashing at the joint.

MATERIALS

- **WHAT YOU'LL NEED.** To install or repair wood shingles, you can use the same basic tools used on asphalt roofs. The main difference is that asphalt is trimmed with a utility knife, while wood shingles are either sawn or split. Many pros use special hammers with a hatchet on the end instead of a standard nail-pulling claw. If you're not used to this tool, it's safer to use a standard hammer.

- **SHINGLE SIZES.** Shingle widths vary from 3 to 9 inches. There are numerous grades and surface finishes, ranging from thin shingles with a smooth surface to thick shingles that are hand split.

WOOD SHINGLES & SHAKES

sheathing or roofing felt (over decking) • plastic mesh • drip edge • step flashing • shingles or shakes

Exposure & Pitch

3 Over sheathing and felt paper, apply plastic mesh to provide air circulation.

4 Nail on drip edge along the rakes and eaves before installing the starter course.

7 Keep about ¼-in. of space between shakes by holding a pencil between them.

8 Keep the exposure consistent by using a homemade spacing jig. (See p. 475.)

In order to install shingles and shakes, the roof must have a steep enough slope. Unlike composition shingles or roll roofing, voids remain between courses of wood shingles and shakes. With enough pitch for quick runoff, this poses no problem, but when installed on a low-slope roof, the roofing is not protected from windblown rain and snow. Wood shingles are not recommended for roofs with less than a 3-in-12 slope. Shakes are not recommended for roofs with a slope of less than 4-in-12. Exposure must also be limited for slight pitches. With a 3-in-12 slope, 16-inch shingles must have a maximum 3¾-inch exposure (5 inches on a 4-in-12 slope). Eighteen-inch shingles may be exposed a maximum of 4¼ inches (5½ inches on a 4-in-12 slope). Shingles that are 24 inches long can have the greatest exposure: 5¾ inches on a 3-in-12 slope roof and 7½ inches on a 4-in-12 slope.

Repairing Shingles & Shakes — MONEY SAVER

USE: ▶ wood chisel • hacksaw blade or nail ripper • hammer ▶ roofing cement • nails

1 To replace damaged shingles and shakes, remove the damaged piece by first splitting it with a wood chisel. **2** Then wedge up the upper course, and cut the nails with a hacksaw blade or nail ripper. **3** Finally, after trimming the edges of a new shingle or shake to size, nail it in place, and cover exposed nailheads with roof cement.

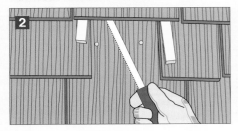

WOOD SHINGLES

18 Roofing

METAL ROOFING

The look of corrugated metal covering a Quonset hut at an army barracks is still most people's idea of a metal roof, and few would consider that suitable for their home. Recent advances in sheet metal, however, have turned the metal roof into an attractive option for certain types of homes. It costs as much as three times more than composite shingles but can last up to 50 years with almost no maintenance—a longevity surpassed only by slate and tile. Unlike the old galvanized tin roofs, metal roofing today is made of steel with a coating of aluminum or a durable polymer.

Metal Roof Options

Many different profiles are available, but simple standing-seam metal roofing (where overlapping ridges run perpendicular to the eaves) is the most appropriate for older homes that might have had a tin roof at one time. Standing-seam panels, being a light material (just one pound per square foot), may even be used over three layers of composition shingles (local building codes permitting). For roofs that have irregularities, narrow, textured, and dull-finish panels work best. Metal roofing (even individual shingles) can also cover low-pitched roofs that have a slope of at least 3-in-12; some metal roof systems can handle slopes as slight as ¼-in-12 (usually a job for built-up roofing).

The installation process involves laying 12- to 16 ½-inch-wide panels and joining them at the seams, wall flashing, valleys, and ridges. The panels are precut to the exact length ordered, up to 40 feet long—for this reason, horizontal seams are unlikely on most homes. Metal roofing can be applied over plywood decking with an underlayment of 30-pound felt. Laying and joining the panels is not difficult, but handling eaves edges, rakes, and ridges can be. Most metal roofing manufacturers will provide an installation guide.

Care should also be taken not to walk on the metal roof, as dents, scratches, and depressions can easily occur. Be sure to replace all copper, lead, and other metal roof fittings, which might corrode the metal panels. Metal roofs should also be grounded with lightening rods in the event of an electrical storm.

Standing Seam

Standing-seam metal panels are one of the most expensive ways to cover a roof, but also one of the longest-lasting applications. Panels run vertically up the roof slope and interlock at the seams. They are made from aluminum or galvanized steel and are available in a variety of finishes, including a wide array of factory-applied paints. Panels can be flat or ribbed between seams, and ordered in lengths up to 40 feet.

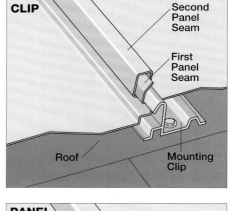

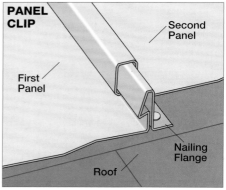

Painted metal roof panels can lace into flashing to shed water at roof openings.

MATERIALS

- Most metal roofing systems are installed by contractors. They fasten mounting clips to the roof over felt paper and attach the panels to the clips. Some panel seams are not pre-formed but joined on-site where contractors use a special machine that travels on wheels down the roof one seam at a time, folding the sections together. The simplest systems for do-it-yourselfers to handle has panels without mounting clips. Instead, each panel has a nailing flange that is fastened directly to the roof. Successive panels simply snap down onto this seam on one side, and fasten with nails on the other as you work across the roof.

METAL

Two Ways to Repair Metal Roofs — MONEY SAVER

USE: ▶ shears • wire brush • soldering gun • flat knife ▶ patch material • flux or roofing cement

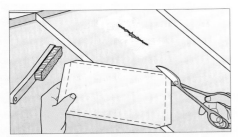

1 Cut a metal patch piece from the same metal as the roof, and nip off the raised edges. Wire brush the repair area.

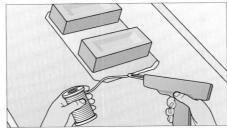

2 Weight the patch in place; then apply flux and heat to make a solder bond around the edge of the patch.

1 Aluminum roofs cannot be soldered. Instead, cut a patch of roofing mesh, and set it in a bed of roof cement.

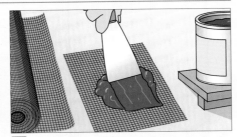

2 Cover the mesh patch with a second coat of roof cement. Work it through the mesh to join with the undercoat.

Metal Edging

In regions with a lot of snow, consider this variation on standing-seam roofs. Instead of covering the entire surface, only the first few courses of shingles along the eaves are clad in metal. The idea is to encourage snow to slide off the roof instead of building up in ice dams at the roof overhangs.

The bargain-basement version of this installation is simply a long roll of sheet metal (aluminum flashing material) rolled onto the roof and tucked under a course of shingles. The drawback is that the sheet metal has to be face-nailed, which creates leak-prone holes.

Corrugated Panels

METAL PANELS

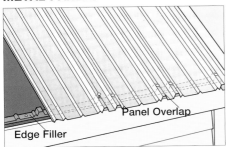

SEALING PANEL SEAMS

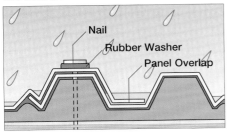

Corrugated aluminum and galvanized steel panels (the kind of roofing you might see on barns), are long-lasting solutions for utility buildings. Corrugated panels made of plastic or fiberglass can provide a watertight yet translucent covering for decks, carports, and greenhouses. Panels of both types are typically sold along with manufacturer-specific nails, filler strips, and caulk. The filler strips fit the contours of the panel, and are installed along the eaves. The nails are compatible with the metal (aluminum or steel). Because these panels do not interlock like standing-seam panels, they have to be face-nailed. To prevent leaks at the nailholes, every nail is set with a rubberized washer. The trick is to set nails just firmly enough to seat the washer without deforming it and causing a leak.

PLASTIC PANELS

Translucent corrugated panels can let light into breezeways and decks but keep out most of the rain. At least one curve of the corrugation along the edge of a panel is covered by the next panel in line to prevent leaks at seams.

18 Roofing

SKYLIGHTS

The extra natural light flooding through a skylight can dramatically change the look of a room, which is one reason why adding a skylight is such a popular home-improvement project. Another reason is that it is fairly easy to do, although the job does involve cutting a hole in your roof. Skylights can be easily added to rooms of a one-story house or on the top floor of a multi-story house. It's a simple matter of mounting the skylight on the roof and building a short light shaft to the room. Where an attic has been converted to a living space, the job is even easier—you don't need the light shaft. (A skylight or two might also eliminate the need for dormers.)

Types of Skylights

As long as you are going to cut that hole in your roof, you might as well pay the extra money for what's called an operator skylight (the industry's way of saying it opens). The added fresh air, as well as the light, is welcome in most kitchens.

Aside from inexpensive all-plastic bubbles, most quality skylights have a glazed section attached by the manufacturer to a frame that raises the glazing several inches above the roof. There are two basic types of operator skylights: bubbles, which are hinged on the high side and open a few inches at the bottom, and roof windows, which are flat frames that pivot about halfway up the frame. In both cases the entire assembly, frame included, is installed in the roof. In addition to being cheaper, clear-bubble skylights expand and contract with changes in temperature. That motion stresses the site-built seam between the roof and the bubble, even if the installer sets the bubble on some type of frame added to the roof. For skylights with an integral frame, the manufacturer takes into account movement at the critical seam between glazing and frame.

Installation

No matter which type you use, the installation consists of attaching the frame of the skylight to the roof. Both fixed and operator frame-mounted skylights should install in approximately the same amount of time and require the same amount of maintenance.

Installing a Skylight

USE: ▶ framing square • pencil • measuring tape • hammer • circular or saber saw • skylight • T-bevel

1 *Lay out openings for the skylight and its light shaft between rafters and joists.*

2 *Drive nails from the inside to mark the corners of the opening in the roof.*

5 *Cut the center rafter, and install same-dimension headers to pick up its roof load.*

6 *Use a plumb bob and measuring stick to mark the ceiling joists for cutting.*

Design Options

Skylights in cathedral ceilings (above) offer direct exposure. The section view of a roof and ceiling (right) shows how a light shaft can be angled on both ends to allow for maximum light entry.

SKYLIGHTS & ROOF WINDOWS

▶ 12d and 16d nails • 2x4s • 2x6s • drywall • drywall screws or nails • joint compound • drywall tape

GREEN SOLUTION
Solar Tube

3 *Reinforce the framing with 2x4s across the rafters before cutting the opening.*

4 *Strip roof shingles, and peel back felt paper to cut the plywood sheathing.*

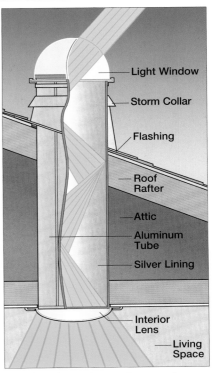

7 *The opening in the ceiling requires double headers where joists are cut.*

8 *With the skylight in place, finish the light shaft surfaces with drywall.*

A solar tube makes only a small opening in the roof and the ceiling, and reflects light down its shaft.

Skylight windows (above) can be opened manually or electronically. Some models have fold-out sections (right), so you can stand outside the roof surface.

Solar tubes concentrate natural light that radiates into the room below through a diffuser.

18 Roofing

FLASHING

One of the most vital parts of any roof is largely invisible—the flashing. Its function is to provide a watertight seal at points where the covering on the roof is interrupted by chimneys, plumbing vents, air vents, and skylights. Without flashing, a roof would leak around all these edges.

Flashing is made from sheet metal, usually copper or aluminum. It can also be made from flat roll-roofing material and plastic. Copper is the most expensive flashing material, but it also lasts the longest. Roll-roofing flashing lasts the shortest time—about 10 to 15 years. If you're investing in a long-term roof of slate, concrete tile, or clay tile, you should use copper flashing.

If your roof has a leak, check the flashing points first. Sometimes it's only a matter of corroded or punctured flashing that needs nothing more than a coat of roofing cement.

Types of Flashing

Chimney flashing is the most complicated of roof flashing jobs because it involves at least three types of flashing: base, step, and counterflashing. Counterflashing is tucked into the mortar joints between layers of brick. The function of counterflashing is to keep water away from the point where step flashing meets with the chimney. Sometimes the freezing and thawing of ice loosens this flashing, which must then be recemented in place.

Other flashing sections that need periodic checking and occasional minor maintenance are valley flashing and sidewall junctures. Check for punctures and corrosion, and repair them with roof tar.

Installing Step Flashing

USE: ▶ metal snips • rubber mallet • vise • hammer ▶ step flashing • shingles • roofing nails

1 Cut flashing from rolls of aluminum into squares with shears or metal snips.

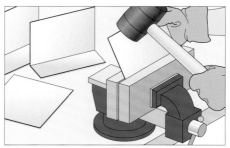

2 To bend flashing, place it between boards, and tap it with a rubber mallet.

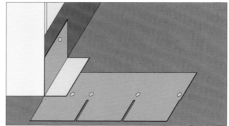

3 The first piece of flashing is laid on top of a course of shingles.

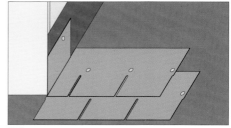

4 The next course of shingles is laid over the flashing, covering its lower half.

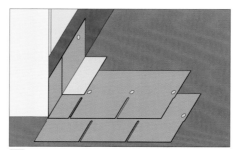

5 The second piece overlaps the first, with the same exposure as the shingles.

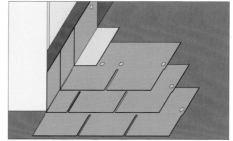

6 Shingle courses are staggered to keep their seams from aligning.

Valleys

Woven valleys don't have any exposed flashing. They use interlaced shingles from both sides of the valley to keep water from penetrating.

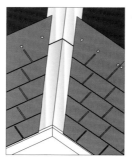

Open valleys have shingles that are trimmed back on both edges of a metal channel. Extra protection is provided by a wide strip of roll roofing beneath the metal.

W-flashing, held to the deck with nailed clips, has a center ridge to keep water from flowing sideways and bent-up edges that prevent water from flowing under the shingles.

FLASHING

Flashing & Counterflashing

Base flashing seals the roof seams, and a cover piece seals the flashing.

The covering pieces, called counterflashing, are sealed into mortared joints.

Even on low-slope roofs, flashing has a bend to deflect water past the chimney.

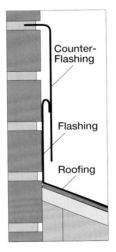

Counter-Flashing
Flashing
Roofing

Counterflashing is installed over flashing to further protect chimney joints. The top lip of the counterflashing is inserted into a mortar joint and secured with mortar. It should overlap the base flashing by at least 4 in.

Pipes & Vents

Vents are best sealed with formed flashing. A rubber collar seals the pipe.

Hoods are flashed similar to vents, with upper and side edges under the shingles.

MONEY SAVER
Flashing Repairs

USE: ▶ chisel • brush • jointing tool
▶ brick mortar • new flashing if needed

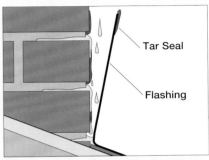

Tar Seal
Flashing

1 *Chimney counterflashing sealed only with tar will eventually break away from the masonry. Then water seeps behind it, into the roof.*

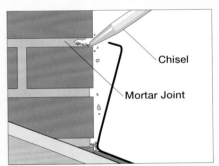

Chisel
Mortar Joint

2 *To repair leaks, chip out mortar from the joint between bricks, and fold the flashing into the seam.*

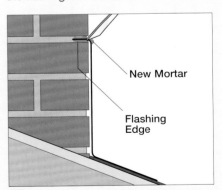

New Mortar
Flashing Edge

3 *Brush out the old mortar; tuck the flashing into the space in the joint; and secure it by filling the seam with fresh mortar.*

18 Roofing

WINTER PROBLEMS

When snow, ice, and freezing rain land on the roof where you live, chances are the coat of frozen stuff will last a while, melting and refreezing several times before finally disappearing. This is not a problem unless your roof overhangs freeze solid, creating an ice dam, which can push water underneath the shingles and cause leaks.

If you can get at the roof edge safely, break off icicles before they become heavy enough to dislodge the gutter. Some homeowners attack ice-choked gutters with a hair dryer or pour hot water in gutters and downspouts to hasten melting. Start at the critical joint between gutter and downspout. Once this area is clear, melting ice will have a drainage channel as you work backward along the rest of the gutter.

Preventing Ice Dams

There are several possible solutions to ice-dam problems. Working from the outside, you can install a strip of sheet metal over the shingles covering the overhang. This method is fairly common in the rural Northeast—it is most effective on steeply sloped roofs, where gravity and the slick metal surface encourage ice and snow to slide off the roof.

Another approach is to install heat cables in a zigzag pattern along the shingles on the overhang. The resistance wiring, which looks like a long extension cord, is attached with small clips tucked under the shingles and even can

Ice Dams

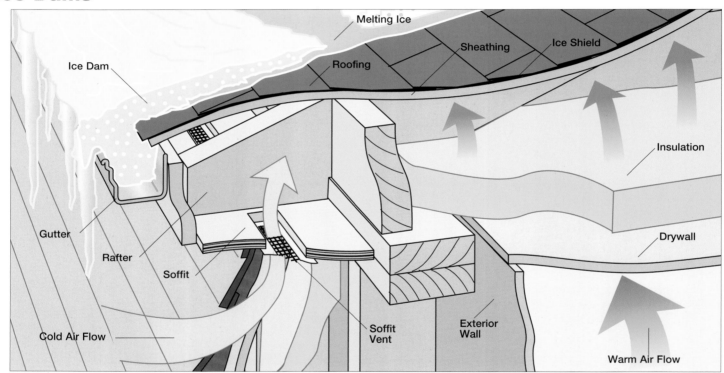

To install an ice shield, peel the paper backing on the rubberized sheet.

How Ice Dams Form

Ice dams form as snow on the roof melts and then refreezes along the eaves. Even in houses with insulation in the ceilings, enough heat can rise through the blankets or batts to gradually warm the bottom of the roof over the attic. In the right conditions, the heat causes the snow blanket to melt from the bottom up, and water trickles down toward the gutter. It may be cold outside, but the trickle is protected from freezing by the snow above. However, when the melted water reaches the roof overhang, there is no longer a heat source from below because the overhang is outside the exterior wall. That's where the water begins to freeze. It forms a dam, and the water above can back up under shingles.

CLIMATE CONTROL

be extended into gutters to help them remain unfrozen. These cables are designed to produce enough heat to prevent freeze-ups.

Because warmth rising through the ceiling or attic is often the cause of ice dams, you can also alleviate the problem by working from the inside to reduce the heat flow with extra insulation. In a typically constructed wood frame attic floor, for instance, the spaces between floor joists should be filled with insulation. An additional layer, even the 3½-inch batting used in walls, can be set on top of and perpendicular to the joists for more protection. At the same time you can increase the vent size in the attic or crack a window at each end. This will make the bottom of the roof colder and closer to the temperature outdoors, which will prevent melting, while the extra insulation will retard heat flow from living spaces below.

On new construction jobs or reroofing projects, consider installing a rubberized ice shield membrane on the roof deck. It should cover the overhang and at least a few courses of shingles over living space. This provides a backup barrier just in case an ice dam does form and works under the shingles.

Usually, these membranes are made of waterproof, rubberized asphalt and polyethylene in self-adhering sheets that bond directly to the roof deck and to each other at overlaps. The material is installed beneath the shingles, and it seals itself around punctures from nails protruding through the shingles above.

Ice Control

There are many ways to keep ice from forming on roofs and in gutters where it can leak and damage shingles and flashing:

◆ Draw enough outside air through the attic so that heat rising through insulation in the ceiling of your living space will be diluted and carried away before it can warm the roof and melt snow. You can ventilate with plug vents, strip-grille vents, or perforated panels in the roof overhang. (See "Soffit Vents," p. 317.)

◆ Where ceiling insulation extends toward the overhang, cut it back at the exterior wall, and make sure that it does not block vents.

◆ Keep gutters and leaders clear of debris and free-draining so any melting water won't be trapped.

◆ Install a rubberized barrier under shingles on the overhang. (See photo at left.) This self-sealing membrane closes around nail shanks driven through shingles and protects the overhang from water that may back up and seep through the overlapped courses.

Temperature Extremes

The best policy is to stay off the roof during any period of extreme temperature. But the weather won't always cooperate when you need to make repairs. When the weather is hot, bear in mind that asphalt shingles become soft. You are more likely to slip and to dislodge surface granules. Also remember that working in extreme heat taxes you as well as the roof. You should be exceptionally careful—work only in shoes with non-slip soles, and only from a safe position. Protect yourself further by wearing a hat, drinking a lot of water, and taking regular breaks. In extreme cold, even with only a few scattered patches of ice and snow, the roof surface may be so treacherous that you should call in a professional. If you must make emergency repairs, use a trowel or drywall knife to spread roof cement under and over shingles in the leaking area. If you want to clear ice-filled and icicle-laden gutters (which can be surprisingly heavy), start work at the corner of the house out from under the main gutter. Some homeowners try a hair dryer or a propane torch, which might clear a relatively small frozen bottleneck. Electric heat cables that can prevent freeze-ups in leaders and gutters should be used with extreme caution. Use only UL-approved units installed exactly to manufacturer's specifications.

When it's very hot outside, the best policy is to work on the roof either early or late in the day. Under midday sun, asphalt shingles soften and mar easily.

When it's very cold outside, let cap shingles warm up inside, and bend slowly into shape. Bending them outside can cause cracks that lead to leaks.

WINTER PROBLEMS

18 Roofing

MAINTAINING GUTTERS

Many old roofs have no trouble shedding water—as long as the water continues to flow off it. The trouble starts when it backs up in the gutters and drains. Leaves, twigs, animal nests, and other debris can block drain outlets, clog gutters and downspouts, and stop up underground drains that take water away from the building where it won't cause any damage.

There are many products designed to prevent blockages, such as wire baskets and gutter screens. In theory, wet leaves are supposed to pile up on them, then dry out and blow away, leaving a clear path for drainage. In reality, the screens and guards often clog themselves, particularly on flat and low-slope roofs. So instead of cleaning out the gutters, you have to clean off the gutter guards: same dreary job; slightly different location.

You may have seen advertisements for a supposedly clog-proof gutter, basically a strip of louvers. It simply breaks the flow into a series of small streams that drop off the edge of the roof, including down your neck while you're fumbling for the door key. They don't clog like a gutter because the louvers don't collect water like a gutter. They aren't really gutters at all.

Blocked Downspouts

If water collects in a cleaned-out gutter instead of draining freely or you hear dripping in the downspout, some debris may have gotten hung up at one of the fittings. The most likely bottleneck is the S-shaped piece of pipe that carries water from gutters at the edge of the roof overhang and curves back toward the house wall to a downspout.

Most downspout systems have enough play that you can take them apart to get at a clog. Some are simply pressure-fitted together; some are joined with small sheet-metal screws you have to remove. But try flushing the debris out with a garden hose first. If you are lucky, you will be able to clear the blockage without taking the pipes apart.

Gutter Hangers

Spikes and ferrules are the standard hanging system. The ferrule (a tube around the spike) prevents crimping.

Brackets nail into the fascia board and clip into the gutter edge. Space brackets about 3 feet apart.

Straps wrap around the gutter and are nailed to the roof deck under the first course of shingles.

System Assembly

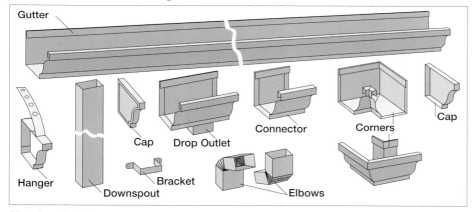

Metal and vinyl gutter hardware includes a full line of fittings. In addition to U-shaped gutters (and end caps), there are inside and outside corner pieces, connectors to link sections of gutters, drop outlets, downspout pipes, and a variety of brackets and hangers that hold the system to the house.

Slope

Gutters should slope about 1 in. for every 10 lin. ft.—more for better drainage. Place downspouts every 35 ft.

GUTTERS & LEADERS

Gutter Cleaners

Keeping gutters clean will help to prevent backups and leaks in bad weather. Clean gutters in the fall after all the leaves have fallen, and in the spring. Start by pulling out all debris by hand; then flush the entire system with water from a garden hose.

A wire basket that rests in the gutter over the downspout opening will keep debris out of the pipe.

Screens that cover the entire gutter are designed to trap wet debris where it can dry and blow away.

Some systems replace standard gutters with louvers that disperse the flow onto the ground.

Repairing Gutters

USE: ▶ wire brush • power drill • pop-rivet gun ▶ patch • rivets • caulk

1 Use a wire brush around the hole to clean and scuff the metal. Cut a patch from the same metal as the gutter.

3 Caulk the area covered by the patch and the back of the patch with silicone, and press the patch into place.

MONEY SAVER

2 Set the patch piece in position, and drill pilot holes for pop rivets through the patch and the gutter.

4 Secure the patch with pop rivets. Install as many as you need to make the caulk ooze out on all sides.

Downspout Extenders

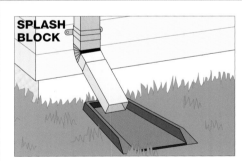
SPLASH BLOCK

ROLL-OUT HOSE

Splash blocks are designed to prevent water leaving the downspout from draining directly down along the foundation wall. Some innovative fixtures, such as a roll-out hose, extend to carry off water during heavy rains, then coil up again, out of the way.

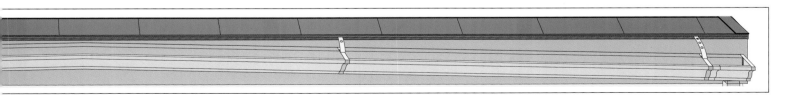

19 Siding

478 TOOLS & MATERIALS
- Siding Choices ◆ House Wraps
- FIBER-CEMENT SIDING ◆ Materials

480 PANEL SIDING
- Installing Panels ◆ Panel Materials ◆ Trim

482 CLAPBOARD SIDING
- Material Options ◆ Corner Design
- Installing Clapboard

484 SHINGLES & SHAKES
- Installing Shingles ◆ Shingle Corners ◆ Staining

486 WOOD-SIDING REPAIRS
- ROUTINE REPAIRS ■ REPLACING SHINGLES
- REPAIRING PANELS ■ REPAIRING CLAPBOARDS

488 BRICK & STONE
- Setting Veneer Stone ◆ Setting Face Brick
- Stonework Types ◆ Brick Bond Patterns

490 STUCCO
- Stucco Pros & Cons ◆ Stucco Textures
- Installing Stucco ◆ EIFS Problems

492 MASONRY-SIDING REPAIRS
- REPAIRING STUCCO ■ REPLACING BRICK

494 VINYL & METAL
- Materials ◆ Installing Vinyl ◆ Details & Finishes

496 VINYL & METAL REPAIRS
- REPAIRING VINYL ■ REPAIRING ALUMINUM
- Painting Aluminum

498 RE-SIDING
- Re-siding Over Existing Siding
- Asbestos Shingles

500 FINISHING NEW SIDING
- Paint vs. Stain ◆ Applicators

502 FIXING PAINT PROBLEMS
- Diagnosing Surface Problems ◆ Stopping Moisture

504 REPAINTING
- Repainting Basics ◆ Wall Cleaning ◆ Stripping

19 Siding

SIDING CHOICES

Siding problems usually can be fixed by making spot repairs—such as replacing a rotted clapboard or cutting out a cracked piece of vinyl siding. But when it comes to completely re-siding the house, you'll have to decide whether to use vinyl, aluminum, wood, masonry veneer, or fiber-cement siding. Every material has its proponents, but it's important to remember that the siding you finally pick will change the look of your home for years to come and may require alterations in trim details that can loom as large as the re-siding job itself.

Many people think vinyl siding looks synthetic, even when it is embossed with a simulated wood grain. But vinyl is usually the least expensive option because it is so easy to install. Aluminum siding tends to look more like painted wood clapboards, but the patter of rainfall takes on a metallic tone and a wayward baseball can leave a dent—a repair you should need to make on cars but not houses.

Wood siding looks like wood, and for many homeowners holds a special appeal. But it needs regular repainting or restaining—a big job that has convinced a lot of people to switch from wood to vinyl or aluminum.

Getting to the Top

Tackling siding jobs means spending time on ladders and scaffolds, so it's worth going over some basic safety tips. For starters, you need a Type I extension ladder, rated to carry up to 250 pounds per rung. Set the ladder's bottom feet far enough out from the wall so that it won't tip backward, but not too far out—about a quarter of the ladder height is a good rule of thumb. Put pieces of wood under the feet if the ground is soft or uneven. When you're on the ladder, wear shoes or boots with heels for the best grip, and don't overreach—leaning over to the side can tip a ladder.

For more ambitious siding jobs that cover more than one story, you'll want to use scaffolding. You can rent ladder jacks that hold a scaffold plank, pump jacks that rise vertically, or pipe scaffolding equipment. Pipe systems provide the most stable work surface, but it takes time to set up the pipe framework and wooden decks. (See page 451 for more information on ladders and scaffolding.)

House Wraps

Building wrap, or house wrap, is an improvement on the felt paper that was once commonly installed over wood sheathing. It is designed to be a one-way material, somewhat like the fabric used on rainwear and parkas, called Gore-Tex. It is woven tightly enough to prevent air infiltration, which improves energy efficiency and comfort by cutting drafts at leak-prone seams in corners and around windows and doors. But the fabric will allow interior moisture to escape, something that tar paper does not do. This is an important distinction, because if interior moisture is trapped in the wall, it can soak insulation, reducing its thermal value and causing rot in the plywood sheathing and framing.

Lightweight house wrap is easily installed with staples. Run wide rolls horizontally to cover one story at a time, with edges tucked into the openings.

GREEN SOLUTION

FIBER-CEMENT SIDING

Fiber-cement siding is made from Portland cement, fly ash, and wood fibers. The products look like wood, are durable (some have 50-year warranties), won't rot or wrap, are resistant to salt spray and UV light, are noncombustible, and are termite and mold proof. Many products come prefinished and are available in a variety of colors.

Their long life means less construction waste, and the use of fly ash, which is a byproduct of coal combustion in electric-power plants, makes them an ecologically friendly choice. Many manufacturers boost environmental benefits by using recycled products and wood from sustainable sources.

TOOLS & MATERIALS

Materials

WOOD SHINGLES AND SHAKES
Siding in these small sections is easy to work with because the joints don't fit tightly like cabinetwork, and the layout does not have to be precise. The main drawback is that the installation is labor-intensive, which can drive up a contractor's price. But repair is easy—split damaged shingles into small pieces for removal, and weave replacement shakes into the wall so there is no noticeable line around the repair.

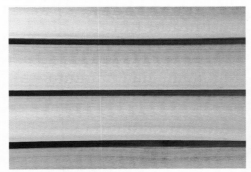

WOOD BOARDS
Solid wood siding can be installed vertically, horizontally, and even on an angle. Cost ranges from moderate to very expensive, depending, of course, on the wood species you use. Although wood siding will need periodic restaining or repainting, it's difficult to match its natural beauty. Stagger joints by at least one stud course to course and when you make repairs. Cut out a damaged section between studs, and conceal the new piece with paint or stain.

PANELS
Panel siding can be made from hardboard or plywood. It's usually less expensive than other types of siding and easy to install (at least on the first floor) because each sheet covers so much area. Some panels are made to resemble materials ranging from shingles to stucco, while others are available in a variety of finishes, including smooth surfaced, rough-sawn, and with grooves every 4 or 8 in. to resemble planks.

ALUMINUM
Aluminum siding is moderately expensive and somewhat difficult to install. It is a lot stiffer than vinyl, the number-one choice today, but any scratches that expose base metal through the finish are noticeable. Metal also dents, a problem that generally requires replacement instead of repair. But from a distance, aluminum looks more like painted wood clapboards than many vinyl products, and unlike wood does not need repainting every few years.

VINYL
Vinyl siding is typically less expensive than aluminum or wood, is easy to install, and requires little to no maintenance. The trade-off can be a synthetic look, although high-quality products are realistic looking. Vinyl siding is more likely than aluminum to fade in sunlight or crack in cold weather. But damaged vinyl can be patched; you have to unlock the interconnected pieces and make a visible lap joint where the replacement meets undamaged material.

BRICK AND STONE
Brick and stone facing materials are both beautiful and durable, but they cost more than other siding materials, mainly because of the time and skills needed for installation. Standard sizes of bricks are easier to install than irregular stone. Arranging an attractive and functional collection of rocks borders on being an art and requires difficult cutting and shaping. But modern face stone—cast masonry that is colored and textured to look like rocks—makes the job easier.

19 Siding

PANEL ADVANTAGES

Because labor accounts for so much of the final bill in a re-siding job, plywood panels can be an attractive, cost-saving alternative. Typically, just two workers can panel an average-size house in about a weekend, and the skill needed is within the range of many DIYers. Also, panels are often available in 4x9- and 4x10-foot sheets that reach from foundation to roof edge.

Many lumberyards carry one of the most popular plywood panels, called Texture 1-11. These sheets have grooves cut into the face 4, 8, or 12 inches apart to simulate separate planks. But many other styles and surface treatments are available, although not all plywood can be used for siding—only sheets rated for exterior use, assembled with special glue that can withstand the exposure.

While the panel surface will be coated with stain or paint for appearance and protection against the weather, panel edges often are not coated. They are the weak links, because layers of thin plywood laminations are exposed along the edges. If they soak up water, the panel is likely to delaminate, which can pop nails and create an array of repair problems. You can protect against this deterioration by brushing a primer coat on the edges prior to installation or by concealing the edges with trim, such as vertical corner boards. It's also important to caulk or flash seams around windows and doors, and on two-story projects where one sheet rests on top of another.

Panel Materials

SURFACED PANELS come in a variety of styles. The final appearance is mainly a product of the wood species, of course, but also depends on how the wood is sawn. The surface texture can be smooth, rough-sawn, striated, or brushed. These panels are often stained or covered with clear sealer for a rustic look, and their mating edges are covered with trim.

COMPOSITE PANELS are made of engineered materials, often including sawdust, wood chips, and other parts of the tree that used to be considered waste. Without a natural grain that could twist one way or another, the panels are stable. Many manufacturers sell either solid-wood or engineered trim pieces matched to the panels to finish off the job.

GROOVED PANELS offer a variety of looks, depending on the groove profile. A wide spacing between grooves generally looks best on a big wall, while narrower spacing fits the scale on smaller surfaces and smaller houses. These panels are built to join in a lap at the last groove with a thin edge of one panel crossing over the other panel so seams aren't noticeable.

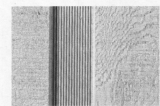

BOARD-AND-BATTEN PANELS have a wide, flat groove cut at regular intervals along the length. This configuration is designed to resemble a twist on the typical board-and-batten design, called a reverse batten, where a narrow board is set behind the simulated planks. In traditional board-and-batten siding, plank seams are covered by narrow boards.

Installing Panels

USE: ▶ circular saw or saber saw • sawhorses • ladder • power drill/driver • combination square • 4-ft. level • hammer • C-clamps • plumb bob • work gloves

1 Establish a level line on the foundation, and nail a 2x4 ledger in place to support the panels.

2 Starting at a corner, make sure the sheet is plumb, and nail into studs using fasteners specified by the manufacturer.

3 On panels with built-in laps, tack the edge of the last panel; then install the next sheet, and nail through the lapped section.

PANEL SIDING

Trim

Panels can be joined edge to edge on an uninterrupted surface and mitered at corners. But the most-efficient installations use trim to frame the house and all openings in the walls, and siding panels to fill the spaces closed in by the trim. This plan will conceal plywood edge grain, giving the job a more finished look. Most panel manufacturers produce trim to match their siding materials. Some offer trim made of composites bonded together with plastic resins. Composite trim is typically less expensive than solid wood and less likely to warp and twist.

Choose trim pieces to suit the style of the house and siding. Plan the installation to minimize cutting and maximize the coverage of the panels—for example, by using trim between stories on high walls.

Install ornamental base trim before the siding panels. Outside corner trim is often made by butting together two 1x4s. You also can use one board or a piece of cove molding to trim an inside corner. Windows and doors can also be trimmed, but most often exterior casing is used. Drip-cap trim and flashing over windows and doors keep water from getting under the siding.

Panel Laps

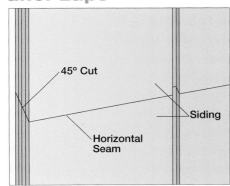

On horizontal joints between sheets of siding on two-story jobs, there are two good ways to protect the seams from the weather. One is to cut 45-degree angles along the mating edges and install them with a bead of caulk. Water would have to run uphill to get behind the siding. The other method calls for a piece of Z-flashing tucked up behind the top panel and extended onto the face of the bottom panel. No 45-degree cuts are needed, but you will see the flashing.

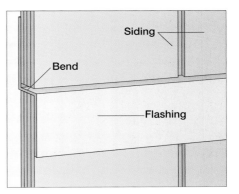

▶ building wrap • plywood siding panels • 2x4 ledger boards • Z-flashing • corner boards • galvanized nails • screws

4 To install corner boards over panels, mount the first one flush and the second one lapping the first, covering its edge.

5 To accommodate protrusions, measure out from the corner and up from the level base line to mark your cutout.

6 You can create double corner boards at inside corners, too, or conceal the plywood seam with molding.

19 Siding

PLANK SIDING

Clapboard, or plank, siding is well suited to a wide range of architectural styles. Real wood offers flexibility and an attractive look—so much so that there are many imitations available in vinyl, aluminum, and other synthetics. The biggest disadvantage of wood siding is that it must be repainted about every 7 years or restained every 3 to 5 years, depending on conditions at your site.

Cedar and redwood are preferred woods for siding, but Douglas fir, larch, ponderosa pine, and local species are also used. You will pay more for cedar and redwood, but these two are naturally more resistant to decay than most other woods and are available in prime-quality grades that look good enough to protect with a clear sealer, instead of coats of paint or stain.

Like panel siding, plank siding can also be made of hardboard. It's cheaper than solid wood but less durable. You also must be careful during installation to prevent damaging moisture from getting in. Hardboard plank siding is generally available in two forms. One has splines that hook over the course below and allows blind nailing. The other is rabbeted along the bottom edge.

Level & Plumb

Though plank siding usually is installed horizontally, it can be applied vertically or diagonally as well. For a good-looking, professional-quality job, horizontal siding must be level, and vertical installations must be plumb.

The layout of the first board is critical, as this board is the base for all successive rows. Even a slight error in measurement during the early stages can lead to noticeable problems after several courses. To keep the installation on track, double-check horizontal installations after every fourth or fifth course—or more often if you haven't installed siding before. Don't measure from course to course, but go all the way back to the first one. After measuring the vertical spacing at the corners, snap a chalk line at the point where the top of the next piece will be installed.

If you find that boards are beginning to run out of level (or plumb on vertical installations), make several small adjustments in the next few courses instead of one big fix that stands out.

Material Options

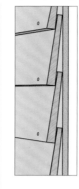
Plain bevel (clapboard) siding is used for horizontal applications. It comes in clear and knotty grades for a more rustic look. The boards are thick along the bottom and taper toward the top.

Shiplap siding provides the weatherproof security of a lap, and a decorative bend that looks like quarter-round molding below each seam. Overlaps can absorb movement of the house frame without opening.

Rabbeted bevel siding (Dolly Varden) is used only in horizontal installations. It is thicker than beveled siding and has a rabbeted overlap. It comes with a smooth or saw-textured face.

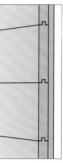
Tongue and groove siding is available in a variety of patterns and sizes from 1x4 to 1x10. Some versions have a chamfer or bead along the edge to create a more finished appearance.

Corner Design

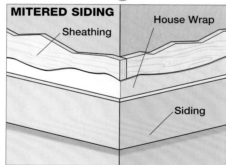

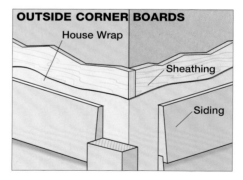

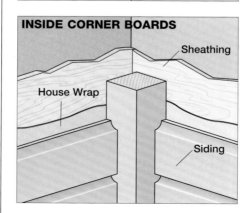

Corner boards protect leak-prone siding joints at the corners of a house. Mitered joints look neat and elegant but aren't as durable as other options, and the technique requires laborious hand-fitting. Butted outside corner boards are easier to install—you nail them in place, then cut the siding to butt squarely against them. Also, at inside corners, you can butt clapboards or simply nail trim boards in place on flat siding.

CLAPBOARD SIDING

Spacing Jig

You can speed up the installation of clapboard or shakes and increase the overall accuracy of the project at the same time by relying on a simple siding jig that you can build yourself. (Jigs save time on clapboards, shakes, and any siding installed in horizontal courses.) The idea is to create a moveable measuring tool that duplicates the overlap on each course. Every few rows you still should measure back to the base course, and check the current course for level. But you won't have to stop work and check each piece if you use a jig. To construct the jig, screw a small cleat to a rectangular piece of wood in a square T-shape. (1x4s work well.) Use a square to check alignment; then, clamp the pieces and screw them together. Be sure to use screws that won't protrude through both pieces and scratch the siding underneath. To use the jig, slide the cleat section along the bottom edge of the last piece of siding you installed, and make a pencil mark, or simply set the next course in position on top of the jig. The long riser of the upside-down T will gauge the amount of exposure on the next piece and keep your clapboard installation uniform.

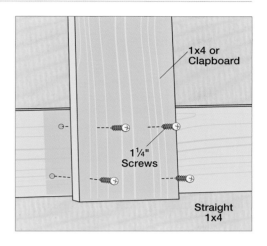

Installing Clapboard

USE: ▶ circular saw • drill • spacing jig • ladder • 4-ft. level • hammer • measuring tape • work gloves ▶ clapboard • building wrap • corner boards • nails

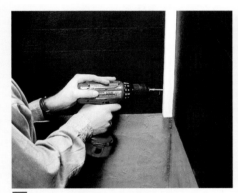

1 Install prefinished inside and outside corner boards to provide square edges against which the siding can butt.

2 Snap a level line for the base course, and nail a starter strip of lath along the bottom edge of the sheathing.

3 Overlap the starter strip with the first board, and drive nails high enough to be covered by the next course.

4 Periodically check the weather exposure at the ends and middle of each row as you progress, even if using a jig.

5 Joints should be staggered at every course. Place joints at random intervals spaced a minimum of 16 in. apart.

6 Cut around obstructions as you go. Always overlap boards and fittings so that rainwater will run off, not in.

19 Siding

BUYING WOOD SHINGLES

Wood shingles and shakes are not just for roofs alone—they also make an attractive siding material for many house styles. Although you can buy traditional and fancy-cut wood shingles made especially for siding, it's okay to cover walls with roof shingles and shakes as well. But because weathering on siding is not as severe, you can save by using a lower grade of roof shingle.

Shakes are machine- or hand-split from blocks of wood called bolts. They're thicker than shingles and less uniform along the exposed edges, but they last longer. Wood shingles are sawed smooth. Both types come from woods like western red cedar and redwood, and are available with a fire-retardant treatment. It's wise to allow an extra 10 percent for waste.

Choosing an Exposure

The amount of the shingle surface exposed to the weather is called the exposure. Manufacturers should specify the allowable exposure because it varies depending on the material. But as a general guideline, calculate the maximum exposure by subtracting ½ inch from half the overall length of the shingle. You can reduce the exposure for looks and greater weather protection, but remember that smaller exposures use more shingles. A typical exposure requires about four bundles of shingles per 100 square feet of wall area.

Design Options

- **RANDOM STRAIGHT**
- **RANDOM DROP**
- **SCALLOPED**
- **ARROW/DIAMOND**

Estimating Guide

Square Footage Coverage of Four Bundles of Shingles (Single Coursed)

Length	Exposure								
	4"	5"	6"	7"	8"	9"	10"	11"	12"
16-inch	80	100	120	140	—	—	—	—	—
18-inch	72	90	109	127	145	—	—	—	—
24-inch	—	—	80	93	106	120	133	146	—

Installing Shingles

USE: ▶ block plane • sliding T-bevel • story pole • 4-ft. level • hammer • ladder • measuring tape • pencil • plumb bob • utility knife • work gloves

1 Build up the first course with starter shingles. Snap a chalk line, or string a level line as a leveling guide.

2 The first course covers the starter course. Select shingles by width to cover all joints, overlapping gaps by at least 1½ in.

3 Mark the weather exposure at the middle and ends of each finished course; then snap a chalk line to guide your row.

SHINGLES & SHAKES

Shingle Corners

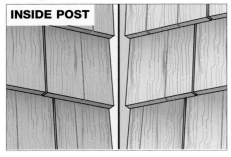
INSIDE POST

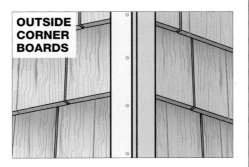
OUTSIDE CORNER BOARDS

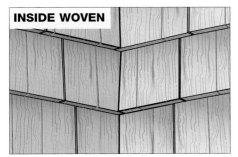
INSIDE WOVEN

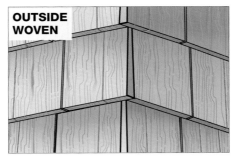
OUTSIDE WOVEN

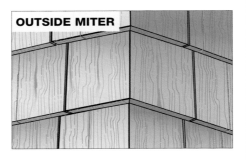
OUTSIDE MITER

At corners, you can butt shingles together or weave them against trim. Mitering looks good, but it's time consuming and doesn't weather well. Woven corners offer better protection but require cutting and fitting. Butted corners offer the most protection.

Staining

Final color depends on the wood hue and the stain color, of course, but also on how long you leave the stain before wiping or brushing out.

The same stain can create different effects, from a wash that resembles a semitransparent coating to a rich tone that conceals most wood grain.

▶ wood shingles (or shakes) • building wrap • edge molding • corner boards • 1x4 ledger board • galvanized nails

4 To keep the courses level and make the installation easier, install a temporary guide board leveled across the wall.

5 Drive galvanized nails about ¾ in. from shingle edges and 1 in. above weather exposure. Space shingles ⅛ to ¼ in. apart.

6 Use a hand block plane to dress shingle edges for fit. Stagger seams so they don't line up course to course.

BUYING WOOD SHINGLES

19 Siding

ROUTINE REPAIRS

Siding, and the roof, are your home's first line of defense against the elements. Wood naturally responds to changes in temperature and humidity, which over a winter can pop nails, open joints, and cause siding to crack or distort. Leaks can do even more damage—to the siding and structural members. It's tempting to smooth over trouble spots with caulk and paint, but if you bury the problem without fixing the cause, it will just work its way back through the refinished surface. Eventually it may require an even bigger repair.

Shingles, Shakes & Clapboards

Damaged shingles and shakes are easy to remove because they come out in small, easily split pieces. This also makes new shakes easy to weave into the wall so that there is no noticeable line around the repair. The most difficult part of shingle and shake repair is removing the nails in a way that won't damage adjacent shingles. New shingles on walls not painted or sealed will weather after a few months and eventually match the surrounding color, so long as the new shingle is made of the same type of wood.

Clapboards are a little more difficult to replace than shingles or shakes, but adding new boards, though a slow process, is well worth the effort, especially if it saves you from having to re-side an entire wall. If damage is caused by rot, check the extent of the problem by poking the wood on each side of the damaged area using a knife or screwdriver. Mark the points where the wood changes from dry and crumbly (punky) to solid. The damage may extend only a few inches, but to make a repair that blends in, you may have to remove some good wood, too.

Repairing Panels

You can patch small punctures or rotten areas of plywood siding with an epoxy compound, but if the damage is major or there is veneer failure, you will have to replace it. Buy the new panels to match the thickness of the existing panels, and prime them with a primer that is compatible with the final finish before installation. Paint or stain the replacement panels as soon as possible to protect them from the weather.

Repairing Bows

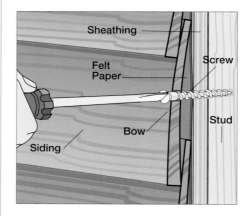

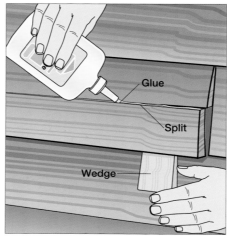

MONEY SAVER

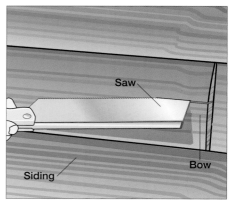

Reattach bowed pieces of siding by fastening them into the studs with long screws. Drill pilot holes for the screws to avoid splits; recess the screwheads, and conceal the heads with putty or caulk. If the siding is heavily bowed, saw a relief cut into each end of the board before screwing it down. To stabilize a large split for caulking, wedge out the split portion; drill pilot holes; apply waterproof glue or construction adhesive; and screw both sides of the split to the wall.

Replacing Shingles

USE: ▶ pry bar • hammer • wood chisel • safety goggles • mini-hacksaw blade • work gloves

1 Drive wood wedges under the course directly above the damaged shingle. Take care not to split the shingles above.

2 Use a hammer and chisel to split the damaged shingle into several narrow pieces that can be removed without pulling nails.

WOOD-SIDING REPAIRS

Repairing Panels

MONEY SAVER

USE: ▶ crowbar • drill • hammer • paint brush or pad • caulking gun ▶ caulk • galvanized screws • wood stain

1 Where panel seams have popped, pull the nails with a crowbar or hammer, and check for water damage underneath.

2 Instead of driving down old nails, use new galvanized screws driven into wall studs through the old nailholes.

3 To conceal the screwheads, set them slightly below the surface; add caulk or filler; and touch up with matching wood stain.

Repairing Clapboards

MONEY SAVER

USE: ▶ pry bar • wood chisel • hammer • paint brush or pad ▶ 1x2 ledger board • exterior paint • exterior wood glue • galvanized finishing nails

1 If the board is split but still intact, hold the split open with a chisel or pry bar while you inject exterior wood glue.

2 Place a 1x2 support beneath the split to hold it closed while the glue sets. Fill the nailholes later with putty.

3 After the glue dries, remove the 1x2 and fill the nailholes; then anchor the repair with finishing nails above and below the split.

MONEY SAVER

▶ replacement shingles • wooden wedges • galvanized nails

3 Cut away hidden nails using a flexible hacksaw blade. Wrap the blade with tape, or wear gloves to protect your hands.

4 Fit the new shingle to fill the damaged area, and gently tap it home. Remove the wedges under the shingles above.

5 Nail the new shingle just beneath the butt of the course above. Set the nails, and fill holes with caulk.

19 Siding

BRICK & STONE SIDING

Few materials are more durable than brick and stone. They don't catch fire or fall prey to rot or insects. And they are so tough that most problems occur at the weak links in the walls—the mortar joints and flashing seams. But they do require considerable skill to work with. Actually building brick or stone walls for a house—and even replacing old siding with brick or stone veneers—are all projects beyond the range of most do-it-yourselfers.

But there are many maintenance and repair jobs on brick and stone walls that DIYers routinely take care of themselves, including cleaning and repointing masonry joints. See "Spot Repairs" in this chapter (p. 484), as well as "Brick Basics" (p. 86), "Brick Maintenance" (p. 58), "Working with Stone" (p. 90), and "Cleaning Masonry" (p. 92) in chapter 6.

Veneers

Masonry veneers add a look of substance to a house and provide a low-maintenance exterior surface. These advantages come at a price, however—brick and stone veneers are the most expensive siding, and they must be professionally installed. Stone veneers are anywhere from 1¼ to 4 inches thick and bricks from ½ to 4 inches. Both individual pieces and preformed panels can be held securely to wall surfaces with metal ties.

To Paint or Not to Paint?

You can paint brick the way you can paint almost anything else. But unless you want to sign up for the periodic repainting job that homeowners with wood siding have to endure, consider other repairs and coatings that let the brick wall be brick.

You can repoint eroded mortar joints to preserve the structure and its appearance. Once the wall is sound, you can seal it with a clear, silicon-based sealer. There are several types of clear coatings that help the wall shed water and dirt and make the surface easier to clean. Also, most clear coatings tend to wear away gradually without becoming the kind of eyesore you can get with flaking and peeling paint. If you must paint, the National Decorating Products Association suggests an alkali-resistant primer coat, followed by a top coat of latex house paint.

Setting Veneer Stone

USE: ▶ mason's trowel • pointing trowel • jointing trowel • brick hammer • grout bag • pitching chisel

1 Start by making a dry layout, placing the stones in approximate position on the ground to see how the joints will fall.

2 Install wire mesh on the wall to help support the mortar. Trowel on a base coat that bonds into the metal weave of mesh.

Stonework Types

 RUBBLE

 RUBBLE MORTARED

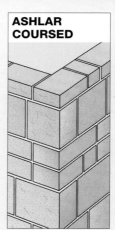 ASHLAR COURSED

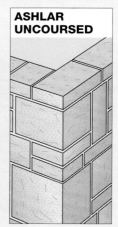 ASHLAR UNCOURSED

Setting Face Brick

USE: ▶ joint spacers • notched trowel • jointing trowel • string level • hacksaw • hammer • plumb bob

1 For the adhesive mortar to bond on the wall, start the job by using galvanized nails to attach sheets of wire mesh.

2 Install an embedding coat of adhesive mortar by troweling the mix onto and into the mesh. Work on one area at a time.

BRICK & STONE

• mason's twine • 4-ft. level • measuring tape • plumb bob • wheelbarrow • work gloves ▶ stone • mortar mix • welded or woven wire mesh (or metal ties)

3 *Butter the backs of stones with mortar. You can work in long rows or build up short sections of several courses at once.*

4 *Fill seams with mortar. To avoid discoloring the surfaces, some manufacturers recommend that you use a grout bag.*

5 *Smooth the grouted seams with a tooling blade. Fill all gaps around the edges of the stones to keep out water.*

Brick Bond Patterns

RUNNING

COMMON

GARDEN

▶ face brick or brick tile • adhesive mortar • welded or woven wire mesh • galvanized nails

3 *Follow manufacturer's specs for spacing; measure the space of two or three courses, and string a level guideline.*

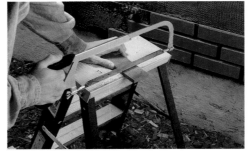

4 *Use a hacksaw to trim bricks and to cut half pieces for alternate courses so that joints will be staggered.*

5 *Mortar the joints to a uniform depth and shape. Tooling blades commonly are concave but are also flat and V-shaped.*

BRICK & STONE SIDING

19 Siding

STUCCO PROS & CONS

Stucco is again popular because of its durability and its uniquely textured appearance. Still, problems can arise in this weather-resistant shell. If the house foundation sinks unevenly or new framing members shift and shrink as they dry out, the movement can create enough stress to crack the rigid stucco walls.

Problems can also arise in the stucco itself. Fresh stucco (and the masonry wall underneath) may contain salt-based compounds that can be carried to the surface of the material. As the moisture evaporates, the salt deposits leave a residue (efflorescence) on the surface. Although the alkalinity of the stucco material normally neutralizes during the curing process, this residue can create discoloration. The presence of alkali may cause expansion and subsequent cracks.

A type of synthetic stucco system (called EIFS) is also widely used for all types of residential applications. The material can be problematic unless applied correctly. (See the facing page for more information.)

Installation

Stucco is made from portland cement, sand, lime, and water. You can mix your own or buy it premixed. (Unless you're experienced, it's advisable to buy premixed stucco.) Because stucco can be tricky to apply—and preparing frame houses for stucco installation can be time-consuming and labor-intensive—consider hiring contractors or at least one professional to work with you during application.

When you're about to begin the job, take care to choose an overcast day to stucco walls with a southern exposure. Excessive heat can dry the stucco prematurely, which causes shrinking and cracking. Conversely, cool temperatures make the stucco too stiff for proper troweling. The ideal temperature for installing stucco is between 50° and 80° F.

Changing Appearances

The finish coat of stucco, if made using white portland cement, can be colored with pigments in a variety of earth colors. The texture results from the technique used to apply the finish. You can create a variety of looks by using different finishing tools.

Stucco Systems

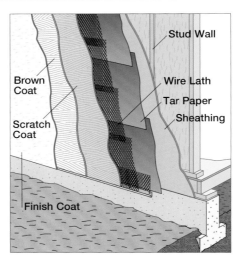

On wood frames, nail expanded metal lath, wire mesh, or woven wire to the sheathing over building paper, and apply the stucco in three separate coats.

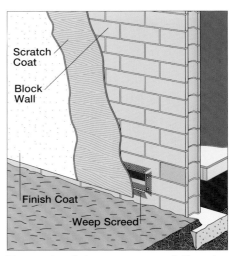

Two coats of stucco are enough if you're stuccoing over a solid, stable substrate like concrete block or other masonry walls (or over old stucco).

Stucco Textures

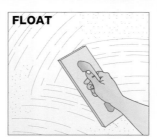

FLOAT — For a smooth texture, trowel on stucco and finish with a wood float.

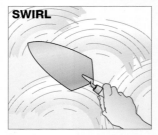

SWIRL — Create a swirled texture the same way as a smooth one, only leave the swirls.

SPATTER — Hit a paintbrush dipped in thinned stucco against a stick to spatter-coat.

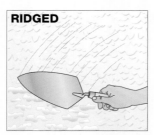

RIDGED — To create a ridged finish, use a steel trowel with a waffling motion.

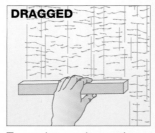

DRAGGED — To produce a dragged finish, scrape a piece of 2x4 down the surface.

LINED — For a lined finish, pounce whisk broom bristles into the final coat.

STUCCO

Installing Stucco

USE: ▶ notched trowel • scratch tool • chalk-line box • hammer ▶ stucco • base strip • nails • wire mesh

1 Most stucco installations are anchored with a base strip. To set it, make level marks at corners, and snap a line in between.

2 Set the base strip along the level line, and fasten it to the wall through the mesh portion of the strip with galvanized nails.

3 Above the base strip, fasten sheets of mesh. Extend the bottom edge of the first sheet onto the base strip.

4 Mix the base coat to manufacturer's specs, and trowel it on, using enough pressure to set the mix into the weave.

5 To maintain a bond between coats, scratch the surface before it hardens. A homemade tool with spaced nails will do.

6 In most applications, the top coat is only ⅛ to ¼ in. thick. You can try to make a smooth finish or create a texture.

EIFS PROBLEMS

▶ **EIFS** is a type of synthetic stucco. It stands for Exterior Insulation and Finish System. The components of the system typically are an adhesive, insulation board, a base coat of cement, fiberglass mesh, and a finish coat. When all the components are installed correctly by contractors (this is not a DIY job), EIFS can be a cost-effective and low-maintenance finish. However, many applications have caused problems to house structures—mainly rot stemming from leaks and moisture trapped in the wall. The most modern EIFS systems attempt to remove these problems by incorporating a system for evacuating moisture—for example, by using insulating panels with narrow grooves on the back.

STUCCO PROS & CONS

19 Siding

SPOT REPAIRS

Make it a point to fix cracks in brick, stucco, and stone as quickly as you detect them. Even small cracks in either the material or the mortar joints in between will let water seep into the underlying structure, where it will eventually cause damage. Over time, water and the winter freeze/thaw cycles will turn minor cracks into major problems, which are both more difficult and costlier to fix.

Brick

Brick repairs usually consist of either replacing crumbling areas of mortar or removing and replacing a cracked or damaged brick. If you shop around, you can usually find replacements for damaged bricks or facing stones that closely match the existing material; mortar, however, is far more difficult to match. Typically, it's necessary to try out a few test batches before you get one that will blend in with the rest of the house. Add powdered colorant, if necessary, to duplicate the existing mortar. Always wait for the mortar mix to dry before making any decisions so that you can make valid comparisons.

The first step in replacing a broken brick is to chisel it—and the surrounding mortar—out of the wall. After cleaning out the hole and spritzing it with water, you should first spread some mortar on the bottom. Then spread mortar on the top and sides of the brick, and slide it into place.

Stucco

Patch large cracks and gaps in a stucco surface with the same stucco mix that was used on the walls. Fill small cracks with all-acrylic or siliconized-acrylic sealants. If you spot large cracks but don't have time to make a thorough repair, at least seal the openings with a bead of silicone caulk to keep out water. The caulk can be peeled away when you get around to making permanent repairs.

Stucco can be difficult to color-match. If the surface requires many patches, it may be easier to cover it with a cement-based paint or an acry-lic exterior paint. (Let the stucco patches cure at least 30 days beforehand.) Dampen the wall with water before painting. Cement-based paint will need a primer coat to prevent blotches, but acrylic paint won't.

Stone

When the joints between stones are the problem, follow the color-matching advice discussed for brick repairs. Chip out the old mortar to between ½ and ¾ inch deep, forming a groove with square sides, one side of which goes down to bare stone. After cleaning and misting the joint with water, add mortar to the groove. Tool the joint when the mortar has cured enough to retain a thumbprint.

For a loose stone, remove it and clean the joints. Butter the back with mortar, and then replace the stone and repoint the mortar joints around it.

Replacing Brick

USE: ▶ cold chisel • hawk • jointing trowel

1 Use a cold chisel and heavy hammer to chip away the damaged brick. Work from the joints in, wearing safety glasses.

Chalking Paint

Many contemporary exterior paints are specially formulated to shed dirt: the topmost surface of the paint breaks down into a chalklike powder and sloughs away slightly with each rainfall or cleaning. This chalking keeps the surface looking clean, but it will not allow a new coat of paint to properly adhere to it. Before painting over a chalking type of paint, scrub the old surface with detergent, and rinse well.

Repairing Stucco

USE: ▶ cold chisel • hawk • mason's trowel • short-handled sledgehammer • safety glasses • work gloves ▶ stucco • straight-edged board • welded wire mesh

1 Use a cold chisel and heavy hammer (while wearing gloves and safety glasses) to chop out loose material.

2 If the underlying wire mesh is damaged (or missing), cut a new piece and attach it to the wall with galvanized nails.

3 Mix more than enough patch material to fill the hole, and trowel it on the mesh so the patch is thicker than adjacent stucco.

MASONRY-SIDING REPAIRS

• mason's trowel • mixing trough • safety goggles • short-handled sledgehammer • wire brush • work gloves ▶ mortar mix • replacement brick • water

MONEY SAVER

2 *Break away remaining chunks of mortar, and use a wire-brush to sweep out the surrounding surfaces to remove debris.*

3 *Mix enough mortar to thoroughly coat the mating surfaces of the replacement brick, a process called buttering.*

4 *Force more mortar into the joints as needed (some will likely fall off); then tool the seams to blend in the repair.*

Wall Cleaning

Power washing uses a pressure washer to remove tough dirt. Some grime may need a nonabrasive detergent. Test the treatment on a patch of wall using a low-pressure setting (under 700 pounds per square inch). In older buildings, mortar may break away under high pressure. Protect nearby shrubs and flowers with drop cloths or plastic sheeting. When using a pressure washer, keep the spray moving. Most power washers come with nozzles offering both a wide spray at a lower pressure and a narrow jet at a higher pressure—you risk less damage by using the nozzle with the wider arc.

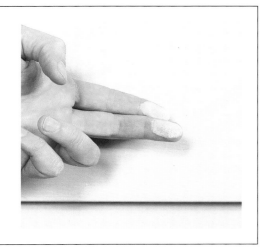

MONEY SAVER

• galvanized nails

Excessive cleaning pressure can erode mortar joints and pit the brick surface.

4 *Use the surrounding wall as a level guide, and smooth the patch surface by sliding a straightedge back and forth.*

Cleaning can reveal a sound brick wall that requires only minor repointing.

Spray at an angle to keep the water jet from digging into the mortar joints.

SPOT REPAIRS 493

19 Siding

VINYL & METAL STYLES

Siding textures in vinyl and aluminum are typically made to resemble wood clapboards or shingles. They come in a variety of colors and styles with matching trim and architectural details. Both vinyl and aluminum siding can be installed horizontally or vertically. When you get a siding estimate, it will likely come with an insulation option. Because both aluminum and vinyl are preformed sheet materials (not solid like wood), the space behind them can be filled with molded panels of insulation, called backer boards. Some vinyl siding comes with an insulated backing attached.

Aluminum siding is paintable, and many vinyl siding manufacturers now claim their products can be painted, too. But painting usually isn't necessary for many years, which makes aluminum and vinyl truly low-maintenance materials. But you will have to wash accumulated dirt and grime off the siding from time to time.

Installation

Both vinyl and aluminum sidings are installed in much the same way, using lock-together components designed to expand and contract with temperature swings. Trim pieces cover expansion joints, give the job a finished look, and help hide minor mistakes. Mounting systems differ slightly depending on the brand, so it's important to use the tools and techniques specified by the manufacturer.

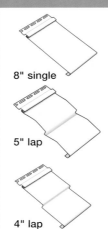

8" single

5" lap

4" lap

Materials

Aluminum siding covered millions of homes through the 1950s and 1960s. It was sold as the last siding you would ever need, and it would never need painting. Most homes now are covered in vinyl—it's less expensive and easier to work with, which saves labor costs on contracted jobs and allows handy do-it-yourselfers to tackle siding projects without contractors. But some vinyl can crack in cold weather, distort if not installed properly, and fade over time. While aluminum is structurally more durable, it dents and scratches easily—and even minor damage is difficult to conceal. Both vinyl and aluminum are sold in complete interlocking systems that cover everything from the roof line to the foundation, including matching systems for covering soffits and fascias.

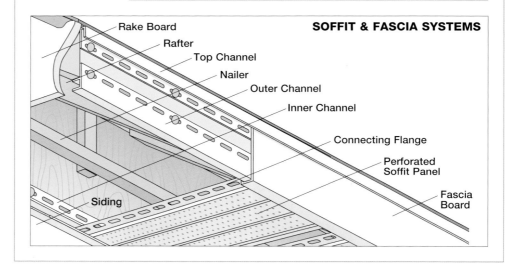

SOFFIT & FASCIA SYSTEMS

Installing Vinyl

USE: ▶ nail-slot punch • snap-lock punch • zip tool • metal snips • 4-ft. level • hammer • ladder • straightedge • utility knife ▶ vinyl siding • vinyl trim (corner trim,

1 Install preformed corner trim first. Siding ends fit into vertical channels on this piece to seal out the weather.

2 At window and door openings, vinyl J-channels butt against the trim. Their outer edges conceal the edges of siding.

3 Level and nail a starter strip for your first course. The first full row of siding locks into the lip on this preformed strip.

494 **SIDING** / VINYL & METAL

VINYL & METAL

Details & Finishes

The pros of vinyl siding are its low cost, easy installation that even some do-it-yourselfers can handle, and exceptionally easy cleanup (with no painting or staining), which every DIYer can appreciate. The downside—at least for some people—is the synthetic appearance, which is no match for real wood. Using vinyl with embossed wood grain may help. Installing clean-lined architectural details in vinyl, such as complete door surrounds, helps even more. Modern trim systems are available to suit many house styles and can dress up the siding behind them.

Before...

...and after

Use a plain siding on the bulk of the house, and cover small, independent sections with a decorative scallop.

Detailed cornices with dental block and surround trim for windows and doors can tie into wall siding above.

House vents are available in many styles, including triangles and circles that fit into siding systems.

Detailed trim pieces, like fluted corner posts, add a custom touch to a house clad in vinyl or aluminum.

fascia panels, F-channels, J-channels, starter strips, under sill trim, etc.) • rigid insulation (for retrofits) • soffit panels • building wrap • galvanized nails

4 *Push up on the siding butt to lock it in; then nail into place through the top flange. Overlap panel end joints by 1 in. minimum.*

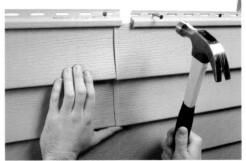

5 *Nail through the center of flange slots at a 90° angle, leaving nails loose enough to allow siding to expand and contract.*

6 *Cut around windows, using aviation snips for vertical cuts and a utility knife to score and snap horizontal cutouts.*

19 Siding

FIXING DAMAGED VINYL

Vinyl is tough, but not indestructible—it often cracks under impact. This is especially true at low temperatures. Patch small cracks by removing the damaged piece, cleaning the crack with PVC primer, and gluing a patch of scrap siding from behind with PVC cement.

If large areas are damaged, the entire piece may have to be replaced. This isn't a major undertaking, because replacement pieces are fairly easy to install. But vinyl siding does fade with time, so your patch or replacement piece may not be a perfect color match—at first.

It's easy to remove a damaged section because each course of vinyl locks into the course below or beside it. The siding is nailed through a flange molded into the top of the course. All you need is a simple tool, called a zip tool, to unlock the courses. You wedge the device under the lower edge of siding and pull it horizontally to unlock the pieces. When working with vertical siding, you pull the zip tool down the seam.

Restoring Aluminum Siding

Although aluminum siding provides good weather protection and is easy to maintain, a stray baseball or large hailstones can dent it. Slight damage can be patched—the procedure for repairing dents in aluminum siding is a lot like repairing a dented car fender. It consists of pulling out the dent, sanding the area, and applying two-part auto-body putty (or auto-body filler for very slight imperfections). Once the surface is dry, you can sand, prime, and paint it. For minor surface imperfections, you can use steel wool. It also helps to know the manufacturer and color of the siding, because you may be able to purchase a touch-up kit in the original color.

You may need to replace areas with more serious damage, although you should try other repairs first for two reasons. First, aluminum isn't as flexible as vinyl; it's more difficult to weave one piece into an existing wall. Second, you may not find replacement pieces that match your old siding. If the repair is in a conspicuous place, you may want to consider removing and using a piece from a less obvious part of the house and then replacing that piece with the new material.

Repairing Vinyl

USE: ▶ zip tool • metal snips • hammer • nail set • pry bar • utility knife ▶ replacement vinyl • wedges

MONEY SAVER

1 Use an installer's zip tool to unlock siding above the damaged piece. Pry down and slide the tool along the edge to free it.

2 Fit temporary wedges under the loose siding course to hold it out of the way while you work on the damaged piece below it.

3 Carefully pry out nails directly above the damaged section. If necessary, protect the piece with tape or scrap beneath the flat bar.

4 Cut out the damaged area with snips or a razor knife. Avoid cutting the siding or locking edge of courses above and below.

5 Cut a replacement section 2 in. larger than the damaged piece you removed to allow a 1 in. overlap on both ends.

6 Fasten the replacement through the top flange, using a nail set to reach under the course above. A zip tool will relock all edges.

VINYL & METAL REPAIRS

Repairing Aluminum

MONEY SAVER

USE: ▶ drill • pliers • sanding block • screwdriver • drywall knife ▶ auto-body putty • emery cloth • exterior spray paint • metal primer • sheet-metal screws

1 Drill one or more holes at the center of the dent; run a screw into the siding only. Pull the screw to pop out the dent.

2 Remove the screws; roughen surface with sandpaper; mix two-part auto-body filler; and smooth it onto the dented area.

3 After the patch dries, sand filler, and paint it with primer. When dry, spray on two coats of finish paint to match siding.

Painting Aluminum

USE: ▶ caulking gun • bucket • 4-in. paint brush • steel wool pad ▶ exterior paint • metal primer • caulk • mild detergent • water

1 For the best adhesion, use steel wool to smooth rough spots, and wash with a mild detergent and water.

2 Caulk around all openings in the siding. Use a flexible caulk that can maintain the seal when siding shifts.

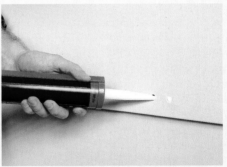

3 Fill small punctures with caulk. Also cover the heads of nails or screws used to make repairs prior to painting.

4 Seal leak-prone seams around windows and doors where aluminum J-channel trim laps onto wood casing.

5 Once aluminum siding is prepared, treat it like any other wall, and use a latex acrylic or oil-base exterior paint.

6 To help disguise the repaint job, carefully cut in around openings, and paint the aluminum trim as well as the siding.

19 Siding

RE-SIDING OPTIONS

When it comes to siding renovation, you have three options. Removing old siding costs time and money, but it lets you open up your exterior house walls. This gives you the opportunity to replace outdated plumbing pipes or damaged framing, string new electrical wiring, or add insulation. The second option, re-siding over existing material, saves you the cost of demolition and carting but adds the cost of coping with new trim work to accommodate an extra layer of siding.

When making your decision, don't overlook the obvious third option, cleaning, which can save you a bundle of time and money. Too often homeowners confuse excessive dirt and dinginess with more permanent deterioration. A thorough washing with soap and water can actually reveal bright paint beneath and make re-siding unnecessary.

If you do decide to re-side directly over the old material, invest some time and effort in surface preparation. Before covering old clapboard siding, for example, wash it with a mild solution of household bleach and water to kill off mold. If major sections of the old siding are deteriorated, don't bury the problem. Where wooden clapboards are rotting, twisted, and split, and many nailheads have popped, leaks or condensation are the likely culprits. Unless you find and fix the source, the problem will only work its way through your expensive new siding.

Vapor Barriers

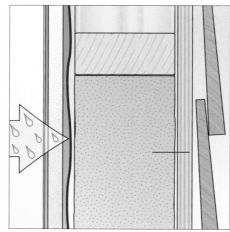

A vapor barrier is a layer of thin material, typically thin plastic sheeting and sometimes foil, that prevents moisture generated inside the house from seeping into exterior walls and condensing. It lies over the wall frame, under the drywall. Without this barrier, which is standard in modern homes but lacking in many built before 1960, moisture working through the wall can meet a surface that is below the dewpoint temperature. That's when moisture becomes water that can soak insulation, cause rot in the wall frame, and peel paint. If you are installing new siding, don't add a vapor barrier over the outside wall. It will stop moisture from escaping. And much of the damage can be hidden until the problems mushroom into major repair projects. House wrap is the right material to use outside the house under new siding because it allows the house to shed any moisture that might get by the vapor barrier inside. Moisture passes out through this fabric, but drafts can't get in.

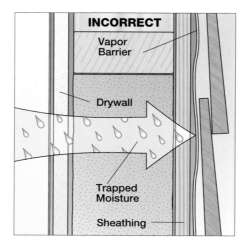

Re-Siding over Existing Siding

USE: ▶ nail-slot punch • snap-lock punch • zip tool • metal snips • 4-ft. level • hammer • ladder • staple gun • straightedge • utility knife ▶ vinyl siding • vinyl trim

1 If you're re-siding over existing siding, secure split, loose, and warped boards to ensure a solid nailing base.

2 Re-siding gives you the option of adding insulation through wall openings or applying insulating foam to the wall surface.

3 Building wrap protects interiors from drafts and heat loss. It can be stapled directly over existing siding.

RE-SIDING

Asbestos Shingles

When asbestos siding was popular decades ago, no one knew that it was carcinogenic; now we know. Asbestos does a good job of protecting houses, but the striated, often powdery, and brittle shingles contain cancer-causing fibers. The material is not a hazard until it is disturbed—for example, by sawing shingles to install a new window. In theory, you could cover old asbestos siding with new vinyl siding. But in practice, nailing on the vinyl would fracture the older shingles. The piles of pieces would get in the way and create a health hazard. On small repairs you can install a fiber-cement board made to look like asbestos (above). On large projects, the safest plan is to have shingles removed by contractors licensed to handle asbestos and who have permits to dispose of it.

Flashing Windows

Above a window you need strip flashing to shed water. Tack the top to the old siding and bend the leading edge over the window frame. Use this flashing as a back up even with aluminum or vinyl siding that has its own trim. There are two ways to flash the sides of a window. When the wall is built out with extra insulation, you may need to install J-channel on the frame (below left). It looks better to butt J-channel against the side of the frame (below right).

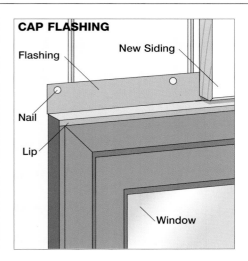

CAP FLASHING — Flashing, New Siding, Nail, Lip, Window

Extend frame with trim to add a layer of insulation with new J-molding for siding.

Insert siding into the J-molding slot to cover cut ends as you add each course.

(F- and J-channels, starter strips, etc.) • building wrap • rigid insulation board (optional) • soffit panels • furring strips • galvanized nails • heavy-duty staples

4 Furring strips create a smooth, even nailing base where walls are uneven or out of plumb. Use shims under strips as needed.

5 Install inside and outside corner posts first, hanging them from top nails so that they are vertically plumb before fastening.

6 Add each course of siding, tapping it upward to lock it onto the course. Nail through the flange into furring strips.

RE-SIDING OPTIONS

19 Siding

PAINT VS. STAIN

Most people stain floors, cabinets, furniture, decks, and just about every other piece of wood, except the siding. Siding gets painted—despite the fact that while you're in the middle of the long, tedious, and expensive job, you know that in a few years you'll be doing it all over again. That's because paint lies on the surface of siding like a blanket. Stain, on the other hand, seeps more deeply into the wood grain and becomes a part of the siding. So what makes more sense: paint or stain?

Color & Coverage

Most homeowners use slightly muted shades, but paint is available in every color in the spectrum. Your house can be bright white, of course, or fire-engine red if you can stand the intensity. But penetrating stains offer almost as wide a variety. There are dark wood hues, brilliant blues, and potent reds, but also very light whites. And if you prefer that one shade you can't find on a paint chip, you can mix stain to a custom shade the same way you mix paint.

Typical stains let some of the wood grain show through, while most paint applications are designed to provide a solid, opaque surface. Both materials can be applied in thick coats or thin coats. You can thin paint to create special effects (generally inside the house), and you can buy heavy-bodied stain that creates the same appearance as full-strength paint on siding. If you choose paint, you'll be covering the wood completely, which may not be the best approach if you have elegant redwood or cedar underneath. In that case, the best choice may be a clear sealer that protects the wood but lets the natural beauty come through. If you choose stain, you can use a full-bodied material to conceal an average grade of pine, or a semitransparent mixture that gives the wood an overall color hue and lets some of the natural beauty come through.

Maintenance

On a modern home with vapor barriers in the wall, paint may last as long as stain. On older homes without vapor barriers (sheets, usually thin plastic, that stop interior moisture from entering the wall), stain will last longer. Here's why:

Moisture from cooking, washing, and other household operations can eventually reach the solid paint film and push it off the wall. But moisture is more likely to seep through a semi-transparent stain. Also, when paint finally deteriorates, the results are obvious and sometimes an eyesore. However, as stain deteriorates, it tends to fade. More wood grain will show but the siding will look basically the way it did a few years ago—just a bit more washed out and weathered.

MATERIALS

◆ **CLEAR COATINGS** are solvent-based and they contain no pigments. Both the wood color and the grain remain visible.

◆ **SEMITRANSPARENT COATINGS (STAINS)** have a solvent-base of oil, oil/alkyds, resin, and/or emulsions. They are lightly pigmented to reveal the grain of the wood.

◆ **OPAQUE COATINGS (PAINTS OR STAINS)** can be water- or solvent-based. They conceal the natural color and grain of wood yet allow the texture to show through. **Latex stains** retain color well and are available in a range of finishes. They clean up easily with soap and water. **Oil stains** are more durable than latex stains because they penetrate into the wood. The finish and solvent used for the cleanup, however, create toxic fumes.

◆ **STAIN KILLERS,** such as pigmented white shellac, are used to conceal blemishes and discolorations, such as bleeding knots.

Finishing New Siding

USE: ▶ 4-in. exterior wall brush • 1½-in. sash brush • caulking gun • drip rags • dropcloths • ladders • putty knife • sanding block • scaffolding (optional)

1 Wrap sandpaper around a block of wood to smooth out minor splits and other small imperfections.

2 Fill holes with putty. Deep holes will need two coats. Then prime the dry patch so that it doesn't show under paint or stain.

3 Apply a weather-resistant caulk around window and door openings and where siding butts against corner trim.

FINISHING NEW SIDING

Applying Paint

A good paint job on bare wood begins with the primer coat. The primer coat seals the surface, hides imperfections and stains, and improves paint adhesion. Use a primer that is compatible with the final top coat you will be using. An oil-based primer works with both water- and oil-based top coats. A water-based primer is generally covered only with a water-based top coat. Be sure the surface being primed is clean and dust-free to ensure good adhesion.

Two coats of paint over a primer or over existing paint provide the longest-lasting, most durable results, particularly on the south and west sides of a house, where the sun is most intense. One coat of paint will last about 5 years; two coats lasts up to 10 years, but much depends on the paint quality. Apply the first coat no later than two weeks after priming and the second coat when the first coat is dry.

Following a few rules will help make your paint job look professional. First, never paint when rain is in the forecast. The surface must be clean and dry before you begin, and a new coat of paint can be washed off or seriously affected by rain before it dries. Second, use a mildewcide before painting. Don't paint over mildew, as it will grow under the paint and cause blistering and discoloration. Lastly, paint at temperatures no lower than 50° F for latex paint and 40° F for oil-based.

Paint can be applied with brushes, rollers, paint pads, or a power sprayer. With any of these, make sure to coat the underside of the siding laps, and always be sure to check the undersides of the courses from the ground. It's easiest to detect drips and skipped areas from below.

In the last century, there was no precut siding but plenty of cheap lumber. This New York farmer saved some measuring by trimming the clapboards right on the house.

Applicators

♦ **WALL BRUSHES** are usually 4, 5, or 6 in. wide. Use smaller brushes for trim—angled bristles are good for cutting in around window glass and edges of another color.

♦ **ROLLERS** apply paint and stain quickly and easily. They leave a stippled surface that becomes more pronounced with deeper naps.

♦ **PAINT SPRAYERS** apply finishes to siding in a fine mist. Using sprayers requires some practice, but they can give you a fast, easy, run-free paint job.

▶ primer • exterior paint (or stain) • caulk • sandpaper • wood putty

4 There is no set application rule for paint or stain. But it works well to cover seams before brushing out the surface.

5 On the finish coat of paint or stain, work on small sections between trimmed borders. Start by cutting in the edges.

6 Work high to low so you can pick up any drips before they dry or scrape them away with a taping knife.

19 Siding

EVALUATING PROBLEMS

Preparing the surface of siding before repainting is probably one of the most important aspects of the job. Paint won't adhere to an improperly prepared surface, and you can be sure that any underlying problems that caused damage to the original paint job will soon reappear to ruin the new coating.

Of these problems, moisture is perhaps the biggest culprit. Water vapor from cooking, washing, and other normal household operations can migrate through walls until it reaches the building skin. On many homes the outermost skin is paint. It is applied to beautify and protect the house on the outside, but it can be cracked, bubbled, and peeled by moisture breaking through from the inside.

Another problem results from painting over thick layers of paint. After 20 or 30 years, many houses have been repainted several times. Not every homeowner scrapes and sands the siding each time it needs new paint. Even when carefully done, the paint coat may be only scraped away in small patches, leaving most of the old paint on the walls. The multi-layer skins are the most likely to crack, often in a pattern resembling an alligator's skin. This condition can quickly resurface in a fresh layer of paint. In this case, rather than scraping off the paint—a task which is laborious, and costly—many people choose to re-side the house with new material.

Water Spots

These gray, milky areas appear on exterior wood near the ground. They're usually caused by sprinkler systems used to water shrubs or flowers near the foundation. Any minerals in the water then get deposited on the siding and cake in place as the water evaporates.

Although hard-water deposits can be softened with white vinegar and other solvents, it's often easier to scrape the surface with a draw-type, razor-blade paint scraper, followed by sanding and refinishing. Since it isn't economical to use softened water on garden plants, the only way to prevent this problem from recurring is to shield the siding during watering or to redirect sprinklers away from the siding. If after scraping and sanding some deep stains linger, cover them with two coats of pigmented white shellac.

Diagnosing Surface Problems

◆ BLISTERING
Blistering appears as a series of bubbles in the paint. These blisters form when paint is applied to a wet surface. It can also occur when paint is applied in direct sunlight. Moisture migrating outward from inside the wall or in the house itself can also cause paint to blister, even though the surface was dry when it was applied.

CAUSES
Cut open a blister to diagnose the cause. If raw wood is under the blister, the problem is moisture. If paint is under the blister, the top coat was probably applied in the hot sunlight.

SOLUTIONS
Scrape off the blistered paint, and sand the area smooth. If the blistering is caused by moisture migration from the interior, there is no point in repainting until the source of the moisture problem has been fixed.

◆ PEELING
Peeling paint is characterized by paint flakes or strips that don't stick to the surface. Several factors can cause peeling, including the application of a very thick layer—the two-coats-in-one approach. But the problem is most commonly due to interior moisture working through the exterior wall.

CAUSES
Paint peels when it does not bond sufficiently to the surface. A dirty surface, a surface coated with many paint layers, or moisture migration from the inside can all cause peeling.

SOLUTIONS
Scrape the paint to the surface of the siding, and then recoat with high-quality paint. Where the peeling has been caused by migrating moisture, fix the problem before repainting.

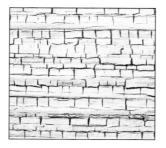

◆ ALLIGATORING
Alligatored paint has small connected cracks that resemble the skin of a reptile. It often occurs on siding where multiple layers of paint have been applied, making it difficult for the outermost coat to adhere properly. The newer paint may have been incompatible or applied to a moist surface.

CAUSES
This results when one layer of paint doesn't adhere to another. It also happens when incompatible paints are used, the surface is poorly prepared, or the previous coat has not dried.

SOLUTIONS
Scrape the paint down to the raw wood; then prime and recoat the surface with high-quality paint. Where many layers of old paint are the cause, you may want to consider replacing the siding altogether.

FIXING PAINT PROBLEMS

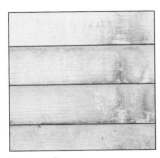

◆ MOLD
Mold and mildew are fungal growths that appear on siding, causing it to look dirty. Growth of the spores is encouraged by soiled, moist, or warm surfaces and lack of adequate sunshine and fresh air. Once the green and gray-green growths take hold, they can discolor most of the wall, prematurely age the paint, and make repainting more difficult.

CAUSES
Mold and mildew thrive on damp surfaces that are shielded from the sun and prevailing winds. Problems often are worst under wide roof overhangs and wherever siding is concealed by shrubs.

SOLUTIONS
Wash mold and mildew with a scrub brush and a solution of up to half household bleach and half water. Try the solution on an inconspicuous area first because the bleach could discolor the siding.

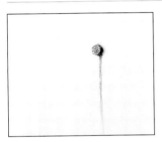

◆ RUST
Rust is the reddish-brown coating formed on iron or steel by oxidation when the material is exposed to water and air. Metals that are unprimed or poorly primed are most likely to develop problems with peeling paint, which exposes the metal and promotes rust that can leave a telltale streak.

CAUSES
Rust stains can be spread onto the siding by rainwater when metal components such as flashing, gutters, ungalvanized nailheads and shutter hardware begin to oxidize.

SOLUTIONS
Try a 50–50 bleach/water solution. If that doesn't work, use oxalic acid, a chemical that must be handled with care. You can also try commercial cleaning products for cleaning rust stains off surfaces.

◆ BLEEDING
Some materials can show through two coats of paint or stain, even when most of the surface is smooth and even in color. This bleeding can come from many sources, including rusting nailheads, resin seeping from wood, and knots that are too hard to accept much stain or paint. The most common fix is to coat trouble spots with pigmented white shellac.

CAUSES
Bleeding comes from uncoated steel nailheads or nails that are coated with a low-quality zinc plating. It also can stem from knots and tight sections of wood grain that produce resin.

SOLUTIONS
Prevent bleeding by using noncorrosive stainless-steel or aluminum box nails on siding. To remove existing stains, scrub with a 50-50 bleach/water solution. Coat discolorations with a stain killer.

Stopping Moisture

Problems with moisture build-up can damage a paint job. Inadequate venting is often the cause, and permanent cures are possible only when venting is improved. Moisture around windows and doors can also cause problems—gaps should be caulked as part of routine maintenance.

When insulating, install foil or asphalt-impregnated kraft-paper vapor barriers toward the heated areas.

Staple through the insulation batt edge flaps every 6 in. or so to create the best possible seal along the studs.

Installing a layer of plastic sheeting is the easiest way to build a complete vapor barrier over paper-faced batts of insulation.

Use clear tape (or duct tape) to seal all tears or gaps in the vapor barrier, and around all pipes and vents.

19 Siding

REPAINTING BASICS

Don't decide to repaint your house until after you have washed the siding and exterior trim. Scrubbing with a soft brush and a solution of warm water and nonabrasive household detergent (about 1/3 cup detergent to a gallon of water) can sometimes be as effective as repainting. As dirt accumulates on siding over several years, you may not notice the gradual increase in dinginess—until you wash one of the dull spots and find surprisingly bright paint underneath.

If repainting is called for, there are a few guidelines that can help make the finish more durable. Scuff the painted surface by sanding lightly. Washing the siding is also a must, because the new paint won't adhere over dirt, mold, or other surface deposits such as grease from a kitchen exhaust fan.

Prepare thoroughly before repainting. On a two-story project or any place you need ladders and scaffolds, it may be tempting to prepare the surface and paint in one step. But chips of old paint, dabs of caulk, and sanding dust may settle onto areas you're about to paint or have just painted.

Hairline cracks can be filled with paint, but caulk larger openings, such as where ends of siding butt against the trim. Use a flexible butyl or paintable silicone caulk there and where dissimilar materials like wood and masonry meet. Don't use old paint, which can be damaged by improper storage and will not match with new paint. And don't try to get by with one thick coat of paint—that causes sagging, wrinkling, and peeling. Wherever you sand, scrape, set nails, or make other repairs that expose raw wood, prime the area before painting. If you don't, these raw patches will pull too much water out of the surface paint, creating dull spots and weakening adhesion.

In addition, limit yourself to working on one manageable area at a time. Trying to cover too much surface from one spot is unsafe and undercuts job quality.

Lastly, always work from high to low. The paint you are using will probably cause drips and splatters, which you can fix on your way down. If they have solidified by the time you get there, scrape them off instead of trying to blend them in.

Wall Cleaning

With extension wands on hose-end cleaners, you can cover large walls.

HOSE-END WASHERS: While power washers use a high-pressure water stream that can sometimes strip paint and cause leaks, hose-end cleaners use standard household water pressure. Many extension wands and cleaning heads are available to suit almost any job, although tough dirt may require vigorous scrubbing by hand. Some systems use cleaning tablets sold by the manufacturer; many allow you to add whatever cleaner you need to the water stream.

Stripping

There are several ways to strip layers of old paint. You can use chemicals that soften the paint so you can scrape it away, a heat gun that does about the same thing without chemicals, sandpaper, sharp scrapers, and more. The trick is to pick the most practical method for the job at hand. On large surfaces like the side of a house, for example, applying gallons of caustic chemicals would make a mess and be difficult to handle. Those surfaces should be scraped. Chemical strippers and heat are better on small jobs such as taking layers of paint off molding. Where grooves and other details in the wood are almost filled with old paint, you may need several applications. On flat surfaces, you can use a putty knife to clear the softened paint. In tight spots, use a shaped scraper blade. There are hand-held models and tools with interchangeable heads designed to dig into all kinds of beading and channels. Before stripping, be sure that the paint is not lead-based. Have a sample checked if you're not sure. Also bear in mind that your town may have restrictions on disposing of the waste. If you use a chemical stripper, wear rubber gloves and safety glasses.

Use chemical strippers in small areas. Allow the reaction to take place (you'll see the paint wrinkle); then scrape.

Heat guns don't add any potentially toxic chemicals to the job but can make paint hot enough to burn you.

REPAINTING

Stain Over Paint

Besides giving siding a natural look, stain has a major advantage over paint. Instead of cracking, peeling, or blistering, it tends to age gracefully and just fades away to a paler but recognizable version of what it once was. That means you're likely to get more years without major maintenance if you use stain instead of paint. So how about switching in midstream and covering paint with stain? It's worth a try on an old house, particularly one without vapor barriers where interior moisture works through the wall and disrupts the paint film every few years. It's wise to try it on an out-of-the-way section of wall. Ideally, you should scrape the siding down to bare wood. But you can get by simply by removing any loose material and leaving a thin paint film that is so well-adhered, you can't easily scrape it away. Pressure-washing is another option. Sand the entire surface with medium to rough paper, and wash the wall clean (if you haven't pressure-washed). Then apply at least one coat of stain. A semitransparent exterior stain that is a close color match for the underlying paint color often works best.

Repainting

USE: ▶ 4-in. wall brush • 1½-in. sash brush • caulking gun • hammer • nail set • scrapers • sanding block ▶ primer • exterior paint • caulk • putty • sandpaper

1 Use a nail set to recess popped nailheads below siding surface. If additional repairs are needed, do them now.

2 Fill nailholes with putty; allow to dry; and then sand the patch smooth. Also fill any gaps, nicks, and other damaged areas.

3 Scrape paint that is blistered or peeling to provide a smooth surface and good adhesion base for the new coat.

4 Recaulk around windows, doors, and openings for pipes and vents. Use a quality, nonshrink, elastic acrylic caulk.

5 Spot-prime sanded and scraped areas, places that have been caulked and puttied, and inside and outside corners and edges.

6 Paint from top down to catch drips and missed spots. Pros apply one primer coat, followed by two finish top coats.

20 Windows & Doors

508 WINDOW BASICS
 ◆ Styles ◆ Window Anatomy ◆ Dividers

510 ENERGY EFFICIENCY
 ◆ Glazing ■ WINDOW LABELS

512 INSTALLING NEW WINDOWS
 ◆ Drip Cap ■ CAULKING ◆ Edging

514 WINDOW TRIM
 ◆ Installing Window Trim ◆ Exterior Details

516 REPLACEMENT WINDOWS
 ◆ Joining New & Old Siding
 ■ INSTALLING AN UPGRADED REPLACEMENT WINDOW

518 BASIC REPAIRS & IMPROVEMENT
 ◆ Unsticking Windows ■ REPLACING GLASS
 ■ IMPROVING WINDOWS ◆ Removing Sash

522 STORMS & SCREENS
 ■ INSTALLING HEAT-SHRINK PLASTIC ■ REPLACING SCREENS

524 DOOR BASICS
 ◆ Types ◆ Anatomy ◆ Hinge Swings

526 INSTALLING DOORS
 ◆ Exterior Details ◆ Glass vs. Security ◆ Installing a Prehung Door

528 WEATHERPROOFING
 ◆ Installing a Door Shoe ■ SILL & JAMB WEATHERSTRIPPING

530 INTERIOR DOORS
 ◆ Interior Details ◆ Installing a Door

532 DOOR REPAIRS
 ■ ADJUSTING A CLOSET DOOR ■ REFINISHING METAL
 ■ LOOSE & BOUND HINGES ■ REPAIRING A SLIDER

534 DOOR HARDWARE
 ◆ Locks ◆ Special Hinges ◆ Closers

536 GARAGE DOORS
 ■ MAINTENANCE ◆ Installing an Opener

20 Windows & Doors

WINDOWS, INSIDE & OUT

Most homeowners just accept their windows as they are, even if they are too small or poorly positioned in a room. Remodeling windows definitely involves more work than some other DIY projects—you can't just rip holes in the side of your house without doing some basic structural work—but the effect of new windows can be dramatic. The extra sunlight can brighten drab rooms and make them seem larger. Repositioning a window may open up an eye-catching view of a backyard garden, warm up part of the house by letting in more sunlight during winter, or improve ventilation by catching prevailing summer breezes.

You do have to consider other things before redoing windows, however. Privacy is one—most people don't want a picture window in their bedroom or bathroom, and it is best to avoid placing windows so people can easily see into the house. Fire codes generally require that at least one window in each bedroom be suitable for emergency escape. You also have to think about how the window will affect the appearance of the outside of the house. The size, style, and position of the new window should be compatible with the rest of the house.

The ideal height of the window off the floor varies according to the use of the room, however—for example, the window's bottom end should be about 3 feet 6 inches above the floor in a kitchen, to allow for counters and cabinets; 2 feet 5½ inches in the dining room; and as little as 1 foot above the floor for a picture window in the living room. You should always consult your local building codes before determining the size and location of a new window.

Before You Begin Work

After deciding on the style and positioning of the new window, the next step is to determine what is lurking behind the drywall. It could be a jungle of wiring, plumbing and heating pipes, or ductwork that would have to be rerouted. Moving the new opening may be easier.

When you order the new window, note the dimensions of the required rough wall opening. You'll need to include them on a scale drawing of the wall, with the location and size of existing and new wall framing. Take the drawing when you apply for a building permit.

Styles

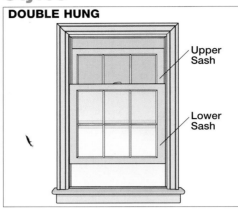

Double-hung windows have a traditional look that is suitable for most home styles. New models have tip-out sash for cleaning.

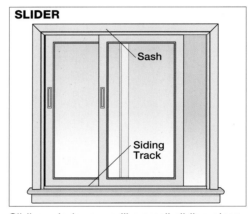

Sliding windows are like small sliding glass doors. For ventilation, one sash slides in a track slightly in and past the other.

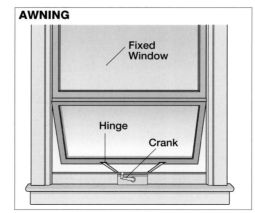

Awning windows are like casements turned sideways. The design can shed rain and still provide ventilation.

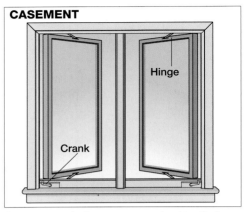

Casement windows are attached to their frame with side-mounted hinges. A crank or slide allows the window to swing out.

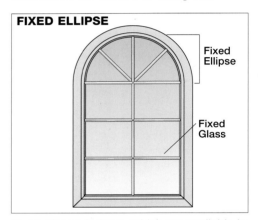

Fixed-glass windows, which are available in many styles, can be used by themselves or paired with operating windows.

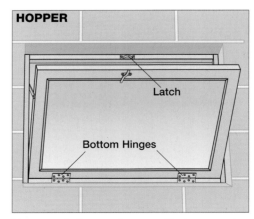

Hopper windows, often used in basements, are hinged at the bottom and open from the top—venting that works well in fair weather.

WINDOW BASICS

Window Anatomy

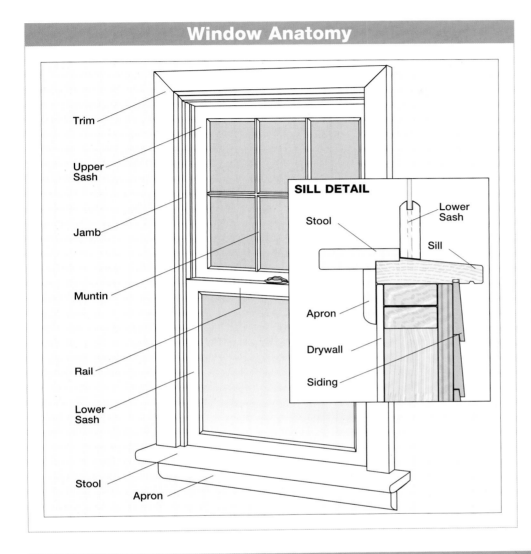

MATERIALS

◆ **WOOD** offers good insulating value. It accepts paint or stain but requires periodic maintenance.

◆ **VINYL** windows generally require less maintenance than wood and are less expensive. It is the most-used material for replacement windows.

◆ **VINYL-CLAD WOOD** windows have the advantage of low-maintenance exterior cladding and the look of real wood windows on the interior.

◆ **ALUMINUM** windows with a baked-on or anodized finish are less energy efficient but are low-maintenance.

◆ **ALUMINUM-CLAD WOOD** offers the same advantages as vinyl-clad units and is more paintable.

◆ **STEEL** makes the strongest window frame, but it can rust and is usually more costly and less energy efficient.

◆ **FIBERGLASS** has strength, durability, stability, and energy efficiency. It's maintenance-free and easily painted.

Dividers

Years ago, when window glass was weaker and more difficult to manufacture, panes had to be small. Muntins—narrow strips of wood—held panes together in a larger sash. Now muntins are no longer needed. In fact, they reduce the energy efficiency of double glazing. (True divided lights in double-glazed windows are costly.) There are several other options, such as a variety of snap-on muntin grilles that simulate divided-light patterns without compromising efficiency.

Many codes today have energy standards that require double glazing. To maintain a dead-air space between panes, use interior grilles made of plastic or wood, a combination of materials inside and out, or true divided lights with individual double panes.

20 Windows & Doors

BETTER WINDOWS

Older homes had single-glazed windows—only one layer of glass separated the inside from the outside. Unless you wanted chilly rooms and high heat bills, every fall you got out the ladder and put up the storm windows. They provided a second layer of glass over the window and slowed the heat loss. Then, every spring, you took down the storm windows and put up screens for ventilation.

New energy-efficient windows eliminate all that hassle with double glazing—two panes of glass separated by an air space are built right into the window frame. The added glass and the air space provides enough insulation to slice anywhere from 10 to 25 percent off your heating bill.

Double glazing is a good compromise between cost and energy efficiency in most parts of the country. But if you live where winters are very cold, you may want triple-glazed windows, which have three layers of glass. They cost more and are heavier, but you may save enough on heating to make up the difference.

It was such a novel idea that C. G. Johnson, who invented the uplifting garage door in 1921, promoted his space-saving design on the back of a Model T—complete with tracks, pulleys, and springs.

Some other high-tech options will conserve even more of your energy dollars. You'll pay about 15 percent more, but windows with low-emissivity (low-E) glass have a microscopically thin metallic coating that blocks heat loss. You may also want to try double-glazed windows with argon gas between the panes. With an insulating value equivalent to 2 inches of fiberglass, some high-tech windows can reduce heating bills in cold climates by 30 percent or more.

Window Frames

When choosing a new window, consider the framing material as well. Wood is a better insulator than many other window frame materials. But wood must be protected from the weather with paint or a coat of vinyl or aluminum.

Framing made from solid vinyl and aluminum generally needs less maintenance, but these materials are not good insulators. To limit the rapid heat loss through these frames, quality windows must have a thermal break between the inner and outer halves of the frame.

Glazing Options

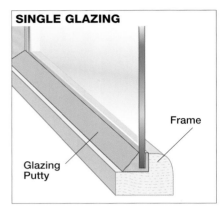

Single glazing is the easiest to install, but it has an insulating value of only R-1, and it fosters condensation.

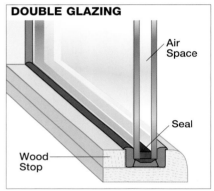

Double-glazed windows provide about an R-2 by trapping a small area of dead air between the panes.

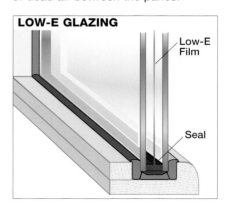

Low-E glazing can more than double the efficiency of double glazing with a reflective film between panes.

Fogging

One by-product of high-tech windows is less condensation, particularly on windows with warm-edge technology—low-conductance spacers that lower heat transfer near the edge of insulated glazing. When it's 20°F. outside, single glazing can sweat when the indoor air has only 20-percent relative humidity; a double-glazed low-E window won't sweat until the interior air has 70-percent relative humidity.

ENERGY EFFICIENCY

GREEN SOLUTION

WINDOW LABELS

The National Fenestration Rating Council (NFRC)—an industry group that tests windows and provides standardized information about window performance—has created a performance label that rates windows for energy performance and for such things as wind and impact resistance.

In the upper right part of every label, you will find the manufacturer, model, style, and materials used in its construction (such as "aluminum-clad") to indicate the outward-facing and inward-facing component that make up the window. How the window opens is also indicated, for example, "vertical slider." That information isn't as important as the characteristics that can affect performance, so take a close look at the information below. For target numbers, a quality window will have a U-factor of around 0.32, a SHGC of around 0.27, and a visible transmittance of around 0.46. Look for a 20-year warranty (or more) on the window. Anything less is a signal that the window isn't of the highest quality.

U-FACTOR. The U-factor is the rate at which the window will lose heat as it passes from the building out into the air—for example, when it's relatively hotter inside the house during the winter. Think of the U-factor as the reverse of an R-value in insulation. The lower the U-factor, the greater resistance the window will offer to heat flow. In other words, the lower the U-factor, the better the window is as an insulator. In areas where you spend most of your energy dollars on heat, the U-factor is the important number.

VISIBLE TRANSMITTANCE (VT). The VT rating tells you how much visible light will be transmitted through the glass. Visible transmittance is a number between 0 and 1. A higher number is better. The higher the visible transmittance, the more light will come through the glass.

AIR LEAKAGE (AL). This is a rating of the window's ability to resist heat loss or gain through cracks in the window assembly. The measurement is expressed as cubic feet of air passing through a square foot of window area. The lower the AL, the tighter the window construction. This is an optional rating.

CONDENSATION RESISTANCE. A rating on how well the window resists the formation of condensation. The rating is a number between 0 and 100; the higher the rating the better. This is an optional rating.

SOLAR HEAT GAIN COEFFICIENT (SHGC). The solar heat gain coefficient tells you how well the window will block heat from the sun. SHGC ranges from 0 to 1, and the lower the number, the better. A window with a low SHGC will transmit less heat than a window with a high SHGC. In areas where you spend most of your energy dollars on cooling, the SHGC number is the one that is most important to you.

NFRC Certified	World's Best Window Co. Millennium 2000+ Vinyl-Clad Wood Frame Double Glazing • Argon Fill • Low E Product Type: **Vertical Slider**	
ENERGY PERFORMANCE RATINGS		
U-Factor (U.S./I-P)		Solar Heat Gain Coefficient
0.35		0.32
ADDITIONAL PERFORMANCE RATINGS		
Visible Transmittance		Air Leakage (U.S./I-P)
0.51		0.2
Condensation Resistance		—
51		

Manufacturer stipulates that these ratings conform to applicable NFRC procedures for determining whole product performance. NFRC ratings are determined for a fixed set of environmental conditions and a specific product size. NFRC does not recommend any product and does not warrant the suitability of any product for any specific use. Consult manufacturer's literature for other product performance information. www.nfrc.org

High-Efficiency Windows

GREEN SOLUTION

When the need for energy efficiency is paramount and you have taken the many less-costly steps to make your home energy efficient, it may pay to upgrade to one of the super glazing systems. Spectrally selective coatings are the latest low-E technology. Glass with these coatings can block from 40 to 70 percent of the heat normally conducted through glass without reducing transmitted light.

Low-E, or low-emissivity, uses one or more invisible metallic coatings suspended between two or more sheets of glass. The glass can be fitted into standard or curved frames. Heavy, triple-glazed units (right) may require special installation.

20 Windows & Doors

CUTTING OPENINGS

Installing a new window in an existing exterior wall is not easy work, even if you don't have a brick wall to contend with. Cutting a hole in the exterior wall of your house can cause irreversible damage. Be sure to locate the new window in a wall section free of wires, pipes, or ducts, and never start the work until after you've actually received the new window.

Begin by marking the exact position of the rough opening—two vertical lines from floor to ceiling (using a long level or plumb bob) and one horizontal line across the top of the rough opening. Cover the floor with a tarp, and use a handsaw to cut through the drywall. (Power saws could cut into buried wires or pipes.) Remove the drywall and insulation underneath.

Before cutting existing studs in the rough opening, build a temporary support for the affected ceiling joists. Make it with 4x4s or doubled 2x4s, and set it up no more than 2 feet from the existing wall. This support should extend at least 2 feet beyond each side of the rough opening. Cut existing studs to be removed into sections to make them easier to pry out. Don't cut through the siding at this point.

Framing the Opening

With the existing studs out, toenail full-length king studs into the soleplate and top plate on either side of the rough opening—or drive in angled screws. Cut and nail up the shorter jack studs to the inside face of the king studs. Trimmer studs support the header, which forms the rough opening's top.

You'll have to make the header by sandwiching a piece of ½-inch plywood between two sections of 2x10 and nailing them with staggered 16d nails about 16 inches on center. The required width of the two-by varies with the length of the span between trimmer studs, however. Check local building codes.

Nail up the header and the short cripple studs above it. Cut and nail in the sill and the jack studs supporting it. Add insulation in stud cavities before nailing up drywall around the opening. If you have wood siding, drive nails through the corners of the rough opening to transfer the outline to the exterior siding. Cut the opening from the outside. Tack a board to the siding as a guide for the power saw.

Drip Cap

A piece of flashing called a drip cap, made from either aluminum or plastic, must be installed to form a barrier between the window and the sheathing to prevent water from seeping inside the wall. One flange of the drip cap is installed underneath the siding above the window; the other folds down over the window frame. Install the drip cap with a downward slope to deflect rainwater.

Although windows generally come supplied with flashing, you can cut and bend your own out of aluminum.

Caulking

Seal the gap between the window casing and the sheathing with an exterior-grade polyurethane caulk. This bead of caulk, along with flashing, will prevent air and water from seeping between the window and the wall, where it can cause rot. After you've installed the window trim, you can also run a second bead of caulk between the trim and the siding. It pays to check the caulk around the windows periodically. By filling cracks (or replacing old caulk), you can cut down on drafts and save on heating costs.

GREEN SOLUTION

On a new window with an integral nailing flange, add a liberal bead of caulk to help seal the window perimeter.

Edging

To seal side seams between windows and wood siding, you need to rely on caulk. With vinyl siding, use molded J-channel. It has a lip that wraps around the ends of the vinyl and a nailing strip where you can attach it to the wall. No matter what type of siding you install, first staple up a layer of felt paper or air-stopping house wrap. Trim the corners, and tack the overage back into the frame opening.

Vinyl J-channel fits against the side edges of the window and is nailed through its perforated flange.

INSTALLING NEW WINDOWS

Installing a Window in New Construction

USE: ▶ measuring tape • circular saw • staple gun • caulking gun • level • hammer ▶ new window • exterior-grade caulk • nails • shims • insulation • trim

1 You can cut sheathing around window openings, but it is often easier to sheathe the entire wall and cut the openings later.

2 After marking the rough opening, set a circular saw to the depth of your sheathing (typically ½ in.) to make the cuts.

3 When you install felt paper or house wrap, leave enough to tuck back and staple onto the sides of the framed opening.

4 To make a weathertight seal, add a bead of exterior-grade caulk to the back of the nailing flange before installing the unit.

5 Set the window unit in place, resting it on the sill so that you can tip it in place. Add a few temporary nails for security.

6 Use pairs of tapered shingle shims to adjust the window on all sides in the opening, checking for plumb and level.

7 Cut drafts and improve energy efficiency by stuffing insulation into gaps between the studs and the window frame.

8 Once the window is plumbed and leveled, fasten it by nailing on all sides through the perforated flange.

9 With the window fastened, you can add caulk and trim: J-channel for vinyl siding or a variety of wood trim for clapboards or shakes.

20 Windows & Doors

FINISHING TOUCHES

Once a new window has been leveled, shimmed, and nailed in place, it's time to begin working on the trim. Factory-built windows won't have anything beyond the bare essentials, and what's there might not match the style used on your house.

Exterior trimwork begins with the factory-supplied exterior casing. It is usually one of two types of bare-minimum molding: brickmold, a narrow molding with a little detail; or a flat ⁵⁄₄ casing, which is just a narrow, flat strip alongside the window's outer edge. By themselves, they don't look like much. Probably the best way to fix factory casing is to just add more molding over or around it. For example, you can widen the flat casing by adding ⁵⁄₄ stock; then cover both pieces with stock molding to add detail to the profile.

You could leave the side casing alone and replace the top piece, called the head casing, instead. This works well with brickmold casing, which is harder to add to because the face is not flat. Replace the factory head casing with a ⁵⁄₄x6 board nailed over the window frame, and apply trim to it—quarter-rounds, coves, drip caps, and other shapes to make the trim as elaborate as you like. When planning your new trim design, it's probably best to tack samples to a wall so that you can step back and take a look. But mixing and matching from stock molding shapes available at lumberyards will surely produce a combination that does the job.

Interior Trim Options

PLAIN — Side Casing, Mitered Corner, Reveal

DECORATIVE — Rosette Corner Block, Stop, Stool, Apron, Fluted Casing

Some windows come with exterior trim already mounted. That system works because windows are installed from the outside. Inside, you have to add trim. Flat or clamshell shapes generally are best suited to casements and fixed glass, while double-hungs can be trimmed with beaded casings and corner blocks.

Installing Window Trim

USE: ▶ power miter saw or miter box and backsaw • power drill/driver • hammer • saber saw • eye protection ▶ side trim • stool • apron • return pieces

1 Cut top and side pieces at 45 deg., add glue to the joints; predrill to avoid splitting; and secure with finishing nails.

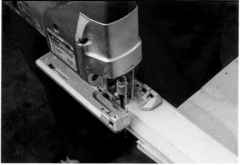

2 Use a saber saw to cut the deep, interior sill, called a stool, where it extends beyond the window frame and side trim pieces.

3 Add glue, predrill the stool, and drive finishing nails at an angle into the window frame. You can fill and sand the holes later.

WINDOW TRIM

Exterior Details

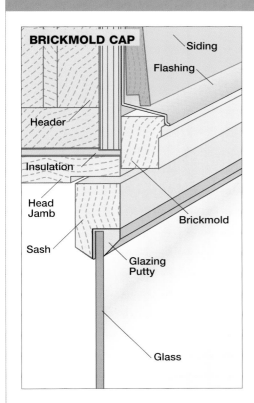

BRICKMOLD CAP

To protect the top edge of windows with standard brickmold trim, tuck flashing under the siding and across the top of the frame.

BUILT-UP CAP

Provide extra protection for windows under a shallow roof overhang by adding an extension of angled cap trim above the window trim.

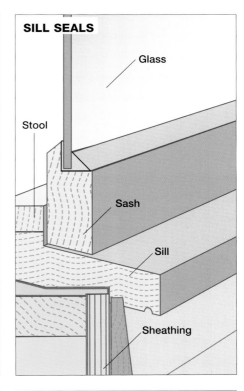

SILL SEALS

For the best seal at windowsills, siding should tuck into a slot in the sill. The small outer groove is a capillary break to shed water.

• finishing nails • wood glue • wood putty • sandpaper

4 Use a power miter saw or backsaw and miter box to cut a miter at each end of the trim that supports the stool, called the apron.

5 After predrilling, drive finishing nails at a slight upward angle through the apron and into the wall and window frames.

6 Make and glue on small return pieces to complete the apron miters. Without this step, you would see the apron end grain.

FINISHING TOUCHES 515

20 Windows & Doors

OLD WINDOWS

After years of use and layer upon layer of paint, your windows may not be broken, but they won't work as they did when they were new. On older homes, single-pane glass and worn jambs aren't doing much to keep the heat in during winter, and that old paint can make windows nearly impossible to open and close.

You could spend hours scraping the windows down to bare wood and wind up with the same old energy-inefficient model. Replacing them makes better use of your time and money. These days you can get inexpensive, custom-made replacement windows with double glazing that drastically improves the unit's insulating value. You could also opt for low-emissivity glass or even triple glazing if you live in an area with very cold winters. (See pp. 502-503 for more on window glazing.)

Replacement Options

You have three basic choices—sash kits, replacement windows that fit into existing frames, or entirely new windows that have their own frames and casing. With sash kits, you get new standard-size sashes—the window glass and the frame around it—plus new jamb liners, which are the tracks that hold the sash in place. Sash frames can be wood or vinyl with double glazing. Your existing window frame, casing, and trim all stay in place, and all installation work is done from the inside. On the downside, a sash kit typically is not as energy-efficient as a new window.

Replacement windows are entirely new, energy-efficient units custom-made to fit into an existing window frame (usually at no extra cost). These windows have less glass area than the old ones, but if cost is a major consideration, vinyl replacement windows are the way to go.

New windows give you a wider range of options—you can go to larger sizes and different shapes—but the installation involves a lot more work because you have to rip out the old window entirely. If the new one is only a little larger, you may be able to take advantage of extra space in the rough frame that was formerly occupied by the sash weights. For larger windows, it means cutting a bigger wall opening and installing new framing. (See pp. 504–505 for more on installing a new window.)

Cutting Back Siding

To cut back vinyl siding, use a zip tool, which unlocks the joint between courses, and trim with metal shears.

Wedge wood siding to make room for your saw blade, and finish the cuts at the top with a sharp wood chisel.

Curved Windows

To set and flash windows with curved frames, you need special flashing. Generally, it is supplied by the window manufacturer, often as a premolded flexible vinyl strip that conforms to the curve. Curved shapes are difficult for DIYers to make because the edge must be snipped and tucked every few inches to create the curve—and every cut is a potential leak. It's better to have flashing made up by a metal shop.

Joining New and Old Siding

You can butt seams where you stagger joints around a new window, but cuts at 45 deg. offer more protection.

Should new siding shrink and open the joint, you'll see more wood instead of a gap. Predrill at the ends of boards.

REPLACEMENT WINDOWS

Installing an Upgraded Replacement Window — GREEN SOLUTION

USE: ▶ pry bar • hammer • chisel • reciprocating saw • level • plumb bob ▶ replacement window • 2x4s • ½-in. plywood • nails • shims

1 To remove an old, single-glazed window, use a pry bar, hammer, and chisel to pry loose the surrounding trim.

2 Use the same techniques inside to remove interior trim. If you want to reuse it, pry gradually as the trim may be brittle.

3 To release the window, cut through the nails that extend from the frame into the house-wall studs.

4 Once the nails are cut, you can pry out the window. First, you may need to release shims at the sides of the frame.

5 Pack out the old opening as required, using 2x4s or other lumber to make a rough opening matched to the new window.

6 The new 2x4 will not be as thick as the adjacent wall. You need to add drywall inside and plywood sheathing outside.

7 Add an overlapped strip of felt paper or house wrap, and plumb and level the replacement window.

8 Tack the window in place; add pairs of tapered shingle shims from the inside; and finish nailing the exterior flange.

9 You can cut back alternate courses of siding so that joints won't align as you fill in around the new window.

20 Windows & Doors

UNSTICKING WINDOWS

Once spring arrives and the weather warms up, you'll want to open all your windows all the way and let the fresh air in. But moving parts that have been locked in place over the winter may refuse to budge. So start small: just try to get the windows cracked. If the frame is swollen from moisture or the window sash and the surrounding frame are sealed by paint, you could be in for a wrestling match.

If a double-hung window won't move, first try a sharp push with the heel of your hand on the sash at the center of the window—by the window lock. A good crack may break a thin paint bond. Do not use a hammer and wood block—they may break the window.

Before going too far, make sure the window is unlocked and that no one has set a finishing nail through the sash or installed some other security device. It's embarrassing to curse at a window that won't move when it's locked.

Usually, the trick is to slice apart the painted seams between the window sash and the trim piece, called a stop, that makes a track for the sash. You can use a utility knife, a tool called a paint zipper, or even a sharp pizza cutter. Don't forget to do the outside seams, too.

If all else fails, try forcing the window with a flat pry bar set between the windowsill and the bottom of the sash. The tool will likely gouge the wood, but the only alternative is prying off the stop, which holds a double-hung window in place.

Cutting Glass & Plexiglas

USE: ▶ scoring tool or wheel cutter • straightedge • eye protection • work gloves ▶ glazing panel

1 Use a scoring tool (left) on plastic glazing panels and a wheel-type tool (right) on single-thickness glass.

2 Work on a flat surface with a square or straightedge to guide the cutter. Sheets of newspaper provide a cushion.

3 Press down firmly as you draw the cutter across the glass. A small wheel scores the glass surface.

4 Set a narrow wood dowel under the score line, and apply pressure on both sides to snap the glass or plastic cleanly.

Replacing Glass

USE: ▶ putty knife • paintbrush • glass cutter (if needed) • eye protection • work gloves ▶ new glass • primer • glazing compound • glazier's points

1 On an older window, the exterior putty may pop off the glass and frame easily when you scrape it with a putty knife.

2 On some frames, you may need to dig out the putty. Once the glass is free, also scrape the wood underneath.

3 Prime raw wood where new compound will rest. This keeps moisture from seeping into the wood and weakening the bond.

BASIC REPAIRS & IMPROVEMENT

Freeing a Stuck Sash

To free a sash stuck in place by many coats of paint, use a utility knife to make several scoring cuts between sash and stop.

On older windows, particularly those that are not regularly opened, layers of paint can seal the sash in its track. Instead of forcing the sash or prying at the base and damaging the wood, release the painted-on seal. You can slice along the seam with a utility knife, making repeated passes to cut through the paint. A circular cutting wheel also works well and is easier to keep in the seam. Sometimes the only permanent solution is to pry off the stop, and remove the window for a thorough scraping and sanding.

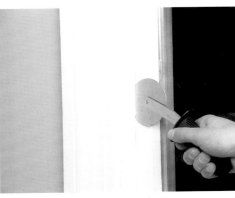

A wheel cutter rides along the seam between the sash and the stop. You may also need to cut the paint film outside.

Glazing Compounds

Traditional compound is an oily putty. You need to work it smooth until it's soft like dough, form a ball, and roll the ball into strips.

At seams between dissimilar materials, such as glass and wood, which expand and contract at different rates, you need glazing compound. To form the compound, work it in your hands or roll it out on a board. If you need to make a glass repair in cold weather, first prepare the compound inside so that it's pliable when you work it along the window. The wood against which the compound rests should be primed. Otherwise, the compound can dry prematurely and crack.

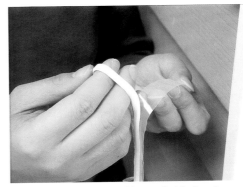

Preformed compound is available in plastic-backed strips. Peel the backing, and push the strip of compound in place.

MONEY SAVER

4 Roll out a rope of fresh compound to back up the glass; set it against the sash; and press it in place with your fingers.

5 To secure the glass, set small holders, called glazier's points. Use a putty knife to force the points into the sash.

6 The exterior layer of compound covers the points. Spread this layer with a putty knife, and use the edge to trim any excess.

20 Windows & Doors

MAKING WINDOWS GLIDE

A window doesn't have to stick to be a pain. Some windows still move but demand a workout to get opened or closed. On a wood window, built-up paint layers can simply make the sash too big for the sliding track. The solution is to scrape or sand off enough paint so that the sash will be a hair thinner. Make sure all exposed wood has been covered with paint (or sealer) to minimize swelling on humid days.

An easier way to make the sash move more freely is to increase the clearance between the sash and stop. Do this by scoring the seam between the sash and the stop with a utility knife, then a metal-cutting jigsaw blade, then a slightly thicker wood-cutting blade.

Adjusting the Friction

On many windows, tension that holds the sash in place is supplied by some type of spring clip, usually in a tight V-shape, that is built into the window frame. It presses against the sash. To decrease the tension, place a 1x2 over the spring clip, and give it a few shots with a hammer. You may find that you have to use some really firm blows before noticing any difference. You can increase the tension by slipping a screwdriver into the V-clip and prying it open a bit—use only a little pressure.

Improve old double-hung windows that use the weight-and-pulley system by installing new friction channels. (See below.) Your sash will operate more smoothly and leak less air.

Removing an Old Sash

USE: ▶ utility knife • wood chisel • pry bar • work gloves

1 To remove an old sash for scraping, painting, or other repairs, start by scoring the painted seam along the stop.

2 Use a chisel to break the bond between the stop molding and the window casing.

3 Use a pry bar to gradually pry off all the stop. An old stop, which you can reuse, is likely to be brittle and snap easily.

4 With the stop removed, you can lift the sash out of its tracks. On some units, you need to disconnect slide systems.

Improving Windows by Using Friction Channels

USE: ▶ pry bar • hammer and wood chisel or small saw • block plane or scraper • paintbrush • plumb bob • power drill/driver • work gloves ▶ friction channels

1 To improve an old window with a loose sash, you can install friction channels. Start by removing the inside and outside stop.

2 Set the new channel in place to mark its location on the existing top trim. You'll have to notch the trim to install the channel.

3 Use a small saw or a hammer and chisel to trim away the end of the trim so that the molded side channels will fit.

BASIC REPAIRS & IMPROVEMENT

New Flashing

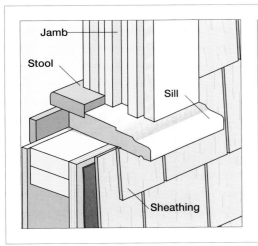

When a window leaks, often the problem stems from the most exposed seam along the top. It should be protected by flashing, which sheds water running down the siding. When flashing deteriorates, particularly synthetics that can become brittle due to constant exposure and the sun's UV rays, you can replace it. The trick is to gently pry up and wedge shakes or clapboards above the window, and pull or cut the old flashing nails. Then you can caulk, and slip in a new piece.

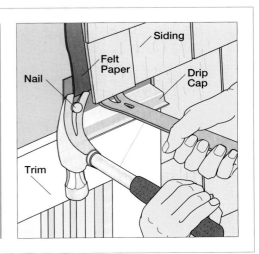

New Sills

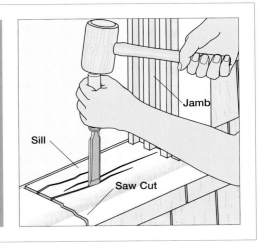

Windowsills can rot on the outside from exposure and on the inside from condensation that runs down the glass. When the wood is beyond repair, you can buy a replacement sill to install without replacing the entire window. To remove the old sill, cut all the way through to make two big pieces. Then use a chisel to split each piece into small sections that you can slide out from under the jamb. You may also need a hacksaw to cut nails.

MONEY SAVER

- primer
- screws

4 You probably will need to scrape or plane down the sides of the sash so that they move smoothly in the new channels.

5 After priming any raw wood, fit the sash into the appropriate sides of the channels, and reinstall them as one unit.

6 Plumb the channels; slide the sash out of the way; and permanently fasten the channels to the window frame with screws.

MAKING WINDOWS GLIDE

20 Windows & Doors

THE CASE FOR STORMS

With so many high-tech alternatives, you may wonder why there is still a market for storm windows. They're old-fashioned, a nuisance, and easily replaced with energy-saving double-glazed windows (two sheets of glass sandwiching an insulating "dead air" space).

In homes with single-pane glass, new double glazing certainly makes a noticeable difference in your comfort and utility bills. However, you can also make a difference for far less money and effort by putting up storm windows. There are many types available. And in almost all cases, adding a removable storm, even triple-track storms with screens, is less costly and less work than ripping out old window sashes and frames and installing completely new units—with trim, touch-ups, and all the other work that goes along with opening a large hole in the side of your house.

And for less than $10 a window, you can add a layer of clear plastic film on the inside during the winter. It is almost unnoticeable, just as energy-efficient as storms and double glazing, and an easy do-it-yourself project. Replacement windows cost more—a lot more.

If you prefer ⅛-inch thick rigid plastic, you can make custom frames for interior-mounted storms using U-shaped plastic or aluminum channels. Many casement windows have interior sash slots that allow you to clip framed storm panels in place.

Installing Heat-Shrink Plastic   GREEN SOLUTION

USE: ▶ hair dryer • utility knife ▶ heat-shrink plastic • double-faced tape

1 Interior clear plastic traps a layer of dead insulating air next to the glass. It mounts with double-faced tape.

2 Unfold the plastic; spread it evenly over the window; and press onto the tape. There are kits for windows and doors.

3 Use a hair dryer to apply heat. This causes the film to shrink, which removes the wrinkles and leaves a clear film.

4 Trim excess plastic using a utility knife. If the material is well installed, it can last through a heating season.

Replacing Screens

USE: ▶ small pry bar • pliers • screwdriver • C-clamps • staple gun • utility knife • straightedge • hammer ▶ blocking • new screening • staples • finishing nails

1 To replace damaged screens on a wood frame, start by prying off the trim pieces that hold the screen edges.

2 Use pliers and a screwdriver to pull the old staples and remove the old screen. You may want to sand and paint the frame, too.

3 Use clamps to create a downward bow in the center of the frame. When the raised ends are released, the screen will tighten.

STORMS & SCREENS

Glazing Connections

There are many types of storm windows, including the old-fashioned sash frame that is glazed like a regular window. To replace or repair those glazing joints, scrape off old putty; prime raw wood; and spread a new layer. Most aluminum storm units have a rubber gasket to hold the glazing in place. You need to disassemble the frame and strip out the gasket to make repairs.

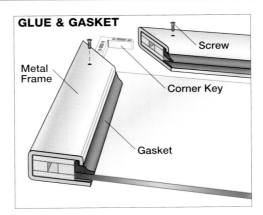

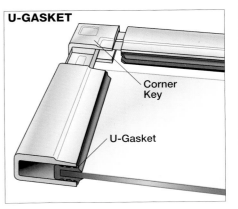

Screening Frames

Use shears or a utility knife to cut new screening with a 2-inch overage on all sides. Lay it over the frame, and roll the screening into the channel with the wheel of a screen roller. Then roll the spline into the channel to tighten and hold the screen in place. Keep the screen tight as you work. With fiberglass or metal screens, it works best to install opposite sides in order.

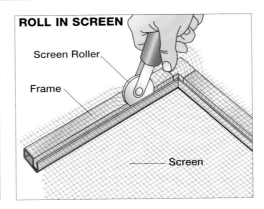

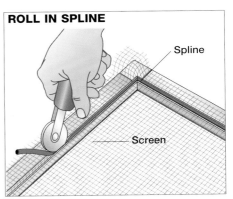

MONEY SAVER

4 Spread a new piece of screen over the bowed frame, and staple it in place, starting at the center and working out.

5 Use a utility knife and a straightedge to trim away excess screen. You have to staple in a straight line to get a neat edge.

6 Finally, nail down the trim pieces that help to hold the screen tight and cover the rows of staples.

THE CASE FOR STORMS

20 Windows & Doors

DOOR BASICS

Doors, like windows, serve various purposes. On the functional side, they provide a way into and out of the house, give us privacy when we want it, and keep out unwanted noise, cold drafts, bugs, and people who bug us. But doors (and doorways) also figure in home design. The front door is an important architectural focal point, and interior doors help express the style of the house or complement a room's decor.

A new front door with sidelights can make your entry hall seem brighter and more inviting, while switching from a painted to a wood-finish door adds a touch of elegance. You can improve traffic flow between parts of the house by installing a new door or two, but it comes at a cost. Because the traffic lanes take up space, the more doors a room has, the smaller it seems. Every new door also needs enough space to swing open and closed.

Don't count on just adding or moving a door in just one afternoon, because there is a lot to the job. You have to locate and cut a hole through an existing wall, remove the old wall framing, frame the rough opening, and install the door itself. You'll need good carpentry skills and a building permit in many locales, but the job is within the range of many DIYers.

Buying Doors

The wood doors you'll find at building-supply centers are either panel doors, which have anywhere from one to ten panels set in a solid wood frame, or flush doors, formed by covering a solid wood core or a lightweight hollow core with thin sheets of wood veneer. While panel doors come in a wide range of wood types, flush doors usually are faced with lauan mahogany or birch, and they are less expensive.

Doors are sold prehung (where the door is mounted in a frame, sometimes with holes for the lockset predrilled) or as individual components. Unless you have hung a door before, spend the extra money for a prehung unit. It's not that the skills required to assemble a frame and hinge a door are all that difficult, it's the time you'll save. On an exterior door, you will also be assured of a tight weather seal. If you want to replace or upgrade an old door, just buy the door alone and use the original as a template for mounting hinges and the lockset.

Types

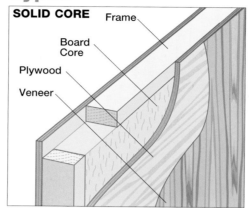

Solid-core doors are used on exterior openings for security and durability. They often need three hinges due to their weight.

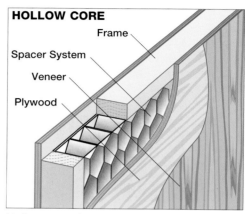

Hollow-core doors are used on interior openings. They are much lighter because the core is an air-filled spacer system.

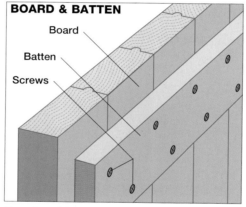

Board-and-batten doors generally are used on sheds and other outbuildings. The batten ties together several boards.

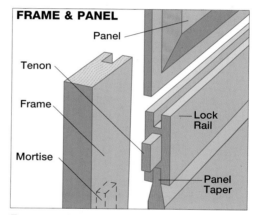

Frame-and-panel doors typically are used on cabinets. Interlocked rails are permanently fixed, and the panel floats inside.

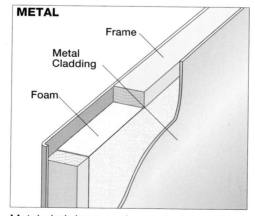

Metal-clad doors are increasingly popular because the metal needs little maintenance, and the foam core is energy efficient.

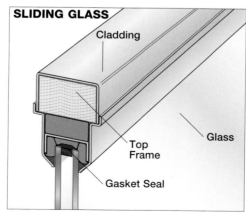

Sliding glass doors may be wood- or metal-framed. One panel is fixed with hardware; the other slides in a track.

DOOR BASICS

Anatomy

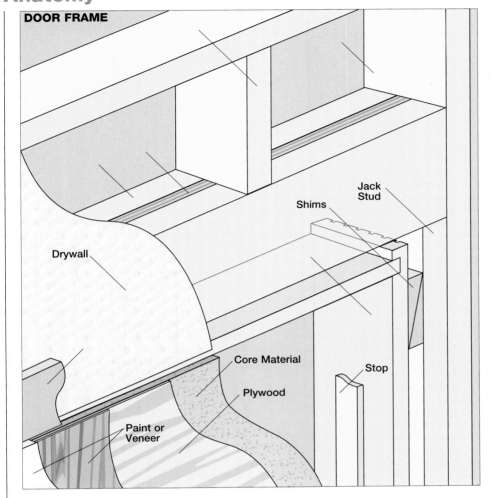

Door Actions

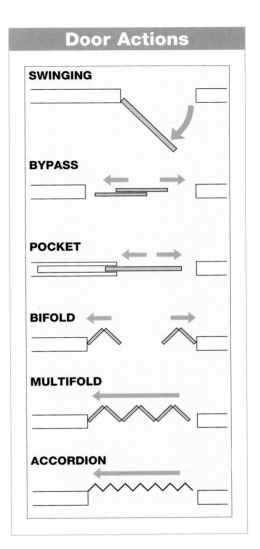

Hinge Swings

When you order a prehung door, you need to specify how it will swing. The convention is to order a door by its hand, which means that the door has either a left-hand or right-hand swing. This is determined by the side of the door that is hinged when you stand on the outside of the room or street side of an exterior door. When you open the door away from you, right-hand doors will swing to the right, left-hand doors will swing to the left. Reverse models are available for doors that swing toward you.

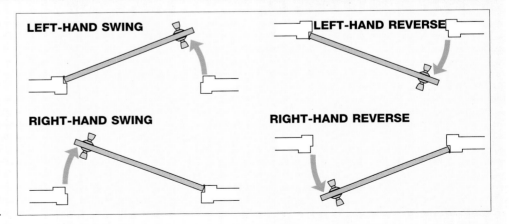

20 Windows & Doors

MORE THAN A DOOR

Doors don't just hang by themselves; other structural pieces surround the door and provide support. Buried beneath drywall and trim is the door frame, which forms the door's rough opening. It is composed of vertical 2x4s and a horizontal piece called a header. The header is needed for extra support over doorways in load-bearing walls, because some wall studs have been removed to make room for the door.

The doorjambs—two side jambs and the head jamb across the top—form the finished opening for the door itself, as well as the mounting point for the hinges and the lockset. The door sill, or threshold, lies below the door, and the door stops—narrow strips of wood nailed to the jambs—keep the door from swinging beyond the closed position. Trim, also called casing, covers gaps between the rough opening and the jambs. The gaps provide room to level and plumb the door and jambs.

Installing a Prehung Door

To start, don't remove shipping braces from the door—they keep the frame square. If the floor is not level, cut one leg of the frame. Prehung doors are built to allow for thick carpeting, so you may need to cut both legs if the bottom of the door is too high off an uncarpeted floor.

Center the unit in the opening, and check that the top is level. Insert shims in the gaps between the doorjambs and rough framing to square and plumb the door opening. Use pre-packaged shims sold for this purpose, tapered wood shingles, or homemade shims.

Remove the stops, and set a pair of shims with tapers opposing between the frame and stud at each hinge location—and if there are only two hinges, in the middle. Increase or decrease the overlap of the shims to adjust the frame until it is plumb. Drive a finishing nail through the jamb, each shim set, and partially into the stud. Then install three sets of shims on the other side jamb and one set above the head jamb. When all shims are in place, the frame is plumb and square, and there is a uniform gap between the door and the jamb, then add a second nail at each shim and drive all nails home.

On an exterior door, stuff insulation behind and above the jamb before installing the casing.

Exterior Details

Exterior door frames are constructed differently from interior doors. Because exterior doors provide protection from rain and seal out drafts, the perimeter must be weatherstripped, even at the sill. The sill must slope down away from the house so that water runs off. It should also have a drip edge to prevent water from seeping underneath. A drip cap above the door and caulk around the casing completes the weather protection.

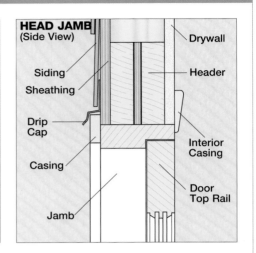

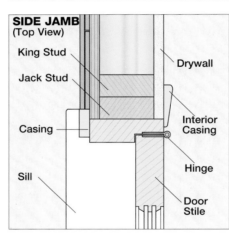

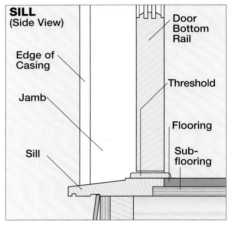

Glass vs. Security

Glass will let in light, but a large panel makes an exterior door vulnerable to break-ins. To minimize your risk, you can use tempered or safety glazing that is harder to break. (Double glazing is also more resistant than single glazing.) The problem is that by breaking the glass, a burglar can simply reach inside to undo the locks. To prevent that easy entry, use glass panels on top of the door or small panes that don't allow access to the locks.

INSTALLING DOORS

Installing a Prehung Door

USE: ▶ utility knife • staple gun • level • drill • hammer • putty knife ▶ prehung door & lockset • caulk or flashing • 2x4 brace • shims • nails • wood putty

1 Cut the felt paper or house wrap across the door opening, and staple back the excess against the sides of the framing.

2 You can install flashing under the sill, although many manufacturers suggest using a waterproof caulk instead.

3 You set a prehung exterior door from the outside of the house. The door already has exterior molding attached to the frame.

4 Working from the inside, use a level to plumb the door. Put a brace across the outside to keep the door in the opening.

5 As you check for level, insert shingle shims in the gaps between the door frame and the 2x4 wall studs.

6 When the door is correctly positioned, predrill and nail through the frame (and hidden shims) into the wall framing.

7 Also drive finishing nails through the face of exterior molding into the wall framing. Set the heads, and fill with putty.

8 You can order most prehung doors with locks already installed or with the holes predrilled so you can install your own.

9 A lockset plus dead bolt provides extra security. Use long screws in the keepers that reach through to the house framing.

MORE THAN A DOOR

20 Windows & Doors

KEEP COLD OUT

Weatherstripping an entry door used to be a job for pros because interlocking metal channels had to be routed into the door and jamb and mated precisely for the door to close. Sealing up doors today is child's play by comparison, and the weatherstripping itself does a much better job of keeping the heat in and the cold out.

As long as the door itself is solid and there are no cracks in the wood panels, your main concern will be sealing up the narrow space around the door's outer edge. Jamb weatherstripping takes care of the space along the sides and top of the door. Sill weatherstripping seals the bottom to keep out cold drafts and water.

Weatherstripping a Door

Home-supply centers sell nail-on and self-adhesive weatherstripping for the sides and tops of doors. Both must be cut to length and mounted on the jamb so that when the door closes, it presses against the rubber or foam sealer on one edge of the strip. Neoprene rubber sealers last longer than foam, and nail-on weatherstripping stays put better than the self-adhesive type. You need a perfectly clean, smooth surface for self-sticking weatherstripping, and it may not last more than a season.

Vinyl, wood, and aluminum door-stop trim, edged with a sealer, comes precut with coped joints on the side pieces. You only have to square-cut to length the top piece and the bottoms of the two side pieces. Then, with the door closed, nail the trim to the jamb with 4d finishing nails so that the weatherstripping compresses very slightly against the door. If you pack in so much that it compresses a great deal, it does no good.

Manufacturers offer a greater selection of weatherstripping for the door sill than for the jambs. Simple door sweeps attach to the inside face of the door and hang down to make a seal with the sill. Usually you don't have to cut down the door itself. Sweeps and any other type of weather seal attached to the door may not work if you have a thick mat inside.

Other types of sill weatherstripping include a U-shaped brass frame and an aluminum threshold with a vinyl sealer. Mount the U-shaped sealer with screws, adjusting to seal the space under the door. On the aluminum threshold, install it and set the height of the vinyl sealer.

Installing a Door Shoe

USE: ▶ chalk-line box • circular saw • paintbrush • hacksaw • screwdriver • drill ▶ door shoe • sealer

1 Exposure to the weather sometimes can cause extensive rot along the bottom edge of an exterior wood door.

2 Snap a chalk line, and cut off the damaged base of the door. An inch or more can be concealed with most door shoes.

Sill Weatherstripping | GREEN SOLUTION

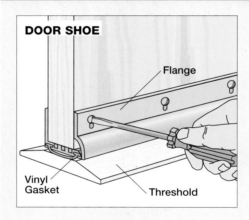

DOOR SHOE — Flange, Vinyl Gasket, Threshold

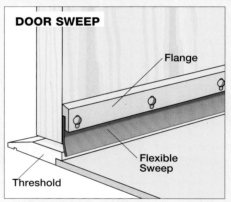

DOOR SWEEP — Flange, Flexible Sweep, Threshold

Replacing a Threshold

USE: ▶ saw • wood chisel • pry bar • caulking gun • putty knife ▶ threshold • waterproof caulk or flashing

1 When the weather causes cracking and rotting in an exposed door threshold, you can buy and install a replacement.

2 To remove an existing threshold without removing the door casing and trim, cut through the middle of the board.

528 WINDOWS & DOORS / WEATHERPROOFING

WEATHERPROOFING

- screws

3 Apply at least one coat of sealer to the raw wood along the bottom edge of the door before installing the shoe.

4 Cut the extruded aluminum shoe to length with a hacksaw; set it onto the trimmed base of the door; and drill pilot holes.

5 Fasten the shoe with screws. This shoe covers the trimmed base and provides a flexible weatherstripping seal.

Jamb Weatherstripping

Most modern exterior doors are sold ready-to-install on their hinges in a frame with integral weatherstripping. To bring an older door up to modern standards, you can add jamb weatherstripping to the existing frame. J-strip is a flexible aluminum strip that fills the gap between the door and the frame. Gasket weatherstripping has a flexible vinyl tube to provide a seal when the door closes.

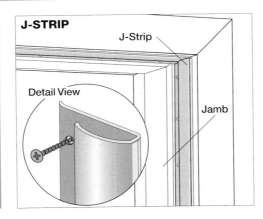

GREEN SOLUTION

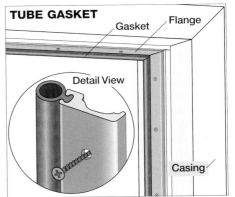

- wood plugs or wood putty

3 With the threshold in two pieces, it's easier to split the shorter sections using a chisel and pry them out with a pry bar.

4 To keep water from seeping in, you can install threshold flashing or use a double bead of waterproof caulk.

5 Notch out for the threshold extensions; slide the new piece in place; and fasten it using screws that are plugged or puttied.

KEEP COLD OUT

20 Windows & Doors

SAVING SPACE

Where there is room to swing open and closed, standard doors work as well as any other for interior rooms, closets, and storage areas. But in a hall without much room, you can gain space by reducing the swing.

You could use several single doors in small openings, but each door requires framing, trim, and hardware, which takes time and money. At the other extreme are pocket doors, which slide sideways into the wall. They save the most hall space but provide limited access because, for every foot of open wall, you need a foot of closed wall into which the door can slide. (And you have to find wall space free of pipes, wires, and ducts.) Another drawback is that the pocket section of the wall has skimpy surface framing instead of full-depth studs. So, unlike other partition walls, if you push on the pocket section, the framing may give a bit.

Another option is to make one large opening and hang two or more sliding doors. But they provide limited access because one door width always stays in the opening. So the best bet for a hall may be a compromise—either double doors or bifold doors. Double doors that meet in the middle of the opening reduce the swing by half. Bifold doors swing over on themselves like an accordion and use even less floor space.

For more money but less work, use prehung doors already hinged in the frames. Don't remove the bracing until the unit is plumbed, shimmed, and nailed.

Interior Details

Interior doors also are often bought prehung. If the unit comes from the supplier with at least one cross brace to keep the door and frame aligned, leave it on until the door is set. On partition walls that do not bear structural loads, you can use 2x4s on the flat across the top of the opening. Load-bearing walls require a structural header. On each side of the opening there is one full-height stud and a shorter stud that helps support the header.

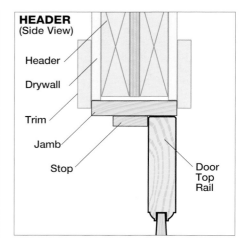

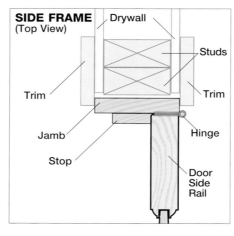

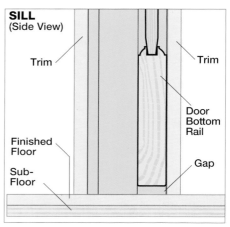

Pocket Doors

Although pocket doors require special framing and more installation time than a standard door, they save floor space by sliding directly into the wall instead of swinging through an arc and taking up wall space when open. New units come with a double-sided frame around the sliding pocket space. Because this framing is not as thick as normal 2x4 wall studs, you have to install it carefully, often with screws instead of nails. The frame must be plumb for the door to slide smoothly on its track.

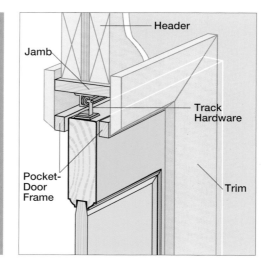

INTERIOR DOORS

Installing an Interior Door

USE: ▶ power drill/driver • hammer • framing square • level ▶ prehung interior door • 2x lumber for header • 2x4s • cross brace • shims • nails • trim

1 To fit an interior door into a non-load-bearing partition, lay out the rough opening, allowing for two jack studs.

2 A jack stud nailed onto a full-height stud helps to support the header. Pack out vertically set headers with ½-in. plywood.

3 With the door framing in place, install drywall panels (typically ½ in. thick) using wide-threaded drywall screws.

4 Although this prehung door is hinged in its frame, it pays to check for square and lock the position with a cross brace.

5 Tip the door into place, and hold it temporarily with shingle shims. The cross-brace keeps the frame flush with the drywall.

6 Working from the inside of the wall, use more shingle shims and a level to plumb the door in the framed opening.

7 When the door is plumb, drive 10d finishing nails through the jamb and shingle shims into the 2x4 wall framing.

8 Select trim for the outside face that matches other trim in the room. Cut mitered corners; add glue; and nail.

9 It's wise to order doors predrilled. You can buy the hardware you like and easily fasten it in the factory-drilled holes.

20 Windows & Doors

STICKING DOORS

Hot, humid summer days don't have to mean a wrestling match with sticking doors. You can minimize this seasonal headache by making sure all edges and door faces are sealed with varnish or paint. If the door rubs only slightly in summer, leave it alone. But if it sticks or needs a real shove to open or close it, it's better to adjust the fit. Eventually all that sticking and shoving may cause the hinges to loosen, which will only aggravate the problem. Bear in mind that a small adjustment will probably do the trick and that if you cut off too much in summer, the door may be too loose in winter. Often you will need less than a 1/16-inch clearance.

Resetting Hinges

Make sure that a loose hinge is not causing the door to bind. Tightening the screws on a loose hinge pulls the door edge closer to the frame on the hinge side. That widens the gap on the handle side, and your binding problem disappears.

If the screws you try to tighten spin in their holes instead of digging in, you have to fix the hole. Dip slivers of wood or toothpicks in glue, and drive them into the hole in a tight pack. When the glue dries, the screw threads will have something to bite into. You can also substitute longer screws that will extend into the framing behind the jamb. This is a good idea for the top hinge of a heavy door and will often solve your problem, even if the hinge is already tight. The longer screw actually draws the jamb closer to the studs, producing extra clearance on the handle side of the door.

Shaving It Down

Unless the binding problem is clearly caused by built-up paint, save sanding or planing the door edge for a last resort. Take off only a little at a time, but remove enough to make room for a new coat of paint. Medium-grit sandpaper is best for small adjustments. If you can't tell where the door is rubbing, sprinkle colored chalk on the edge of the open door. Closing the chalked door will leave smudges on tight spots.

If all else fails and you have to remove a significant amount of wood from the door edge, it's usually best to take the hinges off the door and plane the hinge side. Be sure to deepen the hinge mortises by the same amount.

Tightening a Loose Hinge — MONEY SAVER

When hinge screws on a door work loose, you can solve the problem by installing longer screws with thicker shanks. The idea is to use screws that bite into fresh wood and provide more holding power, with an inch or so more of thread. You can also increase a screw's holding capacity by filling in the existing holes. There are several DIY fixes, including wood filler, a short length of dowel, or even a bunch of toothpicks packed into the old screw-hole to provide material for the screws to bite into.

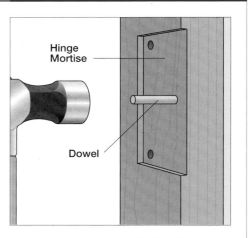

Repairing a Slider

If a sliding door no longer glides, remove it by lifting the panel straight up and edging the bottom out of the lower track. Then examine the wheels, and replace them if they are badly worn or broken. Also check the ribbed channel on the door sill, and straighten out any dents or bends. If the door still rubs after you reinstall it, see whether the ends of the panel are low. Sliders have an adjustment screw recessed in the end of the door that raises and lowers the wheels.

USE: ▶ whisk broom • pry bar • screwdriver

1 To help the door glide smoothly, sweep debris from the ribbed channels of the track, and add a few drops of oil.

Adjusting a Closet Door

Most closet doors have top-mounted wheel assemblies that travel on a ribbed aluminum track. Bypassing doors also have a plastic guide on the floor to keep the doors in alignment. When doors don't operate smoothly, first check the track to make sure that it is tight against the header. Then adjust the wheels to raise or lower the doors as needed. To adjust bifolding doors, loosen the bifold pivots at the bottom and top of the door; adjust as needed; and retighten.

USE: ▶ screwdriver

1 Bypassing or bifolding doors will not operate smoothly unless the aluminum track is securely screwed to the header.

DOOR REPAIRS

Adjusting Bound Hinges

When door hinges close but the door doesn't, even though the door is not rubbing against the frame, the hinges are binding. This problem can occur when hinges sit too deeply or not deeply enough in their mortises. The face of the hinge leaves should be flush with the surface of the door and jamb. Check to see if the hinge is properly aligned and that the screws are tight and don't protrude above the face of the hinge. If that does not work, remove the hinge. Deepen the mortise or add a cardboard shim as needed.

MONEY SAVER

MONEY SAVER
Refinishing Metal

USE: ▶ sanding block • putty knife
▶ sandpaper • auto-body filler • paint

1 To repair surface damage on metal-clad doors, lightly sand the area and fill with auto-body filler.

MONEY SAVER

▶ lubricating oil • wood block

2 To gain more clearance for the door, first unload its weight by lifting the door slightly with a pry bar on a wood block.

3 Hold or brace the door in its unloaded position; then use a screwdriver to turn the adjustment screw and raise the wheels.

2 When the compound cures, smooth the finish with fine sandpaper. Use a block to keep the repair flush.

MONEY SAVER

2 To raise or lower bypassing doors, use a screwdriver to turn the adjustment mechanism attached to the track wheels.

3 To keep bypassing doors from banging into each other, you may need to adjust or tighten the plastic floor guide.

3 You can prime and paint the area with a brush, but a sprayed finish will closely match the factory paint.

STICKING DOORS

20 Windows & Doors

LOCKSETS

The term lockset describes the whole mechanism of a door handle, whether it locks the door or latches it. A lockset includes the door knobs, a latch bolt or locking mechanism, and decorative plates (escutcheons or roses) that cover the lock mechanism. Latch bolts are spring-loaded and may have a locking mechanism. A dead bolt is not spring-loaded and locks and unlocks only with a key or thumb turn.

Locksets fall into two basic categories, although you'll find many variations. Cylindrical locksets are usually installed in 2⅛-inch-diameter holes drilled into door faces. Their latch bolts fit in ⅞-inch-diameter holes drilled into the edges of doors. The other type includes the rectangular, full-mortise lockset, which is mounted in a deep mortise (hole) dug into the edge of a door. Small holes are cut into the door faces to accommodate the spindle on which doorknobs are mounted. Cylindrical locksets are the more common and stronger of the two and are easier to install because you don't have to dig out the deep mortise cavity.

Entry-door locks can be locked and unlocked from both sides of the door. One type will lock automatically when the door closes and unlock with a key from the outside or by turning a knob from the inside. For added security, exterior doors are often fitted with a separate dead-bolt lock, located above a door's key-in-knob lockset. Interior doors may have only a latch or a button that when pushed locks the door.

Common Locks

This combination lockset includes a keyed latchset with a traditional handle and a keyed dead bolt with thumb lever.

Standard keyed locks have a key knob outside and a thumb-lever knob inside that engage the strike assembly.

Passage sets have a thumb-lever knob inside. In an emergency, open them by pushing a long nail into the outside hole.

Installing a Privacy Lock

USE: ▶ power drill/driver • hole saw • spade or Forstner drill bit • utility knife • wood chisel • screwdriver ▶ doorknob with privacy lock • paper template

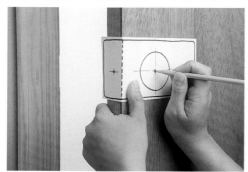

1 Locate the hole centers on the door face and edge using the paper template supplied by the lock manufacturer.

2 To cut clean-edged openings, use a hole saw. Bore through until the pilot bit emerges; then drill from the opposite side.

3 Make the connecting hole for the latch with a spade or Forstner bit. Keep the drill level as you drill through the door edge.

DOOR HARDWARE

Hinge Types

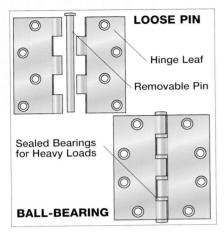

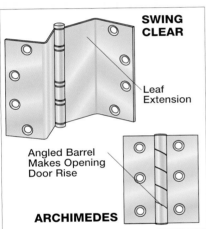

Closers

The same kind of pneumatic piston closer used on exterior screen and storm doors also works inside.

Typical interior closers have a door-mounted piston and a roller arm that travels on a trim-mounted track.

Bumpers

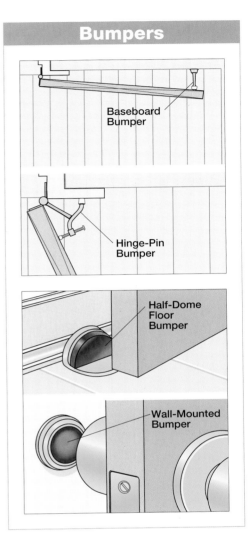

4 Mark the latch mortise; trim around the edges with a sharp utility knife; and clean out the mortise with a chisel.

5 Repeat the mortising sequence for the strike. You can color the end of the sliding bolt to accurately locate the strike.

6 When the bolt and strike are aligned, tighten the screws that hold the latch mechanism and both knobs to the door.

LOCKSETS 535

20 Windows & Doors

OVERHEAD DOORS

Unless you have lived in a house with the old-fashioned swinging doors on the garage, you probably take the convenience of your overhead garage doors for granted. Springs and rollers make sliding the horizontally hinged sections up into the overhead track relatively easy, and if you have an electric motor attached, it only takes the push of a button to open or close even the biggest of garage doors.

The most common type of residential overhead door has an extension spring mounted above the horizontal section of each track. A torsion-spring design has a horizontally mounted spring parallel with and above the door.

Installing an overhead door, or installing a new door that opens automatically, is certainly trickier than most other kinds of doors, but it's not beyond most do-it-yourselfers. The key is to identify and organize the dozens of parts and follow the step-by-step instructions that come with the door. You'll need a drill and a socket wrench set in addition to a level, measuring tape, and some other basic carpentry tools.

Overhead doors are typically trouble-free, provided you keep the hardware tight, clean, and lubricated. It's also important to maintain the exterior finish, particularly on raised-panel-style wood doors, which are subject to rot. Water inevitably seeps between the panel and the frame and has no way to drain. Check these areas annually for cracks in the paint, and caulk seals along joints.

Door Maintenance

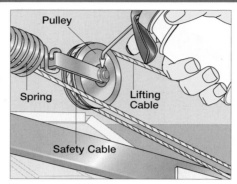

To make the lifting operation smoother (and quieter), add a few drops of machine oil to the spring-mounted pulley.

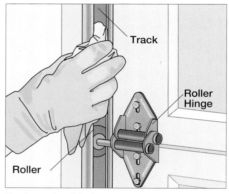

Reduce the friction and clatter of door rollers in their tracks by cleaning the tracks and wiping them with silicone.

MONEY SAVER

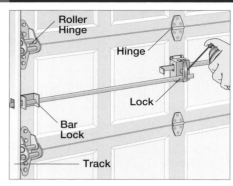

To reach inside the crossbar locking mechanism, use a spray lubricant with an extension straw.

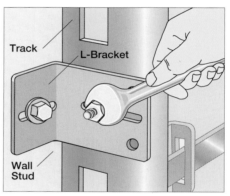

Check and tighten bolts on the frame-mounted L-brackets that hold the door to the frame and overhead track.

Automatic Opener Safety Features

Since 1982, when the Consumer Product Safety Commission began keeping accident records, automatic garage-door openers have caused many injuries and over 100 deaths. Several improvements have been made over the years. They are designed to meet an Underwriters Laboratories standard, which includes these four key provisions. **First,** garage-door operators must reverse a downward moving door within 2 seconds after the door contacts a 2-inch-high test block in the door's path. **Second,** door operators must reopen the door within 30 seconds of the start of downward movement if the door does not fully close to the garage floor. **Third,** once the door is moving down, it must stop, and may reverse, if the control button is pushed again. If the door is moving up, pushing the control button must stop the door and prevent it from moving downward. **Fourth,** door operators must have a manually operated way of detaching the operator from the door.

Following manufacturer's directions, install a UL-approved device that reverses the door when an obstruction is in its way.

GARAGE DOORS

Installing a Garage-Door Opener

USE: ▶ measuring tape • wrench set • power drill/driver • hammer • level ▶ garage-door opener • carriage bolt • 2x4 cleats • angle irons • nails • screws

1 The heart of the system is an electric motor that is mounted on the drive track along the centerline of the door.

2 The chain or rubber drive belt fits around the drive sprocket on the motor. It moves the trolley (and door) back and forth.

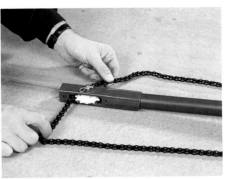

3 The drive belt loops around the plastic idler sprocket that is built into the other end of the main drive track.

4 Bolt a 2x4 cleat to the ceiling frame, and secure the motor with angle irons. A diagonal piece prevents lateral motion.

5 Mount a cleat over the door, and mark a centerline for the header bracket that holds the other end of the drive track.

6 The track (with the drive belt threaded around the idler sprocket) is held in the bracket with a clevis pin.

7 A two-part arm (predrilled so you can adjust the length) reaches from the drive track trolley to a bracket on the door.

8 A terminal block with a keypad, installed on the door frame, can control opening and closing, locking, and lighting.

9 Follow the manufacturer's directions for adjusting the up and down limits and the force applied to the door.

21 Decks, Patios & Walks

540 DECKS
- Planning ◆ Deck Anatomy
- Choosing Wood ◆ Special Tools
- Deck Framing ◆ Post Assembly Options
- Joists ◆ Joist Hangers ◆ Stairs ◆ Railings

552 WALK DESIGN & LAYOUT
- Design Basics ◆ Soils ◆ Slopes
- Front Yards ◆ Side Yards & Backyards

554 CONCRETE WALKS
- Designing a Walk ◆ Forming Curves
- Reinforcing Walks ◆ Forming Walks
- Finishing ◆ Grading & Drainage
- Edges & Joints ◆ Control Joint Spacing
- ■ CRACKS & BREAKS ■ STAMPED CONCRETE

558 BRICKS & PAVERS
- Brick & Paver Basics ◆ Paver Patterns
- Edging Options ◆ Installing a Paver Patio

560 STONE & INTERLOCKING SYSTEMS
- Interlocking Paver Patterns
- Laying Stone Pavers ◆ Cutting Stone
- Stone in Sand ◆ Stone in Mortar

562 GRAVEL & SURFACE SYSTEMS
- Backyard Boardwalks
- Forming a Gravel Walk
- Building a Boardwalk
- Installing In-Ground Steps

564 DRIVES
- Driveway Basics ◆ Design Options
- Slope & Drainage ◆ Material Options
- Asphalt Problems
- ■ REPLACING EXPANSION JOINTS
- ■ CLEANING CONCRETE
- ■ REPAIRING ASPHALT ■ SEALING ASPHALT

21 Decks, Patios & Walks

PLANNING A DECK

Building a deck is usually a satisfying and hassle-free do-it-yourself project. With an outdoors project like this, you don't have any of the difficulties associated with projects inside the house—installing wiring, plumbing, and cabinetry in bath and kitchen renovations. Unlike interior alterations, decks are built in unencumbered free space, allowing you to swing around 12-foot joists without worrying about damaging anything.

With a deck, you expand your living space and increase your home's value for a tiny fraction of what it would cost to build an addition. Homeowners who don't have the time, skill, or inclination to build it themselves might hire a contractor. Although contracting out the job is more expensive than doing it yourself, decks are highly rated features in surveys of home buyers, so the cost is usually justified.

Design Decisions

There are several design decisions common to every deck project. Although most decks are simply platforms raised above the yard, they are an extension of the house—more like living space than yard space. Replacing a solid wall with sliding-glass doors leading to an expansive deck is a quick and relatively inexpensive way to make a room seem larger. Considering the connected inside and outside space as one unit, deck additions are likely to be most useful when added onto living or family rooms and kitchen/dining areas.

Because wooden decks are exposed to the elements year-round, the wood must have weather-resisting characteristics built in or brushed on. Common framing material such as construction-grade fir is more than strong enough and can be treated with penetrating wood stain, clear wood preservative, or both to resist deterioration. In most regions, particularly the Northeast, this protective coating should be reapplied every two or three years. For example, even in wet New England a 20-year-old fir deck that has been drenched in clear preservative five times over the years will show few signs of rot.

Another option, pressure-treated lumber, has chemical preservatives injected throughout. It may take several seasons for the characteristic greenish tint to weather to a more natural tone. This type of decking has a long life and requires only periodic sealing or staining.

For more money (sometimes a third or even half again the cost of fir), you can use redwood or cedar, both of which are more naturally resistant to the elements. Although these woods have an extraordinary appearance when first installed—some select redwood has a unique cinnamon hue—they will weather. Most woods used for decking eventually turn a driftwood gray color. Redwood may turn much darker unless protected with a clear sealer. Even then it gradually changes to an elegant silvery tan.

Deck Levels

Most decks have four structural levels. From the top down they are: the surface (often 1x4s, and generally 2x6s on large decks); the joists (at least 2x6s and as large as 2x12s), usually set 16 inches on center at right angles to the surface boards; one or more beams such as a 4x10 at right angles to and supporting the joists; and posts or piers that support the beams and transfer loads to the ground.

The deck surface is usually the controlling feature of a deck plan. Because it ties into the house, it should be built even with or a few inches below the house floor. The joist and girder dimensions are controlled by the structural requirements specified by an architect or contractor and verified by a building inspector. The posts are the only easily adjustable element, taking up as much or as little space as required between the beam and the ground.

Drafting Your Design

Decks are also among the easiest home improvements to commit to paper—even if

Deck Anatomy

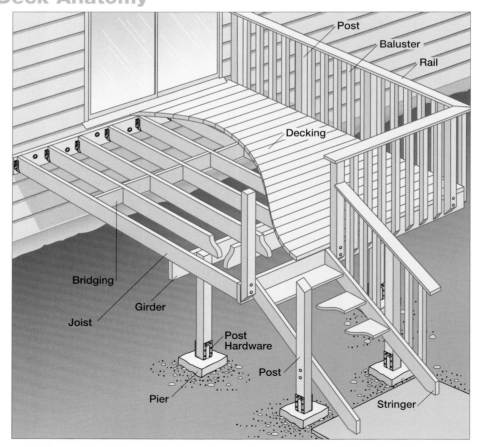

540   **DECKS, PATIOS & WALKS** / DECKS

DECKS

you don't have drafting experience. A few preliminary drawings on graph paper will help you decide what deck design will be most attractive and useful, and what will fit in best with the layout of your house and yard; it will also aid contractors in estimating the job. Many lumberyards will take even a rough plan and do a take-off, or estimate of materials and costs.

However, bear in mind that areas defined on uncluttered scale drawings tend to look larger than they actually will be. To avoid disappointment, transfer the scale drawings to the actual building site. Use stakes driven into the ground connected with string to outline the deck. Deck labor is two to three times as costly as the materials, so extending the deck using 12-foot instead of 10-foot joists has little effect on labor costs and increases material costs only marginally, but may increase the usefulness and sense of space dramatically.

Nails vs. Screws

Most decks are fastened with galvanized common nails. It's still an adequate way to build, but the convenience and fastening power of cordless drill/drivers and exterior-grade screws have upped the ante. Screws are more work and slower to use, but they are better in many ways—nails are more likely to pull loose under any load; screws won't back out of deck boards after a few years of service. Professional deck framers still use pneumatic nailguns, though, because nothing is faster for nailing off large areas of decking.

Air-powered nailguns take time to get used to but speed the work of decking. You can rent the gun and a compressor; you also need special nails for the gun.

Raised
STRUCTURE: Raised decks stay cooler on summer days and are easier to keep clean. To meet building codes, you need a solid foundation, sound attachments to the house itself, and a sturdy railing.
ADVANTAGES: A deck with a view adds an entirely new dimension to a home. Upper-story decks retain valuable space below.
DISADVANTAGES: The higher you go, the higher the costs. Raised decks can be a danger in homes with small children.

Roof
STRUCTURE: Like raised decks, roof decks provide owners with a unique perspective. Access may be from inside the house or by an outside stairway.
ADVANTAGES: Roof decks offer maximum privacy, make excellent outdoor living areas, and utilize wasted space.
DISADVANTAGES: Perhaps the most difficult to build. Areas below must be reinforced to carry the added load of deck materials and occupants. Extra waterproofing is needed.

Ground Level
STRUCTURE: The simplest deck to build is situated on or slightly above grade. Often, foundations and posts are unnecessary.
ADVANTAGES: Ease of construction tops the list, but on-grade decks also improve access and can make a smaller home appear larger.
DISADVANTAGES: Without an enclosed yard, ground-level decks offer little privacy. Also, wood touching the ground is more susceptible to insect and rot damage.

Multilevel
STRUCTURE: The most complex, and usually more costly, type of deck. Generally not a project for the novice builder.
ADVANTAGES: The different levels may be used for privacy, to add interest to larger decks, negotiate a steep building site, or connect the home with a pool.
DISADVANTAGES: Too many levels can look busy and can be hazardous after dark if not adequately illuminated. Maintenance is more difficult and time-consuming.

21 Decks, Patios & Walks

CHOOSING WOOD

There was a time when the word redwood was synonymous with "deck." But when decks became more popular, the supply of redwood couldn't keep pace with demand, and not everyone could afford what was available. Other woods were used in its place—cedar was a good substitute, also naturally decay and insect resistant, but fir and other softwoods didn't last as long or weather as nicely. So chemical treatments were developed to give ordinary woods the resistance they needed to withstand insect infestation, decay, and sometimes even ultraviolet degradation.

This hierarchy of wood remains unchanged, with redwood regarded as the premier deck material, followed by cedar. Pressure-treated woods make up the bulk of the lumber sold for decks today, though. Before you choose any one for your deck, it pays to know the differences not only between these woods but also the grades of each that are available. There is also a growing category of composite material—plastic and composite plastic and wood—now being sold for deck surfaces.

Redwood & Cedar

The range of grades and prices within these related species is so broad that you'll want to be sure to choose the correct grade for your application, and that you get what you pay for. Ask to see a grading chart before you make any purchase. The most expensive, highest-quality woods, for example, are architectural grade Clear All Heart, which are typically used indoors where their decorative beauty shows to its best advantage. This grade may be used for decking because it is also the most decay resistant, but the cost is usually too steep to be practical. Farther down on the chart, but still superior for most outdoor purposes, are construction or garden grades, selects, and common lumber. You won't pay as much for these woods, which contain more knots and less-resistant sapwood, but you'll still get most of the natural benefits of redwood and cedar.

Those benefits include not only long-lasting appearance but wood that is structurally strong and can be used in contact with the ground. Redwood is generally stronger than cedar, and both were once the preferred woods for deck posts and other underpinnings most susceptible to decay. Today, because of their higher cost, redwood and cedar are often reserved for decking and decorative features such as benches, while less expensive pressure-treated wood carries the load and remains hidden from view.

Pressure-Treated Wood

Pressure-treated lumber has been used to build decks for years. In that past, most lumber was treated with chromated copper arsenate (CCA), which gave the wood a greenish tint that faded over time. However, CCA contains arsenic, a known carcinogen, and safe use of the product required special handling procedures. The industry has elected to stop using CCA as a preservative in most residential

Special Tools

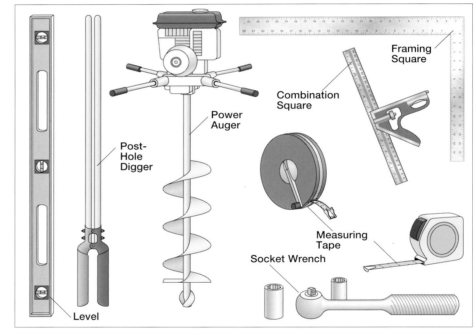

Key tools of the trade: leveling tools include a carpenter's level, line level, torpedo level, water level, and post level. You will also need a combination square, angle square, framing square, hammer, saws, drill clamps, and ratchets for lag screws. Measuring is done with tape rules and reels. A posthole digger, chalk-line box, wheelbarrow, and plumb bob are used for layout and excavation.

Water Levels

A water level is made from a long hose, like a garden hose, with two transparent tubes at each end. It can pinpoint two level spots spanning long distances.

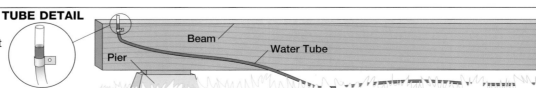

DECKS

applications, including its use on decks. A variety of copper-based compounds are now used to treat lumber.

Newer treatments affect the look and structural characteristics of the lumber—the chemicals used in treating the wood corrode some fasteners, for example. Check with the producer for fastener requirements and specific uses.

Pressure-treated deck lumber can be any one of several softwood species, most often southern yellow pine, fir, or hemlock. Although newly treated wood is soft at first, the chemical treatment leaves it difficult to nail and cut. This wood is also prone to warping and twisting, and you should use it as soon as possible after delivery to minimize these problems. All chemically treated wood must be labeled by grade and chemical content. The label provides a good deal of information about the wood and the treatment process, but the general rating is the most important piece of information. If it says "above ground," use it only where it will not get wet for long periods of time, such as a deck surface. Use "ground contact" lumber for pieces that will be placed near or in the ground, such as support posts.

Buying Lumber

For all but the smallest decks, you will probably want to have the lumber trucked directly to your home. Most lumberyards are happy to deliver large orders for free or for a small fee, and the obvious advantage is that you don't have to load and unload the wood.

But you are pretty much stuck with whatever boards they put on the truck—at least some undesirable pieces are sure to come your way. Often you can find uses for all but the worst pieces. But to protect yourself, ask what the yard's policy is on returns.

Common Lumber Options

Douglas fir is one of the more common types of construction lumber but is not suitable for areas where wood meets the ground. Also, fir needs application of protective coatings to prevent wood rot.

Pressure-treated wood is infused with chemical compounds and will resist rot and insects. PT will turn from green to gray after several seasons and can be stained with an oil-based decking stain to retain color.

Redwood is considered the premier decking wood. It resists rot and insects and has a desirable salmon tint that will turn to gray after several seasons. Select, or knot-free, redwood is rare and expensive.

Cedar is also insect- and rot-resistant and will turn gray after two or more years of exposure. Cedar can be stained in a semitransparent or solid color stain. It's more expensive than PT wood but less costly than redwood.

Fastener Options

Nails or screws? Beams, joists, and posts should be secured with nails. Galvanized wood screws are used to secure decking, stair treads, and railings. Nongalvanized fasteners will bleed rust through the wood.

Lag screws and carriage bolts secure ledger boards to the house and beams to the support posts. Lags are first tapped in an inch and screwed in with a ratchet. These fasteners come in a variety of sizes and diameters.

T-braces, strap ties, and post caps function to keep large pieces of lumber securely tied to each other. Use approved fasteners for this application. Consult the manufacturer's installation guide or your local building department for specific fastener types and lengths.

Joist hangers support joists between beams and ledgers. They give better load support than nailing joists straight into beams. Use of the correct hanger type and size, including the approved fastener, is required. Consult the manufacturer's installation guide or your local building department.

TUBE DETAIL

Electronic units sound a tone when the water levels out, which helps when working alone. You can also make your own water level from clear plastic tubing.

CHOOSING WOOD

21 Decks, Patios & Walks

DECK FRAMING

House framing can be cratered with misguided hammer blows and studded with bent nails because you won't see the mistakes under siding and drywall. But you will on a deck. Because it is out in the open, you want framing to have both strength and a finish-quality appearance. So consider some of the following tips on both form and function.

Posts on Piers

Keep the support system neat by limiting concrete piers to a couple of inches aboveground—even if the ground is uneven—and make a uniform transition to wood by topping the piers with galvanized post anchors. Often, one end has a pin that embeds in the concrete while it's still wet, and the other has a U-shaped bracket that secures the post. (See "Concrete Forms," pp 78–79.)

There are many post-anchor varieties—some that hold 4x6s or 8x8s if your deck needs the support, even some with a threaded rod and nut so you can adjust deck height where a section settles slightly out of level. Although this hardware keeps posts out of the dirt, soak the end grain with wood preservative.

This combination of piers and posts is durable and the easiest way to take up uneven slack between ground and deck. You don't have to establish precise levels and cut posts to exact size ahead of time. You can let them run long and trim to final length when the girder is placed. Posts are more practical than high, aboveground concrete piers and are better-looking, too. While a deck seems perched on top of high concrete piers, posts reaching to grade level seem to anchor it to the site.

Extended Posts

If you are taking the time to set posts that rise from the ground to a girder, you may want to get more for your efforts with longer posts that extend past the girder to anchor the railing, too.

In a typical deck plan, posts reach up to a pair of spiked-together 2x10s or 2x12s that carry the joists and decking above. With minor span adjustments (checking your plan with the local building department), you can make way for the post extensions by assembling the girder in two parts—one on each side of the posts, bolted in position. Figure the approximate post height you need to reach the deck floor, plus another 3 feet or so for railing.

This elegant detail provides many options and makes finishing railings easy because there are solid vertical supports in place every few feet along the deck edge before you start

Deck-Post Anchors

J-bolts and post anchors are attached to the concrete footing, and the anchor bracing is nailed or screwed to the post. Soak post ends in preservative before attaching.

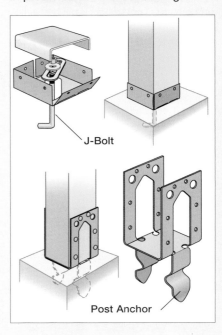

Installing Girders

USE: ▶ circular saw • C-clamps • 4-ft. level • hammer • work gloves ▶ ½-in. pressure-treated plywood • pressure-treated 2x4s and 2x10s • 3-in. galvanized nails

1 Sight a 2x10 beam for crowning or any arching along the length of the wood. Install the beam with the crown side up, not down, to compensate for stress loads.

2 Double up support beams and increase their width to 3½ in. by adding ½-in. pressure-treated plywood between beams. Beams will sit flush on top of posts.

3 Nail beams together using 3-in. galvanized nails. Set nails 16 in. apart in two separate rows. Make sure both beams are crowned upward and nailed flush.

DECKS

adding railings and balusters. And no nailed-on, screwed-on, or even bolted-on railing system applied to the finished deck platform can provide the security of a row of full-height, solid 4x4s.

Dressing Posts

On low decks where supports are recessed, the scruffiest lumber will do as long as it's solid. But on some decks, post supports loom large from the yard. You don't gain anything structurally with a solid 4x4 versus two 2x4s spiked together, but you won't see a seam between boards or nailheads or discolored dings where you swung the hammer and missed.

Also, solid posts can be dressed up with nice touches of carpentry that once were used on exposed beams in early colonial houses. To make tree-size timbers appear more finished and graceful, reduce the hard edges with a chamfer—a 45-degree slice across the corner. Run it across most of the timber length, and taper it away near the top and bottom to make the beam square-edged near the joints.

Another option is to clad the wood posts with one-by facing. You can use quality wood to match your siding or choose a rustic, rough-sawn grade for a contrasting look. Clad 4x4 posts with 1x6 boards and 6x6 posts with 1x8s.

Post Assembly Options

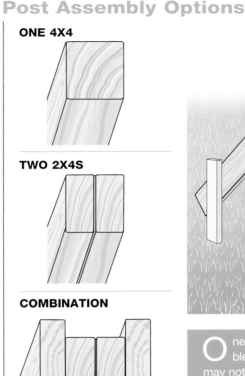

ONE 4X4

TWO 2X4S

COMBINATION

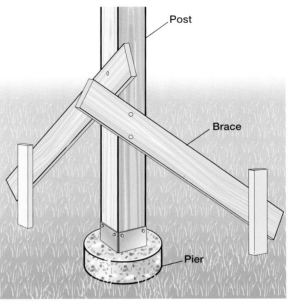

One 4x4 post is easier to erect than doubled 2x4s, which require assembly and may not cost less in the long run. Using larger posts may save you money, but they should be through-bolted to secure them. Two angle braces will hold each post plumb in both directions.

- wood shims

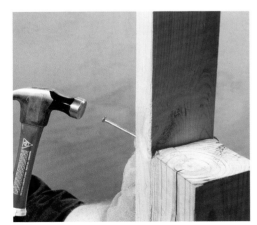

4 To temporarily position the girder, face-nail a 2x4 to the back of the post, extending up several inches. When you set the girder in place, it will rest against the extended 2x4.

5 To keep the girder from falling while you work on the installation, clamp the face to the 2x4 brace. Another option is to install two braces and set the girder between them.

6 Place a level on top of the beam, and use wood shims on posts to make the beam level. Also, shim any gaps in post-beam joints as needed.

21 Decks, Patios & Walks

CHOOSING JOISTS

Most deck designs call for 2x8 or 2x10 joists set 16 inches on center. The length of a joist's span from ledger board to beam also depends on the species of wood and the width of the joist. Naturally, thicker boards can span longer distances and support greater loads than thinner boards. To find out what the allowable measurement should be between joists and how far they can span, consult a span table—your local building department should have them available for your reference.

Wood & Fastener Options

Your deck's joists, beams, and ledger should be made from pressure-treated wood to stand up to the elements, particularly any timbers near the ground. But you can also use standard construction grades if you first coat them with wood preservative. To build a deck economically without sacrificing either strength or durability, you might want to consider using pressure-treated wood for the joists, ledger, and beams, as these parts of the deck generally are out of sight.

For maximum strength (and to satisfy most local building codes), joists are attached to supporting timbers with metal brackets called stirrups or joist hangers. These hangers will support more weight than nails can and reduce the chance of wood splitting at joints. Joists are often doubled up for extra support or to accommodate designs where surface boards require increased nailing support. But there are double-wide hangers to cover those situations—and hardware designed to reinforce just about every structural connection you can make on a deck. Even where hardware isn't required by code, you may find that bridging joints with metal brackets makes construction easier.

Hanging Joists

One good way to hang joists is to first build the deck's outer perimeter, with outside joists nailed to a header joist at the outside of the deck and to a ledger board at the house. Pros who can easily envision the complete layout ahead of time can stack up the components and mark them prior to installation. But for many do-it-yourselfers, it's easier to take one step at a time: build the basic post- and-girder support

Deck Lumber Options

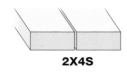
2X4S

2x4s will give a deck a busy look, due to their relatively thin widths. If not pressure-treated, 2x4s must be coated with a preservative; they will also need periodic maintenance. Soak 2x4 ends in preservative before nailing. When you can, check wood for crowns, cups, and warps before purchasing.

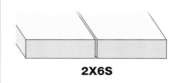
2X6S

2x6s are a better decking option than 2x4s, as they require less time to install. These boards can also be used as joists if the span is short, but check first with building codes. Use two nails over each joist to minimize cupping. 2x6s are heavier than 2x4s; an extra set of hands will help for long boards.

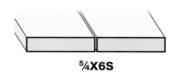
⁵⁄₄X6S

1x4s, ⁵⁄₄x4s, and ⁵⁄₄x6s are some of the more popular choices for decking. Joists may have to be on 12-inch centers, depending on wood species and grade. This decking size is sold in many different grades and different species. Pressure-treated, cedar, and redwood are the most popular.

TONGUE & GROOVE

Tongue-and-groove boards fit into each other for a tight seam. Used mostly on porches with a roof, tongue-and-groove decks should be pitched to shed water and waterproofed against the elements to prevent swelling. They are not recommended for use on open decks where the wood will be exposed to the elements.

Cantilevers

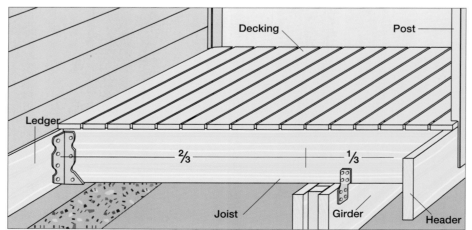

As a rule, no more than one-third of a deck's framing should extend past its support.

DECKS

system; box in the deck space; and then fill in that frame with joists.

As mentioned, most deck designs call for joists on 16-inch centers. But this rule depends on the thickness of your decking material and deck design. If your deck has a herringbone design, for example, which calls for a diagonal pattern of decking boards, the joists may have to be on 12-inch centers. Your plans will only be approved by a building inspector if they specify the placement and thickness of joists.

After you've marked the modular layout of the deck joists on the beam and ledger, check to make sure your frame is square overall by measuring the diagonals from inside corner to inside corner. If the deck is square, the measurements will be equal. You can clamp or temporarily tack the outermost joists in position to complete the deck perimeter. But don't fix them permanently in place until the perimeter is square.

To install the joist hangers, slip one of the brackets around a short section of joist, and set the sample assembly in position. The trick is to position the bracket so that the top of the joist it holds will be flush with the adjacent ledger. If it's not flush, the deck boards—even beefy 2x4s and 2x6s—will ride up and down over their supports. Once you are sure of where the bracket should be, you can use the position of the sample to fix all the brackets in place.

Your joists should fit snugly between the ledger and header. Again, the best bet is to cut a sample joist and test it in place. Then you can use it as a template to cut the other joists.

Setting Joist Hangers

USE: ▶ joist template • combination square • hammer • marker ▶ joist hangers • galvanized nails

1 Make a template to fit just below the flashing, and then mark reference lines for plumb and horizontal placement.

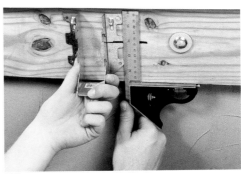

2 Size hangers correctly for joists and ledger, and set them plumb to avoid twisted or uneven joists.

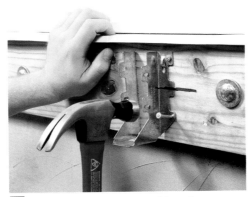

3 Use nailing clips stamped into the hanger to position it and galvanized hanger nails to install it in place.

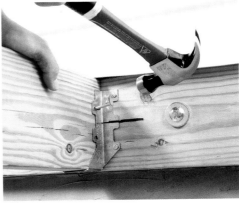

4 With the hanger fastened to the ledger, set the joist, and nail it in place. Use approved hanger nails that won't protrude.

House-Mounted Ledgers

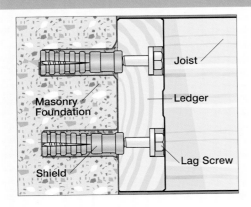

When attaching ledger boards to masonry, predrill the ledger, and use holes as guides for masonry holes. Place board onto the masonry wall, level it, and use a marker to indicate drill points. Then drill and insert shields or anchors. Don't confuse masonry with stucco; ledgers on stucco must be bolted into the house's frame. On other exteriors, strip siding to expose the sheathing; bolt the ledger into the frame. To preserve this crucial timber, use pressure-treated wood and cover the top edge with flashing.

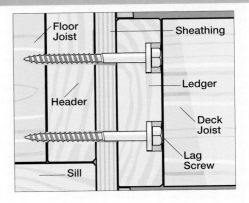

21 Decks, Patios & Walks

LAYING OUT STAIRS

Stairs can be a simple matter of a few steps, or they can be as complicated as a convoluted, two-story spiral. But all stairs and ramps are based on a couple of simple rules that can help you calculate how many steps you need and just how far out the staircase must extend to connect the upper and lower levels.

Because most decks are fairly close to ground level to start with, you can probably get by with a simple, straight staircase that has just a few steps. You may need to remeasure and refigure to get exactly the right angles on stair stringers, but the assembly is a project most do-it-yourselfers can handle.

Stair Anatomy

There are three basic parts to a stair: the stringer, the riser, and the tread. The dimensions and relationships of all three come together to form a typical staircase. But on many decks, stairs are basically platforms that are built without stringers along the sides.

The tread is the flat board that your foot steps on. It is usually at least 11 inches deep from the outer edge to where it meets the riser, but it can be wider to create platform steps.

Not all stairs have risers, the back board which rises up from tread to tread. All stairs do have a rise, however, which refers to the height from step to step. The rise must be the same for every step because people will trip if they encounter even a slight variation in height. Generally, the height of the rise, combined with the thickness of the tread, is about 6 or 7 inches, but you should check this dimension (and details of stairway hand rails) with your local building department.

To find out how many treads and risers your staircase needs, you must first determine what the total rise and total run are for your stair area. After you've measured rise and run, you can calculate the unit rise and unit run, and divide the space into steps.

Measure your total rise in inches, and divide this figure by the 6- or 7-inch riser height you want. Then round off the result to the nearest whole number. This is the number of steps. Next, take the vertical rise in inches, and divide by the number of steps. This will give you the exact riser height.

Stair Layout

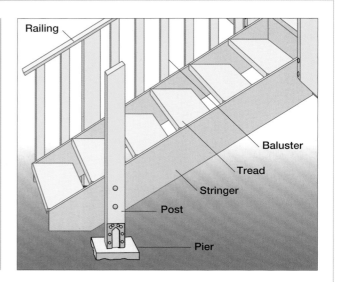

Exterior stairs are generally not as steep as interior stairs because of the more-hazardous conditions that exist outdoors. Wider treads and lower risers make safer exterior steps. Tread and riser tables are available to help you determine the correct number of stairs and angle of the staircase.

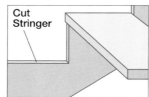

Precut stringers can be purchased at most home centers.

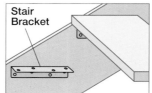

Metal brackets can replace cutout portions of the stringer.

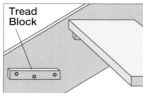

Wood blocks can also support stairs if metal brackets are unavailable.

Installing Stairs

USE: ▶ circular saw • combination square • framing square • spirit level • drill/driver • hammer • pencil

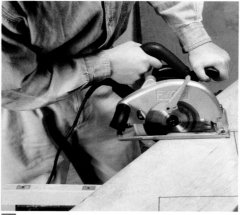

1 Measure twice and cut once is a carpentry rule, especially on stair stringers that can be difficult to calculate.

2 Mark the location of the stringers on the header or rim joist. Measure rise and run from the deck to where stringers will end.

548 **DECKS, PATIOS & WALKS** / DECKS

DECKS

Platform Design

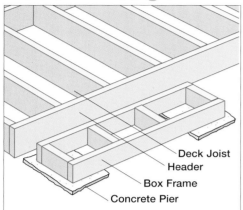

Platforms break up the monotony of a large, flat surface and provide areas for planters, seats, and furniture. They can replace a few stairs and give a deck a classier look. Building a platform area is done by first framing out the intended area. Use the same lumber species and sizes for joists, decking, and stringers as on your original deck, but use pressure-treated framing on the piers. The stairs should rise no higher than 7 inches and no less than 5 inches from the deck. Before you build, check your local building codes for loads on decking.

Accessible Ramp Layout

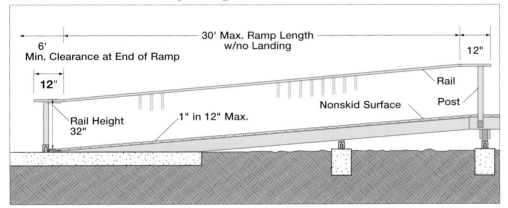

Standards established in the federal Americans with Disabilities Act (ADA) are designed to facilitate access while preventing injury to persons with limited mobility. All public buildings must conform to these design standards, but many homes also are being retrofitted with barrier-free devices, such as lever door handles and access ramps instead of steps and stairs. Because ramps can pose a hazard if improperly designed, it's a good idea to follow the ADA model if you plan to build one.

• measuring tape ▶ pressure-treated 2x lumber • angle brackets or sloped hangers • stair brackets • galvanized nails or ¾-in. lag screws • J-bolts with nuts

3 *Nail or screw on an angle bracket or sloped hanger to strengthen the connection between each stringer and the rim joist.*

4 *Install stair brackets if the stringers are not cut out. Angles should be slightly shorter than the overall tread depth.*

5 *Fasten treads to brackets with nails or screws. Avoid splits on ends of treads by predrilling nailholes.*

LAYING OUT STAIRS

21 Decks, Patios & Walks

FINISHING TOUCHES

Because decks are relatively easy to design and build, even for amateurs, exercise your creativity when planning it out. Decks can conform to slopes, wind their way around building angles, incorporate seats and planter boxes, or have interesting patterns underfoot. They can be whimsical or practical, exotic or prosaic, but there is really no excuse for a deck to be boring.

Curved Railings

A long piece of wood beefy enough to serve as a deck railing (even against the house wall) is difficult to bend. So the best option is to use several thin pieces of wood that, taken individually, can be bent easily and then laminated into one beefy railing. If you use multiple pieces of thin stock—only ¼ or ⅜ inch thick—each piece can be bent to shape in place. For example, you could install the railing hardware, bend one piece into place, clamp it, and then do another, and so on.

Deck Seats

It's not difficult to add seating to your deck for convenience and comfort. To support a seat on the perimeter railing of a deck without adding legs that will block the edge, use the posts to support the seat the same way they support the deck joists—as the core of a structural sandwich that takes the overall shape of a triangle. One leg is the existing post, the other is the horizontal seat support, and the third (the hypotenuse) is an angled brace.

The brace can slope back from the seat to the post at the edge of the deck—out of the way. At its top, the brace is buried between two horizontal 2x4s that support the seat platform. Each pair of 2x4s can be bolted around the 2x6 railing post at one end and around the main brace at the other. To assemble each triangle, sandwich the pieces between clamps; make minor adjustments as needed; and then drill through for bolts. If the deck is square and level, you should be able to make one triangle and use the pieces as templates to cut others. To make the triangle frames blend in, use finishing details such as recessed, round-headed carriage bolts instead of machine bolts, and comfortable, rounded-over square edges on seating that might otherwise catch you under the knees.

Seats & Planters

There is no limit to what you can create when you build a wood deck. A simple box can be enlarged to become a planter or railings extended into built-in seating. Level changes are also easy to make and create decks that look interesting and inviting.

Deck Details

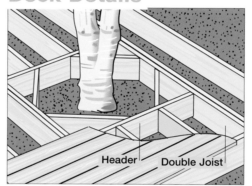

Tree boxes are framed with headers between double joists. Tree holes should be planned and built before decking is laid.

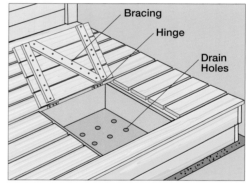

Access hatches leading to sewage pipes or storage areas can be cut out after the decking is constructed.

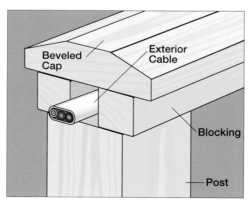

Wiring for outlets and lights can be hidden within the railing. Be sure the wiring conforms to local codes.

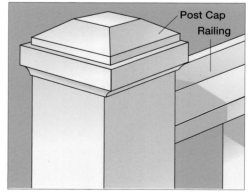

Details such as fancy post caps and railings can be purchased at millwork shops or created with a router and plane.

550   DECKS, PATIOS & WALKS / DECKS

CONCRETE WALKS

- ⅛-in. hardboard • deck screws • nails

3 Use a level long enough to span the walkway to set consistent stake heights through the curving corner.

4 Attach ⅛-in.-thick hardboard to the inside corners of the form frame to create a curve. Drive 3-in. deck screws at all corners.

5 Reinforce the curve against the force of poured concrete by driving support stakes every few feet outside the form.

Finishing Concrete

To add a non-slip surface to your walkway, finish the surface with a broom, working with more or less pressure to create a light, medium, or heavy texture. Use the broom after the concrete is troweled smooth but before it sets up. To make a pebbled surface, embed aggregate (colored pebbles or gravel) into the poured concrete before it sets. You can also add pigment to the concrete or throw rock salt onto the surface, which leaves behind an interesting texture of small holes.

▶ 2x4 lumber for edging • concrete • gravel • rebar or wire mesh (optional)

4 If you can't use forms to screed the surface, embed pipes in the pour to use as a level guide.

5 When you lift out the screed pipes, fill the void with extra concrete. Also tap the forms with a hammer to settle the mix.

6 Smooth screed lines and the entire surface with a trowel or float, making large arcs with the front edge slightly raised.

21 Decks, Patios & Walks

GRADING & DRAINAGE

Your site's terrain, soil conditions, and drainage requirements, as well as the walkway materials to be used, determine how you must prepare the ground for your walk. Virtually every walk project will require some excavation to create a level surface and provide a stable base for the paving materials. This is especially true of concrete. If poured concrete or concrete pavers are not laid on a firm, well-compacted, and well-drained base, they will likely buckle, crack, or sink. A base consisting of 4 inches of compacted gravel or crushed stone topped by 2 inches of builder's sand should suffice. Soils that drain poorly—or those subject to frost heave, settling, or erosion—may require a thicker base of 6 to 8 inches of gravel or crushed stone.

Drainage Systems

There are two basic ways to drain soggy soil: construct a surface drain system or install a subsurface system consisting of area drains, catch basins, trench drains, dry wells, or drain tiles. In wet soils, you may need to bury perforated drainage pipe in a gravel subbase. However, due to the expense and work involved, consider a subsurface system only as a last resort.

Surface drain systems consist of shallow drainage ditches, called swales, and built-up mounds that direct runoff, called berms. After you've identified the area where runoff enters your site, the next step is to decide where to channel the water. Generally, you'll want to direct it to an existing storm sewer located in the street, channeling the water with swales, berms, and retaining walls. If it's impossible to channel storm water to the street system, look for an alternative outlet. But take care to respect your neighbors' property, and don't divert your water into their yards.

Don't install walks that cross swales or run across slopes. Such walks can act as dams; they'll impede natural drainage patterns in the yard and even cause problems with flooding.

In the good old days when wood was cheap, this hotel in Bermuda thought nothing of building a walkway to the beach that was wide enough for an early-1930s sedan to drive across it.

Edges & Joints

USE: ▶ edging and jointing trowels • metal float

1 *Form the perimeter of the slab with an edging trowel. Run it slowly back and forth to smooth the mix and release the form.*

Control-Joint Spacing

Slab thickness	JOINT SPACING	
	Aggregate < ¾ in.	Aggregate > ¾ in.
4 in.	8 ft.	10 ft.
5 in.	10 ft.	13 ft.
6 in.	12 ft.	15 ft.

Control joints reduce surface cracking in concrete. The thicker the slab, the fewer you need.

Cracks & Breaks

USE: ▶ whisk broom • cold chisel • hammer • paintbrush • mason's trowel • safety glasses ▶ concrete repair caulk • masking tape • bonding adhesive

1 *To make minor repairs of small cracks, start by cleaning out dirt and debris with a whisk broom.*

2 *Concrete repair caulk provides a quick fix for minor cracks. It prevents further damage but is only a temporary repair.*

3 *To patch larger areas and edges, clear out loose debris with a cold chisel. Add a form board to contain the patch material.*

CONCRETE WALKS

- trowel ▶ poured concrete walk

2 Form control joints with a jointing trowel run against a squared-up 2x4 for a straight-edged guide.

3 The jointing trowel leaves a smooth groove in the surface, but you may need to clear out the interior seam with a trowel.

4 Use a float to smooth the surface and any marks left by edging or jointing. Try not to overwork the surface and puddle water.

Stamped Concrete

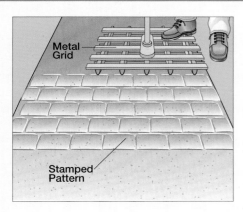

If you like the look of a patterned masonry surface but prefer to use one big pour of reinforced concrete instead of pavers, talk to a contractor about stamping. On a driveway, for instance, workers come back over the surface with a metal grid tool to make an imprint of squares or other shapes in the mix—a little like turning soupy pancake batter into waffles. As the surface wears and weathers, these recesses tend to darken and give the impression of individual tiles.

MONEY SAVER

MONEY SAVER

- vinyl-reinforced patching compound

4 Use masking tape to protect any adjacent surfaces where you don't want any fresh masonry to bond.

5 Apply a thin layer of bonding adhesive, which will help create a strong bond with the patch material.

6 Fill the damaged area with vinyl-reinforced patching compound in thin layers. Wait 30 min. between coats.

21 Decks, Patios & Walks

BRICK & PAVER BASICS

Bricks and concrete pavers make ideal materials for do-it-yourselfers. Because they're packed into a sand base without mortar, installation is forgiving—you can make minor adjustments along the way as needed. Edging keeps the paving material and sand in place.

Bricks and pavers produce beautiful, durable walkways and patios. When selecting brick, choose a type rated for exterior use, with a slightly rough surface that will provide traction in wet weather. (See "Brick Basics," p. 86.) Check with your building department to find out the type of brick that works best in your area.

Concrete pavers come in a variety of shapes and sizes. Most are modular, meaning they fit together in a variety of geometric patterns. Unlike bricks, concrete pavers usually have rounded edges; spacers on the sides keep you from setting them so closely together that there's no room for sand.

Estimating

It will take an average of four-and-a-half 4x8-inch bricks or pavers to cover a square foot of surface area. For example, a 12 x 20-foot patio (240 square feet) would require approximately 1,080 pavers. Order 5–10 percent extra to allow for miscuts, breakage, and future repairs. Some brick patterns, such as herringbone and basket weave, require a paver type whose length measures exactly twice its width.

Calculate the sand you'll need at a rate of 9 pounds per square foot of surface. Filter fabric goes under the sand to keep down weeds and prevent sand erosion, so you'll need to buy some of this material as well.

Installation Tips

The most difficult step in laying bricks or pavers is creating the gravel and sand bed to set them in. The gravel must be firmly compacted: add one layer, compact it; add a second layer, and compact that, too. Add the edging, which must be deep enough to contain both the paving material and the sand layer. Otherwise, the sand can wash under the edging, causing the pavers above to sag. Shovel in the sand, and tamp it down. Then wet it; fill in any low spots; and wet it again. Smooth and level the sand by pulling a notched 2x6 board along the edging.

Paver Patterns

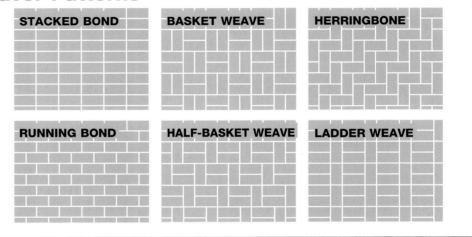

Edging Options

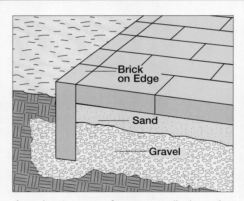

A perimeter row of pavers, called a soldier course because it is set vertically, contains the supporting base.

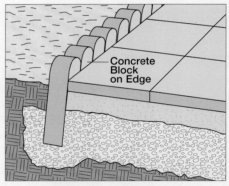

Molded blocks in several shapes also can form a soldier course to prevent undermining from drainage.

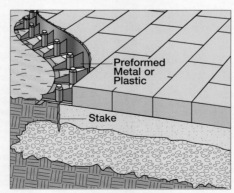

To contain curved edges made of smaller, less stable blocks, use flexible plastic forms staked into the ground.

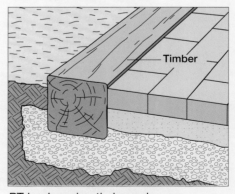

PT landscaping timbers also can provide edge support. They should be nestled in gravel and staked or spiked.

BRICKS & PAVERS

Installing a Paver Patio

USE: ▶ power tamper • circular saw & masonry blade • sledgehammer • rubber mallet • broom ▶ pavers • sand • gravel • filter fabric • pipe • edging • 2x4

1 Once you dig out the sod, rake away stones and twigs. Use a 2x4 to create a roughly level subbase for the patio.

2 To minimize shifting and settling, rent a power tamper to vibrate and compress a 4 to 6-in. bed of gravel over the dirt.

3 Prevent erosion from drainage through pavers by covering the compacted base with filter fabric.

4 There are many ways to contain the edge blocks. One of the most efficient is plastic trim secured with oversize stakes.

5 Shovel a bed of sand over the prepared base. Use pipe set in the sand as screed boards to level the sand bed.

6 Pull the screed pipes; fill in the narrow troughs with sand; and start laying pavers. Use a level to check the surface.

7 You can try to break pavers with a cold chisel and hammer, or make precise edge cuts with a saw and masonry blade.

8 Use a level or straightedge to check the surface level, and tap raised blocks down into the sand with a rubber mallet.

9 To help the interlocked blocks stay in position, dump some sand on the surface, and sweep it thoroughly into the joints.

BRICK & PAVER BASICS

21 Decks, Patios & Walks

STONE PAVING SYSTEMS

Like bricks, both stone and interlocking pavers can be dry-laid—that is, installed on a simple sand bed. This creates an attractive, intricate-looking walkway that requires little preparation. Large flagstones and fieldstones can also be installed directly on compacted soil in warmer areas where frost heave is not a problem. Sand or dry mortar swept between the embedded paving materials keeps them from shifting, as does the edging on each side of the walkway. Sand is best used between interlocking pavers because the units fit closely together. Sand can also be used to fill joints between regular-shaped stones, but wide joints between irregular shapes require joint stabilizer.

Interlocking pavers are installed in the same manner as bricks and concrete pavers. Because it's important to compact the sand and gravel subbase thoroughly, plan on renting a power tamper. These machines work better and faster, and require less effort than tamping by hand.

Depending on the shape of the pavers, you may end up with voids or chinks along the edges of the walk. You can buy special edging pieces to create a straight edge if your local dealer sells them. Otherwise, you'll have to cut the pavers to fit the spaces.

Modular Concrete

In this process, concrete is poured into molds to make big foundation blocks. The concrete can also be worked into a variety of smaller, stonelike shapes and sizes using a special form. Manufacturers offer a range of earth-tone colors that you can mix and match to create subtle gradations, giving your walkway an appearance more like that of real stone at far less cost.

Concrete slabs or smaller areas of fake "stones" should be supported on a 2-inch bed of sand, or 2 inches of sand on 4 inches of compacted gravel. When making areas of stone, it's a good idea to experiment with color and practice using the form before starting on the real thing. You place the form, fill it with the colored concrete, lift the form, and then move on to an adjacent section. After pouring each section, clean up the edges of each stone with a pointing trowel. As a final step, sweep sand into the joints between the fake stones.

Interlocking Paver Patterns

Interlocking pavers come in a variety of sizes and shapes, but most of them measure 2⅜ to 2½ inches thick, about the same size as a standard brick. Grass pavers have an open-grid shape for planting grass or other ground cover. They help to make a durable and natural-looking walk, and the turf itself (or structural edging) holds the pavers in place. Depending on the walkway design—for example, if you lay the stone in a grid or randomly—they can look casual or formal. Keep in mind that curved patterns will require a lot of difficult cuts on the edge blocks.

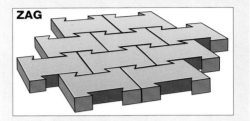
ZAG

HEXAGONAL

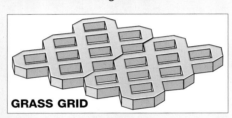
GRASS GRID

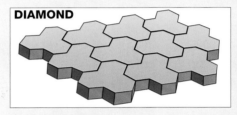
DIAMOND

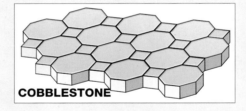
COBBLESTONE

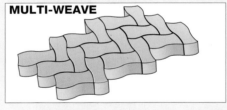
MULTI-WEAVE

Laying Stone Pavers

USE: ▶ power tamper • shovel • mason's twine • rubber mallet • short-handled sledgehammer • cold chisel

1 You can use forms to level a sand base, or temporarily set pipes in the bed and level the sand by dragging a 2x4 over the pipes.

2 Use accurate sod cuts on the border or string guides to position the first few stones. Set them using a rubber mallet.

560 DECKS, PATIOS & WALKS / STONE & INTERLOCKING SYSTEMS

STONE & INTERLOCKING SYSTEMS

Cutting Stone

Mark your cut line across the stone, and score it with a cold chisel and hammer, working back and forth.

With the score line etched in the surface, set a board under the stone for leverage, and break the stone.

Stone in Sand

A 2-inch sand bed is adequate for stones of similar thickness; use a thicker sand bed for stones of varying thickness. To create a level walk surface, remove as much sand as needed. Finish by sweeping sand between the joints with a stiff broom, working a 5- or 6-foot section at a time. A light spray of water will pack down the sand and wash it off the surface. Allow the surface to dry; then repeat the process until all the joints are filled and compacted. Replenish the sand as necessary, usually once a year.

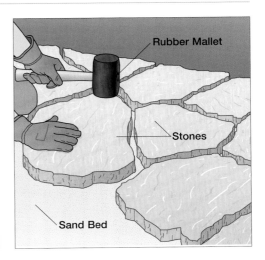

Stone in Mortar

To make a mortar-bed walkway, pour a 3½- to 4-inch concrete slab over a compacted gravel subbase. Then, embed the stones in the concrete slab. The concrete provides support for the stones and keeps them from separating as the ground shifts. As with sand-bed walks, the base must be set on a firm, well-drained bed—usually a 4-inch layer of gravel. The joints between stones are typically filled with mortar but may also be filled with sand or even topsoil.

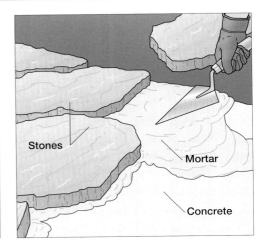

• 4-ft. level • push broom • garden hose ▶ stone pavers • mason's sand • gravel (optional) • pipe (optional) • 2x4 scrap • 2x lumber for forms (optional)

3 Check the surface with a level, and run a straightedge across the seams to identify any high spots.

4 Brush sand across the surface, working your broom back and forth to fill the joints between stones.

5 Spray the surface with water, and brush on another layer of sand. Repeat the process until the joints are full and firm.

STONE PAVING SYSTEMS

21 Decks, Patios & Walks

BACKYARD BOARDWALKS

Boardwalks transplant the advantages of wood decks to the turf, a place normally reserved for more cumbersome concrete and stone. By setting pressure-treated (PT) timber supports and covering them with redwood, cedar, or PT 2x4 decking, you can make an elegant, ground-hugging walkway.

Wooden walks can be built with concrete foundations that extend below the frost line, but most soils can support a simple, less expensive structure. These economical ground-level installations can rise and fall with the frost heaves. In nearly all cases, that's acceptable; wooden walks are flexible enough to bend without breaking.

Basic Walkway Construction

The easiest approach is to embed in the ground parallel pressure-treated boards (called sleepers), setting them on edge 3–5 feet apart, depending on how you plan to use the walk. (See "Designing a Walk," p. 546.) The top edges of these boards, like joists on a deck, carry short lengths of 2x4 or 2x6 decking—the surface you walk on.

To help secure the embedded sleepers, you should add stakes. One way is to drive pointed, pressure-treated 1x4s a foot or two into the ground every 2 feet inside the 2x10 sleepers. Trim them flush with the top edges, and nail them to the sleepers. They'll add strength and help keep the sleepers from tipping. To add even more strength, nail on cross braces, embedding them at right angles to the sleepers (and at the same depth) every 4 to 6 feet. To finish, nail on the decking perpendicular to the sleepers—just the way you would on a deck or porch.

Other Design Options

Some situations may call for more elaborate construction methods. For more stability in sandy soil, replace some of the stakes with 4x4 posts sunk in holes that reach the frost line. If the land rolls up and down, go with 2x12 joists. Use these to help level the terrain, allowing them to sit completely aboveground for a stretch (which may require a railing or even steps) and run belowground where necessary. Keep in mind that walkways needn't be level.

Forming a Gravel Walk

USE: ▶ garden hose • shovel • wheelbarrow • garden rake • hand tamper ▶ gravel • mason's twine

1 Use a length of garden hose to lay out curves in the walk. Cut several sticks the same length to keep the width consistent.

2 Spray-paint the grass outside of the hose, and remove the hose and wood to start digging out the sod.

4 Dig a trench along the edges with a shovel, deep enough so that the edging will rise a few inches above grade.

5 Set edging blocks into the trenches. This perimeter course will contain the surface material and prevent erosion.

Building a Boardwalk

USE: ▶ shovel • wheelbarrow • garden rake • tamper • measuring tape • 4-ft. level • framing square

1 After laying out your walk with string, dig a trench at least 4 in. deep and wider than the walk by 4-6 in. on each side.

2 Lay down landscape fabric to prevent weed growth, and backfill the trench with 4 in. of gravel raked level and tamped.

GRAVEL & SURFACE SYSTEMS

Installing In-Ground Steps

USE: ▶ power drill/driver • sledgehammer • saw ▶ PT lumber • rebar • screws

If your steps will be on a slope, you often can build them directly into the ground. Carve the steps into the hillside, and build them with landscape timbers and brick or other walk materials. These steps work well on a hillside where the run equals 24 to 32 inches per foot of rise. (See p. 344 for how to calculate rise and run.) To determine the number of steps, divide the rise by the height of a landscape timber. Treads on these types of steps usually measure about 15 inches deep, but they can be as deep as 17 inches.

3 Dig up the grass between the painted lines along with 2–3 in. of soil, and rake the ground free of stones and twigs.

1 Use pressure-treated lumber to build an edge for the steps, and drill landscape timbers retaining the steps.

6 Spread the gravel onto the walk in layers, and tamp each layer until it's about ¾ in. below the edging.

2 To minimize shifting, drive a piece of rebar down through the timbers and into the soil.

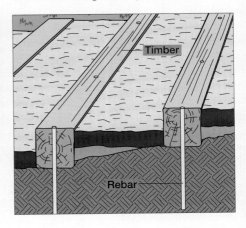

• power drill/driver or hammer • router (optional) ▶ gravel • stakes • mason's string • PT lumber for sleepers • decking • decking screws or galvanized nails

3 Lay pressure-treated sleepers (anything from 2x4s to 2x12s) on the gravel, and screw them to 2x4 stakes in the ground.

4 Cut decking boards to length, and attach them to the sleepers with 3-in. deck screws or 10d galvanized nails.

5 For a more finished look, use a router with a roundover bit to give the rough edges of the 2x4s a smooth, uniform shape.

BACKYARD BOARDWALKS

21 Decks, Patios & Walks

DRIVEWAY BASICS

Sand and gravel make up the basic ingredients of driveways. Sometimes these materials are combined with cement to make concrete. More often, they are bound together with liquid asphalt to make blacktop. Bricks, pavers, stone, or crushed gravel can also be used to construct driveways.

Driveway Design

The driveway serves as the main access to your home. It should accommodate not only the family's everyday vehicle use—including work if you run a home business—but also special purposes such as deliveries and entertaining. As you plan your driveway, keep in mind accessibility and visibility. Offer as clear a view of the road as possible, and get rid of any overgrown plantings that could make it difficult to see oncoming traffic.

Size the main part of the driveway between 10 and 12 feet wide for a one-car garage, and 16 to 24 feet wide for a two-car garage. (If the drive will form part of a walkway, add another 2 feet.) While a narrower drive (6 to 8 feet wide) may suffice, it's better to go wider so that people can step out onto the pavement and not onto grass or flower beds. An apron that tapers outward to meet the street makes it easier to back your car in or out. Extra parking that doesn't obstruct normal activities is extremely useful; allow 12 feet per vehicle for cars, more for trucks and RVs. You can screen these parking areas with plantings or fences, but make sure that the design is proportional to the rest of your yard. For safety's sake, don't crowd the walkway from the parking area to the entry door with plantings or other obstructions. Keep this area fully visible to maximize security and deter trespassers.

A concrete driveway need be only 4 inches thick for normal car traffic. However, if you expect heavy trucks, you should increase the thickness to at least 6 inches. Because a variety of vehicles might use the end of the driveway to turn around, you should make that part at least 8 inches thick. Keep in mind that proper drainage is critical. Any water that collects on the driveway will produce dangerous ice in the winter and will puddle near your house during heavy rains.

Design Options

Safety is the most important thing to consider when designing your driveway. If your house is on a busy street, take extra yard space for a turnaround so you don't have to back out. The figures given in these diagrams should be considered minimums; always consult with your building inspector before beginning work; local codes may be different.

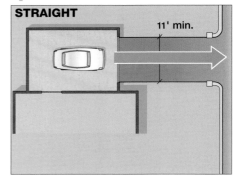

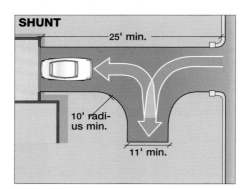

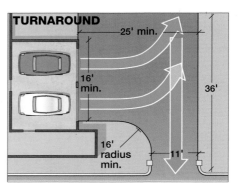

Slope & Drainage

If there is no natural slope to your driveway, it must be shaped to allow for drainage. Three possibilities illustrated at right include the crown (top), concave (middle), and cross-slope (bottom). The ramp can have a slope of anywhere from 4–8 percent; the steeper the ramp, the gentler the slope of the apron. Ramps with 6–8 percent slopes should have aprons with 2 percent slopes or less.

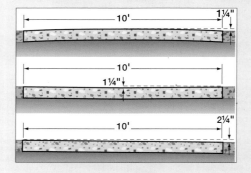

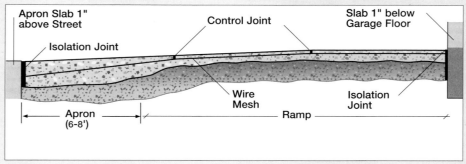

DRIVES

Material Options

◆ **CONCRETE SLABS** make strong, long-lasting driveways as long as the ground underneath is relatively stable and has been prepared thoroughly. The material can be slightly commercial-looking and without much visual interest—unless it is stamped with a pattern when poured. Light-colored concrete will show oil stains. Concrete generally costs more to install than asphalt.

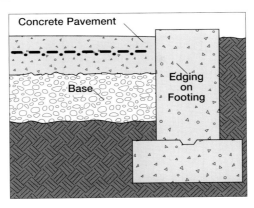
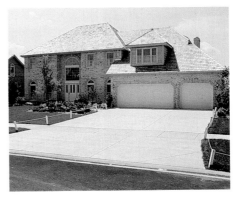

◆ **ASPHALT** doesn't offer the strength of concrete, but it is more malleable, so you can form a shallow swale to carry water away from a garage door or mound up a speed bump. Asphalt also repairs easily. While concrete cracks tend to drift apart no matter how you patch them, asphalt can be made whole, even regraded, and sealed to look new.

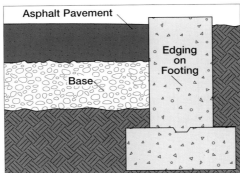

◆ **CONCRETE PAVERS** have the same durability as concrete; some blocks can withstand up to 2,000 psi of pressure. They come in a wide range of colors and styles, so it's easy to complement the look of your home. While fairly simple to install, pavers cost more than plain con-crete or asphalt. They are cheaper than stamped concrete, however.

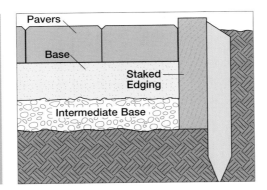

◆ **STONE AND BRICK** make durable drives, but they're not a good choice for steep driveways; they tend to get slippery when wet and ice easily in cold climates. In terms of installation, if you're looking to save money, brick set on a sand bed costs less than brick set on a concrete slab. Stones can be set in a mortar or sand bed or placed directly in the ground.

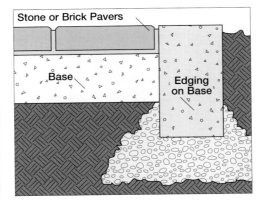

DRIVEWAY BASICS 565

21 Decks, Patios & Walks

ASPHALT PROBLEMS

Although asphalt driveways are fairly durable, they do require some maintenance, especially in cold areas where freeze/thaw cycles are the norm. Small cracks let in water that will freeze and expand, causing existing minor damage to become major. Periodic sealing will help keep these cracks from starting, but they won't prevent damage caused by settling of the ground under the driveway or improper installation.

Minor Repairs

Before you apply any filler or sealer, brush out loose material or vacuum it up with a wet-dry vac. Use blacktop crack filler, which comes in cartridges like caulk, applying it in a continuous bead with a caulking gun. Check the instructions for set time (about 10 minutes). When the material settles, compact and smooth out the surface with a putty knife. Fill deep cracks with sand, compacting it to within ½ inch of the surface before applying the filler.

For small holes, clear away loose chunks and break off any unsupported edges of old blacktop. A wire brush will dislodge debris. Then, use pastelike asphalt patching compound. This material contains fine aggregate and is malleable enough to be feathered to the edges of depressions. It works well where asphalt is broken out along curbing. In the middle of the driveway, a repair will last longer if you dig out around the edges of the hole and make the patch uniform in thickness. Lay in the patch material with a trowel to create a mound about ½ inch higher than the driveway, and then tamp it with the end of a 2x4.

Filling Potholes

Use cold patch for street-size repairs. It has larger aggregate than paste patches and comes in 60- to 70-pound bags. This material resembles the asphalt your driveway is made of, but it's treated with chemicals that keep it workable. If the temperature stays above 50° F, you don't have to heat it the way highway crews heat fresh asphalt—but you do have to roll it.

After clearing loose debris, remove jagged edges around the hole with a hammer and cold chisel. Then, pile on enough cold patch to leave a slight mound after tamping. Fill deep holes in two stages, tamping in between to avoid leaving a water-collecting depression. Next, apply the weight of your car, driving it slowly over a piece of ¾-inch plywood or a layer of sand spread over the cold patch.

Surface Sealing

To determine whether blacktop needs resurfacing, pour a bucket of water on it on a hot day while the sun is out. If the surface water dries leaving a dark circle that takes much longer to dry, the water has soaked in, which indicates that resurfacing is needed.

A stable asphalt surface that was properly installed needs only occasional sealing. New drives won't need it for at least a year. Older drives should be washed using a household cleaner and water—about a quart per 1,000 square feet of asphalt—and then rinsed.

Sealing will make grayish, drying asphalt look better and keep out water that causes erosion and cracking. However, don't expect sealing to take the place of resurfacing, which involves topping an old asphalt drive with at least 2 inches of new material. And no coating can rescue one of those jobs that consists of a 1- or 2-inch layer of asphalt applied over loose gravel.

Sealer comes in 5-gallon cans, enough to cover about 250 square feet. Older, porous drives will soak up more, newer drives less. For extra protection, apply two thin coats, and allow 36 hours between applications.

Replacing Expansion Joints — MONEY SAVER

USE: ▶ wet-dry vacuum • putty knife • caulking gun ▶ foam backer rod • urethane sealant

1 Fiberboard and lumber used to form expansion joints in concrete eventually wear away in the weather.

2 To replace the joint, dig out old pieces of lumber and debris, and clean out the crack with a wet-dry vac.

3 To allow for some shifting and to minimize cracking, pack a strip of foam backer rod into the joint.

4 Cover the backer rod and seal the joint against the weather with a liberal layer of self-leveling urethane sealant.

DRIVES

Cleaning Concrete

USE: ▶ brick • bucket • stiff-bristled brush • 4-in. paintbrush ▶ commercial absorbent or other material • concrete cleaner • clear concrete sealer

1 Use a commercial absorbent, sawdust, or kitty litter to lift deep oil stains. Work the absorbent around with a brick.

2 Use a stiff brush and concrete cleaner to scrub out stains. It also helps to wash the entire surface before sealing.

MONEY SAVER

3 A clear concrete sealer improves the appearance of a concrete driveway and helps it shed water.

Repairing Asphalt

USE: ▶ mason's trowel • caulking gun • utility knife • rubber gloves ▶ cold patch • asphalt-based caulk • asphalt rolls

1 Use cold patch, a workable blend of asphalt and aggregate, to fill small holes. Clean out debris before mounding the mix.

2 Patch asphalt cracks so that they don't let in water by sweeping them clean and filling with asphalt-based caulk.

MONEY SAVER

3 Asphalt rolls are a convenient way to patch small cracks. Peel off the backing, and press the self-adhering mat in place.

Sealing Asphalt

USE: ▶ garden hose with hose-end sprayer • pressure washer (optional) • push broom or squeegee • work gloves ▶ cleaner • asphalt sealer

1 Wash the surface to improve sealer adhesion. Use a hose, hose-end sprayer with detergent, or a pressure washer.

2 Use full-strength cleaner on tough stains. Most solutions can be diluted (1 cup to a gal. of water) for general cleaning.

MONEY SAVER

3 Most sealers can both fill and coat cracks up to 1/8 in. deep. Spread them with a squeegee or an old push broom.

22 Unfinished Space

570 ATTICS
- Renovating an Attic ◆ Meeting Codes
- Dormers ◆ Basic Stairway Variations
- Inspection Checklist
- Structural Alterations ■ ATTIC VENTILATION
- Strengthening Attic Floor Joists
- Collar Ties ■ 7 WAYS TO VENT AN ATTIC

574 BASEMENTS
- Finishing a Basement
- ■ SEALING A FLOOR
- Covering Basement Walls
- Window Wells
- Finishing Touches
- Ceiling Options
- Boxing In Columns, Beams, Pipes
- ■ FIXING WATER LEAKS
- 14 Ways to Eliminate Water
- Sump Pumps

578 GARAGES
- Converting a Garage
- Installing a Garage Entry Door
- Patching & Leveling Floors
- Interior Finishing System

580 SMALL SPACES
- Finding More Space ◆ Space Savers
- In-Wall Hazards ◆ Concealed Space
- Cabinet Storage ◆ Understair Storage
- The Undiscovered Space
- Crawl Spaces

22 Unfinished Space

RENOVATING AN ATTIC

Instead of building an addition from scratch, many homeowners can perform the cost-effective trick of bringing unused space they already have to life. One space that is particularly ripe for renovation is the attic.

Converting unused attic space to habitable space isn't much of a trick in a house with a large area under the roof, especially if it already has plenty of headroom, windows at each end, and a solid floor. The project is more challenging in houses with low-slope roofs over an airless jumble of exposed ceiling joists, electrical wires, and insulation.

Some unused spaces need more revitalization (meaning time and money) than others. But some basic design and construction decisions are similar no matter what shape your attic is in—even on the easiest jobs in which the original builders did a lot of the work for you.

Structural Limits

Some older houses are overbuilt—with extra structural capacity built in. When lumber and labor were inexpensive it was no big deal to make joists in the attic as big as the ones on the first or second floor.

Most houses built in the last 10 or 15 years are not overbuilt. To keep costs down, joists in the attic floor were generally smaller than those in the first and second floor. The discrepancy is allowed by building codes because first- and second-floor joists must be large enough to support dead and live loads, but attic floor joists may be much smaller—designed to support as little as one-third of the loads calculated for the living space downstairs.

Of course, you must comply with state and local building codes. But this can be a challenge when you want to renovate, and an architect or building contractor informs you that new attic floor joists must be added. And if you're adding sizable dormers to the top of the house, some of the posts or load-bearing walls in the existing frame—and even the foundation—may have to be beefed up to carry the new loads.

Trusses & Braces

Here is another case where owners of newer homes may lose out. Trusses are triangular shapes made of small-dimension lumber. The bottom of the truss serves as a ceiling to the living space below. The upper sections serve as roof rafters. But to make the truss out of small lumber (such as 2x4s), many cross braces are attached in a maze of zigzags from one side of the house to the other. Unfortunately, you can't pull out some of the braces to make room. Each one adds strength to the overall truss shape. To add a dormer in a truss-framed roof, trusses would have to be replaced by more substantial floor joists and rafters—a complex job.

Even in large, open attics, you may be hindered by cross beams, called collar ties, that tie opposing rafters together—like the horizontal line in the letter A—to help them resist uplift from high winds and the pressure of the rafters against the outside walls. Often they can be raised enough to provide headroom. However, to be effective, they must be located in the top third of the floor-to-ridge height.

Light & Ventilation

Successful attic conversions must overcome limitations of the space. One such limitation: a window at each end wall might be adequate for one large, uninterrupted storage space, but not for subdivided living space.

Enlarging end wall glazing and installing skylights may be the easiest option. For example, you might exchange one small window for large side-by-side double-hung windows with a spacious, fixed-glass crescent above. That could flood two bedrooms with light, while a central area reserved for closets, stairs, and a bath could be fitted with skylights. Such a plan eliminates the need for dormers—a more complicated and costly structural change in the roof.

Skylights and a half-circle window provide much-needed light and ventilation for this renovated attic room.

Meeting Codes

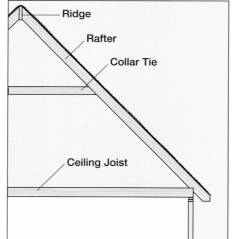
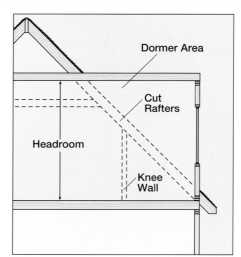

Before you change dead space to living space, check building codes for rules about collar ties, headroom, and knee-wall height above the finished floor.

ATTICS

Dormers

Gable dormers (top) are a good way to create additional space while supplying light and ventilation to an attic room. A continuous shed dormer (bottom) will provide even more space.

Basic Stairway Variations

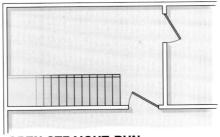
OPEN STRAIGHT-RUN

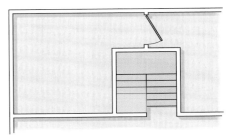
HALF-CLOSED U-SHAPED RETURN

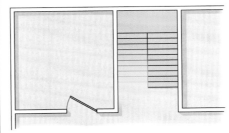
OPEN U-SHAPED RETURN

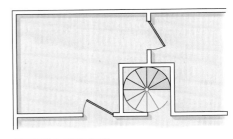
SPIRAL STAIRS

When locating a new stairway, use the option that best fits the space in as well as below the attic. Stairs to any living space must meet building codes.

Obviously, on jobs where dormers must be raised above a low-slope roof to gain headroom, windows can be added in many configurations along the face of the dormer.

The job of renovating unused space in your existing home can be disruptive. The house could be crawling with contractors for weeks. But you'll be saving the time and cost of building foundations and other parts of the building that are already in place.

Access to the Attic

Accessibility is most troublesome in attics reached solely by a trap door or pull-down stairs—local building codes don't allow this for habitable space. There may not be enough room in narrow, second-story halls to install a conventional permanent stairway. Stairs to an attic must be at least 2–6 feet clear between handrails or between a single handrail and the opposite wall. Needed space can frequently be stolen from an existing room or closet. Another option is to gain access with spiral, or library, stairs. If the living space does not exceed 400 square feet, spiral stairs may be permissible. These stairs are available in diameters as small as 4 feet and may fit easily into an alcove or at the end of a hallway.

Installing a Dormer

Dormers serve a number of purposes: they provide light, ventilation, and headroom for renovated attic spaces, and they bring architectural interest to an otherwise nondescript roof. The two main types are gable dormers and shed dormers.

Easiest to build during the initial framing, dormers can, however, be built into existing roofs. The job will be messy, though—involving stripping and cutting out a section of the roof—and will require reshingling at least the roof area around the dormer. Building any kind of dormer requires the application of advanced framing techniques. Gable dormers require some advanced rafter cuts for the valley rafters and valley jack rafters, similar to the cuts used in hip roofs. Shed dormers, with their flat roofs, are easier to frame.

Inspection Checklist

- **HEADROOM.** A minimum height of 7 feet, 6 inches, in at least 50 percent of the floor area, is required.

- **LIVING SPACE.** Each habitable space must be at least 80 sq. ft.

- **VENTILATION.** Natural clear ventilation through windows or skylights must equal at least 4 percent of the floor area of a habitable space.

- **RAFTERS.** Sagging rafters must be evaluated by an engineer.

- **FLOOR FRAMING.** Attic floor joists must be checked by an engineer to determine whether or not they can carry additional loading.

- **WATER LEAKS.** Must be eliminated.

- **PESTS.** Have insects and other pests professionally eliminated.

22 Unfinished Space

STRUCTURAL ALTERATIONS

Structural alterations, as well as improvements and extensions to existing wiring, plumbing, and heating systems, may all be required when you want to make a fundamental change to how a space is used. Below are a few tips on some of the framing changes you may have to make to finish an attic.

Material Access

Instead of trying to haul long, heavy timbers through the finished areas of your home, and inevitably bashing a few 2x4s against the walls, try to deliver joists, studs, and other lumber directly into the attic. New or existing openings for windows or skylights provide a handy entry point for sliding in materials raised up on a scaffold outside the house. If there are no windows or skylights, you might be able to load them through a gable-end vent.

Joist Size vs. Headroom

In some cases, standard joists large enough to span the floor and bring in compliance with local codes for living space may be much deeper than the existing joists—2x12s instead of 2x6s. Sistering these joists would reduce attic headroom by 6 inches—possibly enough to make the attic not legally habitable. To save space and comply with your building code, you may be able to bridge the floor with smaller joists that are supported along their spans by bearing walls or posts in the space below. In extreme cases, you can increase floor strength without increasing joist depth by setting joists closer together than the standard spacing of 16 inches on center—for example, by setting a new 2x6 joist in between old joists spaced at 24 inches on-center, creating a much stronger floor with joists every 12 inches. Always check with a local building inspector beforehand.

Removing Braces

If you have collar ties running across your unfinished attic, they are not there just for you to bang your head—they're an important part of your roof framing. This can be a stumbling block because the collar ties may be the only framing component limiting headroom. If they weren't there, you might be able to save a lot of time and money required to build dormers. It's worth consulting with an an architect or structural engineer, and definitely with the building inspector, about raising them.

The idea is to install shorter collar ties closer to the upper third of the A shape to gain valuable headroom, and only then to remove the existing collar ties. The roof structure will still be rigid and the smaller space above the ties can serve to ventilate the attic, providing a cooling cross flow from one gable-end vent to the other.

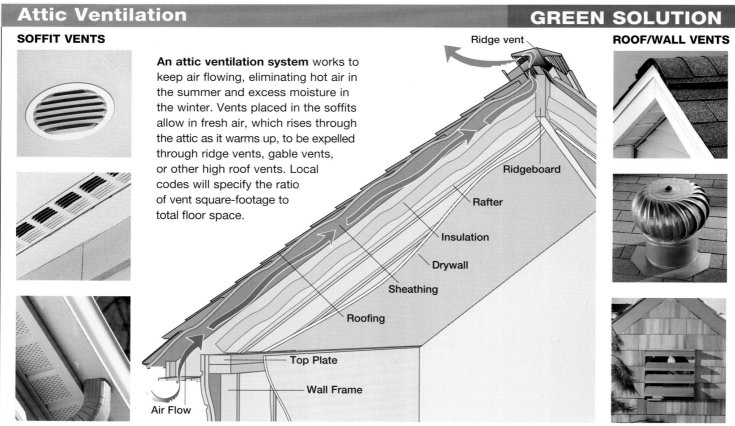

Soffit vents (pictured at left, from top to bottom) include plug vents, strip-grille vents, and perforated vents. Roof vents (pictured at right) include ridge vents, turbines, and motorized exhaust fans. Wall vents include gable vents and motorized exhaust fans.

ATTICS

Ventilation

Because a minimum headroom of 7 feet, 6 inches is required in an attic, it's tempting to expose the ceiling all the way to the roof peak, creating the most expansive A shape possible. But exchanging the vent area you currently have (the entire attic) for living space may create a prohibitively large cooling load on the new living space in the summer and drive up your utility bills. In a finished space directly under the roof, midday temperatures could soar to 120° F or more and require a massive amount of air conditioning. On top of that, when the temperature dropped at night, you would likely have condensation problems where warm, interior air met the underside of the roof.

That's why the best plan may be a compromise: minimum headroom, but more space for insulation and ventilation. Establish this balance by using the collar ties as the floor of a new miniattic. This floor, the attic's new ceiling, may be relatively short, but the miniattic will allow you to vent what remains of the unfinished attic space in two ways. First, air entering the soffit vents in the roof overhangs will vent the eaves, and then it will rise up through the rafter bays to the miniattic above the newly renovated space. Second, the air flow from the soffit vents will join with the continuous flow of air beneath the roof ridge, running from one gable-end vent to the other.

Collar Ties

Sloped roofs have rafters in an inverted V shape. But to tie rafters together, help them resist uplift in high winds, and reduce the force that rafters exert on walls, cross beams called collar ties are often added to make the inverted V-shape into an A-shape. These collar ties are an integral part of the framing. The roof might not collapse if you removed them, but you could probably get enough movement to pop nails, crack plaster, and stress a window or door frame to the point where it would become difficult to open and close. Worse yet, you might violate building codes.

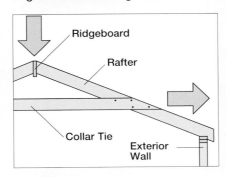

GREEN SOLUTION

7 WAYS TO VENT AN ATTIC

1. **GABLE-END LOUVERS.** Available in many shapes and styles, screened to keep insects out.

2. **STRIP-GRILLE VENTS.** Long narrow grilles set into soffits.

3. **PLUG VENTS.** Small, circular vents used in the soffits instead of a continuous line of strip grilles.

4. **PERFORATED SOFFIT VENTS.** Ventilated panels usually manufactured of aluminum or vinyl.

5. **RIDGE VENTS.** Ideal for attics in which gable cross ventilation is not possible.

6. **ROOF TURBINES.** Designed to improve air flow by turning as warm air rises through the vanes.

7. **MOTORIZED FANS.** Normally mounted on an exterior attic wall and used in conjunction with louvers inside the house to draw air up and through the attic.

Strengthening Attic Floor Joists

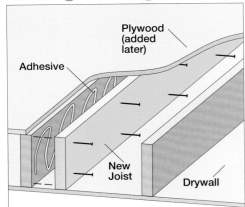

Double up joists as needed for additional strength if the existing attic floor is not adequate to carry the live load. Fasten the new joist with adhesive and nails.

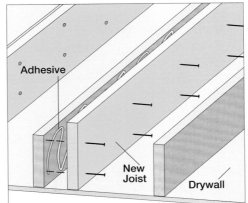

Larger joists such as 2x8s and 2x10s can also be glued and screwed to existing joists. The larger boards will decrease headroom but increase load capacity.

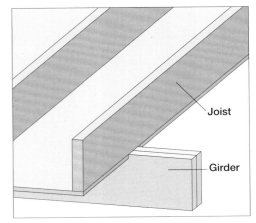

Ceiling joists are supported by beams or girders, usually running along the middle of the house. They may be hidden in the top portion of a load-bearing wall.

22 Unfinished Space

FINISHING A BASEMENT

One of the best bargains you can get is the space gained by converting your basement into habitable space. Homes with full basements already have a floor, walls, and a ceiling, as well as a maze of pipes and wires to supply water and electricity. With these necessities in place, all that is needed is a good plan to turn an unused space into an airy, well-lighted room. Remodeling is often a scaled-down version of doing new construction. Two areas that are critical in remodeling a basement are the concrete floor and masonry walls.

Installing a Floor System

The easiest way to treat concrete floors is with masonry paint. Two coats over a clean, dry floor can make the room much more inviting. If the floor stays dry year-round, a more comfortable solution is carpeting. Even a commercial grade over a foam backing will make the floor softer and warmer; it also absorbs sound that reverberates off hard masonry surfaces.

On floors that are uneven, cracked, or covered with peeling layers of old paint, a sleeper system is often the best solution. Sleepers are treated wood 2x4s laid either flat or on edge on the concrete, usually 16 inches on center in a bed of mastic. A clear waterproof coating over the concrete floor is recommended before installing the sleepers. Rigid insulation is added between the boards, and then a vapor barrier is laid across them. For tile or carpet floor finishes, a ¾-inch plywood subfloor is nailed down. Wood strip flooring may be laid directly over the sleepers, or a second layer of sleepers may be added over the vapor barrier, producing a warm, dry, resilient floor system.

Wall Options

Basement walls are usually unfinished. The simplest alternative is to install a layer of rigid insulation in a mastic bed over the wall and then to nail 1x2 furring strips over this, covered by a layer of wallboard or paneling. Be aware, however, that using masonry nails will create a lot of holes in the wall. Above grade this may not be a problem. But if you have basement walls that don't leak, it makes no sense to puncture them, risking a problem. Nailholes may cause a fault line, inviting stress cracks.

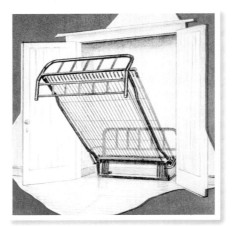

Dancing was the mother of this invention, the Murphy bed. The original Mr. Murphy lived in a small apartment too cluttered to accommodate his afternoon recreation: dancing with his lady friends. Solution? A fold-up bed to make a bigger dance floor.

To avoid this, attach 2x2 furring strips, at 16 inches on center, to a 2x3 sole plate along the floor and a 2x3 top plate along the ceiling. The benefit is that the complete 2x3 frame can be attached to the concrete floor using just a few nails or a bed of construction adhesive, and nailed to the wooden floor joists of the house (the ceiling of the remodeled basement). Individual studs need not be nailed into the masonry wall. Using 2x3s also provides additional room for more insulation than just thin fiberglass batts or rigid polystyrene panels. Cover the assembly with a vapor barrier before applying the new surfacing material. This will keep interior moisture from working through the new wall and condensing on the masonry, where it would come into contact with the frame and the insulation, causing deterioration.

Sealing a Floor

USE: ▶ scrub brush • bucket & sponge mop • pitching chisel or scraper • vacuum cleaner • paintbrush • hand-held sprayer • dust-mist respirator or dust mask

1 *Clean away stains and soiled areas using a solution of trisodium phosphate or a phosphate-free cleaner.*

2 *Scrape away rough and uneven spots in the concrete floor using a pitching chisel or other steel-edged scraping tool.*

3 *Vacuum away concrete dust and other debris in cracks and control joints that could reduce adhesion.*

BASEMENTS

Covering Basement Walls

Masonry paint is an expedient solution for walls, although even a bright color won't disguise the blocks. If the wall was carefully built, a thick coat of masonry paint may just hide the joints, producing a stucco effect. It is more likely that the basement walls were not neatly finished—that's one reason the most common remodeling treatment is nailing furring strips on the wall, covered by a layer of drywall or paneling. The furring strips bridge the little bumps in the wall and the depressions at mortar joints.

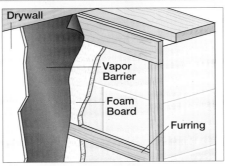

Furring strips can be applied directly onto masonry walls after the walls are painted with waterproofing masonry paint. Nail with 1½-in. masonry nails.

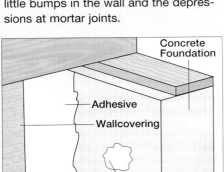

Dry walls can be patched for smoothness and covered with a stiff sisal fiber or fiberglass wallcovering.

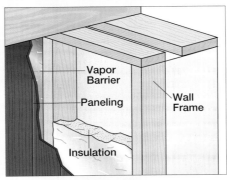

Building 2x4 stud walls is best if you need room to wire for electrical outlets and to insulate your renovated basement.

Window Wells

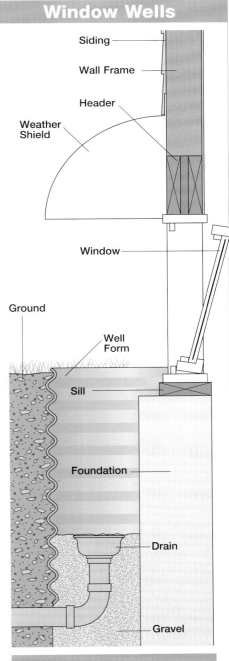

A basement will remain drier if windows are shielded from rain and if drainage is provided by a window well connected to a drainpipe. Waterproof the foundation wall, and then backfill over the drainpipe with gravel.

MONEY SAVER

▶ trisodium phosphate or other cleaner • clear concrete sealer

4 Using a paintbrush, coat around the perimeter of the floor with a clear concrete waterproofing sealer.

5 Spray-apply the sealer to the remainder of the floor for fast, thorough, and even coverage.

FINISHING A BASEMENT

22 Unfinished Space

FINISHING TOUCHES

Basements usually have a combination of ducts, beams, columns, and pipes that present challenges when it comes to converting the area into living space.

Deciding what kind of ceiling to install in a renovated basement depends in part on the amount of existing headroom you have available, and how many pipes, ducts, and wires cross beneath the joists. One of the best ways to hide all of these is by installing a suspended acoustical tile ceiling.

Suspended ceilings can conceal pipes, ducts, and wiring, yet still leave them accessible. Another advantage is that the ceiling is leveled as you install it—the existing joists need not be level or even straight. A suspended ceiling also makes it easy to install lights: simply remove an acoustical tile, and replace it with a fluorescent light fixture of the same size.

Concealing a Beam
Concealing a wood beam is a relatively simple task. Concealing a steel beam, on the other hand, is not so easy because fastening material to it is particularly difficult. The proper way to enclose a steel beam is by building a wooden framework around it and then screwing or nailing gypsum wallboard to the framework, enclosing the beam. The beam enclosure can then be finished to match the rest of the room.

Concealing a Duct
Metal ductwork in a house with hot-air heating typically leads from the furnace to every room. If the ductwork reduces headroom, it is possible to move it, but this is a job for a heating and cooling contractor. In an informal space, you can paint ductwork to match the ceiling. Ductwork may also be concealed above a suspended ceiling or boxed into a framed chase covered by drywall or paneling.

Concealing a Column or Post
If a column or post is not ideally placed for your remodeling plans, try revising the plans rather than moving the column. It is often easy to conceal a post within a new partition wall. If not, you can de-emphasize it with a fabric covering, build a bookcase or shelf unit around it, or frame it into a chase (a continuous enclosure) covered with paneling or drywall.

Concealing a Soil Pipe
Typically, the soil stack is the largest pipe in a house. If possible, enclose it within a chase or soffit. To reduce the sound of running water, wrap insulation around the pipe. Be sure to measure carefully along the length of the pipe because it must slope ⅛- to ½-inch per foot for drainage. If the pipe's cleanout plug will be concealed within an enclosure, build a door for access to the plug.

Ceiling Options

Acoustical ceiling panels are available in a wide variety of colors and textures. To install a tile, angle it through the suspended grid and set it into place. It's a good idea to wear gloves to avoid marring the finished surface of the panels.

Boxing In Columns, Beams, Pipes

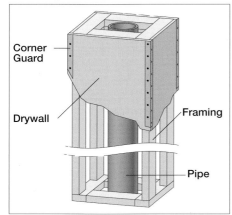

To conceal a column, frame around it using two-by lumber. The frame provides a base for other finishes.

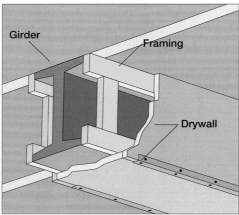

To conceal a steel beam, build and attach a ladder framework to the joists. Cover the framework with drywall or paneling.

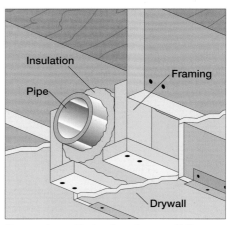

Conceal a soil pipe in a box as if it were a beam. Wrap it in insulation to reduce the sound of running water.

BASEMENTS

MONEY SAVER
Fixing Water Leaks

USE: ▶ wire brush • vacuum • cold chisel • hammer • trowel ▶ hydraulic cement

1 Use a wire brush to remove loose mortar and dirt from the concrete-block wall; then vacuum the wall to remove dust and debris.

2 Chisel out and undercut cracks and small holes to provide a firm anchor for the hydraulic cement. Clean out loose debris.

3 Mix hydraulic cement with water, and use a pointing trowel or a jointing tool to apply the cement to the damaged area.

14 Ways to Eliminate Water

1. Use splash blocks or leader extensions on a functioning gutter system.

2. Grade the area around the house. Grade should drop at least 2½ inches in 10 feet.

3. Repair cracks in masonry walls using hydraulic cement.

4. Seal the space where the house water main and electrical service conduit enter the structure. Use hydraulic cement.

5. Eliminate vegetation close to foundation walls. Plants hold water and reduce evaporation from the soil.

6. Make sure no roof leaks are being carried down to the basement.

7. Keep basement windows closed during rainstorms.

8. Paint concrete floors with two coats of water-locking masonry paint.

9. Insulate pipes and air-conditioning vents to prevent condensation.

10. Install an electric dehumidifier.

11. Install a sump pump to remove water.

12. Hire a contractor to install an interior perimeter drain.

13. Hire a contractor to waterproof the foundation.

14. In extreme cases, hire a contractor to install footing drains around the perimeter of the house to remove any water that manages to accumulate against the foundation wall.

Sump Pumps

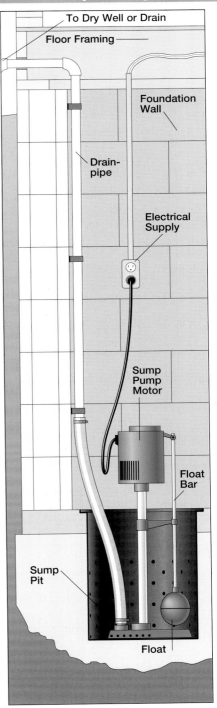

Install a sump pump to keep the basement dry. A check valve will prevent sewer gases from getting in.

FINISHING TOUCHES

22 Unfinished Space

CONVERTING A GARAGE

An unused garage is often an ideal place for a home renovation, offering a large space that contains no more than a lally column. What's important is how the garage relates to the rest of the house. If it is off the kitchen, for example, it may be a good place for a family room but not a master bedroom. You may even be able to convert an unattached garage, connecting it to the main house via an enclosed walkway, which itself may become a studio, home office, or child's playroom.

Converting the Floor

Because most garage floors slope toward the garage door opening, you will probably need to level the floor. One way to do this is to pour a new concrete slab over the existing one. If the floor of the house is set slightly higher than the existing garage floor, this may be the easiest and least expensive option.

Another option is to install a sleeper-and-plywood-subfloor system, using shims to create a level surface. Rigid insulation can be placed between the sleepers.

In homes where the garage is one or two steps down from an adjoining area, you may want to build an elevated subfloor. If you choose not to raise the floor of the garage to the height of the adjoining area, you can make the transition a decorative asset by joining the levels with a dramatic stairway. For safety, be sure that the steps are clearly differentiated by using a contrasting material or color on each level, and by providing handrails.

Finishing

In order to make a conversion appear to be an integral part of the original house design, you must finish the outside to look like the rest of the house. If you remove the garage door and enclose the wall, you might consider building a carport over the existing driveway or even tearing out the driveway and landscaping the area to blend in with the rest of the yard.

The first step to enclosing the garage is to remove the existing garage door. You may need to build a concrete knee wall to match the rest

Installing a Garage Entry Door

USE: ▶ circular saw • hammer • ladder • reciprocating saw (optional) • eye protection ▶ prehung door • door casing • 4-ft. level • 2x4 studs • 2x lumber for header

1 Frame the rough opening for a new door in an existing exposed stud wall before cutting the sheathing.

2 Nail a temporary brace in the stud wall over the proposed door opening to hold the cripples in place as you cut the studs.

3 Cut the studs to make the rough opening for the new door, leaving enough space below the cripples for a new header.

5 Cut away the existing sill plate, as well as the sheathing and siding within the new rough opening.

6 Trim back the siding to accommodate the exterior casing for the new door, and then test-fit the prehung door assembly.

7 Run a bead of caulking around the perimeter of the rough opening; then set the door frame in place and fasten it.

GARAGES

of the foundation and to provide support for the new stud wall. This new wall will be a likely place to put a new door or window because a header already exits here. Position the window or door just below the header.

Once you've installed the wall, attach sheathing over the exterior surface; cover it with house wrap or felt paper; and then install siding to match that already on the house. To make a new door opening in one of the existing walls, follow the sequence shown in the photos below.

Before you finish the interior walls, you will want to eliminate any existing moisture problems. You can then use standard methods to finish the walls. Be sure to include electrical outlets, insulation, and a moisture barrier.

Patching & Leveling Floors

Sweep loose debris and cement dust from the garage floor, and then clean it thoroughly before doing any patchwork.

Dampen cracks; apply a bonding agent; and then trowel patching cement into the crack and smooth it out.

Interior Finishing System

- wood screws or 12d & 16d nails • caulk

4 Cut the new header for the door, and then nail it in place over the trimmer studs. Toenail the cripples to the header.

8 Install the door in the frame, and then nail the exterior casing in place. Caulk between the casing and siding.

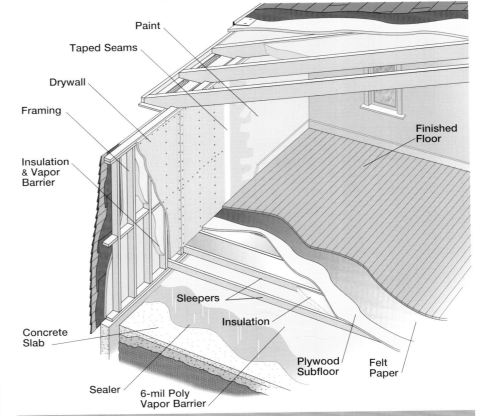

To finish a garage floor, seal the slab with asphalt mastic, and roll out a 6-mil polyethylene vapor barrier. Finished wood flooring needs joists or sleepers and a layer of plywood subflooring. Convert open-frame walls by filling the bays with insulation adding a vapor barrier (either foil or plastic) and a layer of ½-in. drywall.

CONVERTING A GARAGE

22 Unfinished Space

FINDING MORE SPACE

Homeowners with grown children may have the luxury of claiming space that was once a bedroom and converting it into an exercise room, a home office, or the ultimate storage luxury—a walk-in (and walk-around) closet with enough room for bureaus, full-length mirrors, and even a chair so you don't have to balance like a stork to put on your socks.

Yet, even in smaller houses and apartments, you may be able to find more storage space—regardless of whether or not you have a spare room that can be anything you want it to be.

Kitchens

To maximize the use of corner space under kitchen counters—and still be able to get at what's there—store bulk items on multitiered carousels. To better use space in deep but narrow cabinets, install pull-out wire-rack bins.

Many appliance makers offer extensive lines of small appliances and convenience features that can be mounted to the underside of upper kitchen cabinets. These mountings for radios, can openers, recipe and spice racks, coffee makers, and microwave ovens free up counter space. Some companies also offer space-saving larger appliances, such as narrow, 24-inch-wide vertically stacked washers and dryers, which can create additional floor space in your laundry, where it is often badly needed.

Guest Rooms

A time-honored solution for creating a multi-purpose room is the Murphy bed. The Murphy Door Bed Co. was founded at the turn of the century by William Murphy, who designed a fold-up bed to gain space in his crowded one-room apartment.

A modern version of the Murphy bed is constructed with a built-in counterbalance mechanism that allows you to tip up the mattress, bedding, and frame into a prefabricated cabinet. Convertible sofas or trundle-bed couches (with a second mattress on a pop-up frame stored beneath the main cushions) are sometimes able to provide twice the bed space within the footprint size of a single bed, allowing more space for furniture. There are even some oversize chairs available that convert into beds, creating even more options.

Closets

Interior shelves in many closets and cabinets are set back and are therefore not as deep as the space behind the door. Even a few inches of such unused space can be reclaimed by mounting one or more ventilated wire-shelving units on the back of the door. (Make sure that the units are a few inches less in width than the door, to allow for closing.)

A split-closet wall can provide a convenient, space-saving plan. Instead of taking dispersed closet space from bedrooms that share a common wall, make the partition wall the depth of a closet, opening half to one room and half to the other. This will reduce noise transmission between the rooms. For even more sound deadening value, insulate the interior closet walls.

Garages

The roof framing in garages often includes several cross ties, usually 2x4s or 2x6s, that link the ends of opposing rafters. This platform of framing was probably not designed to support heavy loads, but with a few 1x2s to make a supporting lattice, it's an ideal place to store large but lightweight items such as screens. If you want the space for heavy-duty storage, with a solid floor, you probably will have to install a stronger framework with a ⅝-inch plywood subfloor—roughly similar to the floor framing in a house attic.

Entry Halls

Entry halls, or foyers, may contain a surprising amount of potential storage space. If you have

Space Savers

A heater/fan/light saves space by combining the comforts of light, warmth, and ventilation.

A lazy Susan rotates inside a corner cabinet, making good use of a difficult-to-reach space.

Kitchen soffits, otherwise empty, can be replaced with high cabinets, useful for storing infrequently used items.

A recessed toe-space heater is especially welcome in a kitchen where available wall space for conventional heating is scarce.

SMALL SPACES

an older home, commonly built with higher ceilings than contemporary homes, you may find several feet of space between the top of an entry door and the ceiling.

Rescue this space by building a platform, 2 to 3 feet deep or more, supported by ledger strips, one on each side wall. In the simplest form, two strips of ¾-inch or larger molding could support a sheet of ½-inch-thick plywood. If you expect to support heavy loads, upgrade the plywood to ¾-inch, and use heavier ledger strips. This construction may be dressed up to simulate a dropped foyer ceiling. Access to the storage is gained through cabinet doors facing the hall. Painted to match the surrounding walls and ceiling, the structure will recede, leaving the entry door as the focus of attention.

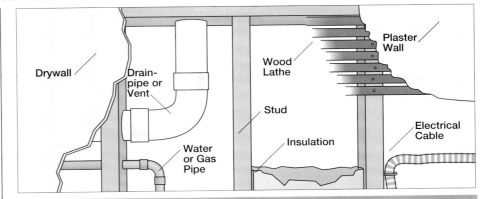

In-Wall Hazards

Do you know what's inside your walls? This illustration shows some of what you might find. Remove drywall or plaster slowly, to avoid hazards or damage.

A Murphy bed can be hidden away into a wall unit. It's a convenient space saver for guest rooms and small apartments.

A pocket door slides between the walls, allowing for furniture to be placed where doors would swing out into the room.

A corner medicine cabinet can make available the use of what would otherwise be wasted space.

A central vacuum system eliminates the need for a freestanding vacuum and helps to remove common allergens from a home.

A pet door in a wall or conventional door can provide your pet with easy passage in time of need, when you don't want to be disturbed, or when you are at work during the day.

An ironing board can fold up onto a door like a Murphy bed, ridding your clothes closets of unnecessary clutter.

FINDING MORE SPACE

22 Unfinished Space

CONCEALED SPACE

Calculating the amount of unused space in a house or apartment is almost as difficult as totaling the interest payments on a 30-year mortgage. The cost of the unused space behind doors, under stairs, and at the back of closets, plus the dead space boxed into kitchen soffits, knee walls, and so on really adds up.

Vanquished parcels of interior real estate may be equal to adding an entire room or moving to a larger apartment, but they can generally ease overcrowding, remove obstacles formed by tight clusters of furniture, even make an unbearably small apartment seem more spacious. And just about everyone can benefit by gaining extra storage space.

Deep Closets

Most clothing hung on hangers is not less than 2 feet wide, yet many closets are only that deep. Roomy closets are luxurious, but they lose their appeal if there is no room for socks and underwear. Without ripping apart walls, you can better use space by resetting the closet pole toward the back of the closet. Gaining even 3 or 4 inches is enough to allow room for wire shelving (plastic-coated steel, like a dish drainer) to be added to the back of the closet door. This narrow shelving is perfect for small items that have a tendency to get lost in bureau drawers. Also consider using coat hooks and small clothes racks on the back of closet doors.

Kitchen Cabinets

Kitchen wall cabinets often do not extend all the way to the ceiling. Because the upper shelves are usually difficult to reach, even with a stepstool, this space is commonly framed into a soffit, covered with wallboard, taped, and then painted. By cutting an opening in the wallboard and adding hinged doors, you can create a substantial storage space for infrequently used items such as turkey platters, large pots and pans, fruit dehydrators, mason jars, and the old toaster your great-grandmother gave you.

Under-counter kitchen cabinets, or base cabinets, sit on a raised framework and project past it to create a kick, or toe, space. Your toes tuck underneath the base cabinets into this space so that you can stand closer to your workspace without leaning. The base may only be only 4 inches high, but the deep, shallow space beneath the base cabinet is a good place to run utilities to other remodeled areas without having to break into the existing stud walls.

It can also be fitted with an electric toe-space heater to warm a chilly kitchen. Electric heat is generally more expensive than heat from a central gas- or oil-fueled furnace, but a toe-space heater may be an economical alternative to new hot-air ducts or hot-water pipes.

The inside face of a cabinet door may also be considered for space-saving devices, such as vinyl-coated wire baskets, cup hooks, towel holders, or even a spice rack.

Eaves

Where dormers will be used to provide light and ventilation for renovated spaces under sloped roofs, there will likely be a lot of dead space created behind the knee walls of the finished room. While not every inch of this triangular space is usable, floor space can be gained by using the space lacking headroom for built-in closets and bureau drawers, leaving more area for living space.

Framing Cavities

You can't use the space within insulation-filled exterior walls for storage, but every interior partition contains a 4-inch-deep space that may be used to provide more depth and, therefore, more storage capacity for a built-in fixture, such as an extra-deep medicine cabinet.

A large custom cabinet can be recessed into a partition wall near the bathroom lavatory. With a ¾-inch-thick door, it may protrude a couple of inches, yet it might enclose 4- or 5-inch-deep shelves—a medicine chest that actually holds more than tiny medicine bottles. Partition cavities can also free up space, for example, by housing pocket doors that slide into the wall rather than swinging into usable room space. Interior wall spaces can also be used to hide small valuables, such as jewelry, or dangerous articles, such as guns. A painting or a mirror may be readily hung to conceal a shallow, personal storage box or cabinet.

Cabinet Storage

A pull-down false drawer front under a kitchen sink is a handy place to store sponges, steel wool, and scrub brushes.

Roll-out trays are a great back-saving device in base cabinets where easy accessibility to large pots and pans is desirable.

Pull-downs can accommodate anything from appliances to cookbooks, freeing up countertops for clutterless meal preparation.

SMALL SPACES

Understair Storage

Space beneath a stair is perfect for storing dry goods and other non-flammable household items. For a straight-run stair, consider building a closet or casework under the high side of the stairs and an access door to a low cabinet on the short side.

Roll-out shelving works well for small-item storage, eliminating the awkward reaching required in a standard cabinet.

The Undiscovered Space

Buried in the maze of joist and studs are many small parcels of unused space. A good eye and a ruler may be the only tools you need to discover hidden space in walls, above closets, beneath stairs, and in many other out-of-the-way places in your house. In older homes, for example, there may be almost a full 4 inches in the wall cavity, enough room for you to recess a large bathroom vanity. In newer houses, high prices per square foot have encouraged builders to use every possible area. And even in locations where some dead space is inevitable—for example, under the eaves of a roof—you can bring some of it back to life by building in bureau drawers and storage closets that free up space in the room.

Crawl Spaces

Crawl spaces aren't suitable for finishing, but they make a great place for long-term storage. Common problems with a crawl spaces are dampness and limited air circulation. A well-insulated and ventilated space can cut down on moisture.

CONCEALED SPACE

23 Canadian Code

586 PLUMBING & ELECTRICAL
- Complying with Canadian Code
- Plumbing Code
- Drain Requirements
- Emergency Requirements
- Electrical Code
- Smoke Detectors
- R-Value Zone Map

588 GENERAL CONSTRUCTION
- Header Spans
- Rafter Nailing
- Wood Moisture Content
- Insulating Crawl Spaces
- Ice-Dam Membranes
- Air Changes
- Radon Control

23 Canadian Code

Complying with the Canadian Code

Just as Canadian and American housing are alike, so too are the building codes used by the two countries. In the U.S., there has been a broad movement to unify building codes under the International Code Council. The council has created the International Residential Code (IRC), which has been adopted by most of the states. However, there are some areas of the country where local codes exceed or differ from the standards of the IRC. In the Canadian provinces, the national code is called the National Building Code (NBC), which was created in 1941. It, too, does not carry the force of law until it has been adopted by the local building authority.

The code enforcement environment in Canada is similar to the environment in the U.S. If a model code has been adopted, local authorities can move to toughen its basic requirements. So no matter what you are building—and the country in which you are building—check the code that applies to your local area and the structure you intend to build. You can determine this by contacting the office that issues local building permits or your area's building inspector's office.

Note that one country's code is not considered stronger or more rigorous than the other, but there are idiosyncratic differences, often inspired by the weather. Also note that the codes can change every year. As many as 800 changes to the Canadian code were introduced the last time it was issued, so be sure you are working from the latest version. (The National Building Code of Canada can be ordered online at www.nationalcodes.ca.)

There are not many differences between the U.S. and Canadian building codes. The differences are notable, however, and not recognizing them may force you to tear out a section of your structure or even rebuild it entirely. An example of how close the two codes are can be shown in "Minimum Requirements for Emergency Escape and Rescue Openings," below. So, as you review the applicable code, keep in mind that code can have the force of law, and there are more than a few stories of people who didn't obey the code and had to correct the problems at a substantial cost.

Plumbing Code

One of the main differences between the Canadian and U.S. code is where ABS (acrylonitrile-butadiene-styrene) and PVC (polyvinyl chloride) piping can be used. PVC is a versatile material, and in the states you will see PVC used in everything from drain pipes to roof gutters.

In Canada, ABS is more popular and can be used for waste and drain lines. ABS is only marginally different from PVC. For example, ABS is lighter than PVC and more expensive to produce, but ABS has twice the thermal expansion of PVC. The adhesives required for joining ABS and PVC are different. You cannot use PVC adhesive on ABS, and vice versa.

Other Differences

In the U.S., it is very typical for the drainpipe from double sinks to drain to a central drain located midway between the sinks; the central drain serves both sinks. In Canada, the code calls for one sink to be plumbed directly to the drain, and the second sink connects to the first sink's drainpipe by means of a T-fitting. This cuts down on the number (and cost) of the plumbing fixtures required for plumbing the sinks, but probably won't affect sinks' performance.

It is extremely common in the U.S. to plumb sinks with a P-traps (a section of plumbing shaped like a P), which fills with water upon first use of the sink. The P-trap keeps gas from

Minimum Requirements for Emergency Escape and Rescue Openings

ITEM	IRC	NBC
Required locations	Bedrooms and habitable basements	Bedrooms
Minimum Area	5.7 sq. ft.	3.77 sq. ft.
Minimum height or width	24 in. height, 20 in. width	15 in. for both
Maximum height of sill	44 in.	59 in.
Minimum clearance in window wells	3 ft. x 3 ft.	Front clearance greater than 22 in.
Additional requirements for window wells	If depth greater than 44 in., a ladder must be provided. Any cover must be removable from the interior side.	

PLUMBING & ELECTRICAL

Drain Requirements

One sink of a double sink must drain directly to the drainpipe.

P-traps are required by the Canadian building codes.

emerging from the sewer into the living space. P-traps are not always required by U.S. codes, but they are required by Canadian code.

No Self-Tapping Valves

When hooking up small copper water lines (⅜ inch) to larger copper feed lines, it is common in the U.S. to use a "saddle T-fitting." A saddle T is screwed in place, tight against a pipe with a bracket that looks like a saddle. Once in place and secured, a self-piercing valve stem is screwed down into the pipe. A needle pierces the pipe, and water flows through the fitting to a valve. This is a common practice in the U.S., especially for plumbing ice makers. But in Canada, the code does not allow this type of fitting. You must use a standard compression fitting and attach the ⅜-inch copper line to it.

Electrical Code

In the service panel (the box that contains the circuit breakers), there are conductive strips of metal called a "bus bars" that serve as conductors to bring or return power to the electrical system. Typically, one bus bar serves the hot side; the other serves the neutral side. In the U.S., the electrical code generally allows a single, combined bus bar to serve both the neutral and ground wires. But Canadian electrical code requires a separate service panel bus bar for the ground wire.

The two codes are similar concerning grounding of the electrical boxes, switches, and receptacles. They all must be grounded by approved means. For metal boxes, this means a short bonding jumper between a metal screw and the box. The bonding jumper should be attached to the system's grounding conductor.

As in the IRC, once-common battery-operated smoke detectors are insufficient for new construction according to the NBC. Currently, each sleeping room and each level of the home must contain a hard-wired smoke detector that is interconnected, which means that when one alarm sounds, they all sound. Battery backup is required for the alarms. Hard-wired carbon monoxide detectors are required in each bedroom or bedroom hallway.

Service-Entrance Panel

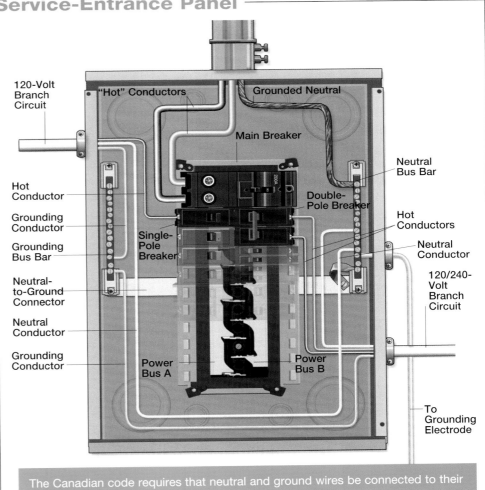

The Canadian code requires that neutral and ground wires be connected to their own bus bars in the panel box. In the U.S., the two types of wires often share a bus bar, although each must be connected to its own terminal screw.

COMPLYING WITH THE CANADIAN CODE

23 Canadian Code

Detector Wiring

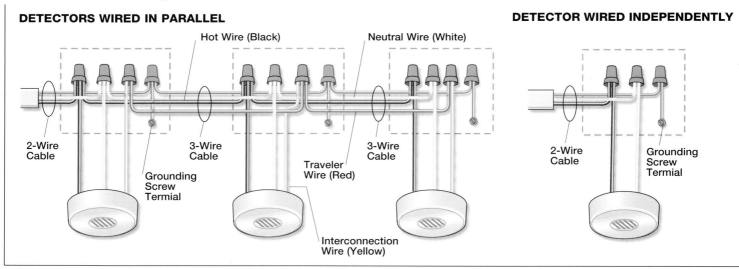

General Construction

A good practice when building a doorjamb is to "block" around the jamb to provide overall stability and a nailing surface for setting the jamb permanently in place. This is especially important when providing a nailing backer around the hinges and the lockset. Though the U.S. code does not always call for this, the Canadian code requires solid block backers behind all hinges and lockset.

Header Spans

A header is typically fabricated by taking 2-by lumber and creating a sandwich with a section of plywood in the middle. The plywood provides structural reinforcement, but it also adds thickness to the header, so the header presents a flush nailing surface for drywall on either side of the wall. A header constructed in this way achieves a longer span rating. Span ratings determine how big an opening the header can span. Because of snow load and the weight of the upper stories of the structure, a header may change size from application to application. U.S. and Canadian codes calculate the allowable header spans differently. Be sure you are using a region-appropriate span table when determining which types of lumber used to make your headers. Note that these differences can also affect engineered lumber as well as dimension lumber.

The sizing of joists also differs between the U.S. and Canadian codes. Be sure you are using a current joist span table when specifying joists. Canada's joist spans tables tend to run a little more conservative (closer spaces, larger dimension lumber) that U.S. code. Other differences include:

Rafter Nailing

Though U.S. code calls for a minimum number of nails at the heel joint with no distinction for the associated pitch, span, or rafter spacing, the NBC contains tables that vary the requirement with slope and span of the rafters.

Be sure to check span tables when selecting and spacing floor joists. The NBC tends to require smaller spans than the IRC.

Wood Moisture Content

The NBC requires that an "alternative construction be used" if wood frame walls must be finished if the moisture content of the lumber is above 19 percent. (This is a condition that will not affect the vast majority of builders.)

Insulating Crawl Spaces

Crawl spaces can be damp locations in any structure because they are often unconditioned spaces that aren't very well vented. Additionally, water can flow across the floor of a crawl space, and crawl spaces are rarely designed with a drainage system. Insulation in a crawl space that runs all the way to the floor can wick that moisture, so the Canadian code requires that insulation stop 2 inches above the crawl-space floor. There is no such requirement in the U.S. code, and insulation can roll down the walls and across the floor, though an increasingly common practice is to waterproof the walls, drain the floor, and even condition the air within the crawl space by including it as a conditioned space within the building envelope.

Ice-Dam Membrane Underlayments

When finishing the roof, Canadian and U.S. codes differ in their treatment of ice-dam membranes. An ice dam is piece of ice that forms near the edge of a roof and that prevents the flow of melting snow, causing a backup of water that can work its way under roofing materials.

GENERAL CONSTRUCTION

R-Value Zone Map

DEGREE-DAY (DD) ZONES

- A — UP TO 3,500 DD
- B — 3,500–5,000 DD
- C — 5,000–6,500 DD
- D — OVER 6,500 DD

ZONE		A	B	C	D
WALLS	RSI	3.0	3.6	4.1	5.3
	R	17.0	20.0	23.0	30.0
BASEMENT WALLS	RSI	3.0	3.0	3.0	3.0
	R	17.0	17.0	17.0	17.0
ROOF OR CEILING	RSI	4.5	5.6	6.7	9.0
	R	26.0	32.0	38.0	51.0
FLOOR (over unheated spaces)	RSI	4.7	4.7	6.7	9.0
	R	27.0	27.0	38.0	51.0

NOTE: Each zone on the map represents an area that experiences a similar number of degree-days. Degree-days are a measure of heating demand based on the difference between the average daily outdoor temperature and 18°C (65°F). Cumulative totals for the month or heating season are used to estimate heating energy totals.

RSI indicates thermal resistance in metric terms.

To control the damage caused by ice damming, U.S. code typically requires a 24-inch ice-dam membrane, measured back from the inside face of the exterior wall; Canadian code requires 900 mm (36 inches). In addition to self-adhering ice-dam membranes, the NBC allows various types of ice-dam membranes to be used, though self-adhering membranes are generally considered superior.

Air Changes

The NBC requires that new homes replace a certain total volume of air with outside air every hour, and the rate of that exchange differs from the IRC; indeed, many local and regional codes have various requirements for air changes per hour. The NBC requires that new homes include a mechanical ventilation system with a minimum exhaust capacity of one-third of an air change per hour.

Radon Control

The latest version of the NBC calls for radon control in all new buildings. Basically, builders must thoroughly seal the foundation from the surrounding earth by installing polyethylene vapor barriers under all slabs and sealing all penetrations in the slab and the foundation walls, including sump pits, floor drains, and water pipes. Cracks where the foundation wall intersects the slab must be filled with caulk. In addition, the builder must provide a rough-in for a radiation extraction system. This consists of a length of pipe that penetrates the slab and rests in gravel that is 100 mm (4 inches) deep. If high levels of radon are discovered in the building, the pipe serves as the connection point for a subslab depressurization system.

The coverage of ice dam membrane in Canada is 36 in. from the edge of the roof. U.S. requirements call for a minimum of 24 in.

Glossary

Actual dimensions The exact measurements of a piece of lumber, pipe, or masonry. See "Nominal dimensions."

Aerator The unit screwed onto the end of a faucet to control splashing.

Aggregate Crushed stone, gravel, or other material added to cement to make concrete or mortar.

Ampere (amp) A unit of measurement describing the rate of electrical flow.

Anchor bolt A bolt set in concrete that is used to fasten lumber, brackets, or hangers to concrete or masonry walls.

Apron Architectural trim beneath a window stool; also, the wider end of a drive that abuts the street.

Ashlar Stone cut at the quarry to produce smooth blocks that are easily stacked.

Auger Flexible metal cable used to clean out drains; also, a tool used for boring holes in the ground.

Backfill Soil or gravel used to fill in between a foundation or retaining wall and the ground excavated around it.

Baluster One of the vertical supports for a handrail.

Battens Narrow wood strips that typically cover vertical joints between siding boards.

Batter board A level board attached to stakes and used to position strings outlining foundations and footings.

Bay window A window that projects from a wall, creating a niche in the interior.

Beam A steel or wood framing member installed horizontally to support part of a structure's load.

Bearing wall A wall that provides support to the framing above.

Bevel An angled surface not at 90 degrees, typically cut into the edge of a piece of lumber.

Bird's mouth The notch cut near the tail end of a rafter where it fits on a top plate or horizontal framing member.

Blocking Lumber added between studs, joists, rafters, or other framing members to provide a nailing surface, additional strength, or as a fire stop to keep fire from spreading.

Board foot A measurement of wood by volume, equivalent to 1 foot square and 1 inch thick.

Bottom plate The horizontal framing member at the base of a wall.

Bridging Lumber or metal installed in an X-shape between floor joists to stabilize and position the joists.

Btu British thermal unit; the standard measurement of heat energy.

Cable Two or more insulated wires inside a sheathing of plastic or metal.

Cantilever Joists projecting from a wall to create a porch or balcony floor without external supports.

Carcass The basic case of a cabinet.

Casing The exposed trim around windows and doors.

Cement board Cement-based backer board used as an underlayment for tile.

Check valve A valve that allows water to flow in only one direction.

Chord The outer framing members of a roof truss. See "Web."

Circuit breaker A protective device that opens a circuit automatically when a current overload occurs. They can be reset manually.

Cleanout A removable plug in a trap or drainpipe, which allows easier access for removing blockages.

Cleat A block used to support wood braces or other members.

Code The rules set down by local or county governments that specify minimum building practices.

Collar tie A horizontal board installed rafter to rafter for extra support.

Column A vertical support in a building frame, made of wood, metal, or concrete.

Conduit Metal or plastic tubing designed to enclose electrical wires.

Control joints Joints tooled into the surface that make concrete crack in planned locations.

Cornice Ornamental trim at the meeting of roof and wall (exterior) or at the top of a wall (interior).

Coped joint A curved cut on a piece of trim that makes the reverse image of the piece it must butt against; made with a coping saw.

CPVC Chlorinated polyvinyl chloride; a plastic used to make water pipe.

Cripple studs Short studs that stand vertically between a header and top plate or between a bottom plate and the underside of a rough sill.

Curing Providing proper moisture to a concrete slab to reduce cracking and shrinkage and develop strength.

Dado A wide, flat-bottomed groove cut at a right angle to the grain of the wood. Called a "rabbet" if cut at the edge of a board.

Dead load The weight of a building's components, including lumber, roofing, permanent fixtures, etc.

Deadman Also called a T-brace; a wooden cross brace at the end of a support used to hold a ceiling panel in place until it is fastened.

Deflection The bending of wood due to live and dead loads.

Dormer A shed- or doghouse-like structure that projects from a roof, built to add space to an attic.

Drip edge A metal piece bent to fit over the edge of roof sheathing, designed to channel rain off the roof.

Drywall Gypsum sandwiched between treated paper, used as an interior wall covering. Also called gypsum board or wallboard.

Dry well A hole in the ground filled with rocks or gravel, designed to catch water and help it filter into the soil.

D-W-V Drain-waste-vent; the system of pipes and fittings used to carry away wastewater.

Easement The legal right for one person to cross or use another person's land.

Eaves The lower part of a roof that projects beyond the supporting walls to create an overhang.

Efflorescence A deposit of soluble salts on the surface of masonry.

Elastomeric Made of a material that does not lose shape when subjected to heat or stress.

Escutcheon A metal plate that covers the hole in the wall around a pipe or faucet.

Façade The exterior face of a building.

Face brick A type of brick used when a consistent appearance is needed.

Face-nailing Nailing perpendicularly through the surface of lumber.

Closers, 535
Closets
    adding space, 442
    building new, 442
    cedar, 446
    deep, 582
    finding space for, 580
    framing, 442–43
    layout of, 442
Closet systems, 443
Coatings
    clear, 500
    opaque, 500
    semitransparent, 500
Cold chisel, 317
Collar ties, 573
Colonial base, 390
Colonial casing, installing simple, 399
Columns
    boxing in, 576
    concealing, 576
Combination bit, 59
Combination locksets, 21
Combing paint, 375
Compact fluorescents, 134
Composite panels, 480
Composite trim, 481
Compression faucet repair, 191
Concealed space, 582–83
Concrete
    cleaning, 567
    diagnosing problems, 76
    edging and jointing, 76
    estimating, 75
    finishes for, 74
    finishing, 555
    finishing tools for, 73
    forming curved corner, 76
    forms for, 78–81
    mixing, 74
    modular, 560
    placing, 76
    pouring patio, 74–75
    reinforcing, 76
    repairing cracks, 76
    stamped, 557
    testing, 75
    types of, 78
    weight of, 72
Concrete block
    building, 82–85
    buttering, 84–85
    cutting, 82
    estimating, 83
    laying, 82–83
    mortar types, 83
    painting, 84
    patching, 84–85
    repairing walls, 84
    replacing, 84–85
    ties, flashing and reinforcement, 82
    tools for construction, 73
    weight of, 72
Concrete pavers as driveway material, 565
Concrete slabs as driveway material, 565
Concrete tiles
    installing, 463
    repairing, 463
Concrete walks, 554–57
    reinforcing, 554
    repairing, 556
Condensation resistance, 511
Conduit, bending and cutting, 107
Conduit connectors, 107
Connections, fishing, 110
Console humidifiers, 298
Construction, installing window in new, 513
Construction adhesive, 68
Contact cement, 69, 423
Continuity tester, 130
Continuous footing forms, 78
Continuous perforated soffits, 325
Contractors, hiring, 38–39
Contracts
    payments on, 41–42
    starting and completion dates on, 41
Control joints, 76, 84
    spacing of, 556
    tooling, 76
Cooling, 302–3
    basic maintenance, 308
    choosing system, 302–3
    cleaner air, 310
    dry air, 312
    efficiency of, 303
    in-wall unit basics, 306–7
    installing, 306–7
    picking window unit, 304
    recharging, 308
    saving energy, 309
    thermostats, 313
    types of units, 302–3
Coped joints, 391, 393
Coping miter, 394
Copper, connecting, 174–75
Copper tubing, 166
Copper water supply lines, 174
Cordless power drill/drivers, 49, 59
Cord of wood, 296
Cords, fixing, 101
Corner boards, 482
    inside, 482
    outside, 482
Corner knife, 366
Corner medicine cabinet, 581
Corners
    bracing, 64
    finishing, of drywall, 366–67
Cornices, detailed, 495
Corrugated panels, 467
Counterflashing, 470, 471
Counters
    ceramic tile, 424–25
    installing post-form, 422
    laminate for, 423
    materials for, 410
    solid masonry, 425
Countersink bits, 59
Countertops
    installing sinks, 180, 422
    laminate, 410
    materials for, 422
    sinks mounted in, 180
    solid surfacing for, 424
Coursed ashlar, 90
Coursed rubble walls, 90
Cove moldings
    attaching, 394
    uses of, 390, 391
Coverage, 450
Cover plate types, 159
CPVC (chlorinated polyvinyl chloride) pipe, 168
Crawl spaces, 583
    insulating, 278–79
    insulating, in Canadian code, 589
    ventilation for, 317
    venting, 317
Crime watch program, 19
Crimping ferrules, 105
Crown moldings
    attaching, 394
    uses of, 390, 391
CSST (corrugated stainless-steel tubing) pipe, 168
Cubic feet per minute (CFM) rating for fans, 318
Curing, 76
Curved corner, forming, 76
Curved railings, 550
Curved windows, 516
Curves, forming, 554
Customized storage, 440
Custom-made cabinets, 414
Cutoff valves, 14, 167

Index

Cutout tool, 49
Cutters, 49, 52–53
Cutting tools, 170
Cyanoacrylate adhesive, 68
Cylindrical locksets, 534

D
Dadoes
    cutting, 412
    dovetail, 437
Dado joint for shelving, 436
Dead-bolt locks, 19, 21
    installing, 23
Deadman, 367
Dead-zone problems, 326
Deck nailer, 66
Deck-post anchors, 544
Decks
    anatomy of, 540
    buying lumber for, 543
    cedar for, 542
    fastener options for, 543
    finishing touches, 550
    framing for, 544–45
    ground-level, 541
    installing girders, 544–45
    joists for, 547–47
    laying out stairs, 548
    levels of, 540
    lumber options for, 543, 546
    multilevel, 541
    nails versus screws, 541
    planning, 540–41
    planters for, 550
    pressure-treated lumber for, 540, 542–43
    raised, 541
    redwood for, 542
    roof, 541
    seats for, 550
    wood for, 542
Deep-set escutcheons, 192
Deep socket kit, 229
Dehumidifiers
    capacity of, 312
    servicing, 312
Deicing, 18
Delta wrench, 184
Dents, filling, with joint compound, 375
Design-build, 36
Design professionals, 36
Detector wiring, Canadian, 588
Diaphragm, removing grit from, 212
Dimmer switches, 130
    middle-of-run, 130
    testing, 130
    types of, 130
    wiring, 130–33
        single-pole, 131
Direct vent, 320
Dirt, moving, 72
Dishwashers, 252
    direct-wiring, 146
    installing new, 253–55

removing old, 252
Disputes, settling, 45
    material substitutions, 45
    time problems, 45
Do-it-yourself plans, 36–37
Doorbell, fixing, 142–43
Doorbell circuits, 142
Doors, 524–37
    actions of, 525
    adjusting closet, 532–33
    anatomy of, 525
    automatic opener safety features, 536
    bifold, 532
    blockouts for, 81
    board-and-batten, 524
    buying, 524
    cabinet, 418–19
    casings for, 400–1
    construction of, 418
    exterior details, 526
    flush, 419
    frame-and-panel, 524
    glass versus security, 526
    hinge swings, 525
    hollow-core, 524
    installing
        garage entry, 578–79
        garage opener, 529
        interior, 531
        prehung, 526, 527
        shoe, 528
    interior details, 530
    lip, 419
    locksets for, 534
    locks for, 20
    maintenance of, 536

metal-clad, 524
    overhead, 536
    overlap, 419
    pet, 581
    pocket, 530, 581
    raised-panel, 418
    replacing threshold, 528–29
    resetting hinges, 532
    saving space, 530
    shaving it down, 532
    sliding glass, 22, 524
    solid-core, 524
    sticking, 532
    strengthening frames, 22
    types of, 524
    weatherstripping, 528–29
Dormers, 571
    adding, 570
    installing, 571
Double-glazed windows, 509
Double glazing, 510
Double-hung windows, 508
Double-pole circuit breakers, 102, 103
Douglas fir for decks, 543
Dovetail balusters, 355
Dovetail cutting, 420–21
Dovetail dadoes, 437
Doweling, 412
Downspouts
    blocked, 474
    extenders for, 475
Dragging, 375
Drainage systems, 556
    in driveway, 564
Drains, 228
    access in bathtubs, 229
    attaching fittings, to new sink, 201

hose for, 252
removing, 194
stubborn fittings, 201
troubles of, 270–71
vents and, 178
Drain-waste-vent system (DWV), 166
Drawers
    anatomy of, 420
    cutting dovetails, 420–21
    installing
        metal guides, 420–21
        wood guides, 420–21
    pull-down false front, 582
    rebuilding, 431
    repairing, 431
    types of, 420
    unsticking, 431
Drills, 49, 52, 63
    bits for, 170
    cordless, 49
Drip cap, 512
Drivers, 52
Driveway
    design of, 564
    material options, 565
Dryer receptacles, 125, 153
    choosing, 153
    installing, 153
Dryers, 252
    fires related to, 321
Dryer vents
    installing, 321
    options for, 321
Drywall, 358, 359
    basics, 364
    cutting for outlets, 365
    finishing, 366

596

corners, 366–67
    panel seams, 366–67
fixing corners, 369
gypsum, 359
hammers versus screwdrivers, 365
installing, 364–65
repairs, 368–69
taping joints, 366
Drywall knife, 376
Drywall saw, 366
Drywall tools, 364
Ducted range hood, 149
Ducts
    cleaning, 310
    types of, 149
Dull/brown spots, 374
Duplex receptacle, 118

E

Earthquakes, need for evacuation plan, 12
Easy-fix flush valve, 215
Eaves, 582
Eaves shelves, 447
Edge flashing, 454
Edgings
    in concrete, 76
    metal, 467
    of windows and doors, 512
Efflorescence, 93
Electrical boxes, 104, 114
    depth gauge marked on, 105
    outdoor, 158
    types of, 104
    wiring, 141, 148
Electrical codes, 100
    Canadian, 586
Electrical hazards
    built-in shock protection, 15
    extension cords, 15
    GFCIs, 15
Electrical metallic tubing (EMT), 106, 107, 108
Electrical safety, overloading circuit, 323
Electrical tester, 100
Electric furnaces, 289
Electrician's pliers, 50
Electricity
    attaching wires, 109
    basic materials, 108–9
    bending and cutting conduit, 107
    bulb types, 134
    capping wires, 109
    ceiling fans, 148
    chandeliers, 138
    childproofing and, 29
    circuit breakers, 102
    conduit and accessories, 106
    creating new circuits, 114
    cutting off power, 100
    direct-wiring dishwasher, 146
    dryer receptacles, 153
    electrical boxes, 104

electric radiant floor heating, 150-52
extending outdoors, 160–61
fishing cable
    into ceilings, 113
    into walls, 112
fixing cord, 101
fixing doorbell, 142–43
generation of, 96
GFCI receptacles, 126
high-voltage receptacles, 124–25
installing
    ceiling box, 135
    circuit breaker, 115
    floodlight, 156
    fluorescent fixture, 135
    outlet, 120
    surface wiring, 111
    transformer, 142–43
low-voltage lighting, 162–63
low-volt circuits, 142
metal-box markings, 105
new wiring options, 110–11
outdoor wiring, 157
plugged in, 118–21
point of use, 96–97
provision of, 96–97
quartz halogen, 140
raceway components, 106
range hoods, 149
range receptacles, 146
receptacles and switches, 159–61
recessed lighting, 141
replacing thermostat, 143
socket safety, 101
splicing grounding wires, 131
stripping cable sheathing, 108–9
switch types, 128–29
telecommunications, 144–45
testing, 100–101
testing fuses, 102
track-mounted fixtures, 137
transmission of, 96
types of smoke detectors, 154–55
typical outlet and wiring layouts, 122–23
typical switch and wiring layouts, 132–33
vanity lighting, 139
waste-disposal unit, 147
weatherproofing, 158
wire connectors ratings, 105
wiring
    appliances, 147–51
    capacities, 107
    carbon monoxide (CO) detectors, 156
    ceiling fixtures, 136
    dimmer switch, 130–33
    GFCI outlet, 126–27
    telephone jack, 145
Electric-powered planers, 413
Electric radiant floor heating, 150-52
Electric systems, 290
Electric water heater, installing, 249

Electronic air cleaners, 311
    installation of, 311
Electronic stud finder, 416
Emergency telephone numbers, 14
End of run dimmer switches, 130
Energy Star, 45
Engineered flooring, 334–35
    advantages of, 334
    installing, 334–35
    preparing, 334–35
    types of, 334
Entrances, universal design and, 32
Entry halls, finding space for, 580–81
Environmental hazards, 30–31
Epoxy, 68
Eroded mortar joints, 89
Escape routes, 14, 26
Escutcheons, 192
    deep-set, 192
Estimates, checking, 40–41
Evaporative humidifier, 298
Excavation, 72
Exhaust fan, choosing, 318
Expansion joints, replacing, in driveways, 566
Extension cords, 15
Exterior keyed locks, 21
Exterior walls, installing tub against, 235
Extinguishers, 14
Extruded polystyrene, 276

F

Face brick, 86
    setting, 488–89
Face-framed base cabinets, 415
Falling hazards, 14
False beam, installing
    in ceiling, 386–87
Fans
    ceiling, 322
    cubic feet per minute (CFM) rating for, 318
    exhaust, 318
    installing whole-house, 322
    motorized, 316
    reducing vibration of, 323
    ventilation with, 322
    whole-house, 322
Fasteners
    frame-wall, 62
    masonry-wall, 63
Faucets
    anti-freeze, 268–69
    anti-scald, 267
    attaching, 202–3
    bottom-mounted, 195
    filters for, 257
    fixing, 180, 228–29
    getting rid of sprayer, 187
    installing new bathroom, 196–97
    lever, 180
    matching types, 228
    removing old bathroom, 195

repairs for
    bathroom faucet and drain, 194–95
    of ceramic-disk, 188
    compression, 191
    fixing
        ball-type, 184
        O-ring stem leaks, 183
        spout leaks, 190–91
    repairing single-handle cartridge, 187
    replacing seat washer to fix leaks, 182
    of single-handle cartridge, 187
    of single-handle kitchen, 185
    of tub and shower, 192
    of two-handle cartridge, 186
    of two-handle ceramic-disk, 189
    working with scald-control faucets, 193
    replacing tub-shower, 232–33
    stem, 180
    top-mounted, 195
    water connections for, 195
Feather board, 411
Fiber-cement siding, 478
Fiberglass insulation
    most common, 275, 276
    R-value of, 275
    in venting soffits, 325
    for windows, 509
Fiberglass mesh tape, 368
Fiberglass wallcovering, 359
Fiction-reducing lubricant, 113
Field tiles, 378
Fill valves, 218–19
    fixing, to cure hissing, 215
    replacing, 206
Filter vent, 320
Finishes, types of, 426
Finish work, 390–91
Firebrick, 86
Fire codes on windows, 508
Fireplaces
    gas, 296
    installing masonry, 297
Fire safety, 14
    emergency numbers, 14
    escape routes, 14, 26
    extinguishers, 14
    preventing and detecting, 26
    smoke detectors, 14, 26
Firewood, 297
Fishing routes, 110
Fish tape and lubricant, 113
Fittings, sink, 181
Fixed-glass windows, 508
Fixed grilles, 324
Fixing ball-type faucets, 184
Fixtures
    connections of, 139
    recessed, 140
Fixture vents, 167

Index

Flanges, 221
Flapper replacement, 214–15
Flared fittings, making, 175
Flashing, 470–71
    base, 471
    edge, 454
    installing step, 470
    new, 521
    repairs of, 471
    types of, 470
    valley, 470
    W-, 470
Flashing windows, 499
Flat pry bar, 333
Flat roofs, 452, 454–55
    common problems of, 455
    details of, 454
    fixing edge leaks, 454
    repairing, 454–55
    typical systems of, 454
Flexible braided-steel supply line, 168
Floating floors, installing, 334–35
Floodlights, installing, 156
Floods
    foundation repairs, 13
    pumping out water, 13
    sump pumps, 13
Floor nailer, 333
Floors
    changing height for toilet, 224
    fishing cable from below, 112
    fixing rotted, 227
    for garage conversion, 578
    installing, in basement, 574
    laminated wood, 331
    parquet, 331
    patching and leveling, for garages, 579
    preparing old, 331
    resilient, 331
    sealing, 574
    setting ceramic tile, 342
    subflooring for, 330
    universal design and level of, 32
Floor sander, 338
Floor trim
    options for, 330
    removing old, 330–31
Flues, metal, 296
Fluorescents
    compact, 134
    installing fixtures, 135
Flush doors, 419
Flush-valve cartridge, 220
Flush valves, 216–17
    replacing, 216–17
Fogging, 510
Forebay, 96
Formed footings, building, 80–81
Formless piers, building, 78
Form ties, 80
Foundation
    insulating exteriors, 278–79
    insulating interiors, 278–79
    repairs of, 13

Foundation anchor, 64
Foundation strap, 65
4-slot outlet, 146
Frame-and-panel doors, 524
Frameless cabinets, 415
Frame-wall fasteners, 62
Framing hardware, 64–65
Friction channels, improving windows by using, 520–21
Front yards, 553
Fuel costs, saving, 249
Fuel supply, 288
Full-mortise locksets, 534
Furnaces, gas-fired, 288
Fuse puller, 104

G

Gable-end shelves, 447
Gable-end vents, 324
Gable roofs, 452
Gable vents, 316
    installing, 326–27
Galvanized pipe, 233
Galvanized-steel pipe, 168
Galvanized wire mesh, 317
Gambrel roof, 452
Garage-door opener, installing, 537
Garages
    converting, 578–79
    finding space for, 580
    installing entry door, 578–79
    patching and leveling floors for, 579
Garden block, 85
Gas, 31
Gas and carbon monoxide (CO) detectors, 14
Gas-fired furnaces, 288
Gas-fired water heaters, replacing, 247–48
Gas fireplaces, 296
Gaskets, 221
Gas piping, 166, 239
Gas space heaters, 294
Gazebo roof tie, 66
Gel stains, 426
General construction, Canadian, 589
Generators, 96
    portable, 13
Girders, installing, 544–45
Glass
    cutting, 518
    replacing, 518–19
    safety, 28
Glass blocks, 18
    finishes, 382
    installation of, 382
Glass-domed meter, 96–97
Glass doors, sliding, 22
Glazed tile, 378
Glazing
    double, 510
    low-E, 510
    single, 510

Glazing compounds, 519
Glazing connections, 523
Grading, 556
Gravel walk, forming, 562
Gravity, defying, 250–51
Gravity-flow toilets, 210–11
Gravity-tank toilets, 178
Grazing, 76
Greenfield, 119
"Green" paints, 370
Green remodeling, 45
Grooved panels, 480
Ground-fault circuit interrupter (GFCI) circuit breakers, 103, 116–17
    installing, 127
    for multiple locations, 126
Ground-fault circuit interrupter (GFCI) outdoor power, 159
Ground-fault circuit interrupter (GFCI) outlets, 116–17, 126, 138
    for electrical hazards, 15
    wiring of, 117, 126–27
Ground-fault circuit interrupter (GFCI) receptacles, 100, 126
    installing, 163
Ground faults, 127
    on appliances, 99
Grounding boxes, 119
Grounding wires, splicing, 131
Ground-level decks, 541
Grout mixes, 344
Grouts for ceramic tiles, 378
Guest rooms, finding space for, 580
Guide hole, drill, 112
Guide wire, locating, 112
Gutter cleaners, 475
Gutter hangers, 474
Gutters
    clearing iced, 15
    maintaining, 474
    repairing, 475
    slope of, 474
Gypsum drywall, 359

H

Hacksaw, 107
Half-circle window, 570
Halogen bulbs, 134
Hammer drill, 63
Hammers, 51, 52
    screwdrivers versus, on drywall, 365
Handle-rotation stop, 193
Hand tools, basic, 51
Hangers
    top-mount, 65
    top-mount masonry, 65
Hardboard, applications for, 411
Hardware
    knock-down, 433
    security, 22
    special-use, 66
Hardwoods, durability of, 332

Hazardous materials, 29, 48
    storing safely, 28
Headers, insulated, 281
Header spans in Canadian code, 589
Heat
    conserving, 13
    movement of, 274
    preserving, 284–85
Heat cables, 269
Heat guns, 504
Heating, 288–99
    air quality, 298–99
    baseboard convectors, 290–91
    comparing systems, 288
    electric furnaces, 289
    electric radiant floor, 150-52
    electric systems, 290
    expanding system, 290
    filter and flue checks, 292
    furnace maintenance, 292–93
    gas-fired furnaces, 288
    heat pumps, 289
    high-pressure oil burners, 289
    hot-air systems, 288, 290
    hot-water systems, 289, 290
    improving efficiency, 293
    maintenance, 293
    moisture, 298–99
    relining flues, 292
    supplemental, 294–95
    with wood, 296–97
Heating element, replacing, for water heater, 246
Heating equipment, regular checkups for, 26
Heat pumps, 289, 303
Heat savings, 288
Heat-shrink plastic, installing, 522
Heavy loads, handling, 72
Help, calling for, 37
Hide glue, 68
High-efficiency windows, 511
High-intensity-discharge (HID) bulbs, 134
High-pressure oil burner, 289
High-voltage receptacles, 124–25
    for appliances, 119
    layouts, 125
    wiring, 124
Hinges
    adjusting bound, 533
    resetting, 532
    tightening loose, 532–33
    types of, 535
Hip roofs, 452
Holding tools, 170
Holes
    filling, with wood putty, 440
    plugging of, 440
Hollow-core doors, 524
Hollow-door anchor, 62
Hollow-wall anchor, 62
Home fire extinguishers, 27
Hoods, 471

Hopper windows, 508
Hot-air systems, 288, 290
    filter on, 292
Hot glue, 69
Hot water, 238–41
    preserving, 284
    typical number of gallons used, 238
Hot-water dispensers, 242–43
    installing an instant, 242
Hot-water systems, 289, 290
Household hazards, 14
House-mounted ledgers, 547
House vents, 495
House wraps, 478
Humidifiers
    central, 298
    console, 298
    evaporative, 298
    installing, 298–99
    portable, 298
    steam-vaporizer, 298
    types of, 298–99
    ultrasonic, 298
Hurricanes, need for evacuation plan, 12
Hurricane tie, 65
Hydroelectricity, 96

I

Ice control, 472
Ice-dam membrane underlayments in Canadian code, 589
Ice dams, 472–73
    formation of, 472
    preventing, 472–73
Ice shield, installing, 472
I-joists, 67
Immersion detection circuit interrupters (IDCIs), 15
Incandescent bulb, 134
In-frame shelves, 447
In-ground piping, installing, 172
In-ground steps, installing, 563
Inside corner boards, 482
Installation, electronic air-cleaning, 311
Insulated ceiling, 141
Insulated headers, 281
Insulation, 274–85
    adding, 280
    for attic, 282–83
    basic configurations, 276
    blowing-in, 282–83, 325
    for ceilings, 280–81
    choosing, 276
    comparing R-values, 274
    for crawl spaces, 278–79
        in Canadian code, 589
    diagnosing wall, 279
    for existing walls, 282–83
    fiberglass, 325
    for foundations, 278–79
    materials, 276–77
    performance of, 275
    preserving heat, 284–85
    rigid board, 274
    for roofs, 280–81
    terminology in, 274
    tools and special handling equipment, 277
    types of, 276
    for walls, 280–81
    water heater, 239
Insulation displacement connectors (IDCs), 145
Interior designers, 36
Interior water supply systems, 169
Interlocked plank, replacing, 362
Interlocking paver patterns, 560
Intermediate metallic conduit (IMC), 106, 157
International Code Council, 586
International Residential Code (IRC), 586
In-wall air conditioners, 302, 306
    basics of, 306–7
    installing, 306–7
Ionization smoke detector, 155
Ironing board, 581
Isolation joints, 76

J

Jacks
    ladder, 451
    pump, 451
Jambs
    assembling, 398
    weatherstripping for, 529
J-bolts, 64, 544
Joinery cuts, 393
Joint compound, 375
    setting-type, 368
Jointer, 410
Joints
    back-cutting, 390
    control, 76
    coped, 393
    coping, 391
    isolation, 76
    mitered, 393
    scarf, 393, 399
    scribing, 363
    tooled, 89
Joist furring rack, 447
Joist hangers, 64
    setting, 547
Joist-hanging rack, 447
Joists
    choosing, 546–47
    hanging, 546–47
    installing pipe through, 172
    strengthening attic floor, 573

K

Keyed window locks, 21
Keyhole saw, 53
Keyless lock, 18
Key-lock cutaway, 18

Kitchens, 320
    finding space for, 580
    installing metal-rim sink in, 202–3
    range venting in, 320
    repairing cartridge faucet in, 185
    replacing sink in, 201
    soffits in, 580
    universal design and, 32
    vent hoods in, 320
    ventilation for, 320
    wall and ceiling vents in, 320
Knife
    corner, 366
    taping, 366
Knock-down hardware, 433
Knockout punch, 146
Knotty-pine planking, 361

L

Ladder jacks, 451
Ladders, 451
Laminate
    for counters, 423
    for countertop, 410
    installing, 423
Laminated wood flooring, installing, 331
Lamp repair, 134
Latches, 418
Latex paint, 370
Latex primer, 370
Lattice
    plastic, 317
    pressure-treated (PT), 317
Laundry
    dryer vents in, 321
    installing freestanding sink, 204–5
    venting, 321
Layout, closet, 442
Lazy Susan, 580
Lead anchor, 63
Lead in water, 30
Lead paint, 30
Leak reservoirs, releasing, 15
Leaks
    fixing faucet spout, 190–91
    replacing seat washer to fix, 182
Leaky flange gaskets, 222–23
Ledgers, house-mounted, 547
Level, 253, 416–17
Leveling, 253
    tools for, 542
Lever faucets, 180
Light fixture boxes, 134
Light for attics, 570–71
Lighting
    abundant, 48
    automatic, 13
Light switch box, 139
Line cord, 144
Lineman's pliers, 50
Lip doors, 419
Liquid adhesive, 175
Locking pliers, 50
Locks, 20
    common, 534
    dead-bolt, 21
    door, 20
    exterior keyed, 21
    fixing common problems, 18

Index

installing privacy, 534–35
keyed window, 21
keyless, 18
rotating cam, 20
rotating lever, 20
sash, 21
sliding-door, 21
window, 20
Locksets
combination, 21
common, 534
cylindrical, 534
for doors, 534
full-mortise, 534
passage, 20
standard keyed, 20
Loose fill insulation, R-value of, 275
Low-E glazing, 510
Low-emissivity, 511
Low-voltage lighting, 162–63
installing, 163
Low-volt circuits, 142
Low-volume toilets, 178
Lumber
applications for, 411
buying, 543
for decks, 543
pressure-treated, 543
Lumens, 134

M

Machine-screw anchor, 63
Main vents, 167
Mansard roofs, 452
Manual conduit bender, 107
Masonry, 63, 72–93, 425
cleaning, 92–93
removing stains from, 93
Masonry bolt, 63
Masonry fireplace, installing, 297
Masonry paint, 575
Masonry veneers, 488
Masonry-wall fasteners, 63
Mason's twine, 73
Material substitutions, 45
Measuring, 174
Media air cleaners, 311
Medicine cabinet, corner, 581
Metal-box markings, 105
Metal bracing, 66
Metal-clad doors, 524
Metal conduit, or tubing, 106
types of, 106
Metal ductwork, concealing, 576
Metal edging, 467
Metal flues, 296
Metal guides, installing, 420–21
Metal-rim sinks, 201
installing, 202–3
Metal roofing, 466–67
materials, 466
options for, 466
repairing, 467
Middle-of-run dimmer switches, 130

Mildew in bathrooms, 374
Mineral deposits in toilet, 211
Mineral wool, 276
Mirrors, mounting, 383
Miter box, 390, 410
Miter chop saw, 410
Mitered joint, 393
Mitered siding, 482
Miter saws, power, 390
Modular brick styles, 87
Modular cabinets, 415
Modular concrete, 560
Moisture, 298–99
stopping, 503
Mold, 503
in bathrooms, 374
Molded-case circuit breakers (MCCBs), 102
Molded trim, installing, 392–93
Molding
common profiles, 390
cove or crown, 390
types of, 390
Mortars, 87
maintaining, 88
mixes of, 344
stone in, 561
types of, 83
Mosaic, 90
Mosaic tile, 343
Motion detectors, 25
Motorized fans, 316
Multilevel decks, 541
Multipart base molding, 335
Muntins, 509
Murphy beds, 574, 581

N

Nailers, power, 49
Nailholes, filling
with joint compound, 375
with putty, 505
Nailing
one-hand, 57
rafter, in Canadian code, 589
specs for, 56
Nail plate, 66
Nails
applications of, 56
concealing, 405
safety with, 57
screws versus, 541
sizes of, 56
tips for using, 57
types of, 56
weights of, 57
Nail set, 399
National Association of Home Builders, 45
National Association of Remodeling Industry (NARI), 39, 45
National Building Code of Canada, 586
National Electrical Code (NEC), 15, 103, 104, 107
National Fenestration Rating Council (NFRC), 511
Natural-gas alarms, 30
Needlenose pliers, 50
Neoclassical fluted casing, 401
Neon circuit tester, 100
Neutral current, 128
NM (flexible nonmetallic sheathed cable), 108

Noise
controlling, 384
reducing transmission, 29
Non-insulated ceiling (NIC), 141
Nonmetallic conduit, types of, 106
Nonmodular brick styles, 87
Nut driver, 52
Nuts, bolts and, 60
Nylon carpet, 348

O

Oak flooring, grades of, 332
Oil-fired systems, filters in, 292
120/240-volt appliance receptacle, 125
One-hand nailing, 57
One-piece baseboard, 396
Opaque coatings, 500
Open valleys, 470
Orbital sander, 410
O-ring packing seals, replacing, 183
O-ring stem leaks, fixing, 183
Outdoor cable, installing, 157
Outdoor circuits, 157
Outdoor electrical boxes, 158
Outdoor receptacles, installing, 160–61
Outdoor wiring, 157
Outlets
connections of, 139
cutting, in drywall, 365
installing, 120
soffit, 319
typical wiring of, 118
Outside corner boards, 482
Overhang options, 453
Overhead doors, 536
Overlap doors, 419
Oxidation, 76

P

Packing washer, repairing, 183
Painters, 371
Painting
- aluminum siding, 497
- basics, 370
- combing, 375
- dragging, 375
- preparing walls, 372
- problems, 374
- ragging, 374
- spattering, 375
- special finishes, 374–75
- sponging, 374
- spot-painting patches, 375
- stippling, 375
- walls, 372–73

Paint mixer, 370
Paints
- application options, 371
- applicators for, 501
- chalking, 492–93
- choices, 370
- "green", 370
- lead, 30
- masonry, 575
- for siding, 500–1
- stain over, 505

Paint sprayers, applicators for painting, 501
Paneling, 358
- advantages of, 480
- board, 404
- lighting dark, 362
- options, 360
- patching, 362–63
- plank, 405
- prefinished, 404
- products for, 404–5
- scribing joints, 363
- solid, 361
- solid hardwood, 361
- touch-ups, 363

Panel laps, 481
Panel lifter, 367
Panel materials, 480
Panels, 479
- board-and-batten, 480
- composite, 480
- finishing seams, 366–67
- grooved, 480
- installing, 360–61, 480–81
- repairing, 487
- surfaced, 480

Panel siding, 479
Panel trim, 360
Paper reinforcing tape, 368
Parquet flooring
- installing, 331
- laying, 335

Particleboard
- applications for, 411
- for shelving, 436

Passage locksets, 20

Patio
- paver, 559
- pouring, 74–75

Paver patio, 559
Pavers, 558–59
- interlocking patterns, 560
- patterns of, 558–59

Paving bricks, 86
Peeling, 374, 502
Peephole, installing, 20
Peg-board, 444
Penetrating sealer as wood floor finish, 347
Penstock, 96
Perforated soffit vents, 325
Personal protection, 48
Pet door, 581
PEX manifold system, installing, 177
PEX pipe, 168
PEX stub-outs, installing, 177
PEX water tubing, 176–77
- tools and equipment for, 176

Photoelectric smoke detectors, 155
Pickling as wood floor finish, 347
Pier forms, 78
Pin drill, 62
Pipes
- boxing in, 576
- cutting, 173
- galvanized, 233
- installing, through joists, 172
- quieting noisy, 269
- supporting, 172
- types of, 168
- wrapping, 285

Pipe-sleeve blockouts, 81
Pipe slope, 179
Piping, gas, 239
Plain siding, 495
Plane, 390
Planking, knotty-pine, 361
Plank-paneling, 405
Plank siding, 482–83
Planters, deck, 550
Plaster, fixing, 199
Plasterboard, 358
Plastic, connecting, 174–75
Plastic lattice, 317
Plastic panels, 467
Plastic pipe adhesive, 175
Plastic supply lines, 174
Plastic toggle, 62
Plate rail, 406
- milling and installing, 406–7

Platforms, 549
Plexiglas, cutting, 518
Pliers, 50, 107
- electrician's or lineman's, 50
- locking, 50
- needlenose, 50

Plugging, 337
- of holes, 440

Plug vents, 325
Plumber's helper, 268

Plumber's putty, 231
Plumbing, 166–271
- aerators, 181
- appliances, 253–55
- Canadian code for, 586–87
- cleaning fouled air inducer, 220
- cleanout, unblocking, 179
- cleanups, 266
- cutoff valves, 167
- cutting pipe, 173
- defying gravity, 250
- dishwasher, installing new, 253–55
- drains and vents, 178
  - low-volume toilets, 178
  - unblocking cleanout, 179
- drain troubles, 270–71
- dryers, 252
- easy-fix flush valve, 215
- faucet repairs, 180, 183–93
- fixing fill valves to cure hissing, 215
- fixing running toilet, 213
- flapper replacement, 214–15
- getting better water, 256–57
- gravity-flow toilets, 210–13
- heat cables, 269
- hot water, 238–41
- hot-water dispensers, 242–43
- installing
  - cast-iron bathtub and shower, 234–35
  - electric water heater, 249
  - laundry sink, 204–5
  - metal-rim sink, 202–3
  - new bathroom faucet, 196–97
  - recirculating system, 250
  - sediment filter, 256–57
  - shower stall, 236–37
  - sink in plywood top, 200
  - toilet flanges on wood floor, 222
- interior water supply systems, 169
- making flared fittings, 175
- one-piece silent-flush toilet, 213
- parts of system, 166
- PEX water tubing, 176–77
- pipes, quieting noisy, 269
- plumber's helper, 268
- pressure-assisted toilets, 220
- private water wells, 264–65
- removing
  - grit from diaphragm, 212
  - old dishwasher, 252
- repairing cast-iron flange, 223
- replacing
  - gas-fired water heater, 247–48
  - kitchen sink, 201
  - leaking pressure tank, 265
  - S-trap, 205
  - tank ball, 213
  - tub-shower faucet, 232–33
- running water, 174
- servicing
  - electric water heaters, 246

- gas-fired water heaters, 244–45
- showers, 228
- sinks, 180
  - fixing faucet, 180
  - installing, 180–81
- sump pumps and wells
  - installing submersible, 262
  - private water wells, 264–65
  - replacing leaking pressure tank, 265
- supplies, 171
- supporting pipe, 172
- toilets
  - flanges and gaskets, 221
  - installing new, 226
  - installing wax-free gaskets, 221
  - leaky flange gaskets, 222–23
  - replacing fill valve, 218–19
  - replacing flush valve, 216–17
  - taking up and resetting, 224–25
- tools, 170–71
- traps and vents, 167
- treating waste, 258–59
- types of pipe, 168
- wall-hung sink, 198–99
- washers, 252
- waste-disposal units, 260–61
- water and waste treatment, 257–61
  - installing submersible sump pumps, 262
  - installing water softeners, 263
- water flow, 166
- water heaters, 243
- water meters, 166
- water-saving toilets, 208–9
- water softeners, installing, 263
- well systems, 251
- workings of toilet, 206–207

Plumbing supplies, 171
Plumbing system, parts of, 166
Plywood
- applications for, 411
- cutting and edging, 412
- installing bathroom sink in top, 200
- for shelving, 436
- on sleepers, 331
- on subfloors, 331
- on wood, 331

Pneumatic cleaner, 271
Pneumatic impact wrenches, 49
Pneumatic tools, 49, 57
Pocket doors, 530, 581
Polyester carpet, 348
Polyisocyanurate, 276
Polypropylene carpet, 348
Polyurethane, 69, 276, 426
- as wood floor finish, 347

Polyurethane foam, 276
Pop-up drain in bathtub, 231
Portable air cleaners, 310
Portable electric baseboard heaters, 295
Portable electric radiators, 295

601

Index

Portable generators, 13
Portable humidifiers, 298
Post anchors, 544
    with bolt, 65
Post beam cap, 65
Post bolt cap, 65
Posthole digger, 78
Posts
    caps and, 64
    concealing, 576
Power drill/drivers, cordless, 59
Power miter saws, 48, 390
Power nailers, 49
Power outages
    automatic lighting and, 13
    conserving heat, 13
    portable generators, 13
Power sanders, using, 338
Power tools
    operating, 48
    secured, 48
Power transmission, 96
Power washing, 493
Prefinished paneling, 404
Prehung door, installing, 526, 527
Prepared, being, 12
Pressure-assisted toilets, 209, 220
    fixing running, 208
Pressure-balanced cartridges, 193
Pressure-balance spool, 193
    cleaning, 193
Pressure switches, 264
Pressure tanks, 250
    replacing leaking, 265
Pressure-tank toilets, 178
Pressure-treated lattice, 317
Pressure-treated lumber for decks, 540, 543
Primer-sealer, 370
Priming, 372
Private water wells, 264–65
Projects
    changes and extras, 43–44
    supervising, 43–44
Prying tools, 51
P-trap, 167
Pull-down false drawer front, 582
Pull-down stairs, 352–53
Pullout platform, installing, 432
Pump jacks, 451
Pump system, submersible, 264
Push stick, 411
Putty, filling nailholes with, 505
PVC (polyvinyl chloride) pipe, 168

Q
Quad circuit breakers, 102–3
Quarry tile, 343
Quartz halogen bulbs, 140
    replacing, 140

R
Rabbets, alternatives for cutting, 413
Raceway components, 106

Raceway wiring, 106
Rack and prefab systems, 445
Radial-arm saw, 48
Radon, 31
    control of, in Canadian code, 589
Rafter bracing, 64
Rafter clip, 65
Rafter-hung shelves, 447
Rafter nailing in Canadian code, 589
Rafters, 64
Ragging, 374
Railings
    curved, 550
    installing, 551
Rails
    chair, 406
    plate, 406
Raised decks, 541
Raised-panel doors, 418
Random-orbit pod sander, 49
Range, venting, 320
Range hoods, 149
    ducted, 149
Range receptacles, 125, 146
Ready-mix, 75
Rebar, 76, 80
Receptacles
    cover for, 101
    dryer, 125, 153
    ground-fault circuit interrupter (GFCI), 126
    high-voltage appliance, 119
    interpreting, 124
    120/240-volt appliance, 125
    range, 125, 146
    240-volt appliance, 125
    types of, 119
Recessed fixture, 140
Recessed lamp housing, 140
Recessed lighting, 141
    installing, 141
Recessed toe-space heater, 580
Reciprocating saw, 48
Recirculating system, installing, 250
Redwood for decks, 542, 543
Remodeling
    contracts for, 41–42
    following codes, 42–43
    green, 45
    settling disputes, 45
Remodeling job
    planning, 36–38
        design professionals, 36
        di-it-yourself plans, 36–37
        general contractor, 38
        management options, 38
    watching budget, 39–41
Remote sensors, 25
Remote switching, 142
Rental tools, 171
Replacement windows, 516–17
Residential circuit breakers, 102
Resilient flooring
    installing, 331

    repairing tile, 341
    sheets and tiles, 340
Resorcinol, 68
Respirator, 370
Resurfacing kit, 428
Retrofit kits, 216
Reverse-osmosis units, 257
Ridge strap, 65
Ridge vents, 316, 324, 452
    installing, 326–27
Right-angle drill, 172
Rigid board insulation, 274
    R-value of, 275
Rigid copper pipe, 168
Rim clips, 202
Rimless sinks, 201
Ringer equivalency number (REN), 144
Rise and run of roofs, 453
Rollers, 371
    applicators for painting, 501
Roll-out shelving, 583
Roll-out trays, 582
Romex, 108
Roof decks, 455, 541
Roofing, language of, 450–51
Roofing tools, 451
Roofing vents, 316
    gable, 316
    motorized fans, 316
    ridge, 316
    soffit, 316
    turbine, 316
Roofs
    anatomy of, 450
    clay tile, 452–53
    climate and, 452–53
    design of, 452–53
    estimating surface area, 450–51
    flat, 452, 454–55
    gable, 452
    gambrel, 452
    hip, 452
    insulating, 280–81
    overhang options for, 453
    rise and run of, 453
    shed, 452
    slopes of, 450
    temperature extremes and, 472
    temporary repairs to, 12, 15
    types of, 452
    ventilation for, 317, 452
Roof vents, 320, 324
    combinations for, 324
    gable-end, 324
    installing, 326–27
    soffit and gable, 324
    soffit and ridge, 324
    soffit and turbine, 324
Room-unit maintenance, 308
Roots, mansard, 452
Rotary drill, 62
Rotating cam lock, 20
Rotating lever lock, 20
Router, 49, 410, 413

Rubble walls, 90
    coursed, 90
Rust, 503
R-values, 274
    comparing, 274
    comparisons of, 275
R-value zone map, 275
    Canadian, 588

S
Saber saw, 48
Safety
    checklist, 27
    equipment, 53
    fire, 14
    gates, 28
    gear, 73, 370
    with nails, 57
    socket, 101
    table saw, 411
    tool and shop, 48
Safety glass, 28
Safety glazing, 14
Safety grates, 28
Saffir-Simpson scale, 12
Sand, stone in, 561
Sanders, 49
    random-orbit pod, 49
Sanding
    alternatives for, 413
    cabinets, 427
    wood floors, 338
Sanding sealers, 426
Sandpaper, 375
Sash
    freeing stuck, 519
    removing old, 520
Sash lock, 21
Saws, 48, 52–53
    circular, 48, 325
    drywall, 366
    keyhole, 53
    power miter, 48
    radial-arm, 48
    reciprocating, 48
    saber, 48
    table, 48
    utility, 366
Scaffolds, 451
    setups, 359
Scald-control faucets, working with, 193
Scarf joint, 393, 399
    making, 399
Scrapers, 51
Screen block, 85
Screening frames, 523
Screens, replacing window, 522–23
Screwdrivers, hammers versus, on drywall, 365
Screws
    filing holes left by, 431
    heads and washers, 58
    nails versus, 541

removing, 59
sizes of, 58
Scupper troughs, 454
Seats, deck, 550
Seat washer, fixing, 182
Secured power tools, 48
Security, vacation, 19
Security hardware, 22
Security programs, 19
Security systems
    auto-dialers, 25
    basic layout, 24
    common components, 24
    motion detectors, 25
    remote sensors, 25
    types of, 24
    whole-house, 24
Sediment
    cleaning, from pressure-balance spool, 193
    removing, 189
Sediment filter, installing, 256–57
Self-rimming sinks, 201
Self-vent, 320
Semicustom cabinets, 414
Semitransparent coatings, 500
Septic and graywater systems, 259
    maintaining, 258
Service drop, 96
Service-entrance conductor (SEU) cable, 146
Service-entrance panel, 97–98
Service entrance round (SER) cable, 146
Service panel, working on, 114
Setting-type joint compound, 368
Shakes, 464
Shaving, alternatives for, 413
Shed roofs, 452
Sheet floorings, installing, 340–41
Sheet materials, applications for, 411
Sheet paneling, installing, 404
Sheetrock, 358
Sheet vinyl, advantage of, 340
Shelf supports, screwing, to wall studs, 439
Shelving, 436–39
    adjustable, 437
    bracket options for, 438–39
    built-in, 440
    eaves, 447
    gable-end, 447
    in-frame, 447
    joinery for, 436
    making wide plywood, 436–37
    materials for, 436
    rafter-hung, 447
    reinforcing, 436
    roll-out, 583
    solid wood, 436
    truss-mounted, 447
    types of, 436
Shims, 399
Shingles
    asbestos, 499

    asphalt, 453
Shock hazards, eliminating, 29
Shower head, replacing, 266
Showers, 228
    faucets in, 228–29
    installing cast-iron, 234–35
Shower stall, installing, 236–37
Sidewall junctures, 470
Side-wired receptacles, 118
Side yards, 553
Siding, 478–505
    aluminum, 478, 479, 494–99, 496
    asbestos, 499
    brick, 488
    choices of, 478
    cutting back, 516
    details and finishes, 495
    evaluating problems, 502–3
    fiber-cement, 478
    finishing new, 500
    joining new and old, 516
    mitered, 482
    paint versus stain for, 500–1
    panel, 479
    plain, 495
    repainting, 504–5
    re-siding over existing, 498–99
    routine repairs, 486–87
    solid wood, 479
    stone, 488
    vinyl, 478, 479, 494–99, 496–97
    wood, 478
Silent-flush toilet, 213
Silicone, 69
Sills
    new, 521
    transitions, 345
    weatherstripping of, 528
Simulated wood boards, 358
Single glazing, 510
Single-handle cartridge faucet, repairing, 187
Single-pole circuit breakers, 103
Single-pole dimmer switches, 128–29
    wiring, 131
Sink-clip wrench, 202
Sinks, 180
    countertop-mounted, 180
    fittings for, 181
    installing, 180–81, 422
        bathroom, in plywood top, 200
        freestanding laundry, 204–5
        metal-rim, 202–3
    metal-rim, 201
    replacing kitchen, 201
    replacing S-traps, 205
    rimless, 201
    self-rimming, 201
    wall-hung, 198–99
Skylights, 468–69, 570
    in cathedral ceilings, 468
    design options, 468
    installation of, 468
    installing, 468–69
    types of, 468
Slate roofing, 460–61
    installing, 460
    materials, 460
    repairing, 461
    selecting, 460
    tips for, 461
Sledgehammer, 51
Sleepers, plywood on, 331
Sleeves, 371
Slider, repairing, 532–33
Sliding-door locks, 21
Sliding glass doors, 22, 524
Sliding windows, 508
Slopes, 552
    of driveway, 564
    of roof, 450
Smoke alarms, wiring, in series, 155
Smoke detectors, 14, 26
    battery-powered, 26
    hardwiring, 155
    hard-wiring, 26
    installing, 154
    ionization, 155
    photoelectric, 155
    types of, 154–55
    typical locations, 26
    wiring, 154
Socket safety, 101
Soffit and gable vents, 324
Soffit and ridge vents, 324
Soffit and turbine vents, 324
Soffit outlet, 319
Soffits
    continuous perforated, 325
    kitchen, 580
Soffit vents, 316, 325, 572
Soft copper pipe, 168
Softeners, 257
Softwoods, durability of, 332

Index

Soil pipes, 178
    concealing, 576
Soils, 552
Solar heat gain coefficient (SHGC), 511
Solar tube, 469
Soldier course, 558
Soleplate, locating, 112
Solid-core doors, 524
Solid hardwood paneling, 361
Solid masonry counters, 425
Solid paneling, 361
Solid surfacing, 424
    for countertops, 424
Solid wood, 424
Solid-wood flooring
    grades of oak, 332
    installing, 331, 332, 333
    ordering and storage, 332
    staining, 332
Solid wood shelving, 436
Solid wood siding, 479
Sound-absorbing walls, 384–85
Sound insulation for shower, 228
Sound-resistant construction, options, 385
Sound transmission, 384
Space, concealed, 582–83
Space heaters, recessed toe, 580
Space savers, 580
Spacing, control-joint, 556
Spacing jig, 483
Spalling, 76, 89
Spattering, 375
Specialty circuit breakers, 102–3
Special-use hardware, 66
Sponging paint, 374
Spot-painting patches, 375
Spot patching for wall-to-wall carpeting, 351
Spray attachment, replacing, 183
Sprayed or foamed fill insulation (per inch), R-value of, 275
Sprayers, 371
Spud wrench, 201
Squeaks, stopping, in wood floors, 336
Stains
    removing from wood floors, 336–37
    sealing over, 372
    for siding, 500–1
Staircase cracking, 89
Stair pullouts, 447
Stairs, 352–53. See also Steps
    anatomy of, 352, 548
    building, 352
    building codes for, 352
    carpeting, 353
    formulas, 352
    installing, 548
    installing railings, 551
    layout of, 548
    pull-down, 352–53
    replacing balusters, 355
    replacing treads, 354

stopping squeaks, 354–55
tightening balusters, 355
tightening posts, 354
Stairway, variations of, 571
Stairwell, scaffolding for, 359
Stamped concrete, 557
Standard keyed locksets, 20
Standing-seam metal panels, 466
Stay-in-place tube forms, 79
Steam-vaporizer humidifiers, 298
Steel as material for windows, 509
Stem faucets, 180
Step-down utility transformer, 96
Steps. See also Stairs
    installing in-ground, 563
    patching, 92
Stippling, 375
Stirrups, 64
Stock cabinets, 414
Stone
    buying, 90
    as driveway material, 565
    laying face, 91
    laying full, 91
    in mortar, 561
    repairs to, 492
    in sand, 561
    shaping, 91
    weight of, 72
Stone pavers
    cutting, 561
    laying, 560–61
Stone paving, 560–61
Stone siding, 488
Stone tiles, 343

underlayment for, 330
Stone walls
    dry-stacking, 90
    patterns for, 90
Stonework, types of, 488
Storage, 440–47
    adding space, 446–47
    built-in shelving, 440
    customized, 440
    eaves shelves, 447
    gable-end shelves, 447
    in-frame shelves, 447
    joist furring rack, 447
    joist-hanging rack, 447
    rafter-hung shelves, 447
    stair pullouts, 447
    truss-mounted shelves, 447
    understair, 583
Storms, 12
    clearing bottlenecks, 12
    de-icing, 12
    returning home after, 15
    survival tips, 12
    temporary roof repairs, 12
Storm windows, 522
Straightedge, 416
S-trap, replacing, 205
Strip-grille vents, installing, 325
Stripping, 504
Structural limits of older houses, 570
Structural timbers, cutting, 172
Stucco, 490
    changing appearances, 490–91
    installation of, 490
    installing, 491

patching, 92
pros and cons of, 490
repairing, 492
systems of, 490
textures of, 490
Stud-and-ranger forms, 78
Stud finder, electronic, 416
Stud wall, 358
Subfloors, 330
    plywood on, 331
    replacing, 227
Submersible pump system, 264
Submersible sump pumps, installing, 262
Sump pumps, 13, 250
    eliminating water, 577
    installing submersible, 262
Supply lines, plastic, 174
Supporting pipe, 172
Surfaced panels, 480
Surface form, 79
Surface-mounted fixtures, 136
Surface-mounted strip vents, 325
Surface wiring, 110
    installing, 111
Surge-protection device, 103
Switch-controlled outlet, 118
Switches
    pressure, 264
    timer, 319
    types of, 128–29
    vanity, 138
Symmetrical arts and crafts casing, 401
Synthetic stucco system (EIFS), 490, 491

T

Table saw, 48, 410, 413
    safety of, 411
Tank-ball replacement, 213–14
Tank condensation, eliminating, 207
Tape dispenser, 367
Taping knife, 366
Telecommunications, 144–45
Telephone numbers, emergency, 14
Telephones
    number of, 144
    wiring jacks for, 145
    wiring of, 142, 144
    wiring plugs, 144
Temperature, reducing attic, 326
Temperature extremes, roofs and, 472
Tester, electrical, 100
Thermal envelope, 274
Thermocouple, replacing, 244
Thermographic picture, 279
Thermostats, 313
    replacing, 143
    for water heater, 246
Thickset adhesive, 342
Thinset adhesive, 342, 379
    for ceramic tiles, 378
Three-piece base trim, installing, 396
Three-piece Victorian-style casing, 400
Three-way switches, 128

wiring, 128–29
Threshold, replacing, 528–29
Tie-down strapping, 64
Tile floors
    cutting and nipping, 345
    repairing, 341
Tiles
    ceramic, 343
    installing, 379
    mosaic, 343
    quarry, 343
    stone, 343
Time problems, 45
Timer switches, 319
    installing, 319
Toenailing, 57
Toe-space heaters, installing, 294–95
Toggle bolt, 62
Toilets, 206–227
    ball float in, 206
    caulking of, 223
    cellar, 250
    checking for clogs, 209
    cleaning bacteria-clogged, 211
    easy-fix flush valve, 215
    eliminating tank condensation, 207
    fixing fill vales to cure hissing, 215
    fixing slow-filling, 212
    flanges and gaskets, 221
    flapper replacement, 214–15
    gravity-flow, 210–13
    gravity-tank, 178
    installing flanges on wood floor, 222
    installing new, 226
    installing wax-free gaskets, 221
    leaky flange gaskets, 222–23
    low-volume, 178
    mineral deposits in, 211
    pressure-assisted, 208, 209, 220
    pressure-tank, 178
    removing bacteria from, 210–11
    removing grit from diaphragm, 212
    replacing fill valve, 218–19
    replacing flapper, 214–15
    replacing flush valve, 216–17
    replacing tank ball, 213
    silent-flush, 213
    supply lines for, 218
    taking up and resetting, 224–25
    water-saving, 208–9
    wax seal in, 206
    workings of, 206–207
Tool and shop safety, 48
Tooled joints, 89
Tools
    air-powered nailguns, 541
    cutting, 170
    drywall, 364
    holding, 170
    leveling, 542
    plumbing, 170–71
    practical, 50
    rental, 171

roofing, 451
shopping for, 170–71
turning, 170
Top-mounted faucets, 195
    removing, 195
Top-mount hanger, 65
Top-mount masonry hanger, 65
Tornadoes, need for evacuation plan, 12
Total circuit demand, 323
Toy box, 446
Track-mounted lighting, 137
    installing, 137
Transformer, installing, 142–43
Traps and vents, 167
Trays, roll-out, 582
Trim, 481
    baseboard, 346
    composite, 481
    floor, 330–31
    panel, 360
    repainting, 373
Trimwork, 390–407
    basics, 396–401
    chair rail, 406
    decorating with, 392–93
    finish work, 390–91
    installing crown molding, 394–95
    paneling products, 404–5
    plate rail, 406
    wall frames, 402–3
Tripped circuit breaker, 100
Tripwaste, connecting, 230–31
Troughs, scupper, 454
Trusses, 453, 570
Truss-mounted shelves, 447
Truss plate, 66
Tub-shower faucets, 192
    fixing single-handle, 192
    replacing, 232–33
Tuft patching for wall-to-wall carpeting, 351
Turbine vents, 316
Turning tools, 170
240-volt appliance receptacle, 125
Two-handle cartridge faucet, repairing, 186
Two-handle ceramic-disk faucets, repairing, 183
Type K mortar, 83
Type M mortar, 83
Type N mortar, 83
Type O mortar, 83
Type S mortar, 83
Type TW (thermoplastic-wet), 157
Type UF (underground feeder), 157

U

U-factor, 511
UF cable, installing, 160–61
Ultrasonic humidifiers, 298
Underlayment, 330
    replacing, 227
Understair storage, 583
Underwriters Laboratories (UL), 156

Unfinished space, 570–83
Ungrounded appliance, 99
Universal desiagn, 32–33
    in baths, 33
    entrances and, 32
    floor level and, 32
    kitchens and, 32
USB outlets/receptacles, 121
Utility boxes, sealing, 285
Utility saw, 366
Utility storage
    assembling wall system, 445
    rack and prefab systems, 445
Utility transformer, step-down, 96

V

Vacation
    checklist, 19
    security, 19
Vacuum system, central, 581
Valley flashing, 470
Valleys
    open, 470
    woven, 470
Valves
    cutoff, 167
    water inlet, 253
Vanity cabinets, 198
Vanity lighting, 139
    installing, 139
Vanity sink, anatomy of, 197
Vanity switch, 138
Vapor barriers, 285, 498
Varnish as wood floor finish, 347
Veneer stone, setting, 488–89
Vent hoods, 320
Ventilation, 314–27
    for attics, 317, 570–71, 572, 573
    for basements, 317
    basics in, 316
    for crawl spaces, 317
    with fans, 322
    insulating ducts, 285
    for kitchen, 320
    paths for, 316
    requirements for, 317
    for roofs, 317
    systems for, 316
    whole-house, 323
Venting
    alleviating problems in, 316
    for baths, 318–19
    crawl-space, 317
    laundry, 321
    roof, 324
Vent pipes, 167
Vents, 471
    bathroom, 318
    direct, 320
    filter, 320
    for foundations, 317
    gable, 316, 326–27
    house, 495
    perforated soffit, 325

placement of bathroom, 318–19
plug, 325
ridge, 316, 324, 326–27, 452
roof, 320, 326–27
roofing, 316
self-, 320
soffit, 316, 325, 572
strip-grille, 325
surface-mounted strip, 325
turbine, 316
wall, 572
wall and ceiling, 320
Vibration, reducing fan, 323
Victorian bellyband casing, 400
Victorian-style casing, three-piece, 400
Vinyl as material for windows, 509
Vinyl-clad wood as material for windows, 509
Vinyl siding, 478, 479, 494–99
    fixing damaged, 496–97
Visible transmittance (VT), 511
Volatile organic compounds (VOCs), 30, 370

W

Wainscoting, 404–5
    installing, 404–5
Waiver of Mechanic's Lien Rights, 42
Walks
    concrete, 554–55, 556–57
    design and layout, 552–53
    forming, 554–55
Wall and ceiling vents, 320
Wallboard, 358
Wall brushes, applicators for painting, 501
Wall cabinets, installing, 416
Wallcoverings
    vinyl, 358
    vinyl-coated, 358
Wall forms, building, 80
Wall foundation forms, 78
Wall-frame assembly jig, 402–3
Wall frames, 402–3
    installing, 402
Wall heaters, installing, 294–95
Wall-hung sink, 198–99
    removing, 198
Wall options for basement, 574
Wallpaper
    hanging, 377
    making spot repairs, 376
    scarifying tool for, 376
    stripping old, 376
    stripping old paper, 376
    types of, 376
Walls, 358–85
    built-up, 360
    ceramic tiles for, 378–81
    cleaning of, 493, 504
    covering basement, 575
    decorative options, 358
    drywall, 364–65
    exterior, installing tub against, 235

Index

field tiles for, 378–81
fishing cable, 112
glazed tile for, 378–81
insulating, 280–81
insulating existing, 282–83
painting, 372–73
patterns, 378
sound-absorbing, 384–85
sound transmission, 384
stud, 358
wiring in closed, 110
wiring in open, 110
Wall studs, screwing shelf supports to, 439
Wall system, assembling, 445
Wall-to-wall carpeting, 348
   caring for, 350
   installing, 348–49
   materials and weaves, 348
   pads, 349
   pad versus cushion-backed, 349
   patching, 350–51
   spot patching, 351
   spot repairs, 350
   tuft patching, 351
   use of stretchers and kickers in installing, 350
Wall vents, 572
Washers, 252
Waste, treating, 258–59
Waste-disposal units, 147, 260–61
   installing, 260
   removing, 258, 261
   restarting jammed, 261
   wiring, 147
Waste-line clog, clearing, 270
Waste pipes, 178
Water
   connections for faucets, 195
   as destructive force, 316
   eliminating, 577
   flow of, 166
   getting better, 256–57
   lead in, 30
   pumping out, 13
   running, 174
   turning off, 183
Water closet, upgrading, 206
Water-hammer, 249
Water heaters
   anatomy of, 245
   anode rods and, 243
   elements of, 238
   enclosing, in thermal blanket, 285
   installing electric, 249
   insulation, 239
   replacing gas-fired, 247–48
   safety and maintenance, 241
   servicing electric, 246
   servicing gas-fired, 244–45
   tankless, 240
Water inlet valves, 253
Water leaks, fixing, 577
Water level, 542

Water meters, 166
Water-saving toilets, 208–9
Water softeners, installing, 263
Water spots, 502
Water supply lines, copper, 174
Water supply systems, 167
   interior, 169
Water-treatment system, selecting, 256
Wax as wood floor finish, 347
Wax-free gaskets, installing, 221
Wax seal, replacing, 206
Weatherproof boxes, 158, 159
Weatherproofing, 158
Weatherproof receptacles, 157
Weatherstripping
   door, 528–29
   jamb, 529
   sill, 528
Web, 90
Wedge-type anchor bolt, 64
Welded or woven wire mesh, 76
Well pumps, 250
Wells, private water, 264–65
Well systems, 251
W-flashing, 470
White glue, 68
Whole-house fans, installing, 322
Whole-house security systems, 24, 25
Whole-house ventilation, 323
Window air conditioners
   installing unit, 304–5
   picking, 304
Window blockouts, 81
Window casings, 400–1
Window locks, 20
Windows, 508–523
   anatomy of, 509
   awning, 508
   basement, 18
   casement, 508
   curved, 516
   cutting glass and plexiglas, 518
   cutting openings for, 512
   dividers in, 509
   double-glazed, 509
   double-hung, 508
   exterior details, 515
   finishing touches, 514
   fixed-glass, 508
   flashing, 499
   frames for, 510
   framing opening, 512
   freeing stuck sash, 519
   glazing compounds, 519
   glazing connections, 523
   glazing options, 510
   half-circle, 570
   high-efficiency, 511
   hopper, 508
   improving, by using friction channels, 520–21
   installing, in new construction, 513
   installing heat-shrink plastic, 522
   installing trim, 514–15

   installing upgraded replacement, 517
   interior trim options, 514
   labels for, 511
   locking open, 20
   making glide, 520
   materials for, 509
   new flashing, 521
   new sills, 521
   removing old sash, 520
   replacement, 516–17
   replacing screens, 522–23
   screening frames, 523
   sliding, 508
   storms, 522
   styles of, 508
   unsticking, 518
Window units, 303
Window wells, 575
Wire connector ratings, 105
Wire connectors, 105
Wire crimper, 105
Wired alarm system, 24
Wire hangers, 172
Wire mesh, 76
   galvanized, 317
Wires
   aluminum, 98
   attaching, 109
   capping, 109
   colors of, 99
   maximum, in a box, 114
   sizes of, 99
   types of, 99
Wire terminals, 108
Wire ties, 80
Wiring
   capacities, 107
   in closed walls, 110
   for electric water heaters, 246
   ground-fault circuit interrupter (GFCI) outlet, 126–27
   installing surface, 111
   in open walls, 110
   phone, 142
   raceway, 106
   surface, 110
Wiring conduit, 106
Wiring layouts, 122
   typical switch and, 132–33
Wood
   heating with, 296–97
   as material for windows, 509
   plywood on, 331
   solid, 424
Wood boards, 479
Wood bracing, 66
Wood dowel, 63
Wood finishes, 346–47
Wood floors
   finishes, 347
   fixing, 336
      dents and cracks in, 336
      stains, 336–37

   installing toilet flanges on, 222
   refinishing, 339
   replacing boards, 336–37
   restoring, 338–39
   sanding, 338
   staining, 347
   stopping squeaks in, 336
Wood-framed standards, hanging, 438–39
Wood guides, installing, 420–21
Wood moisture content in Canadian code, 589
Wood planking for ceiling, 386
Wood putty, filling holes, 440
Wood shingles, 464–65, 479
   buying, 484
   corners, 485
   exposure and pitch, 465
   installing, 464–65, 484
   materials, 464
   repairing, 465
   replacing, 486
   staining, 485
Wood siding, 478
Wood stains, 426
Woodstoves, 296
Wood trim for shelving, 436
Wool carpet, 348
Workshops, ideas for, 444–45
Woven valleys, 470
Wraps
   building, 478
   house, 478
Wrenches, 50, 107
   pneumatic impact, 49
   sink-clip, 202
   spud, 201

Y
Y cleanout fittings, 178
Yellow glue, 68

Z
Z-brackets, 439
Z-clips, 67
Zoning, 42

Photo Credits

All photography by **John Parsekian/CH**, unless otherwise noted.

pages 10–11: Lisa F. Young/Dreamstime.com **page 12:** *both* Brian C. Nieves/CH **page 13:** *bottom right* Merle Henkenius **page 14:** *both* Brian C. Nieves/CH **page 15:** *bottom right* Brian C. Nieves/CH **pages 16–17:** Stanislav Komogorov/Dreamstime.com **page 18:** *top right* courtesy of Interlock Technologies **page 19:** *top left* courtesy of Overhead Door Co. **pages 20–21:** *Basic Door Locks sequence* courtesy of Schlage Lock Co.; *Window Locks sequence* Brian C. Nieves/CH **page 23:** *all* Merle Henkenius **pages 24–25:** *all* courtesy of Interactive Technologies, Inc. **page 26:** *top* Brian C. Nieves/CH **page 27:** *top* Brian C. Nieves/CH **page 29:** *top left & Electricity sequence* Brian C. Nieves/CH **page 30:** *all except bottom right* Brian C. Nieves/CH **page 31:** *top right & bottom right* Brian C. Nieves/CH **page 33:** *top & bottom both* courtesy of GE Appliance, design: Mary Jo Peterson; *middle both* courtesy of AD•AS **pages 34–35:** Fever-pitched/Dreamstime.com **pages 46–47:** Wendy Kaveney/Dreamstime.com **page 48:** *bottom right* courtesy of Craftsman by Sears; *bottom left* Brian C. Nieves/CH **page 49:** *middle & middle left* Brian C. Nieves/CH; *top left* courtesy of Black & Decker; *bottom left & right* Gary David Gold/CH; *bottom middle* Arenacreative/Dreamstime.com **page 50:** *middle row, bottom right* Brian C. Nieves/CH **age 51:** *left column down, bottom row & middle left* Brian C. Nieves/CH **page 52:** *middle left & bottom both* Brian C. Nieves/CH **page 53:** *top left* courtesy of Celotex; *middle right, middle left & bottom row* Brian C. Nieves/CH **pages 54–55:** Crystal Craig/Dreams-time.com **page 57:** *Nail Tips sequence* Brian C. Nieves/CH; *bottom left* Benjamin Gelman/Dreams-time.com **pages 59–60:** *all* Brian C. Nieves/CH **page 61:** *all except bottom row* Brian C. Nieves/CH **page 62:** *all* Brian C. Nieves/CH **page 63:** *bottom* courtesy of Makita U.S.A. **pages 66–68:** *all* Brian C. Nieves/CH **page 69:** *Hot Glue & Common Caulks sequences* Brian C. Nieves/CH **pages 70–71:** Patricia Hofmeester/Dreamstime.com **pages 78–79:** *all* Merle Henkenius **pages 84–85:** *Patching Block & Replacing Block sequences* Robert Anderson **pages 86–87:** *top left* David K. Hand; *Mortaring Bricks sequence* courtesy of Brick Industry Association **pages 88–89:** *Laying Brick sequence* courtesy of Brick Industry Association **page 91:** *all* Robert Anderson **pages 94–95:** Monkey Business Images/Dreamstime.com **page 99:** *all* Brian C. Nieves/CH **page 100:** *all except bottom left* Brian C. Nieves/CH **page 101:** *top row & Socket Safety sequence* Brian C. Nieves/CH **pages 102–104:** *all* Brian C. Nieves/CH **page 105:** *top and bottom left* Brian C. Nieves/CH *bottom middle and bottom right* Charles T. Byers **pages 106–109:** *all* Brian C. Nieves/CH **page 110:** *top* courtesy of Schenectady Museum; *bottom both* Merle Henkenius **page 111:** Merle Henkenius **pages 112–115:** *all* Brian C. Nieves/CH **pages 116–117:** *all* Charles T. Byers **pages 118–120:** *all* Brian C. Nieves/CH **page 121:** *all* Charles T. Byers **pages 122–134:** *all* Brian C. Nieves/CH **page 135:** *all* Merle Henkenius **page 136:** *all* Brian C. Nieves/CH **page 138:** *Install a Chandelier sequence* Brian C. Nieves/CH **pages 140–141:** *all* Brian C. Nieves/CH **pages 142–143:** *Fixing a Doorbell sequence* Merle Henkenius **pages 144–149:** *all* Brian C. Nieves/CH **pages 151–152:** *all* Charles T. Byers **pages 153–155:** *all* Brian C. Nieves/CH **pages 156–157:** *top & Hardwire CO Detector sequence* Brian C. Nieves/CH; *Installing a Floodlight sequence* Merle Henkenius **page 158:** *all* Brian C. Nieves/CH **page 159:** *all except bottom left* Brian C. Nieves/CH **pages 160–161:** *all* Brian C. Nieves/CH **page 162:** *all* courtesy of Malibu Lighting/Intermatic, Inc. **page 163:** *all* Brian C. Nieves/CH **Pages 165:** Merle Henkenius **pages 168–171:** *all* Merle Henkenius **pages 176–177:** *all* Merle Henkenius **page 178:** *top* courtesy of Sloan Flushmate **pages 181–185:** *all* Merle Henkenius **page 186:** *top* Freeze Frame Studio/CH; *bottom row* Merle Henkenius **page 187:** *all* Merle Henkenius **page 188:** *top* Freeze Frame Studio/CH; *bottom row* Merle Henkenius **pages 189–198:** *all* Merle Henkenius **page 199:** *bottom row* Merle Henkenius **pages 200–212:** *all* Merle Henkenius **page 213:** *top left* courtesy of Kohler; *all others* Merle Henkenius **pages 214–237:** *all* Merle Henkenius **page 238:** Charles T. Byers **page 239:** *bottom left & bottom middle* Merle Henkenius **page 240:** Charles T. Byers **pages 241–250:** Merle Henkenius **page 251:** *bottom row* Merle Henkenius **pages 252–255:** Merle Henkenius **page 256:** *left* Merle Henkenius **page 257:** *top right* Brian C. Nieves/CH; *bottom row* Merle Henkenius; *middle left & top left* courtesy of Moen; *middle* courtesy of Honeywell **pages 258–265:** *all* Merle Henkenius **page 266:** *top right, middle right & bottom right* Merle Henkenius **page 267:** *Relining a Tub sequence* courtesy of Re-Bath Corporation; *all others* Merle Henkenius **page 268:** *bottom right* Merle Henkenius **page 269:** *top right, bottom middle & bottom left* Merle Henkenius; *middle left* Brian C. Nieves/CH **page 270:** *bottom right* Merle Henkenius **page 271:** *top left* Brian C. Nieves/CH; *all others* Merle Henkenius **pages 272–273:** Christina Richards/Dreamstime.com **page 276:** *all* Stephen Munz **page 277:** *middle* courtesy of Innovative Insulation **page 279:** *bottom right* courtesy of Infraspection Institute, Inc. **page 282:** *top* courtesy of Nebraska State Historical Society **page 283:** *middle* courtesy of Owens Corning **pages 286–287:** Cynthia Farmer/Dreamstime.com **pages 290–291:** *left* courtesy of Linda L. Riley, Valley Forge Convention & Visitors Bureau; *Baseboard Convectors sequence* Brian C. Nieves/CH **pages 292–293:** *all* Merle Henkenius **page 295:** *middle right & middle left* courtesy of Honeywell; *middle* courtesy of Holmes Products Corp. **page 296:** courtesy of Heat-N-Glo **page 297:** *all* courtesy of Buckley Rumford Co. **pages 300–301:** Lisa F. Young/Dreamstime.com **page 302:** *top left* courtesy of Carrier Corp. **page 303:** *middle* courtesy of Carrier Corp.; *right* courtesy of Adobe Air Inc. **page 304:** *left* courtesy of Carrier Corp.; *Basic Cleaning sequence* Merle Henkenius **page 308–309:** *top left* courtesy of Carrier Corp.; *Basic Cleaning sequence* Merle Henkenius **page 310:** *all* courtesy of Honeywell **page 313:** *top right & middle right* courtesy of Honeywell; *middle left* courtesy of Heating & Cooling Supplies **pages 314–315:** Lisa F. Young/Dreamstime.com **page 317:** *top left* courtesy of Vantage Products Corporation; *bottom left* courtesy of Hickson Corporation **page 320:** *top* courtesy of US Department of the Interior, National Park Service, LBJ National Historic Park; *bottom* courtesy of Broan MFG. Co, Inc. **page 324:** *all* Merle Henkenius **page 325:** *Installing Strip-Grille Vents sequence* Merle Henkenius **pages 328–329:** Michael Pettigrew/Dreamstime.com **page 338:** *top* courtesy of Otis Elevator Company; *Preparing the Room sequence* Brian C. Nieves/CH **page 339:** *all* Brian C. Nieves/CH **pages 340–341:** *all* Merle Henkenius **pages 342–343:** *Materials sequence from top to bottom* courtesy Custom Building Products; courtesy of Metropolitan Ceramics; courtesy of Dal-Tile; courtesy of Custom Building Products; *Installing Tile sequence* courtesy of Custom Building Products; *top middle* Brian C. Nieves/CH **pages 344–345:** *top right* Brian C. Nieves/CH; *middle right & Replacing Tile sequence* Merle Henkenius **page 347:** *Finishes & Staining sequences* Brian C. Nieves/CH **page 349:** *Pad vs. Cushion-Backed & Pads sequences* Brian C. Nieves/CH **page 350:** *top right* courtesy of Crain Cutter Company, Inc. **pages 354–355:** *Stopping Stair Squeaks sequence* Brian C. Nieves/CH **pages 356–357:** Christina Richards/Dreamstime.com **page 358:** *top both* courtesy of Georgia-Pacific; *bottom right* courtesy of Crown Berger; *bottom left* courtesy of Eisenhart Wallcoverings **page 359:** *top both* courtesy of USG; *bottom right* courtesy of York Wallcoverings; *bottom left* courtesy of Tasso **page 360:** *top* courtesy of New England Classic Interiors; *middle* courtesy of George-Pacific **page 361:** *Replacing an Interlocked Plank sequence* Brian C. Nieves/CH **pages 364–365:** *all except top left* courtesy of Celotex **pages 366–367:** *all* courtesy of Celotex **pages 368–369:** *all* Merle Henkenius **page 371:** *top row & middle row* Brian C. Nieves/CH; *bottom right* courtesy of Wagner Spray Tech Corp. **page 372:** *middle right* Brian C. Nieves/CH **page 373:** *Repainting Trim sequence* Brian C. Nieves/CH **pages 374–375:** *top* courtesy of USG; *top right* Brian C. Nieves/CH; *Special Finishes sequence* Don Wong/CH **page 376:** *all* Merle Henkenius **page 377:** *all* courtesy of Brewster Wallcoverings **page 378:** *top both*, upper *middle left* courtesy of Delta; upper *middle right* courtesy of American Standard; *Thinset & Grout sequences* Brian C. Nieves/CH **page 380:** *all* Brian C. Nieves/CH **page 381:** *all* Merle Henkenius **pages 382–383:** *top right* courtesy of Corian; *middle both & Installing Glass Blocks sequence* courtesy of Pittsburgh Corning; *top left* courtesy of Hi-Lite Block Windows **pages 384–385:** *all except top right* Brian C. Nieves/CH **page 387:** *Installing a Suspended Ceiling sequence* courtesy of Chicago Metallic Corp. **pages 388–389:** courtesy of Fypon **page 392:** *top right & middle right* courtesy of White River Hardwoods/Woodworks; *middle left* courtesy of Georgia-Pacific; *top left* courtesy of Elite **page 393:** *top left* courtesy of Focal Points Architectural Products; *middle right & middle left* courtesy of White River Hardwoods/Woodworks **pages 406–407:** *all* Neal

GREYSCALE

BIN TRAVELER FORM

Cut By: Miriam Qty: 15 Date: 07/24

Scanned By: _____ Qty: _____ Date: _____

Scanned Batch IDs

Notes / Exception